Supramolecular Chemistry

NATO ASI Series

Advanced Science Institutes Series

A Series presenting the results of activities sponsored by the NATO Science Committee, which aims at the dissemination of advanced scientific and technological knowledge, with a view to strengthening links between scientific communities.

The Series is published by an international board of publishers in conjunction with the NATO Scientific Affairs Division

A Life Sciences **B Physics**	Plenum Publishing Corporation London and New York
C Mathematical and Physical Sciences **D Behavioural and Social Sciences** **E Applied Sciences**	Kluwer Academic Publishers Dordrecht, Boston and London
F Computer and Systems Sciences **G Ecological Sciences** **H Cell Biology** **I Global Environmental Change**	Springer-Verlag Berlin, Heidelberg, New York, London, Paris and Tokyo

NATO-PCO-DATA BASE

The electronic index to the NATO ASI Series provides full bibliographical references (with keywords and/or abstracts) to more than 30000 contributions from international scientists published in all sections of the NATO ASI Series.
Access to the NATO-PCO-DATA BASE is possible in two ways:

– via online FILE 128 (NATO-PCO-DATA BASE) hosted by ESRIN, Via Galileo Galilei, I-00044 Frascati, Italy.

– via CD-ROM "NATO-PCO-DATA BASE" with user-friendly retrieval software in English, French and German (© WTV GmbH and DATAWARE Technologies Inc. 1989).

The CD-ROM can be ordered through any member of the Board of Publishers or through NATO-PCO, Overijse, Belgium.

Series C: Mathematical and Physical Sciences - Vol. 371

Supramolecular Chemistry

edited by

V. Balzani

and

L. De Cola

"G. Ciamician" Department of Chemistry,
University of Bologna,
Bologna, Italy

Springer Science+Business Media, B.V.

Proceedings of the II NATO Science Forum on
Supramolecular Chemistry
Taormina (Sicily), Italy
December 15–18, 1991

Library of Congress Cataloging-in-Publication Data

Supramolecular chemistry / V. Balzani and L. De Cola.
p. cm. -- (NATO ASI series. Series C, Mathematical and physical sciences ; no. 371)
Includes indexes
ISBN 978-94-010-5099-9 ISBN 978-94-011-2492-8 (eBook)
DOI 10.1007/978-94-011-2492-8
1. Macromolecules--Congresses. I. Balzani, Vincenzo, 1926- II. De Cola, L. (Luisa) III. Series.
QD380.S86 1992
547.7--dc20 92-5813

ISBN 978-94-010-5099-9

TABLE OF CONTENTS

FOREWORD

The first NATO Science Forum was held in Biarritz in September 1990. This Taormina Conference is the second in a series that we wish to be a long one and I believe that it has equalled the success of its predecessor. In setting up these meetings the NATO Science Committee wanted to gather leading experts to review fields of strong present interest. It was intended that presentations and discussions should pay special attention to potential developments. This "forward look" is indeed precious to us in mapping out the evolution of our Science Programme but more importantly, it is an essential part of the progress of Science. I believe that NATO, being able to bring together eminent scientists from both sides of the Atlantic, is in a priviliged position to provide this service to our Scientific Community.

It was only proper that Chemistry should be one of the first areas to be targeted: a central science with many rich borders touching on other disciplines, it deserved the full attention of our Committee. In its vast domain, among many possible topics, the present one was carefully selected and its choice resulted from an extensive consultation of many leading chemists. The large fraction of replies which pointed to Supramolecular Chemistry left us with little doubt about the timeliness of a Forum in this area and the strong interest attached to it.

In fact the field seemed to us important enough to warrant a particular effort from our side. This is why we decided at the same time to support it through a NATO Special Programme. Through this mechanism we shall focus for the next five years part of our activities (Advanced Study Institutes, Advanced Research Workshops, Collaborative Research Grants, etc.) on this highly promising area in order to catalyze the development of international collaborations. We hope that this development will extend beyond the member countries of NATO and will particularly actively involve the scientists of Central and Eastern Europe. I am happy to note that several of our "non-NATO" colleagues were already present in Taormina.

The NATO Science Committee is deeply indebted to all those who assisted us in getting this new initiative started. They are too numerous to be all cited here but special thanks go to Professor Balzani who assumed the heavy burden of organizing this very successful Forum. As for the future I hope that the direction indicated by the Biarritz and Taormina meetings will be followed with the continued support of the NATO Science Programme.

Jacques Ducuing
Chairman of the NATO Science Committee

FOREWORD

Supramolecular chemistry, the chemistry beyond the molecule, is a highly interdisciplinary field, at the meeting point of chemistry with biology and physics.

It extends into organic chemistry and the synthetic procedures for receptor construction, into coordination chemistry and metal ion-ligand complexes, into physical chemistry and the experimental and theoretical studies of interactions, as well as into biochemistry and the biological processes that all start with substrate binding and recognition. The resulting cross-fertilization opens wide horizons that represent a challenge and a stimulus to the creative imagination of the chemist.

Based on the chemical, physical, and biological features of recognition, catalysis and transport processes, novel lines of investigation have developed, concerned with the design of supramolecular photonic, electronic and ionic devices as well as of systems displaying higher forms of molecular and supramolecular behavior such as self-organization, regulation, cooperativity, communication, and replication.

The development of supramolecular chemistry involves a more and more complete control over supermolecules, molecular assemblies and materials. It rests on the elaboration of strategies towards properties and functions rather than towards structures, the goal being to express a desired property in a molecule or a material, whatever its nature.

The attractive power of supramolecular chemistry expresses itself in the many researchers from different fields who recognize themselves in its projects and are shaping its present and its future. Its coming of age is attested by the numerous highly imaginative and active groups that join it, by the many books that provide overviews and of course by the illuminating encounters that report advances and seed new projects.

It is always a great enjoyment to see how much each meeting, each symposium adds to the picture and enriches the perspectives.

The initiation of a NATO Special Program on "Supramolecular Chemistry" is most welcome; it should provide further impetus and facilitate the exchange of ideas and researchers. The first activity held within this scheme was the present symposium, that has brought together many of the most active actors in the field. Judging from the presentations made, it has indeed achieved the goals of imagination and diversity, the keys to creative research.

Both the funding agency and the organizers are to be thanked most warmly for furthering our common good, supramolecular chemistry!

Jean-Marie LEHN

PREFACE

Chemistry is the most central of the scientific disciplines. Any possible link between biology and physics is through chemistry. Material science, chemical engineering, earth sciences, ecology, and related areas are largely based on chemistry. In the September 13th, 1991 issue of Science (p. 1212) there is this beautiful description of chemistry: *Like a species that has moved into open niches, evolved and diversified, chemistry can no longer be seen as a discrete scientific field; its methods, concepts, and practitioners are penetrating virtually every nook and cranny of Science and Technology*. And in an Italian scientific magazine (Sapere, August 1991, p. 34) it was pointed out that *chemistry can now be regarded as a trans-disciplinary science, as is mathematics: a science that provides an essential means and a fundamental "language" to fully understand other scientific disciplines*. Everybody knows that chemical research is a fundamental factor for the progress of humanity. The most important global problems (food, human health, energy, environment) cannot be solved without chemistry. Without chemistry there would be no hope in filling the gap regarding the quality of life between developed and developing countries.

Unfortunately, chemical research is largely underfunded. Two examples taken from our own country, but certainly representative of a general situation, will suffice to demonstrate this issue: (i) the sum spent to buy twelve Harrier vertical-landing bombers for the Italian Navy is as much as four times the sum spent in 1991 by all the Italian public research agencies (MURST, CNR, ENEA, ISS, etc.) to finance chemical research; (ii) the subdivision of the 1991 budget for scientific research in Italy was as follows: space 23%, physics 18%, chemistry 4%. These two examples, of course, show where changes should be made in the allocation of financial resources.

In the present situation, the decisions taken by NATO to reduce by 80% its nuclear arsenal in Europe and to launch a new Special Programme in the field of Chemistry are most welcomed. These two decisions were certainly unrelated and concern, in fact, very different budgetary and political levels. We like to believe however, that the coincidence in time and location between those two events*, albeit casual,

---

*The decision to reduce the nuclear arsenal was taken at the Defense Ministers meeting of NATO Countries at Taormina (Sicily) on October 17, 1991; the NATO Special Science Programme on Supramolecular Chemistry was launched by the II NATO Science Forum on December 15-18, 1991, in the same town.

is of the utmost significance since it offers an opportunity for meditation to scientists and politicians.

Chemical research can certainly be a fundamental factor for the progress of humanity, but it must be driven in the right direction. Past experience shows that scientific and technological progress takes place mainly at the cross-roads of different disciplines. Therefore, as a first action we need to demolish all those artificial barriers which have been created to subdivide chemistry into specialized subdisciplines like organic chemistry, inorganic chemistry, physical chemistry, and so on. Supramolecular Chemistry is an appropriate vehicle for this reunification.

But what is "Supramolecular Chemistry"? As for any other interdisciplinary field of science, a general definition would necessarily be arbitrary. Different meanings can in fact be attributed to Supramolecular Chemistry and the II NATO Science Forum has shown many facets of this interdisciplinary field. The importance of this meeting can be measured, in fact, not only by the number of represented countries (16), but especially by the number of represented disciplines (more than 20).

The conceptual feature that distinguishes a supramolecular from a molecular species is not its size, but the possibility to split the species into individual molecular subunits (components) capable, as they are or with minor modification, of an individual existence. Such a definition is clear-cut in several cases, i.e. when the components are molecules held together by intermolecular forces as in the case of host-guest systems. In other cases (e.g., for covalently linked components) the definition is more subjective. The main characteristic of supramolecular species is certainly the possibility to construct them by combining molecular building blocks in the same way as molecules are obtained by the combination of atoms. A suitable choice of the building blocks and their appropriate location in the supramolecular array are, of course, the key factors in obtaining new materials capable of performing new functions.

In the context of the immaginative and creative area of Supramolecular Chemistry, the II NATO Science Forum held at Taormina in December 1991 offered the opportunity for very interesting reports and discussions on important challenges: the design, synthesis, and characterization of new building blocks and new assemblies with desired structures and properties for (i) molecular recognition, self-organization, and self-replication, (ii) selective catalysis and reactivity, (iii) energy and electron transfer processes related to the elaboration of signals, information storage, and solar energy conversion, and (iv) encapsulation of radicals and unstable molecules. These and other related topics are extremely important both from a fundamental and an applicative viewpoint. But we should not forget another appealing aspect of Supramolecular Chemistry: namely, the intrinsic beauty of most supramolecular species, the aesthetic value of their architectures; some supramolecular species are not only chemical, but also artistic masterpieces.

The main factor for the success of the Forum was, of course, the high scientific level of the presented lectures, which have been collected in this volume.# Another important factor for the success was the enthusiastic involvement of participants and observers during the long time periods dedicated to formal and informal discussions.

We would like to thank Prof. Jacques Ducuing and Dr. Alain Jubier of the NATO Scientific Affairs Division for their constant help and encouragement. Financial support was mainly offered by NATO, with additional contributions from Consiglio Nazionale delle Ricerche, ENEA, Bracco Industrie Chimiche, Regione Sicilia, Comune di Messina.

We would also like to thank the members of our research group, in particular Massimo Guardigli and Luca Prodi, for their assistance in organizing the meeting. A special thank goes to the staff of the Jolly Hotel Diodoro of Taormina who offered us not only outstanding facilities for the scientific sessions, but also a friendly and peaceful atmosphere in a beautiful surrounding (including fireworks from the Etna volcano!).

Vincenzo Balzani, Luisa De Cola
University of Bologna
Bologna, Italy
February 1992

#Besides 27 papers presented at the Forum, this book also contains two papers that could not be delivered due to the last minute problems that prevented the participation of two expected lecturers.

Organizing Committee

Vincenzo Balzani (Director)
"G. Ciamician" Department of Chemistry
University of Bologna,
40126 Bologna, ITALY

François N. Diederich
Department of Chemistry and Biochemistry
UCLA
Los Angeles, CA 90024, USA

Jean-Marie Lehn
Institut Le Bel
University L. Pasteur
67008 Strasbourg, FRANCE

Richard A. Lerner
Scripps Research Institute
10666 N. Torrey Pines Road
La Jolla, CA 92037, USA

William L. Jorgensen
Department of Chemistry
Yale University
New Haven, Conn. 06511-8118, USA

David N. Reinhoudt
Laboratory of Organic Chemistry
University of Twente
7500 AE Enschede, THE NETHERLANDS

Helmut Ringsdorf
Institute of Organic Chemistry
Johannes-Gutenberg University of Mainz
6500 Mainz, GERMANY

Jean-Pierre Sauvage
CNRS, Institute de Chimie
University L. Pasteur
67008 Strasbourg, FRANCE

Fraser J.Stoddart
Department of Chemistry
University of Birmingham
Birmingham B15 2TT, ENGLAND

George M. Whitesides
Department of Chemistry
Harvard University
Cambridge, MA 02138, USA

List of Participants and Observers

Michael V. ***Alfimov*** *(P)*	Institute of Chemical Physics Academy of Sciences USSR 117421 Moscow, USSR
Pier Lucio ***Anelli*** *(O)*	Bracco Industrie Chimiche Via Folli 50 20134 Milano, ITALY
Jun-ichi ***Anzai*** *(P)*	Pharmaceutical Institute Tohoku University Aobayama, Sendai 980, JAPAN
Yasuhiro ***Aoyama*** *(P)*	Department of Chemistry Nagaoka University of Technology Nagaoka, Kamitomioka 940-21, JAPAN
Roberto ***Ballardini*** *(O)*	FRAE-CNR Institute Via dei Castagnoli 1 40126 Bologna, ITALY
Vincenzo ***Balzani*** *(P)*	"G. Ciamician" Department of Chemistry University of Bologna 40126 Bologna, ITALY
Francesco ***Barigelletti*** *(O)*	FRAE-CNR Institute Via dei Castagnoli 1 40126 Bologna, ITALY
Jan ***Becher*** *(P)*	Department of Chemistry University of Odense 5230 Odense, DENMARK
Paul D. ***Beer*** *(P)*	Inorganic Chemistry Laboratory University of Oxford Oxford OX1 3QR, ENGLAND
Peter ***Belser*** *(P)*	Institute of Inorganic Chemistry University of Fribourg, Perolles Fribourg, SWITZERLAND
Carlo A. ***Bignozzi*** *(O)*	Department of Chemistry University of Ferrara 44100 Ferrara, ITALY

Fabrizio ***Bolletta*** *(O)* — "G. Ciamician" Department of Chemistry
University of Bologna
40126 Bologna, ITALY

Henry ***Bouas-Laurent*** *(P)* — Laboratory of Photochemistry
CNRS, University of Bordeaux I
Talence 33405, FRANCE

Sebastiano ***Campagna*** *(O)* — Department of Inorganic Chemistry
University of Messina
98166 Messina, ITALY

Claudio ***Chiorboli*** *(O)* — CNR- Department of Chemistry
University of Ferrara
44100 Ferrara, ITALY

Edwin C. ***Constable*** *(P)* — Department of Chemistry
University of Cambridge
Cambridge CB2 1EW, ENGLAND

Luisa ***De Cola*** *(P)* — "G. Ciamician" Department of Chemistry
University of Bologna
40126 Bologna, ITALY

Javier ***De Mendoza*** *(P)* — Institute of Organic Chemistry
Universidad Autonoma de Madrid
28049 Cantoblanco, SPAIN

Gianfranco ***Denti*** *(O)* — Institute of Agricultural Chemistry
University of Pisa
56124 Pisa, ITALY

Franz C. ***De Schryver*** *(P)* — Department of Organic Chemistry
Katholieke Universiteit Leuven
B-3030 Leuven (Heverlee), BELGIUM

François N. ***Diederich*** *(P)* — Department of Chemistry and Biochemistry
UCLA
Los Angeles, CA 90024, USA

Geerd H.F. ***Diercksen*** *(P)* — Max Planck Institut für Astrophysik
8046 Garching
Munchen, GERMANY

Heinz ***Dürr*** *(P)* — Department of Organic Chemistry
University of Saarlandes
6600 Saarbrucken, GERMANY

Luigi ***Fabbrizzi*** *(P)*
Department of Chemistry
University of Pavia
27100 Pavia, ITALY

Paolo ***Finocchiaro*** *(P)*
Faculty of Engineering
University of Catania
91125 Catania, ITALY

Maria T. ***Gandolfi*** *(O)*
"G. Ciamician" Department of Chemistry
University of Bologna
40126 Bologna, ITALY

Massimo ***Guardigli*** *(O)*
"G. Ciamician" Department of Chemistry
University of Bologna
40126 Bologna, ITALY

Andrew D. ***Hamilton*** *(P)*
Department of Chemistry
University of Pittsburg
Pittsburg, PA 15260, USA

Anthony ***Harriman*** *(P)*
Center for Fast Kinetics Research
University of Texas
Austin, Texas 78712, USA

Ronald E. ***Hester*** *(P)*
Department of Chemistry
University of York
Heslington, York Y01 5DD, ENGLAND

David W. ***Hutchinson*** *(P)*
Department of Chemistry
Warwick University
Coventry CV4 7AL, ENGLAND

Zbigniew ***Jedlinski*** *(P)*
Polish Academy of Science
ul. Marii Curie - Sklodowskiej 34
41-800 Zabrze, POLAND

William L. ***Jorgensen*** *(P)*
Department of Chemistry
Yale University
New Haven, Conn. 06511-8118, USA

Alain H. ***Jubier*** *(P)*
Scientific Affairs Division
NATO
1110 Bruxelles, BELGIUM

Alberto ***Juris*** *(O)*
"G. Ciamician" Department of Chemistry"
University of Bologna
40126 Bologna, ITALY

Dimitris ***Katakis*** *(P)*
Department of Chemistry
University of Athens
10680 Athens, GREECE

John M. ***Kelly*** *(P)*
Department of Chemistry
Trinity College
Dublin, IRELAND

Yannis ***Kovanis*** *(O)*
Department of Chemistry
University of Athens
10680 Athens, GREECE

Michael G. ***Kuzmin*** *(P)*
Department of Chemistry
Moscow University
Moscow, 117234, USSR

Mauro ***Maestri*** *(O)*
"G. Ciamician" Department of Chemistry
University of Bologna
40126 Bologna, ITALY

Antonino ***Mamo*** *(O)*
Faculty of Engineering
University of Catania
91125 Catania, ITALY

Luigi ***Mandolini*** *(P)*
Department of Chemistry
University "La Sapienza"
00185 Roma, ITALY

Thomas J. ***Meyer*** *(P)*
Department of Chemistry
University of North Carolina
Chapel Hill, NC 27599, USA

Fernando ***Montanari*** *(P)*
Department of Organic Chemistry
University of Milano
20133 Milano, ITALY

Sandra ***Monti*** *(P)*
FRAE-CNR Institute
Via dei Castagnoli 1
40126 Bologna, ITALY

Ana L. ***Moore*** *(O)*
Department of Chemistry
Arizona State University
Tempe, Arizona 85287, USA

Thomas A. ***Moore*** *(P)*
Department of Chemistry
Arizona State University
Tempe, Arizona 85287, USA

José J.G. ***Moura*** *(P)* — Center for Chem. and Biol. Technology
Universidade Nova de Lisboa
P - 2780 Oeiras, PORTUGAL

George R. ***Newkome*** *(P)* — Department of Chemistry
University of South Florida
Tampa, FL 33620, USA

Piero ***Paoletti*** *(P)* — Department of Chemistry
University of Firenze
50144 Firenze, ITALY

Marek ***Pietraszkiewicz*** *(O)* — Institute of Physical Chemistry
Polish Academy of Sciences
01-224 Warszawa, POLAND

Fernando ***Pina*** *(P)* — Department of Chemistry
Universidade Nova de Lisboa
P - 2825 Monte da Caparica, PORTUGAL

Vania ***Poulopoulou*** *(O)* — Department of Chemistry
University of Athens
10680 Athens, GREECE

Luca ***Prodi*** *(O)* — "G. Ciamician" Department of Chemistry
University of Bologna
40126 Bologna, ITALY

Maria A. ***Rampi*** *(O)* — Department of Chemistry
University of Ferrara
44100 Ferrara, ITALY

David N. ***Reinhoudt*** *(P)* — Laboratory of Organic Chemistry
University of Twente
7500 AE Enschede, THE NETHERLANDS

Vittorio ***Ricevuto*** *(P)* — Department of Inorganic Chemistry
University of Messina
98166 Messina, ITALY

Helmut ***Ringsdorf*** *(P)* — Institute of Organic Chemistry
Johannes-Gutenberg University of Mainz
6500 Mainz, GERMANY

Enrico ***Rizzarelli*** *(P)* — Department of Chemical Sciences
University of Catania
95125 Catania, ITALY

Sergio ***Roffia*** *(O)* — "G. Ciamician" Department of Chemistry
University of Bologna
40126 Bologna, ITALY

Nanda ***Sabbatini*** *(O)* — "G. Ciamician" Department of Chemistry
University of Bologna
40126 Bologna, ITALY

Jeremy K.M. ***Sanders*** *(P)* — University Chemical Laboratory
University of Cambridge
Cambridge CB2 1EW, ENGLAND

Jean-Pierre ***Sauvage*** *(P)* — CNRS, Institute of Chemistry
L. Pasteur University
67008 Strasbourg, FRANCE

Franco ***Scandola*** *(P)* — Department of Chemistry
University of Ferrara
44100 Ferrara, ITALY

Hans-Jörg ***Schneider*** *(P)* — Department of Organic Chemistry
University of Saarlandes
6600 Saarbrucken, GERMANY

Paolo ***Scrimin*** *(O)* — Department of Organic Chemistry
University of Padova
35131 Padova, ITALY

Scolastica ***Serroni*** *(O)* — Institute of Agricultural Chemistry
University of Pisa
56124 Pisa, ITALY

Galina A. ***Shagisultanova****(P)* — Department of Chemistry
State Pedagogical University
191186 Leningrad, USSR

Ivan ***Stibor*** *(O)* — Department of Organic Chemistry
Prague Institute of Chemical Technology
166-28 Prague 6, CZECHOSLOVAKIA

*Fraser J.****Stoddart*** *(P)* — Department of Chemistry
University of Birmingham
Birmingham B15 2TT, ENGLAND

Theodore ***Tarasow*** *(O)* — Scripps Research Institute
10666 N. Torrey Pines Road
La Jolla, CA 92037, USA

Fujio ***Toda*** *(P)*	Faculty of Bioscience and Biotechnology Tokyo Institute of Technology Midori-ku, Yokohama, 227 JAPAN
Umberto ***Tonellato*** *(P)*	Department of Organic Chemistry University of Padova 35131 Padova, ITALY
Fulvio ***Uggeri*** *(O)*	Bracco Industrie Chimiche Via Folli 50 20134 Milano, ITALY
Rocco ***Ungaro*** *(P)*	Institute of Organic Chemistry University of Parma 43100 Parma, ITALY
Jan W. ***Verhoeven*** *(P)*	Laboratory of Organic Chemistry University of Amsterdam 1018 WS Amsterdam, THE NETHERLANDS
Fritz ***Vögtle*** *(P)*	Institute of Organic Chemistry University of Bonn D-5300 Bonn 1, GERMANY
Michael D. ***Ward*** *(O)*	School of Chemistry University of Bristol Bristol BS8 1TS, ENGLAND
Michael R. ***Wasielewski*** *(P)*	Chemistry Division Argonne Chemical Laboratory Argonne, IL 60439, USA
Itamar ***Willner*** *(P)*	Department of Organic Chemistry The Ebrew University of Jerusalem Jerusalem 91904, ISRAEL
Raymond ***Ziessel*** *(P)*	EHICS, IPCMS 1, rue Blaise Pascal 67008 Strasbourg, FRANCE

SELF-ASSEMBLY IN CHEMICAL SYSTEMS

PETER R. ASHTON, RICHARD A. BISSELL, DOUGLAS PHILP, NEIL SPENCER, and J. FRASER STODDART

School of Chemistry
University of Birmingham
Edgbaston
Birmingham B15 2TT
United Kingdom

ABSTRACT. Molecular and supramolecular assemblies, some of which should be amenable to electrochemical control, have been assembled from molecular components composed of simple building blocks. Self-assembly in chemical systems appears to take place under very precise constitutional control and, in some cases, it is easier to construct the molecular and supramolecular assemblies than it is to make some of the molecular components on their own.

1. OPENING REMARKS

Nature uses self-assembly processes [1] to construct large and ordered molecular arrays for addressing certain prescribed functions. These self-assembly processes are used by Nature in a variety of ingenious ways to create structures and superstructures at the sub-cellular and cellular levels. Just as recognition between molecular components is the key to self-assembly in natural systems, so we can expect [2] that molecular recognition will have to be an integral part of any wholly synthetic chemical system constructed by self-assembly processes. A truly challenging goal for synthetic chemists is to find out how to design and construct nanometre-scale structures and superstructures from appropriate and inexpensive molecular components. If this objective can be reached then new materials, which will allow us to store and transfer information at a molecular level in nanometre-scale devices [3], will result.

We have evolved a synthetic strategy that has relied upon the formation of catenanes and rotaxanes in the beginning. We regard these compounds - with their mechanically interlocked molecular components - as prototypes for the construction of large and ordered molecular arrays. Once we have established for the chemical systems

V. Balzani and L. De Cola (eds.), Supramolecular Chemistry, 1–16.

of our choice, the principles underlying an efficient self-assembly process, then we can relax the initial constraint of mechanical interlocking and replace it by entanglement of a threading or intertwining nature.

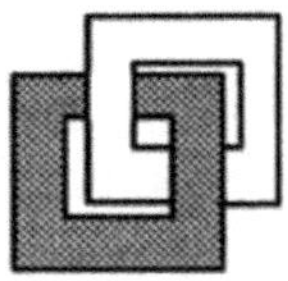

Catenane

Rotaxane

The story is now told in a sequence of graphic displays with some supporting text provided at the top of each display.

2. GRAPHIC DISPLAYS

Display 1 shows how a [2]catenane and a [2]rotaxane can be self-assembled. A [2]catenane can be made by **(a)** clipping an unshaded ring around an already complete shaded ring, or **(b)** *vice versa.* A [2]rotaxane can be made by **(c)** threading a complete shaded ring with an unshaded rod and then covalently stoppering the ends with black stoppers, or by **(d)** clipping a shaded ring around an unshaded rod that is already covalently linked to big black stoppers at each end.

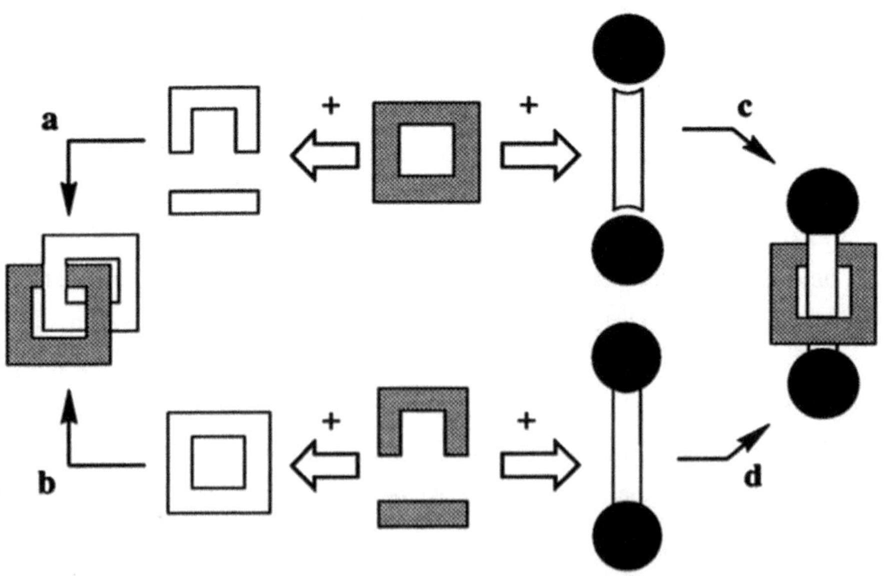

Display 2 shows the final step in the preparation of **BisParaPhenylene-34-Crown-10 (BPP34C10)** and its complex formation with **ParaQuaT** ([**PQT**]$^{2+}$). Thus, the 1:1 complex is stabilised by (i) electrostatic interactions, including [C-H..O] hydrogen bonding between the methyl groups and the hydrogen atoms α to the nitrogen atoms on the bipyridinium unit of the [**PQT**]$^{2+}$ dication and some of the polyether oxygen atoms in the **BPP34C10** macrocycle, and (ii) dispersive forces, including charge-transfer interactions between the π-electron rich hydroquinol rings in **BPP34C10** and the π-electron deficient bipyridinium unit of the [**PQT**]$^{2+}$ dication. The complex is a very deep orange colour both in solution and in the solid state. See reference [4].

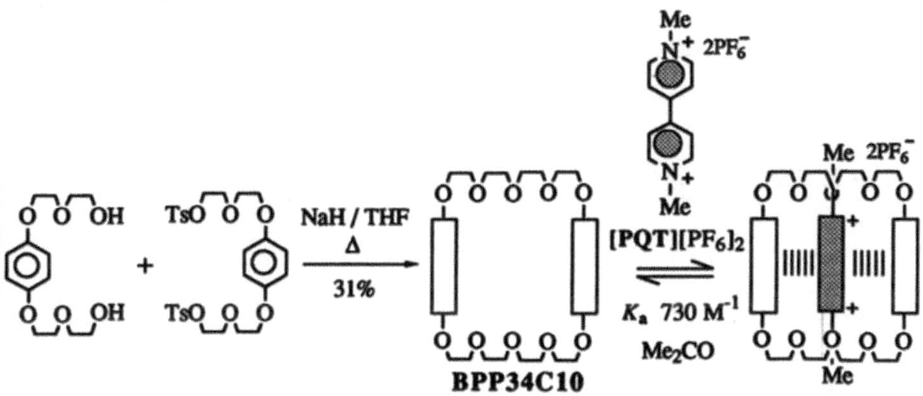

Display 3 shows the final step in the preparation of the **BisBIPY**iridinium **BIsXY**lyl **CY**clophane ([**BBIPYBIXYCY**]$^{4+}$) and its complex formation with 1,4-dimethoxybenzene (**1/4DMB**). This 1:1 complex is also a deep orange colour reflecting its stabilisation by charge transfer and other dispersive forces between the π-donor and acceptor units. See references [5] and [6].

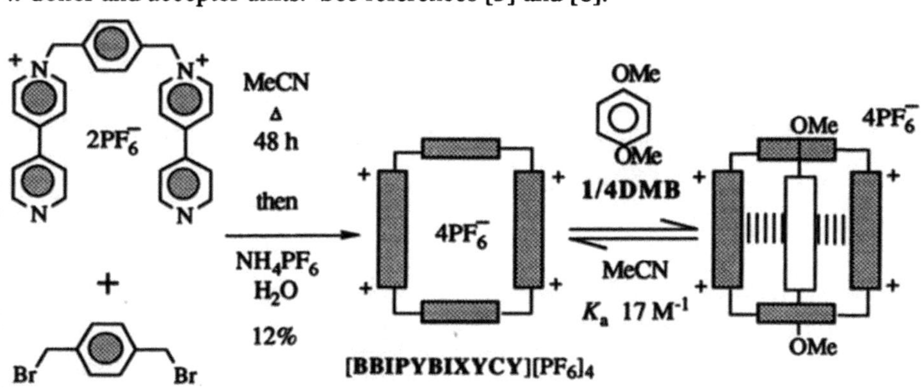

Display 4 shows a [2]catenane being made to order. The **BPP34C10** is present in excess. See reference [7].

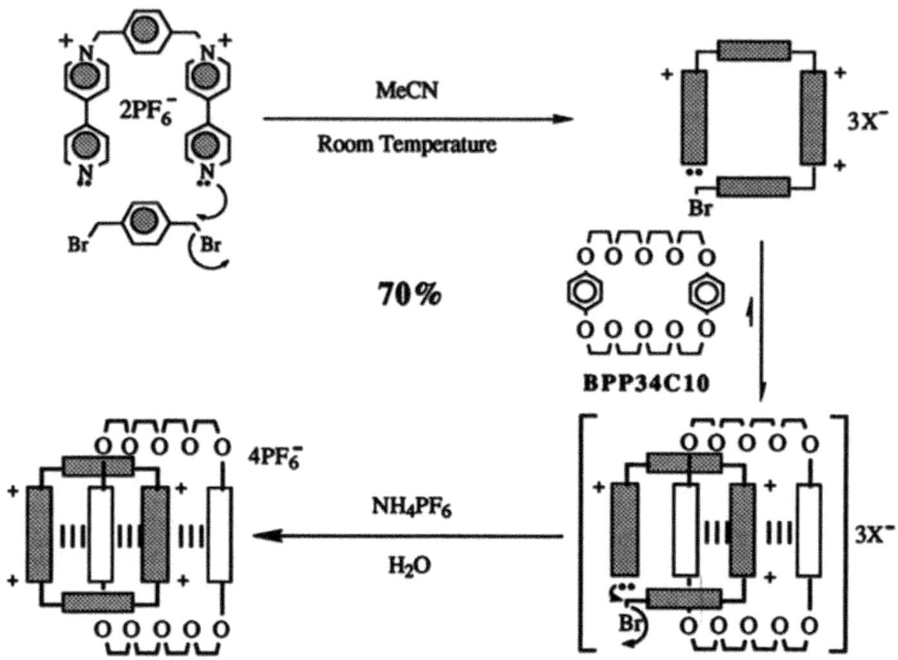

Display 5 shows a [2]rotaxane being made to order by the clipping procedure. See reference [8].

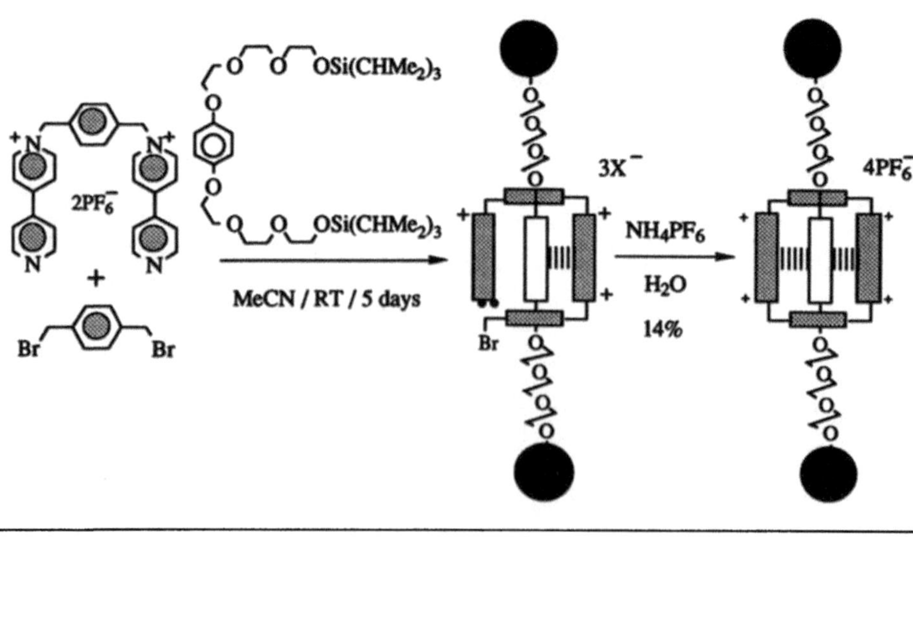

Display 6 shows how the yield of [**BBIPYBIXYCY**][PF_6]$_4$ can be progressively enhanced by templating, by changing solvents, by adding sodium iodide to act as a catalyst, and finally by carrying out the reaction at ultra-high pressure. See reference [8].

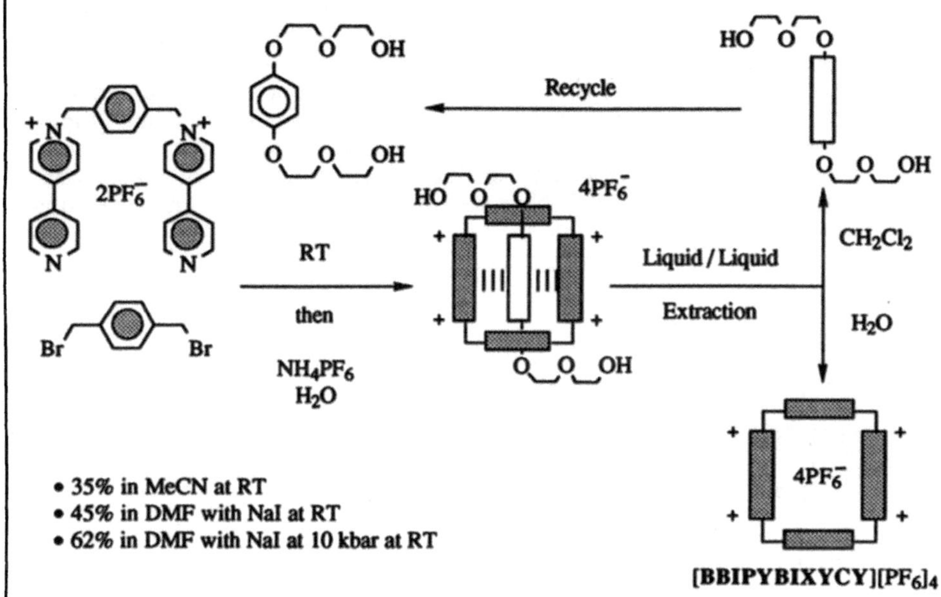

Display 7 showing how the oxidation potential of hydroquinol rings depend on their environment in different compounds dissolved in acetonitrile. See reference [8].

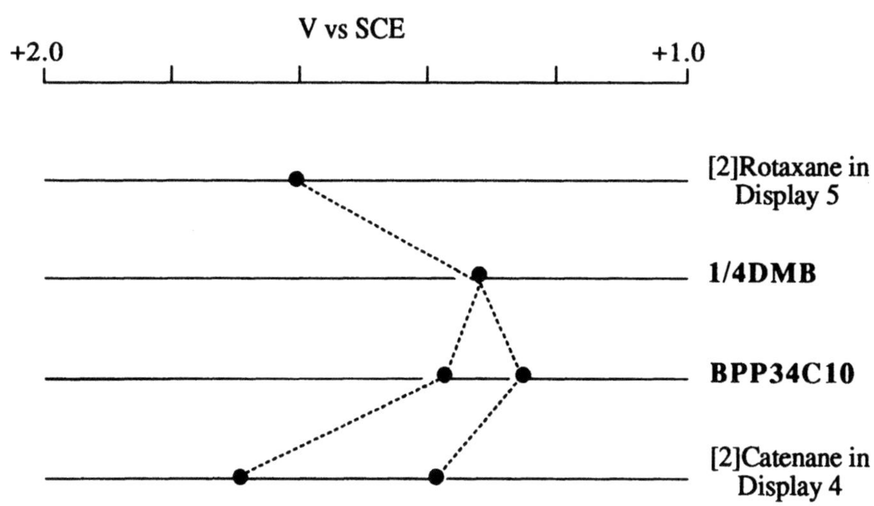

Display 8 shows the step that can be taken conceptually from the [2]catenane - at least as neighbouring molecules exist in the crystal - to a [3]pseudorotaxane. The molecular components of the pseudorotaxane are held together not only by their mutual noncovalent bonding attraction and but also by a mechanical feature as in a rotaxane. The polyrotaxane-like superstructure can be viewed as part of a logical progression from the situation which exists in the solid state structure of the [2]catenane - wherein molecules are positioned side-by-side so as to establish a repeating unit of π-donor and π-acceptor units along one particular crystallographic axis - to the double helical type superstructure in which two polymer chains, one containing π-donor units appropriately spaced and the other containing π-acceptors units appropriately spaced, are intertwined. Such macromolecules are reminiscent of DNA, RNA, and some proteins in the way in which one polymer chain recognises another polymer chain.

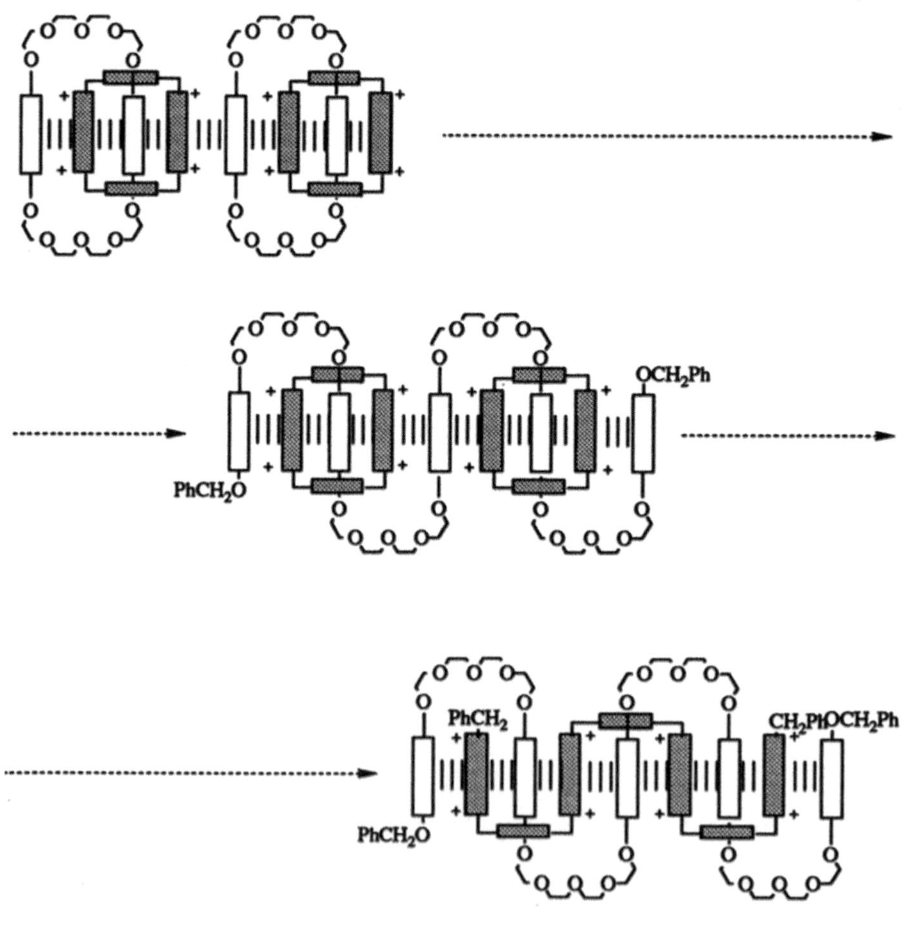

Display 9 outlines the synthetic route by which oligomers containing **three (3HQ)** and **five (5HQ)** hydroquinol residues have been prepared starting from hydroquinol. See reference [9].

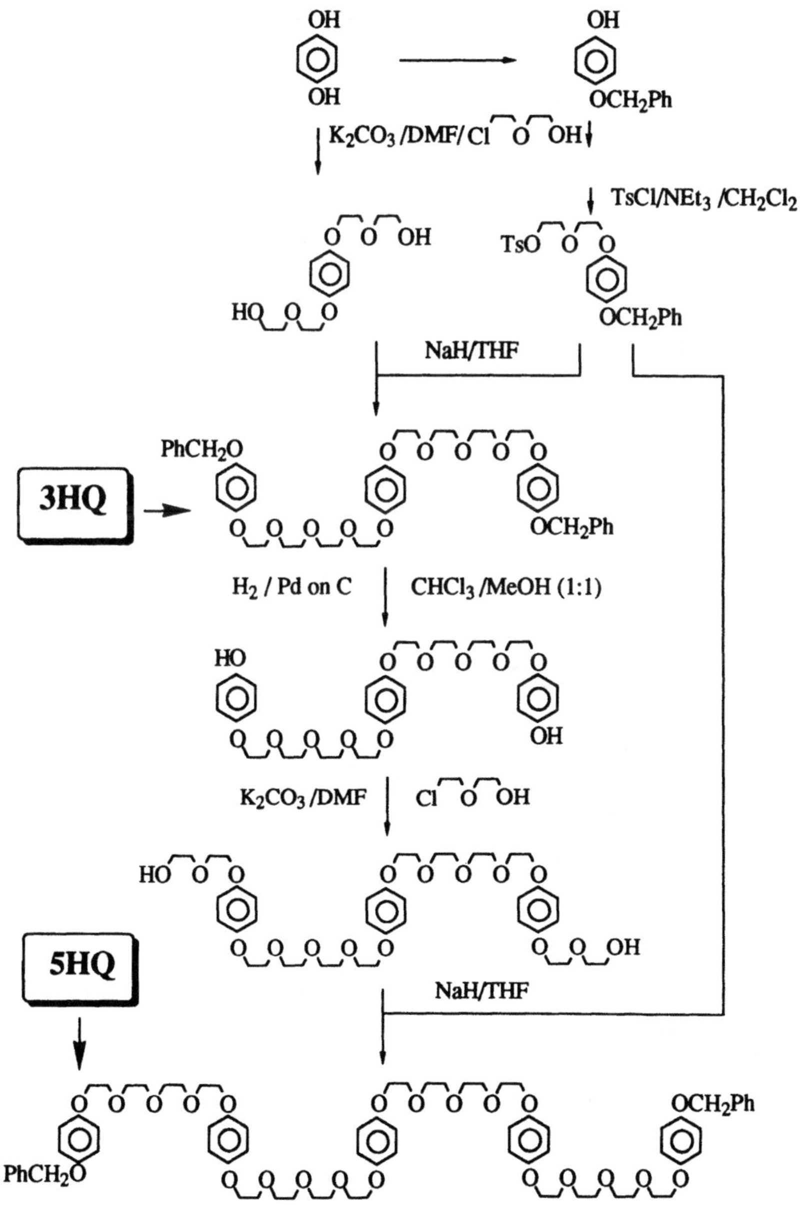

Display 10 outlines the synthetic route by which oligomers containing **two (2HQ)** and **four (4HQ)** hydroquinol residues have been prepared starting from 4-benzyloxyphenol. See reference [9].

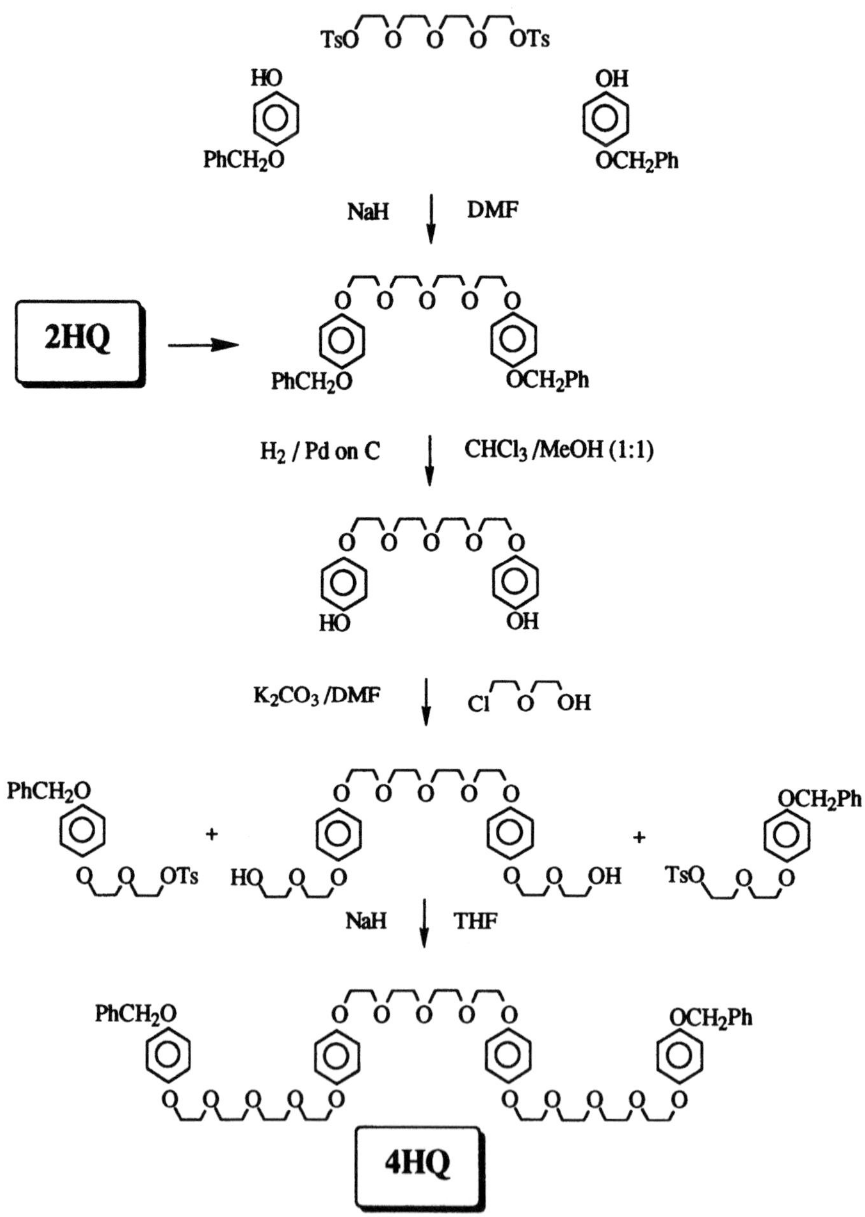

Display 11 shows the self-assembly of a [2]pseudorotaxane from **3HQ** and **[BBIPYBIXYCY]**$[PF_6]_4$ in acetonitrile. The ^1H NMR spectrum recorded in CD_3CN at -40°C reveals a singlet resonating at δ3.43 for the protons on the central hydroquinol ring and AA'BB' system centred on δ6.36 and δ6.58 for the protons on the two outer hydroquinol rings. The X-ray crystal structure of the 1:1 complex indicates a superstructure in which the **3HQ** oligomer is centrosymmetrically threaded through the centre of the tetracationic cyclophane such that the three π-electron rich hydroquinol rings enter into π–π stacking interactions with the two π-electron deficient bipyridinium units. The mean plane of the central hydroquinol ring is separated from the mean planes of the bipyridinium rings on each side of it by *ca.* 3.5Å. This central hydroquinol ring also enters into electrostatic T-type edge-to-face interactions with the *p*-xylyl spacer units in the **[BBIPYBIXYCY]**$^{4+}$ tetracation. See reference [10].

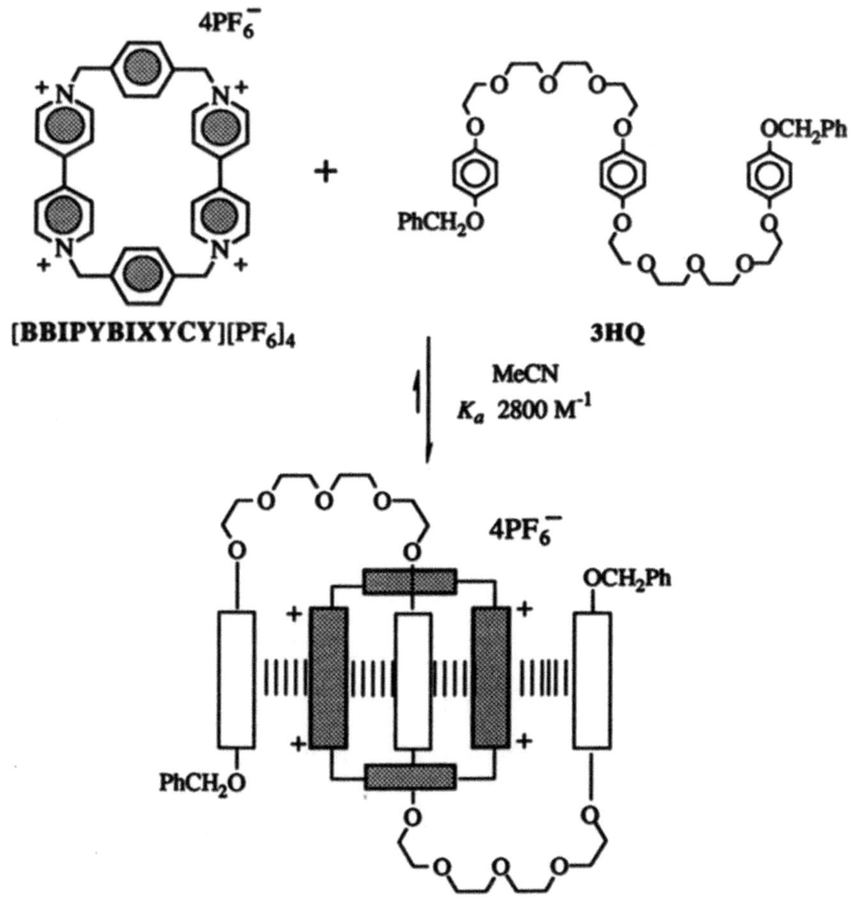

Display 12 shows the self-assembly of a [3]pseudorotaxane from **5HQ** and two molar equivalents of **[BBIPYBIXYCY]**[PF_6]$_4$ in acetone. The ^1H NMR spectrum recorded in CD_3COCD_3 at -40°C reveals a singlet resonating at δ6.22 for the protons on the central hydroquinol ring trapped between the two tetracationic cyclophanes and another singlet resonating at δ3.70 for the protons on the hydroquinol rings located inside the cyclophanes: an AA'BB' system is observed in the range δ6.39-6.62 for the protons on the two outermost hydroquinol rings. Assuming a tightly packed π–π stacked structure for the 2:1 complex, the [3]pseudorotaxane is *ca.* 30Å long with a diameter of around 15Å.
See reference [11].

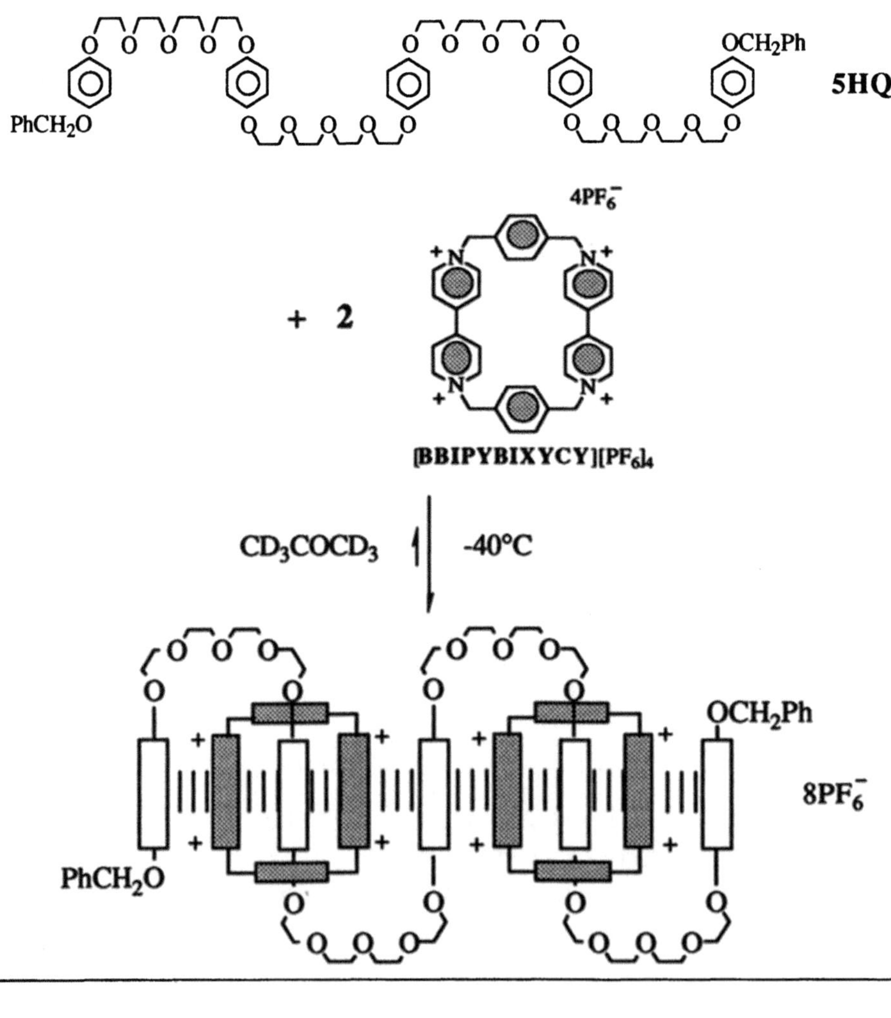

Display 13 shows the self-assembly of a [2]pseudorotaxane from **5HQ** and one molar equivalent of **[BBIPYBIXYCY]**[PF_6]$_4$ in acetone. On cooling the CD_3COCD_3 solution to -40°C, evidence is obtained from the ^{1}H NMR spectrum which suggests that the tetracationic cyclophane spends 40% of its time on the central hydroquinol ring of the **5HQ** oligomer and the rest of its time equally between the hydroquinol rings second and fourth along the oligomeric chain. See reference [11].

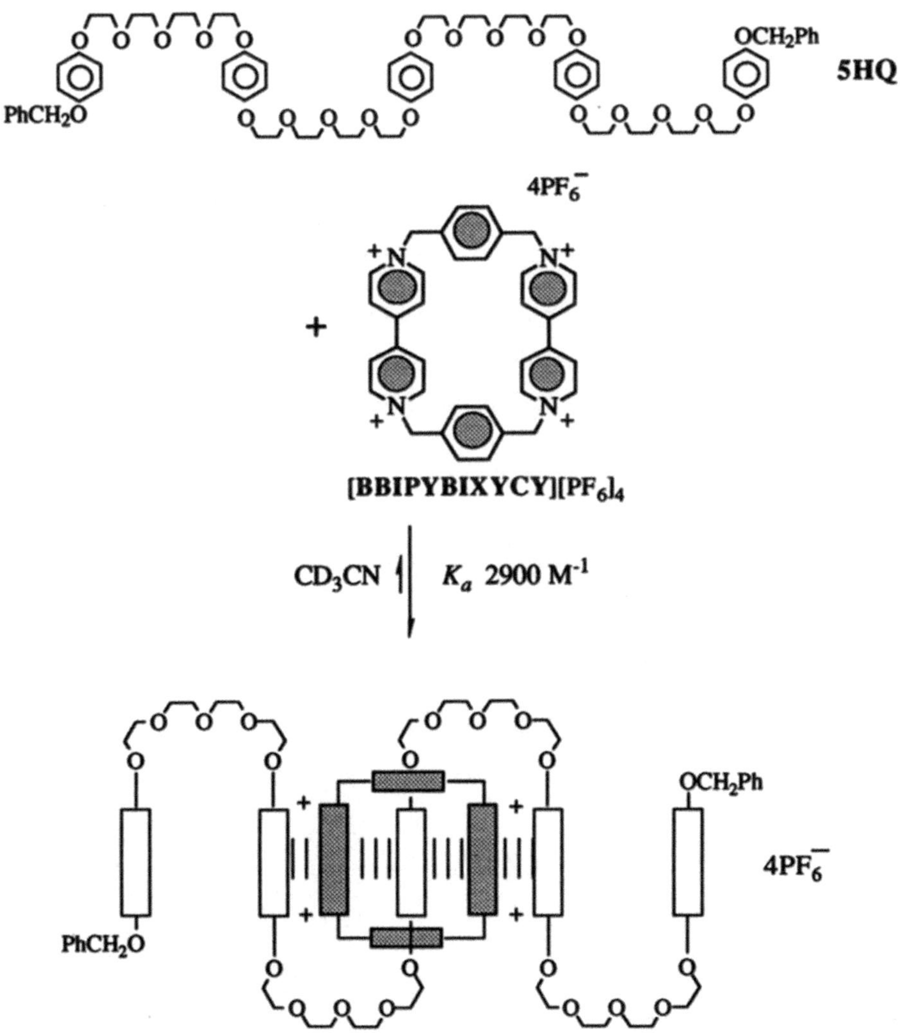

Display 14 shows a [2]rotaxane being made to order by a clipping procedure. Note that the yield of 32% is much higher than that recorded earlier in Display 5. The reason for the increased efficiency of the self-assembly process is undoubtedly the presence of two hydroquinol rings in the thread component of the [2]rotaxane. See reference [12].

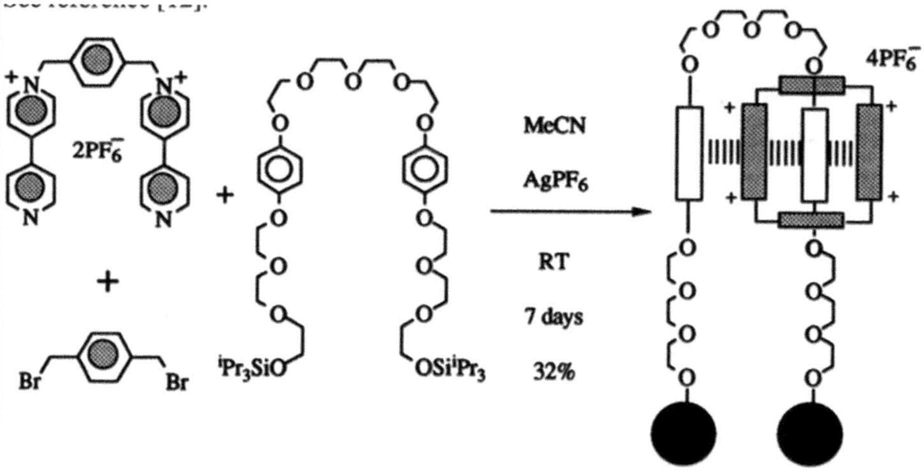

Display 15 reveals that the [2]rotaxane whose preparation is described in Display 14 behaves like a molecular shuttle. Dynamic ^1H NMR spectroscopy in CD_3COCD_3 solution indicates that the tetracationic cyclophane is darting back and forth 500 times per second between the two hydroquinol rings. See reference [12].

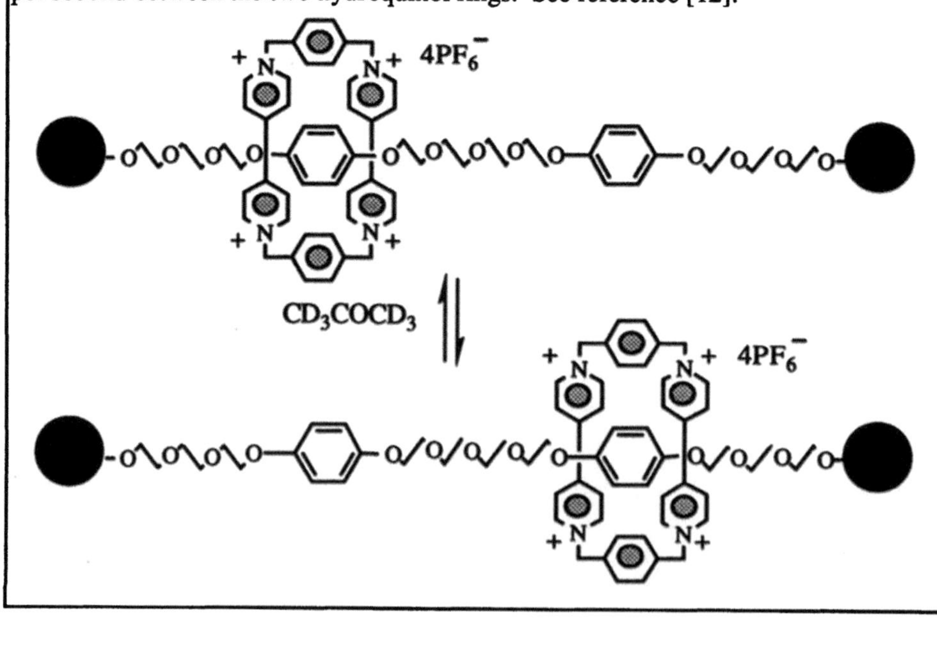

Display 16 considers how a molecular shuttle may be driven electrochemically on the basis of the knowledge that the oxidation potentials of hydroquinol and *p*-xylene are approximately +1.3 V and >+1.7 V, respectively. The theory goes something like this -

(1) The hydroquinol ring is more π-electron rich than the *p*-xylyl residue.

(2) The hydroquinol ring should be occupied preferentially by the tetracationic cyclophane.

(3) The hydroquinol ring should be more easily oxidised than the *p*-xylyl ring.

(4) On oxidation, the π-donating ability of the hydroquinol ring will decrease.

(5) The tetracationic cyclophane should now preferentially occupy the *p*-xylyl residue.

(6) Subsequent reduction should regenerate the hydroquinol ring and the bead will return to its starting position, completing the switching cycle.

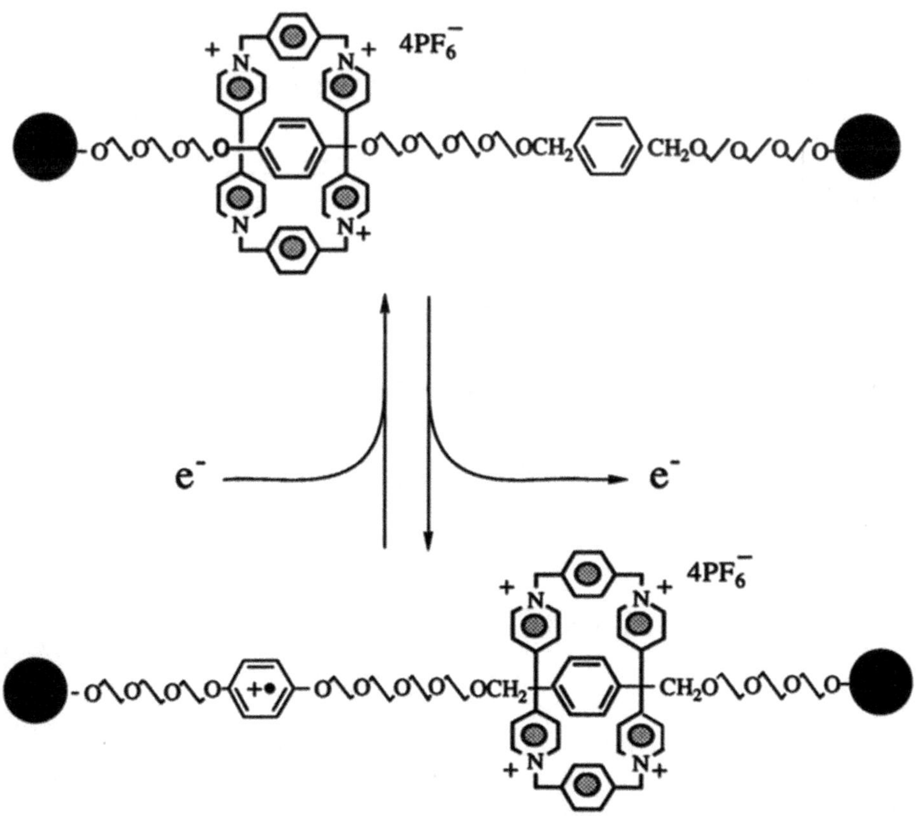

Display 17 outlines the synthetic route for the synthesis of an asymmetric thread containing a hydroquinol ring and a *p*-xylyl residue.

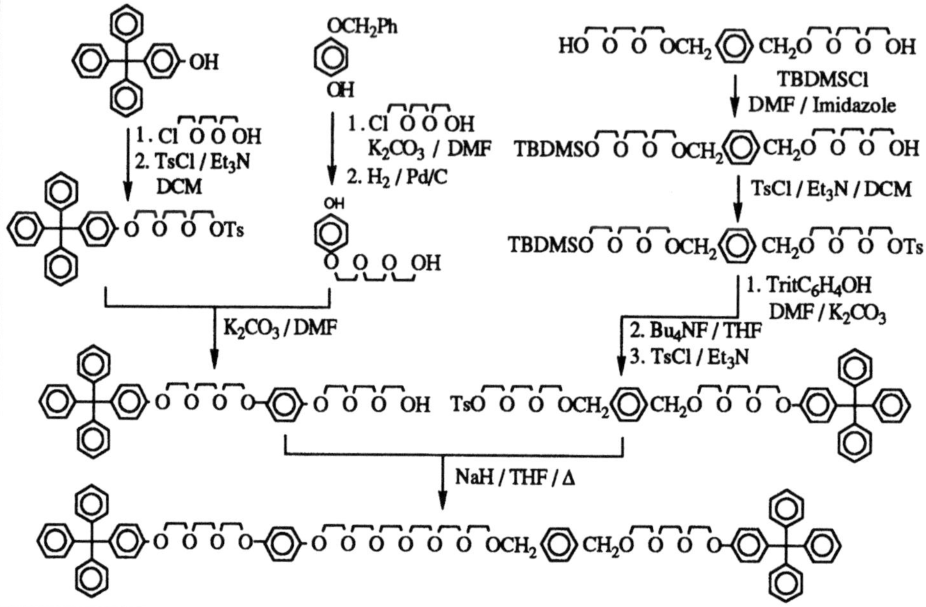

Display 18 describes the preparation of a [2]rotaxane which constitutes an asymmetric molecular shuttle. Preliminary examination of the temperature dependent ^{1}H NMR spectrum of the [2]rotaxane reveals that the tetracationic cyclophane prefers to encircle the hydroquinol ring rather than the *p*-xylyl residue, as predicted. The molecular assembly has still to be investigated electrochemically. The hoop represents the tetracationic cyclophane, **[BBIPYBIXYCY]**$^{4+}$.

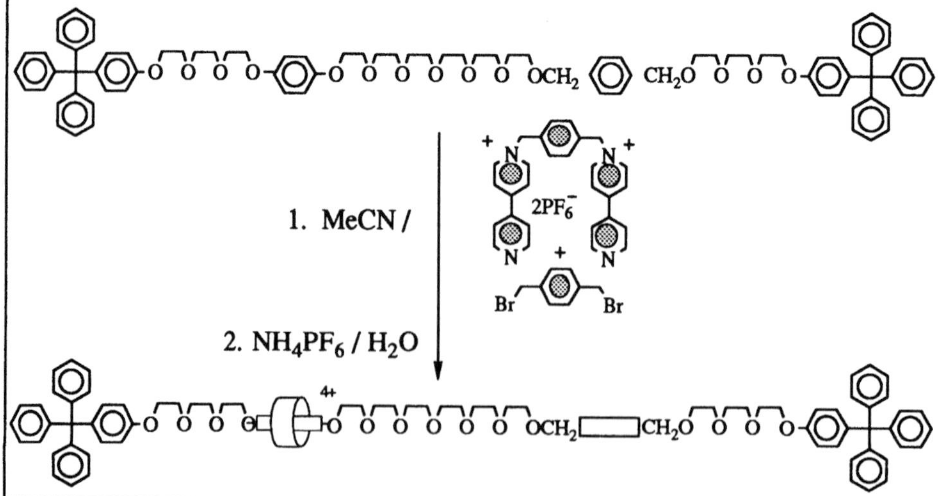

3. CONCLUDING REMARKS

The following conclusions can be drawn on the basis of this programme of research into self-assembly in chemical systems —

- Molecular and supramolecular assemblies have been constructed which should prove to be amenable to being addressed and controlled electrochemically.

- Molecular and supramolecular assemblies have been formed from molecular components comprised of simple and inexpensive building blocks.

- Self-assembly occurs under very precise constitutional control.

- It is often easier to make molecular and supramolecular assemblies than it is to make molecular components on their own.

Self-assembly in chemical systems offers a new and exciting conceptual challenge to synthetic chemists. The irony of what is beginning to happen in the development of synthetic chemistry is intriguing to behold. Just as **natural products** and their analogues have been synthesised more often than not by quite **unnatural** means, so it is that **unnatural products** and new materials will be prepared increasingly by approaches that are reminiscent of **natural** processes.

4. REFERENCES

[1] Lindsey, J.S. (1991) 'Self-Assembly Processes' *New J. Chem.* **15**, 153.

[2] Philp, D. and Stoddart, J.F. (1991) 'Self-Assembly in Organic Synthesis' *Synlett* , 445.

[3] Drexler, K.E. (1990) Engines of Creation, Fourth Estate, London.

[4] Allwood, B.L., Spencer, N., Shariari-Zavareh, H., Stoddart, J.F. and Williams, D.J. (1987) 'Complexation of Paraquat by a Bisparaphenylene-34-crown-10 Derivative' *J. Chem. Soc., Chem. Commun.* 1064.

[5] Odell, B., Reddington, M.V., Slawin, A.M.Z., Spencer, N., Stoddart, J.F. and Williams, D.J. (1988) 'Bisparaquat(1,4)cyclophane. A Tetracationic Multipurpose Receptor' *Angew. Chem. Int. Ed. Engl.* **27**, 1547.

[6] Ashton, P.R., Odell, B., Reddington, M.V., Slawin, A.M.Z., Stoddart, J.F. and Williams, D.J. (1988) 'Isostructural Alternately-charged` Receptor Stacks. The Inclusion Complexes of Hydroquinol and Catechol Methyl Ethers with Bisparaquat(1,4)cyclophane' *Angew. Chem. Int. Ed. Engl.* **27**, 1550.

[7] Ashton, P.R., Goodnow, T.T., Kaifer, A.E., Reddington, M.V., Slawin, A.M.Z., Spencer, N., Stoddart, J.F., Vicent, C. and Williams, D.J. (1989) 'A [2]Catenane Made to Order' *Angew. Chem. Int. Ed. Engl.* **28**, 1396.

[8] Anelli, P.L., Ashton, P.R., Ballardini, R., Balzani, V., Delgado, M., Gandolfi, M.T., Goodnow, T.T., Kaifer, A.E., Philp, D., Pietraszkiewicz, M., Prodi, L., Reddington, M.V., Slawin, A.M.Z., Spencer, N., Stoddart, J.F., Vicent, C. and Williams, D.J. 'Molecular Meccano 1 - [2]Rotaxanes and a [2]Catenane Made to Order' *J. Am. Chem. Soc.* In press.

[9] Ashton, P.R., Philp, D., Spencer, N. and Stoddart, J.F. 'Self-Assembling Synthetic Supramolecular Polymers' *Makromol. Chem.* In Press.

[10] Anelli, P.L., Ashton, P.R., Spencer, N., Slawin, A.M.Z., Stoddart, J.F. and Williams, D.J. (1991) 'Self-Assembling [2]Pseudorotaxanes' *Angew. Chem. Int. Ed. Engl.* **30**, 1036.

[11] Ashton, P.R., Philp, D., Spencer, N. and Stoddart, J.F. (1991) ' The Self-Assembly of [*n*]Pseudorotaxanes' *J. Chem. Soc., Chem. Commun.* 1677.

[12] Anelli, P.L., Spencer, N. and Stoddart, J.F. (1991) 'A Molecular Shuttle' *J. Am. Chem. Soc.* **113**, 5131.

HYDROGEN BONDING IN SUPRAMOLECULAR FUNCTIONS

YASUHIRO AOYAMA
Department of Chemistry
Nagaoka University of Technology
Kamitomioka, Nagaoka, Niigata 940-21
Japan

ABSTRACT. Here are described some newer aspects, in relation to supramolecular function developments, of host-guest complexation based on multipoint hydrogen-bonding interactions in apolar organic media. The following topics are covered. A supramolecular strategy for the functional-structure construction provides a very simple route to a great variety of face-to-face porphyrin-quinone architectures. A supramolecular strategy for the structure elucidation presents an exciton-chirality-induction method for the assignments of absolute stereochemistry of chiral molecules. A supramolecular strategy for the three-dimensional encapsulation leads to versatile polar-guest binding via a polar microsolvation effect. Some future prospects of supramolecular chemistry are also briefly referred to. These include (1) supramolecular control of organic solid and crystal structures by using divergent functional groups and (2) supramolecular complexation of polar guests in aqueous media as a result of balanced cooperation of polar and apolar interactions.

1. Introduction

Molecular recognition mainly deals with selective complexation of the host-guest type. In order to achieve *selectivity*, a host molecule must be rigid so as to allow a preorganized binding site which is complementary in size and polarity to a particular guest [1]. Selective complexation itself is an important object, but more than that. It plays an essential role in the various aspects of supramolecular functions [2].

A rapidly growing area of molecular recognition is concerned with the polar host-guest interaction based on multipoint hydrogen-bonding in apolar organic media [3]. The use of hydrogen bonding as a driving force of complexation has opened the door to a rich field of molecular recognition involving *common* organic molecules which are usually more or less polar. In fact, not only nitrogen-containing molecules [3] but also oxygen-functionalized compounds can now be regarded as good targets. Examples include di(poly)ols and steroids [4,5], sugars [5,6], dicarboxylic acids [7], amino acids [8], phosphoric acids [9], esters [10], and quinones [11].

The present review is intended to provide some new and novel aspects of the supramolecular chemistry. The contents are based on the author's own work on multipoint hydrogen-bonding in apolar organic solutions. Molecular recognition of polar compounds in the solid states and in aqueous media are also introduced in view of potential application of supramolecular chemistry to the material science and technology as well as the life science and technology.

V. Balzani and L. De Cola (eds.), Supramolecular Chemistry, 17–30.

2. Supramolecular Construction of Functionally Significant Molecular Architectures

The charge separation processes in photosynthesis involve photoinduced electron transfer from chlorophyll to quinone derivatives. A popular methodology of biomimetic chemistry for biological active sites is assembly of supposedly essential components via covalent linkage. Thus, a variety of *covalently* linked porphyrin-quinone derivatives such as **1** and **2** below have been prepared and photoinduced electron transfer therein investigated as models of photosynthetic electron transfer [12]. There are two major difficulties in the covalent approach to the preparation of this type of face-to-face porphyrin-quinone derivatives. The first is synthetic problems associated with the preparation of highly substituted quinone precursors and the porphyrin-quinone double coupling reactions. Second, there is little room for the systematic modification of the electronic/steric structures of quinones.

Two-point hydrogen-bonding interactions present a noncovalent *supramolecular* route to such face-to-face porphyrin-quinone architectures [11]. Thus, an octaethylporphyrin derivative **3** having 2-hydroxynaphthyl groups at 5- and 15-meso positions in a cis configuration reversibly binds various 1,4-quinones in chloroform (Scheme 1). The simultaneous two-point hydrogen-bonding was convincingly evidenced by spectroscopic means as well as X-ray crystallography. The binding constant for anthraquinone (**5**) at 298 K is $K = 2.3 \times 10^{2}\ M^{-1}$. On the other hand, *trans*-bis(2-hydroxynaphthyl)porphyrin **4** and anthrone (**6**) respectively as one-point reference host and guest form far less stable

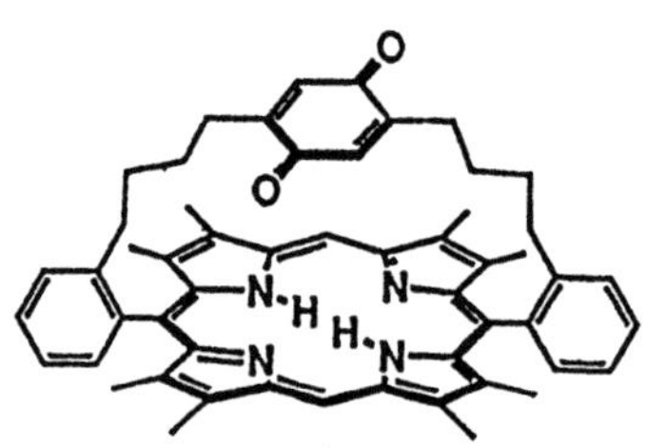

1

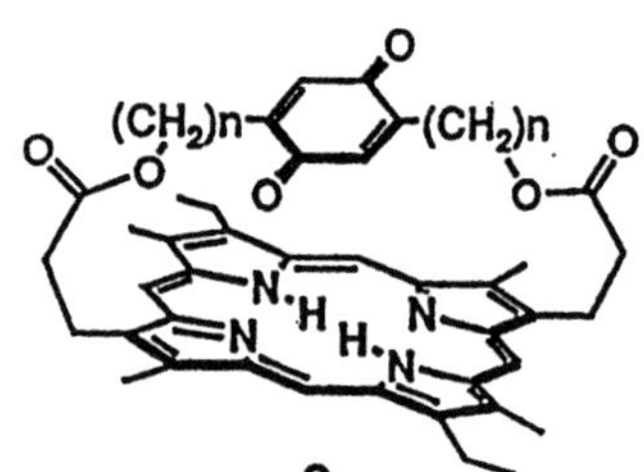

2

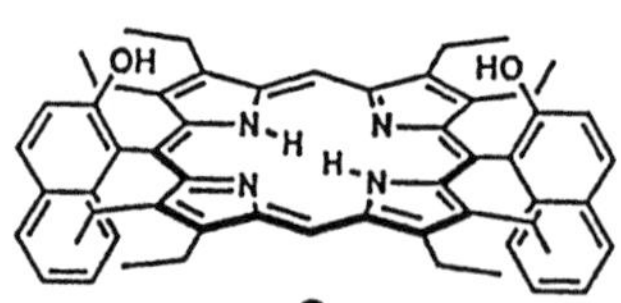

3

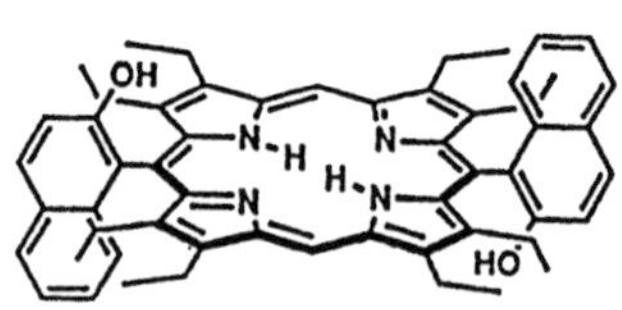

4

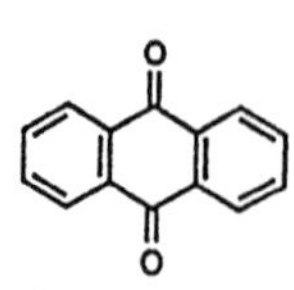

5

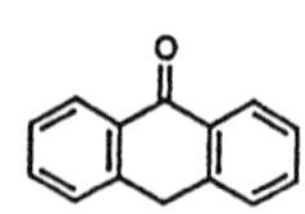

6

Scheme 1

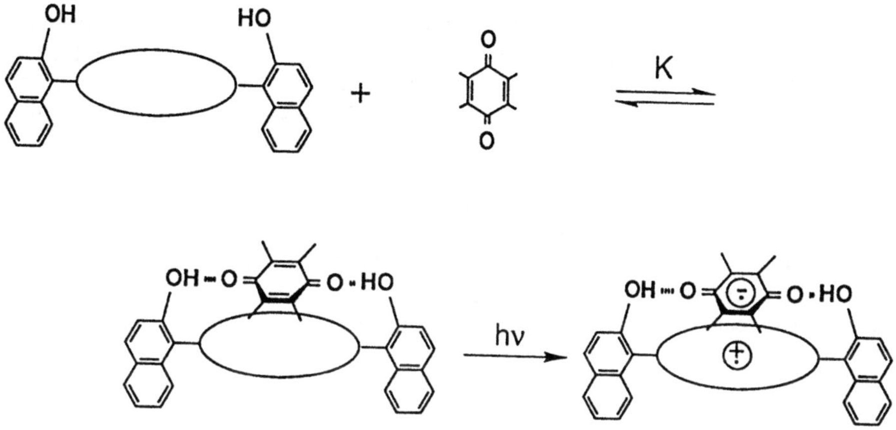

Scheme 2

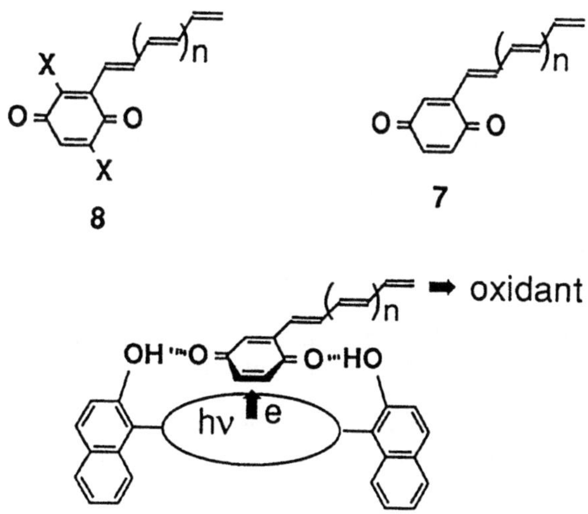

complexes; $K = 4.2$ (**3·6**), 5.9 (**4·5**), and 4.5 M^{-1} (**4·6**). These results clearly point to the importance of two-point host-guest hydrogen-bonding. The stability of a resulting two-point adduct is governed not only by the strength of the hydrogen bonds (as reflected on the complexation-induced shifts in δ(OH) and ν(OH)) but also by the porphyrin-quinone interaction of the charge transfer type. Thus, both electron-donating and electron-withdrawing substituents in a guest quinone promotes host-guest complexation.

Examination of CPK molecular models indicates that the face-to-face separation of the porphyrin and quinone rings is as short as ~3 A. This enforced proximity of the two rings results in their significant π-π electronic interaction. In addtion, the porphyrin fluorescence of the adducts is completely quenched by an efficient photoinduced electron transfer from host porphyrin to bound quinone (Scheme 1). The efficiency of fluorescence quenching by various quinones parallels the binding constants for adduct formation, indicating that the electron transfer is intracomplex in nature. Thus, photoreduction of weakly oxidizing, electron-rich quinones such as duroquinone is readily achieved in this manner.

The present host-guest complexation provides a synthetically very simple route to a great variety of face-to-face porphyrin-quinone architectures. The present method may exert it's maximal usefulness when one considers to construct a suramolecular device for charge separation using a functional quinone such as **7** (Scheme 2). This quinone is functionalized by a molecular wire. All you have to do is prepare a trisubstituted benzene derivative **7**. If you follow the covalent approach, you have to prepare a pentasubstituted derivative **8**, where the substituent X is essential as the site of double coupling with a suitably disubstituted porphyrin derivative.

3. Supramolecular Probing of the Stereochemistry of Chiral Guests

Circular-dichroism (CD) is specific to chiral molecules that absorb light. When a chiral molecule has two or more light-absorbing units (chromophores), exciton coupling therein results in split Cotton effects, the signs of which can be correlated with the absolute structure of the molecule. The stereochemistry of chiral molecules can thus be elucidated via the exciton chirality method [13]. This method has been successfully applied to the determination of absolute configuration of chiral glycols after their conversion to light-absorbing dibenzoate derivatives. An essential point of this method is that the chirality of the two C-OH bonds (either clockwise or counterclockwise on going from the front one to the rear one) is the same as that of the two transition moments located on the benzene rings (Scheme 4, *vide infra*) and that the chirality of the transition moments determines the signs of split Cotton effects.

A similar probing of stereochemistry can also be done supramolecularly based on *interaction* (complexation) with an achiral multibenzenoid host [5]. Resorcinol cyclic tetramer **9** is a multidentate host capable of complexing various polar guests via multipoint hydrogen bonding interactions involving the four independent binding sites (A, B, C, and D) composed of a pair of hydrogen-bonded hydroxyl groups [4-7]. Host **9** is achiral. It forms hydrogen-bonded 1:1 complexes with a variety of chiral glycols in chloroform. The resulting adducts exhibit induced CD with split Cotton effects as a result of exciton chirality induction in otherwise symmetric **9** upon binding of a chiral guest. The actual spectra are shown in Figure 1 for the enantiomeric 1-phenyl-1,2-ethanediol (S-**10** and R-**10**). The signs of split Cotton effects (either positive-to-negative or negative-to-positive on going from longer to shorter wavelengths) for complexes derived from various chiral glycols were found to be correlated with the chiralities or the absolute configurations of the glycols.

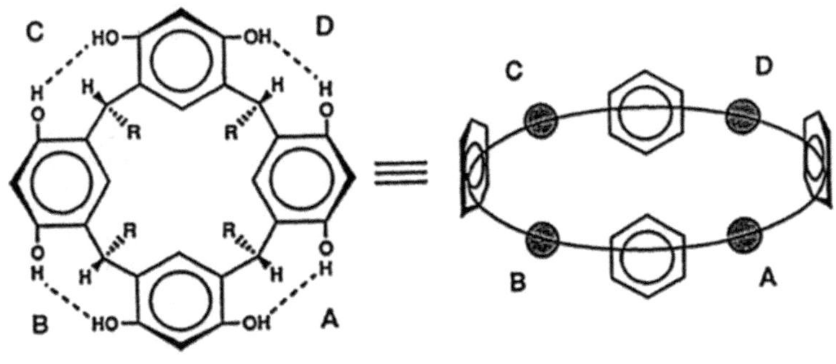

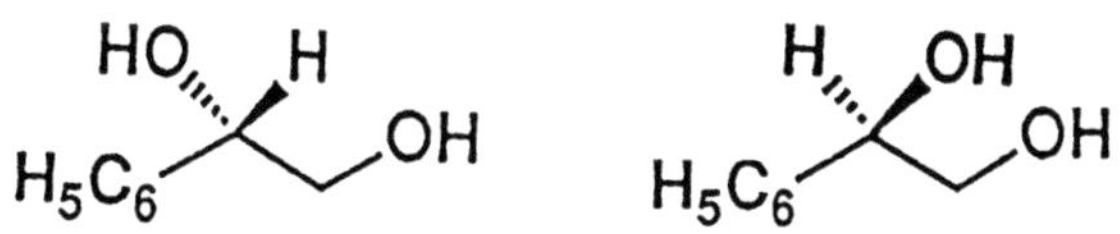

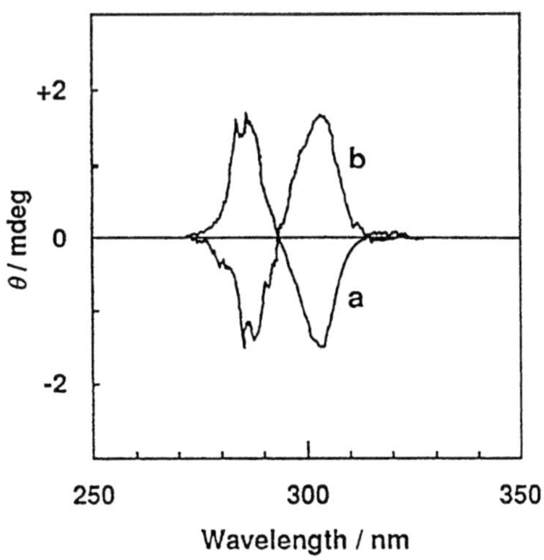

Figure 1. Induced circular-dichroism spectra for $CHCl_3$ solutions (0.1-cm path-length) of host **9** (1.0 x 10^{-2} M) and guest S-**10** (a) or R-**10** (b) (2.0 x10^{-2} M).

Scheme 3

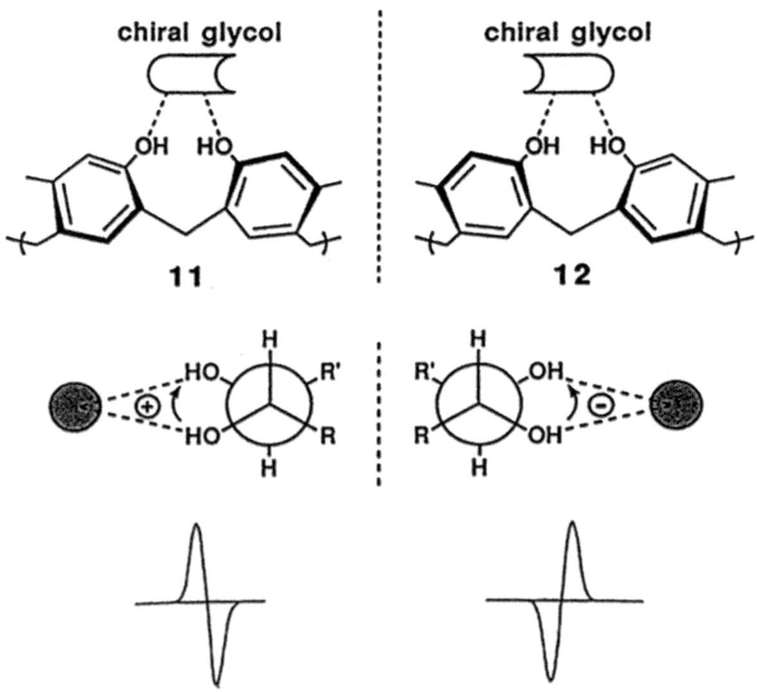

chiral glycol
OH
HO
11
chiral glycol
OH
HO
12
H
HO
R'
HO
R
H
H
R'
OH
R
OH
H

Host **9**, by virtue of the present exciton-chirality-induction method, can also be used as a very novel supramolecular probe for the assignments of ring conformations (either C1 or 1C) of bound sugars [5]. More recent study of ours indicates that the present method is also applicable to chiral monools including 2-alkanols (**13**), menthol (**14**), borneol (**15**), and steroidal monools (**16**) such as cholestanol [14]. The absolute configuration (R or S) of the OH-carrying asymmetric carbon atom of a chiral monool can be supramolecularly assigned on the basis of the signs of split Cotton effects upon compexation with host **9**.

Scheme 4

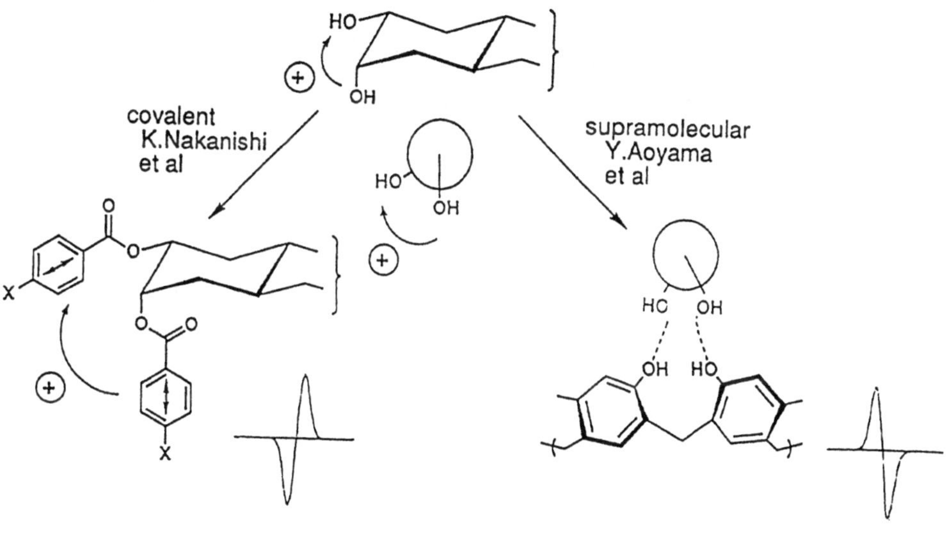

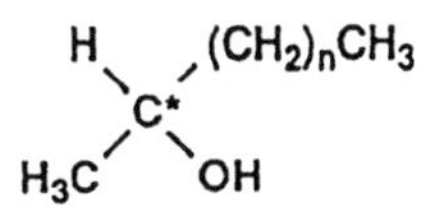

13

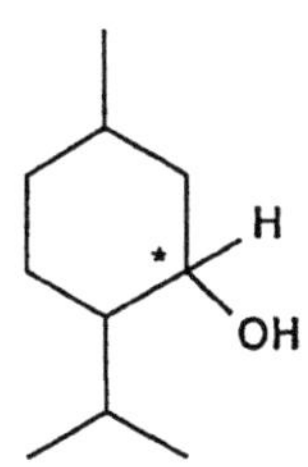

14

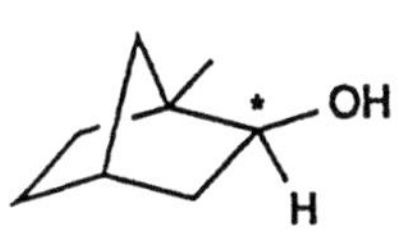

15

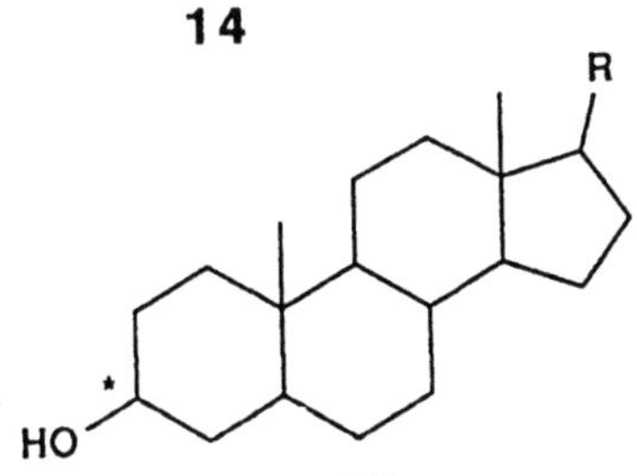

16

4. Supramolecular Solubilization of Polar Guests via Microsolvation

Selectivity is an important aspect of molecular recognition. A guiding principle in this respect is to preorganize a limited but multiple binding site using a rigid skeleton [1]. Resorcinol cyclic tetramer **9** is a good example. Because of the rigid, multidentate, and essentially two-dimensional nature of the binding site, host **9** exhibits remarkable stereo- and chain-length selectivities in the binding of polar guests [4-7]. It also shows a pronounced selectivity in the sugar extraction. The extractabilities of ribose, fructose, and glucose dramatically decrease in this order [6,15]. An alternative approach other than preorganized binding sites is to mimic the solvation shells for a polar solute in water. A remarkable aspect here is the induced-fit adjustment of solvation to the structures of a great variety of polar solutes.

A unimolecular three-dimensional polyamine/polyhydroxy core can accomodate various polar guests via what may be called a flexible intramolecular polar microsolvation effect [16]. Thus, octaazamacrocycle **17** having 8 dihydroxyalkyl chains in carbon tetrachloride accomodates ~40 molecules of water in its polyamine(ammonium)/polyhydroxy core. The resulting supramolecular complex containing a water pool (structure **19**; Guest is ~40 H_2O) is nearly monomeric as such. Glucose as well as ribose and fructose (3 M in water) can be extracted into CCl_4 upon formation of 1:1 **17**-sugar complexes (**19**; Guest is sugar). There is no significant selectivity among sugars, as evidenced by competitive extraction. There is no anomer selectivity, either. The anomer ratio for bound glucopyranose is the same as that for glucopyranose in water, i.e., $\alpha/\beta = 1/2$. Control experiments indicate that the presence of clustering dihydroxyalkyl chains, although not requiring a macrocyclic skeleton, is essential for the sugar extraction. Thus, noncyclic

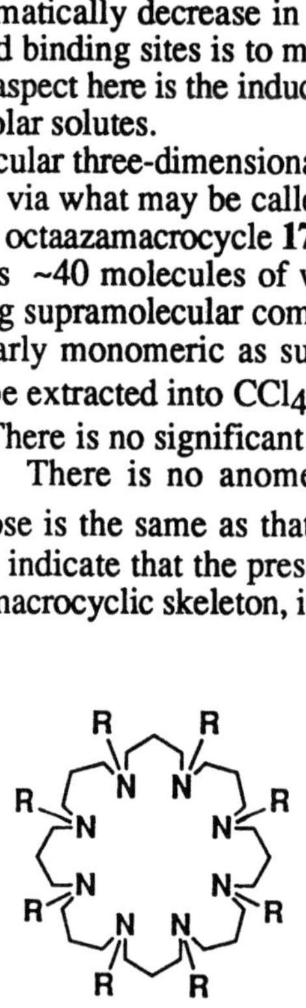

17

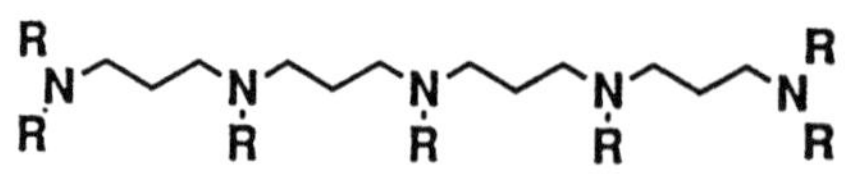

18

$R = CH_2CH(OH)CH(OH)(CH_2)_{10}CH_3$

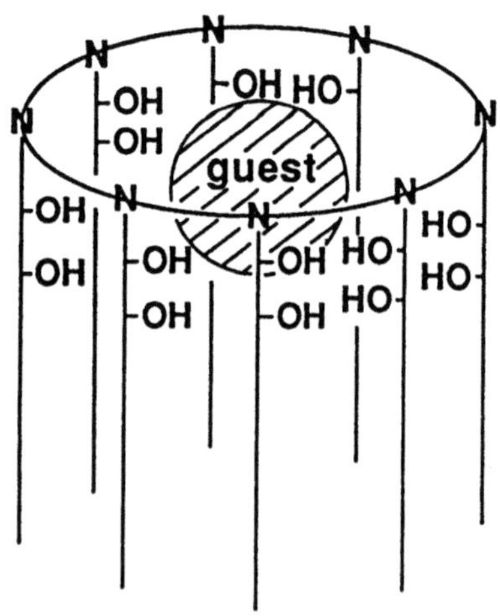

19

reference host **18** works fairly well, too. These results strongly suggest that the sugar encapsulation in the polar core is promoted via a flexible, intramolecular microsolvation effect (**19**). Anions, especially lipophilic ones such as picrate and nucleotides, are also extracted into CCl_4 upon formation of microsolvation-stabilized 1:1 salts with protonated **17**.

The binding site or the polar core of host **17** is an intramolecularly associated polyamine(ammonium)/polyhydroxy cluster. The polar core is adjustable to and hence incorporates various polar guests by the induced-fit mechanism or the microslvation effect, in a similar manner as solvent water dissolves various polar solutes. Thus, flexibility leads to *versatility* [17], as contrasted to selectivity that results from rigidity. This versatility will find many general applications for the reactions and transport of polar compounds in apolar organic media.

5. Supramolecular Synthetic Chemistry of Organic Solids and Crystals

Whatever the type of guest may be, a selective host has to provide convergent functional groups (**20**) that allow a multipoint interaction with the guest [18]. When the hydrogen-bond-forming functional groups are arranged in a divergent manner (**21**), no well-behaved host-guest complexation can be expected. The functional groups, however, would form intermolecular hydrogen-bonds that may result in *polymerization* of the host, as schematically shown in structure **22**.

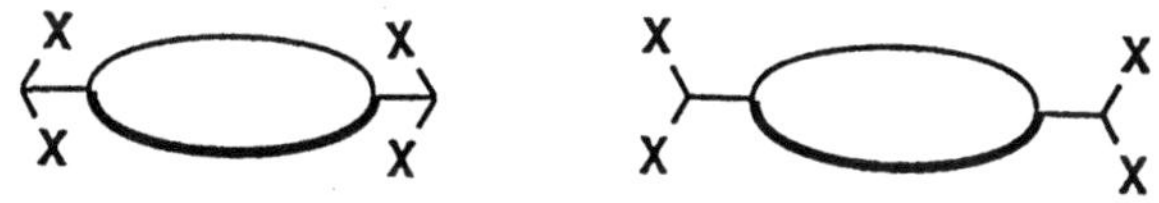

20 **21**

Supramolecular control of the structures of organic solids and crystals is a potential area of molecular recognition based on multipoint hydrogen bonding [19]. By ingenious design of the hydrogen-bonding sites (X), one may be able to control the supramolecular structures of the *functional moieties* (represented as a circle) in a face-to-face stacked polymer (**22**), for example. These approaches may open the door to a supramolecular synthetic chemistry of organic solids and crystals.

6. Supramolecular Complexation of Polar Guests in Water

Apolar organic solvents have been repeatedly used in the study on molecular recognition of polar guests. This is simply because the host-guest hydrogen-bonding can be maximized in such a noncompeting solvent. From the viewpoint of biological molecular recognition, however, much attention should be given to the selective binding of polar guest in aqueous media. Sugars in particular provide one of the most exciting challenges. Especially important are a class of cell-surface oligosaccharides, which are well documented to play essential roles in the intercellular recognition [20]. Contribution of supramolecular chemistry to this field seems to be strongly encouraged for the following

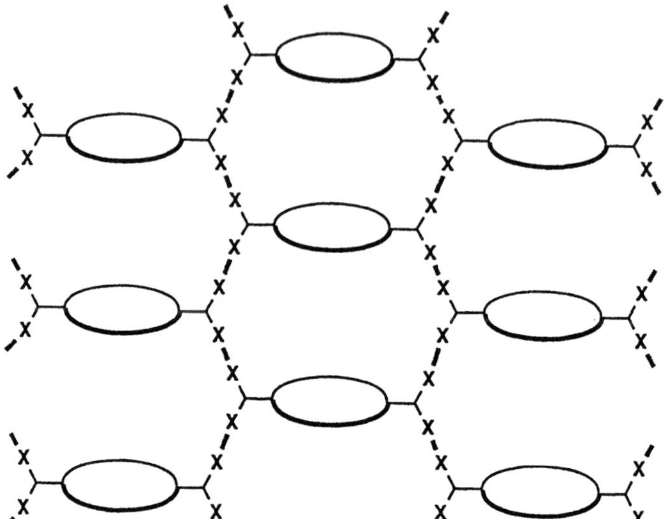

22

reasons. (1) Molecular recognition of sugars in fact is a rapidly growing area [5,6,15,21]. (2) Naturally-occurring oligosaccharides are not under direct genetic control, so that biotechnology can not apply. (3) They are also relatively *small* (molecular weight 1000) and can thus well be a subject of organic chemistry.

A first example of selective (oligo)sugar-sugar interaction has recently been demonstrated by using β -cyclodextrin (β-CD) as host and aldopentoses as guests [22]. β-CD was found to bind aldopentoses in water. The binding constants decrease in the order methyl β-D-ribopyranoside (K = 14) > D-2-deoxyribose (9.4) > D-ribose (5.3) > D-lyxose (4.3) > D-xylose (1.0) > D- and L-arabinose (0.7). This order indicates that hydrophobic interactions play an important role in the present β-CD-sugar complexation. In marked contrast to aldopentoses, aldohexoses and derivatives (glucose, galactose, mannose, fucose, and methyl fucopyranoside) show almost no affinity to β-CD (fucose = 6-deoxygalactose or 5-methylarabinose), even though the last two are significantly more hydrophobic than aldopentoses.

The inhibitory effect of the 5-methyl group of fucose suggests that the inhibition is steric in origin. Thus, an *axial* binding mode (**23**) is unlikely, even though this structure can minimize the steric effect of the methyl group. This is probably because some of the hydrophilic ring-OH groups of a bound sugar are forced to point to the hydrophobic wall of β-CD (CPK models).

It is interesting to consider the present selectivity in the sugar binding in water in light of that observed in the sugar extraction into CC_4 using host **9** [6]. α-Ribopyranose which has all-cis OH groups is readily extracted upon formation of a multipoint hydrgen-bonded complex **24**. Xylose, on the other hand, is hardly extracted. This is probably because one trans OH group of the sugar in the xylose complex **25** is forced to be exposed to

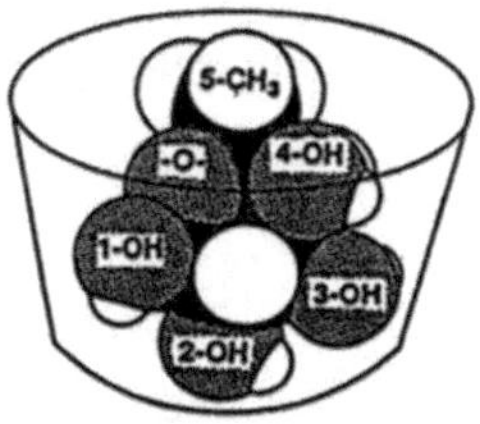

23

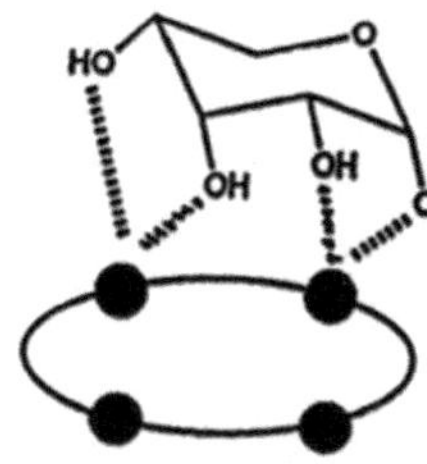

24

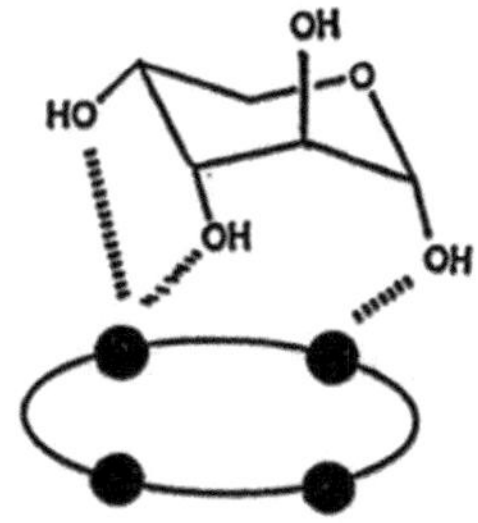

25

apolar solvent CCl_4. In *both* cases of the sugar binding in water and the sugar extraction into CCl_4, the OH groups of a sugar should not be placed in a relatively apolar environment. This requirement leads to remarkable selectivities.

7. Conclusions

Multipoint hydrogen-bonding is a nature's device for the maintenance of structures and functions of informational biopolymers such as nucleic acids, proteins, and oligo- and polysaccharides. Molecular recognition based on such interactions is still in an early stage, but there have already been accumulated many good examples that convince one of its potential utilities in modern chemical science.

First, multipoint hydrogen-bonding widens the scope of host-guest association so as to include common organic molecules as guests in organic media. Thus, molecular recognition can now be a strategy in organic chemistry, as highlighted by the developments of supramolecular methods for the construction and elucidation of complicated molecular structures.

Second, the general concept of preorganized binding sites composed of rigid and convergent functional groups can be modified in two ways. One is to use a polar core or cluster composed of *many* flexible polar groups. This leads to an induced-fit type accomodation of versatile guests via a microsolvation effect. The other is to use divergent functional groups so as to allow extensive intermolecular hydrogen-bonding. This may result in supramolecular control of organic solids and crystals.

Third, the binding of highly polar guests such as sugars in aqueous media is now accessible. Hydrophobic interactions play an essential role at least in the present CD-sugar complexation. However, polar interactions of the OH groups of bound sugar are also of primary importance in determining the structure of the complex and hence the selectivity of sugar binding.

Molecular recognition is the strategy by which a molecule bears supramolecular functions. The choice of interacting molecules is essential in this respect. Supramolecular chemistry would surely occupy the central position in the design of new functional materials with controlled intermolecular interactions. Contribution to life science and technology is also a high possibility; biopolymers, cells, and even organs can be targets of supramolecular chemistry. Multipoint hydrogen-bonding itself is a tremendous interaction. However, a balanced cooperation of polar (hydrogen bonding and electrostatic) and apolar (hydrophobic and van der Waals) interactions would play a key role in both rigorous structure control of materials and elaborate molecular recognition of living systems.

Acknowledgement

I am grateful to all of my colleagues whose names appear in the references cited below. I also sincerely thank the organizing commitee, especially Professor V. Balzani, of the Second NATO Science Forum - Supramolecular Chemistry for inviting me to the Forum and to write this review.

References

[1] Cram, D. J. (1988) 'The design of molecular hosts, guests, and their complexes', Angew. Chem. Int. Ed. Engl. 27, 1009-1020.
[2] Lehn, J.-M. (1988) 'Supramolecular chemistry - scope and perspectives. Molecules, supramolecules, and molecular devices', Angew. Chem. Int. Ed. Engl. 27, 90-112.
[3] Rebek, J. Jr. (1990) 'Molecular recognition with model systems', Angew. Chem. Int. Ed. Engl. 29, 245-255.
[4] Kikuchi, Y., Kato, Y., Tanaka, Y., Toi, H., and Aoyama, Y. (1991) 'Molecular recognition and stereoselectivity: geometrical requirements for the multiple hydrogen-bonding interaction of diols with a multidentate polyhydroxy macrocycle', J. Am. Chem. Soc. 113, 1349-1354.
[5] Kikuchi, Y., Kobayashi, K., and Aoyama, Y. (1992) 'Complexation of chiral glycols, steroidal polyols, and sugars with a multibenzenoid, achiral host as studied by induced circular-dichroism spectroscopy: exciton chirality induction in resorcinol-aldehyde cyclotetramer and its use as a supramolecular probe for the assignments of stereochemistry of chiral guests', J. Am. Chem. Soc. 114, in press.
[6] Aoyama, Y., Tanaka, Y., Toi, H., and Ogoshi, H. (1988) 'Polar host-guest interaction. Binding of nonionic polar compounds with a resorcinol-aldehyde cyclotetramer as a lipophilic polar host', J. Am. Chem. Soc. 110, 634-635.
Aoyama, Y., Tanaka, Y., and Sugahara, S. (1989) 'Molecular recognition of sugars via hydrogen-bonding interaction with a synthetic polyhydroxy macrocycle', Ibid. 111, 5397-5404.
[7] Tanaka, Y., Kato, Y., and Aoyama, Y. (1990) 'Two-point hydrogen-bonding interaction: a remarkable chain-length selectivity in the binding of dicarboxylic acids with resorcinol-aldehyde cyclotetramer as a multidentate host', J. Am. Chem. Soc. 112, 2807-2808.
[8] Aoyama, Y., Yamagishi, A., Asakawa, M., Toi, H., and Ogoshi, H. (1988) 'Molecular recognition of amino acids: two-point fixation of amino acids with bifunctional metalloporphyrin receptors', J. Am. Chem. Soc. 110, 4076-4077.
Aoyama, Y., Asakawa, M., Yamagishi, A., Toi, H., and Ogoshi, H. (1990) 'Simultaneous hydrogen bonding and metal coordination interactions in the two-point fixation of amino acids with a bifunctional metalloporphyrin receptor', Ibid. 112, 3145-3151.
[9] Motomura, T. and Aoyama, Y. (1992) 'Accumulation of hydrogen-bonding and electrostatic binding sites: stabilization of salts in hydroxylic media via intramolecular hydrogen bonding', J. Org. Chem. 57, in press.

[10] Motomura, T., Inoue, K., Kobayashi, K., and Aoyama, Y. (1991) 'Transition-state stabilization via dynamic molecular recognition: a concerted acid-base bifunctional catalysis in ester hydrolysis', Tetrahedron Lett. 32, 4657-4760.

[11] Aoyama, Y., Asakawa, M., Matsui, Y., and Ogoshi, H. (1991) 'Molecular recognition of quinones: two-point hydrogen-bonding strategy for the construction of face-to-face porphyrin-quinone architectures', J. Am. Chem. Soc. 113, 6233-6240.

[12] Borovkov, V. V., Evstigneeva, R. P., Strekova, L. N., Filippovich, E. I. (1989) 'Porphyrin-quinone compounds as synthetic models of the reaction centre in photosynthesis', Russ. Chem. Rev. (engl. Transl.) 58, 1032-1063.

[13] Harada, N and Nakanishi, K. (1983) Circular Dichroic Spectroscopy - Exciton Coupling in Organic Stereochemistry, University Science Books, Mill Valily, CA, USA.

[14] Kikuchi, Y., Kobayashi, K., Toi, H., and Aoyama, Y. 'Supramolecular probing of the absolute configuration of chiral monools via host-guest complexation', to be published.

[15] Tanaka, Y., Ubukata, Y., and Aoyama, Y. (1989) 'Selective sugar binding with a synthetic polyhydroxy macrocycle. A remarkable selectivity for fructose over glucose', Chem. Lett. 1905-1908.

[16] Kobayashi, K., Ikeuchi, F., Inaba, S., Aoyama, Y. (1992) 'Accomodation of polar guests in unimolecular polyamine-polyhydroxy cores: solubilization of sugars in apolar organic media via intramolecular polar-microsolvation', J. Am. Chem. Soc. 114, in press.

[17] Menger, F. M., Takeshita, M., Chow, J. F. (1981) 'Hexapus, a new complexing agent for organic molecules', J. Am. Chem. Soc. 103, 5938-5939.
Vögtle, F., Müller, W. M. (1982) 'Selective inclusion of uncharged guest molecules by neutral hosts with fence-shaped side arms', Angew. Chem. Int. Ed. Engl. 21, 147-148.
Murakami, Y., Kikuchi, J., Ohno, T., Hayashida, O., Kijima, M. (1990) 'Syntheses of macrocyclic enzyme models. 7. Octopus cyclophane having L-aspartate residues as novel water-soluble hosts. Aggregation behavior and induced-fit molecular recognition', J. Am. Chem. Soc. 112, 7672-7681.

[18] Rebek, J. Jr., Askew, B., Killoran, M., Nemeth, D., and lin, F.-T. (1987) 'Convergent functional groups. 3. A molecular cleft recognizes substrates of complementary size, shape, and functionality', J. Am. Chem. Soc. 109, 2426-2431.

[19] Lehn, J.-M., Mascal, M., Decian, A., and Fischer, J. (1990) 'Molecular recognition directed self-assembly of ordered supramolecular strands by cocrystallization of complementary molecular compounds', J. Chem. Soc., Chem. Commun. 479-481.
Zerkowski, J. A., Seto, C. T., Wierda, D. A., and Whitesides, G. M. (1990) 'Design of organic structures in the solid state: hydrogen-bonded molecular "tapes"', J. Am. Chem. Soc. 112, 9025-9026.
Simard, M., Su, D., and Wuest, J. D. (1991) 'Use of hydrogen bonds to control molecular aggregation. Self-assembly of three-dimensional networks with large chambers', Ibid. 113, 4696-4698.

[20] Eggens, I., Fenderson, B., Toyokuni, T., Dean, B., Stroud, M., and Hakomori, S. (1989) 'Specific interaction between Le^X and Le^X determinants', J. Biol. Chem. 264, 9476-9484.

[21] Bonar-Law, R. P., Davis, A. P., and Murray, B. A. (1990) 'Artificial receptors for carbohydrate derivatives', Angew. Chem. Int. Ed. Engl. 29, 1407-1408.

[22] Aoyama, Y., Nagai, Y., Otsuki, J., Kobayashi, K., and Toi, H. 'Selective sugar binding to β-cyclodextrin: a prototype of sugar-sugar interactions in water', to be published.

NEW SHAPES FOR SELECTIVE MOLECULAR RECOGNITION FROM CALIXARENES

A. ARDUINI, A. CASNATI, M. FABBI, P. MINARI,
A. POCHINI, A.R. SICURI and R. UNGARO
Istituto di Chimica Organica dell'Università
Viale delle Scienze
I-43100 Parma
Italy

ABSTRACT. The functionalization of calixarenes at the "lower rim" reduces their conformational mobility creating new shapes which have been exploited to organize binding groups in space, thus producing efficient and selective Molecular Receptors. Calix[4]arene- ester and amide ligands in the fixed cone conformation show high efficiency and selectivity in the complexation of Na^+ and Ca^{2+}; the amide is able to encapsulate lanthanide ions ($Eu^{3+}$and Tb^{3+}) in its polar cavity, giving water soluble luminescent complexes. Calix[4]arene-based siderophores, which strongly complex iron (III) in ethanol/water, have been obtained by the introduction of three hydroxamate chelating chains on the bottom rim of the calix fixed in the cone conformation. The selective 1,3,5-methylation of p-*tert*-butylcalix[6]- arene has been achieved and the subsequent introduction of ester and amide binding groups fixes the calix in a "flattened cone" conformation. The resulting ligands have a C_3 symmetry and are able to transport guanidinium cation through a Supported Liquid Membrane. Water soluble calix[4]arenes in the fixed cone conformation, which selectively bind organic cations and neutral molecules in aqueous solution, have been synthesized, together with lipophilic cage molecules derived from calix[4]arenes which are potentially host for the encapsulation of organic molecules in apolar media.

1. Introduction

Cyclophanes, which are bridged aromatic compounds, are probably the most investigated class of host molecules in Supramolecular Chemistry [1]. Among them, the cyclic oligomers derived from the condensation of phenols and formaldehyde, and recently named calixarenes [2], occupy a special place as proved by several books and review articles devoted to the chemistry of these macrocycles [2-6]. The reason for the interest in calixarenes is based on the fact that these compounds can be easily

V. Balzani and L. De Cola (eds.), Supramolecular Chemistry, 31–50.

synthesized on a large scale and that their sizes and conformational properties offer several possibilities for the design of new Molecular Receptors. The most easily available calixarenes are those derived from p-*tert*-butylphenol for which high yield synthetic methodologies have been developed [7].

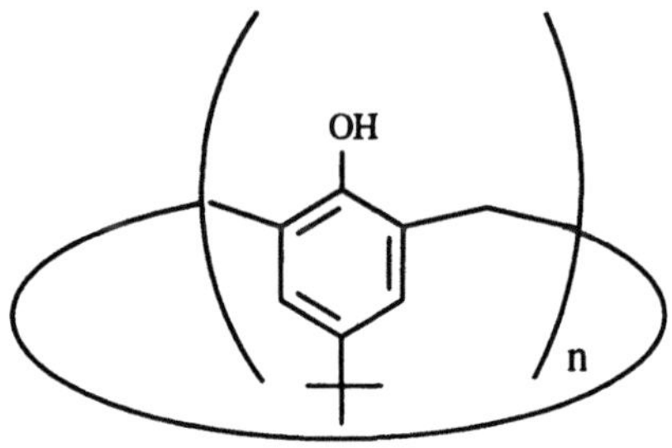

1: n = 4

2: n = 6

3: n = 8

p-*tert*-Butylcalix[4]arene **1** exists in the solid state in a cone conformation [8] mainly determined by strong intramolecular hydrogen bonding, whereas in solution it is conformationally mobile, showing a free energy of activation for the ring inversion process ($\Delta G^{\neq}$) of 15.7 kcal mole^{-1} in $CDCl_3$ [9]. p-*tert*-Butylcalix[6]arene **2**, whose solid state conformational preferences are determined partially by hydrogen bonding and partially by the minimization of van der Waals interactions [10], is conformationally more mobile than the corresponding calix[4]arene **1** showing a $\Delta G^{\neq}$ = 13.3 kcal mole^{-1} for the ring inversion process in $CDCl_3$ [9]. Finally, p-*tert*-butylcalix[8]arene **3** exists, in the solid state, in a "pleated loop" conformation [11], where no tridimensional cavity is present. Surprisingly it shows temperature dependent ^{1}H NMR spectra in $CDCl_3$ very similar to the cyclic tetramer, with the same inversion barrier of 15.7 kcal mole^{-1} [12]. The functionalization of calixarenes at the "lower rim" (phenolic OH groups) or at the "upper rim" (aromatic nuclei) often produces derivatives with reduced conformational mobility thus creating new shapes useful for organizing binding or catalytic groups in space. This article will summarize some of our recent efforts to exploit these shapes for the selective recognition of ions and neutral molecules.

2. Calix[4]arene Cation Ligands in the Fixed Cone Conformation

2.1. ENCAPSULATION OF LANTHANIDE IONS

p-tert-Butylcalix[4]arene **1** has been treated with NaH in tetrahydrofuran (THF)/N,N-dimethylformamide (DMF) followed by the reaction with bromoacetic acid derivatives to produce in high yield the tetraester **4** [13] and the tetramide **5** [14] in the fixed cone conformation. Later studies have shown that the use of NaH in pure DMF or

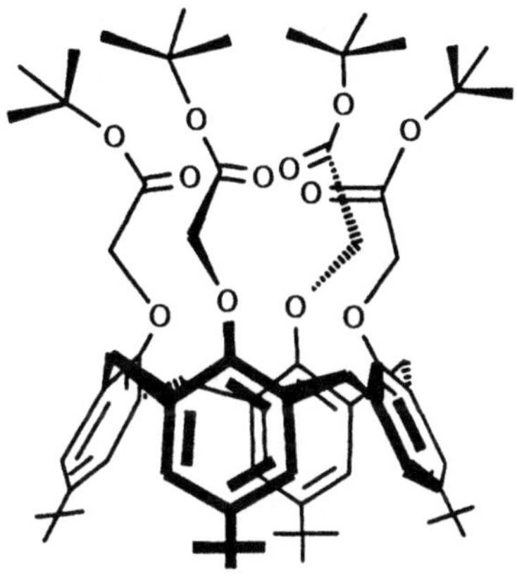

4

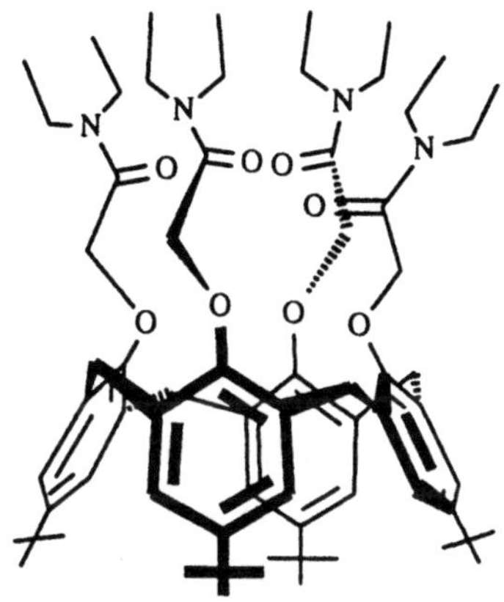

5

acetonitrile is a general method [15] for the obtainement of tetraalkoxycalix[4]arenes in the fixed cone conformation. The convergence of four donor groups on the lower rim of the calix creates a rather polar cavity which interacts strongly with alkali and alkaline earth cations forming complexes of comparable stability with those of cryptands.

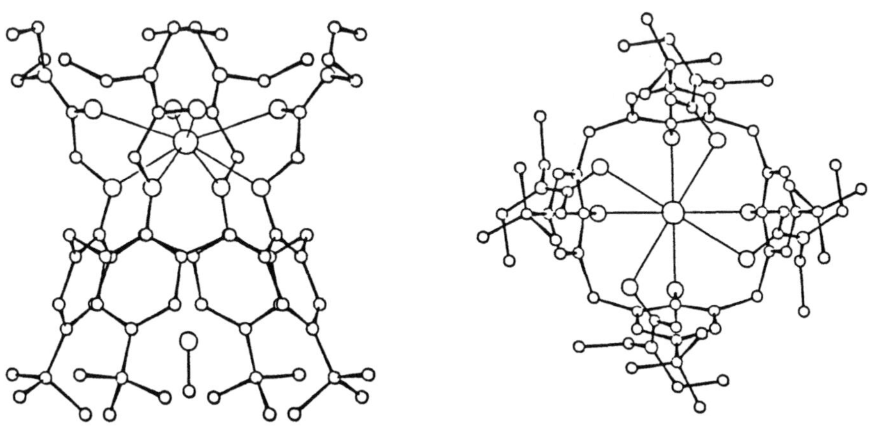

(a) (b)

Fig. 1: X-Ray crystal structure of 1:1 complex between ligand **5** and KSCN. (a) Perspective view; (b) View along the fourfold axis.

The selectivity observed with this type of ligands is always toward the small cations with high charge density such as Na^+ and Ca^{2+}. The general properties of the cation complexes of ester and amide calix[4]arene ligands have been recently reviewed [5,16].

The X-ray crystal structure of the potassium complex (Figure 1) of ligand **5** shows the cation completely encapsulated in the polar cavity consisting of eight donor atoms [14]. The four amide carbonyl groups and the other four ether oxygen atoms lie in two distinct parallel planes forming an antiprism, whereas the alkyl groups on nitrogen efficiently deshield the cation from the counter anion (SCN^- or $Picrate^-$). The latter are dispersed in the crystal lattice, showing no interaction with the K^+ cation. The ability of ligand **5** to deshield the complexed cation from the counteranion and from the solvent has been exploited in the preparation of luminescent Eu^{3+} and Tb^{3+} complexes [17]. It is known that in order to show efficient luminescence the lanthanide complexes have to be deshielded from the water solvent and this property is shown only by few uncharged synthetic receptors. 1:1 Eu^{3+} and Tb^{3+} complexes with tetramide **5** have been prepared and they are both water soluble and highly luminescent. Table 1 provides data from photophysical studies of these complexes together with those reported in literature for some selected neutral ligands.

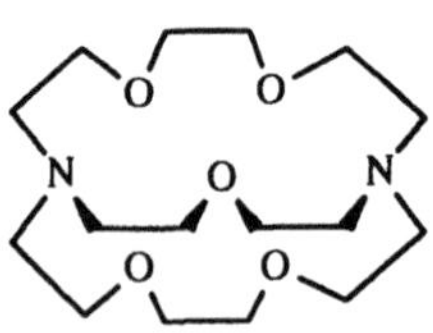

6

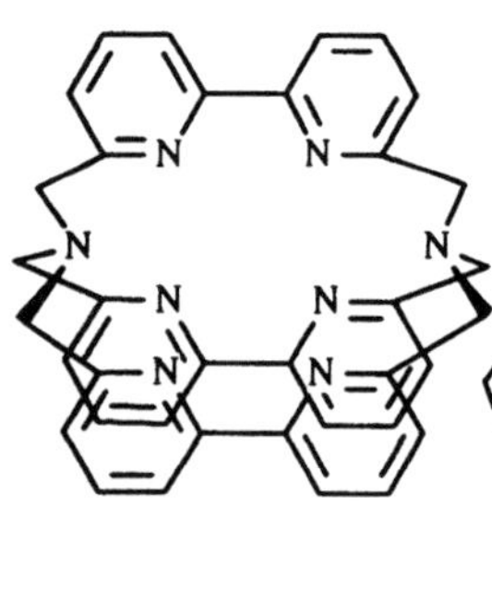

7

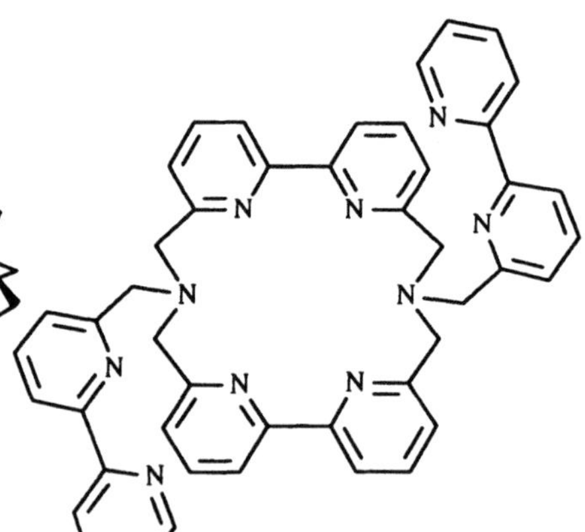

8

Analysis in terms of Sudnick and Horrocks' equation of the lifetimes τ of the luminescent $^5D_0$ excited state of the $[Eu \subset \mathbf{5}]^{3+}$ inclusion complex in H_2O and D_2O allowed to establish that, on average, only one molecule of water ($\pm$ 0.5) is coordinated to the cation. Corresponding values are 9.6 for $Eu(aq)^{3+}$, 3.3 for Eu^{3+} encapsulated into 2.2.1 cryptand $[Eu \subset \mathbf{6}]^{3+}$ [18a], 2.5 in the case of $[Eu \subset \mathbf{7}]^{3+}$ [18b] and 0 for $[Eu \subset \mathbf{8}]^{3+}$ [19]. As a consequence of this efficient shielding the luminescence lifetime and the quantum yield of the $^5D_0$ Tb^{3+} emission upon ligand excitation are greatly enhanced and are comparable with those observed in methanol (at 300 K) for ligand **8** (τ=1.1 ms; ϕ = 1.4 E-1), whose lanthanide complexes are considered to be the most promising luminescent labels for biological substrates [19].

Table 1. Luminescence properties of lanthanide complexes[a].

Complex	Lifetimes (ms)				Quantum yield	Ref.
	300 K $\tau(H_2O)$	300 K $\tau(D_2O)$	77 K $\tau(H_2O)$	77 K $\tau(D_2O)$	300 K $\phi(H_2O)$	
$[Eu \subset \mathbf{5}]^{3+}$	0.65	1.9	1.8	2.0	2.0E-4	[17]
$[Tb \subset \mathbf{5}]^{3+}$	1.5	2.6	1.6	2.8	2.0E-1	[17]
$[Eu \subset \mathbf{7}]^{3+}$	0.34	1.7	0.81	1.7	2.0E-2	[18]
$[Tb \subset \mathbf{7}]^{3+}$	0.33	0.43	1.7	3.8	3.0E-2	[18]
$[Eu \subset \mathbf{8}]^{3+}$	1.5	1.9	1.8	1.9	1.0E-1	[19]

a)excitation in the ligand-centered band.

2.2. CALIX[4]ARENE SIDEROPHORES

In supramolecular chemistry there has been a certain interest in the synthesis of iron (III) complexing agents in the biomimetic approach to reproduce the properties of Enterobactin, Ferrichrome or other natural siderophores [20]. A general tendency has emerged to synthesize trihydroxamate or tricathecolate siderophores with a high degree of preorganization [21].

In order to introduce three chelating chains on the lower rim of calix[4]arenes, two general methodologies have been developed for the selective synthesis of calix[4]arene monoalkyl ethers [22]. The synthesis of trihydroxamate siderophores was achieved through the sequence of reactions reported in Scheme 1. The convergence of the three chelating chains was established by ^1H and ^{13}C NMR [23]. The pKa values of compound **10** were evaluated in ethanol/water 9:1; the first proton dissociation constant

Scheme 1.

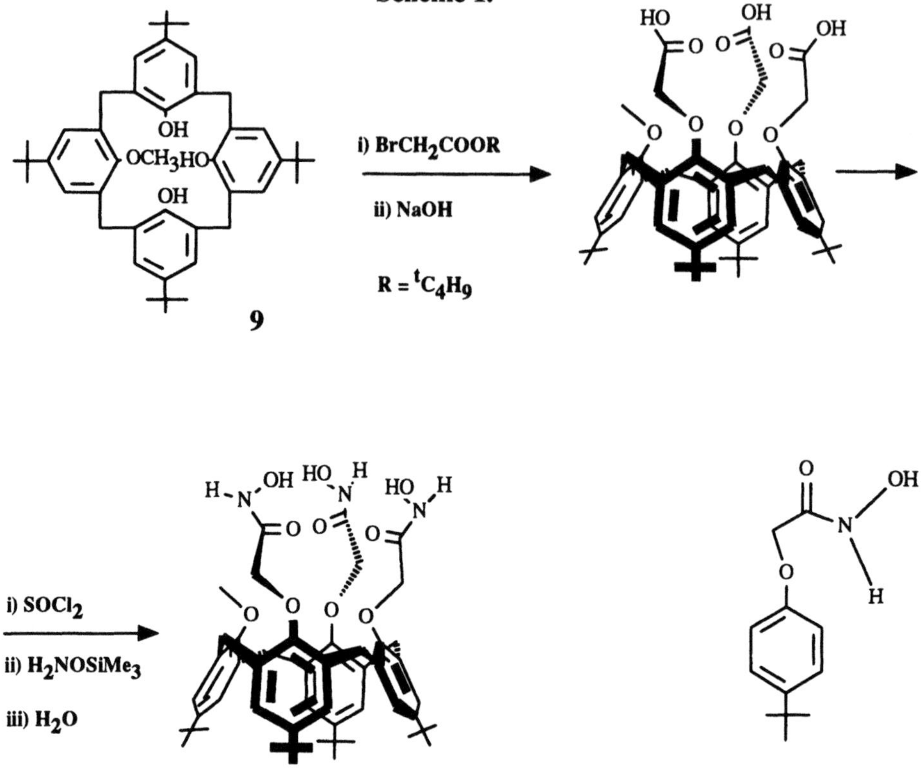

10 **11**

(pKa1 = 8.64) indicates a higher acidity of one hydroxamate unit in compound **10** compared with the monomeric p-*tert*-butyl-N-hydroxy phenoxyacetamide **11** (pKa = 10.29), whereas the other two hydroxamate units have a lower acidity (pKa > 11). UV-Vis and potentiometric studies in the presence of iron (III) indicate that only one type of complex is formed in the pH range 2.4-7.0 which corresponds to the octahedrically coordinated Fe(III) with three ionized hydroxamate units. Due to the high stability of this complex in EtOH/H_2O 9:1 only an estimation of its formation constant (log β_{111}> 39) could be made by direct potentiometric titration [23].

3. Stereochemical Control of Cation Complexation in Calix[4]arene Ligands

3.1. CALIXCROWN ETHERS

The tetraalkylation of calix[4]arenes at the lower rim with groups bulkier than CH_3 results in the formation of a mixture of stereoisomers.

cone **partial cone** **1,2 alternate** **1,3 alternate**

By changing the nature of the base and the solvent a certain control of the stereochemical outcome has been achieved in few alkylation [15] and acylation [24] reactions of calix[4]arenes. Recently, in collaboration with the group of Prof. D.N. Reinhoudt, several stereoisomeric calix[4]arene crown ethers have been synthesized and their ionophoric properties toward alkali metal cations determined by the Cram's extraction method [25]. In all cases it was found that these ligands are very selective for potassium cation and that the compounds in the partial cone structure are more efficient and more selective than those in the fixed cone conformation. The highest difference in

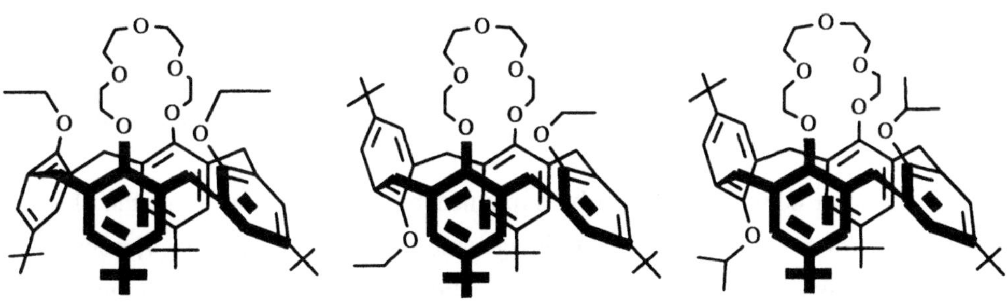

12 **13** **14**

the binding free energy of complexation of potassium cation ($\Delta\Delta G° = 3.9$ kcal/mole) was observed between the two stereoisomers of 1,3-diethoxy p-*tert*-butylcalix[4]arene crown-5 **12** and **13**. Examination of the two available X-ray crystal structures of K^+ complexes of compound **12** in the cone conformation and 1,3-diisopropoxy p-*tert*butylcalix[4]arene crown-5 in the fixed partial cone structure **14**, reveals some interesting features (Figure 2) [26]. In the 1:1 K^+ complex of ligand **12** the cation is coordinated by all seven oxygen atoms which show O...K^+ distances shorter than the corresponding distances in the K^+ complex of ligand **14** in the partial cone structure, where only six of seven oxygens are involved in complexation. Nevertheless ligand **14** forms stronger complexes with alkali metal cations than ligand **12** by 3.0 kcal mol^{-1} in $CDCl_3$ [25].

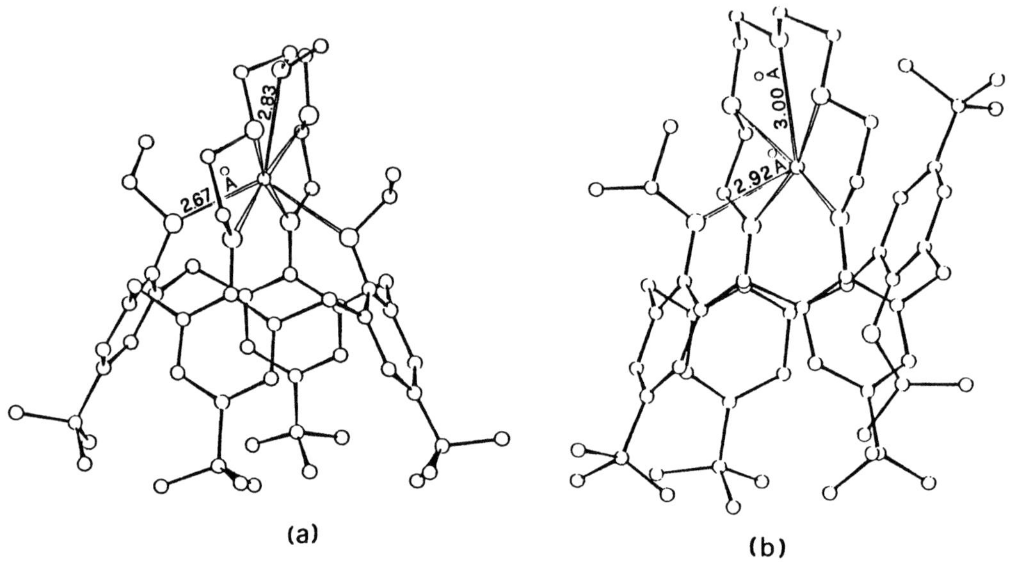

Fig. 2: X-Ray structures of K^+ complexes of (a) calixcrown **12** and (b) calixcrown **14**.

This apparent contradiction can be explained by taking into account the shielding effect as an important factor in stabilizing cation complexes of calix[4]arene crown ethers. In the partial cone structure one face of the complex is occupied by the aromatic nucleus and by the *tert*-butyl in the *para*-position whose shielding effect contributes to the stability of the K^+ complex more than one ethereal binding site which does not partecipate to complexation in ligand **14**.

3.2. IONIZABLE CALIX[4]ARENE DICARBOXYLIC ACIDS

Another example of stereochemical control on cation complexation has been observed in calix[4]arene dicarboxylic acid derivatives **15-17** which were designed for the selective extraction of divalent metal cations.

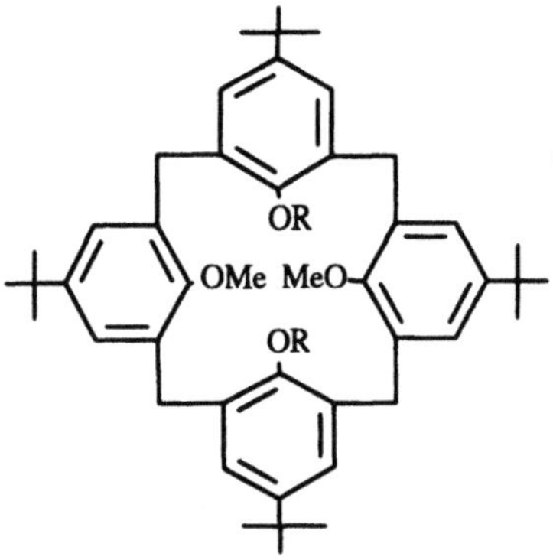

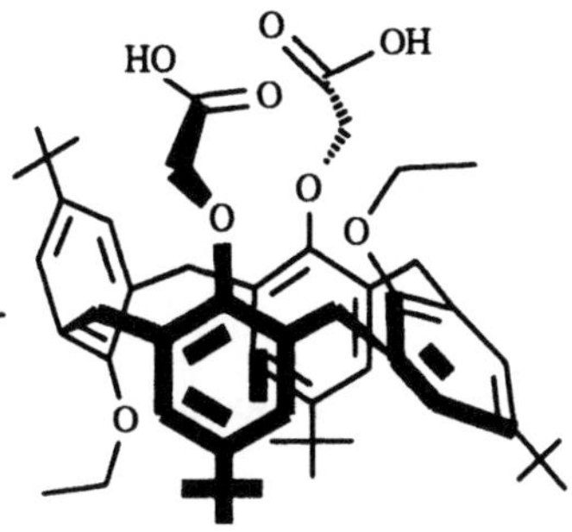

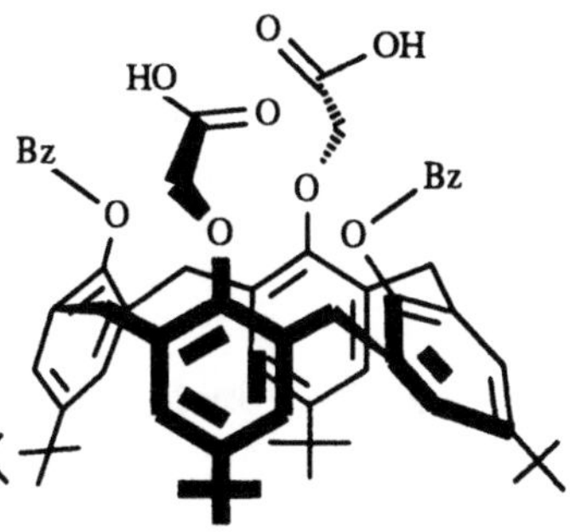

15: R = CH_2COOH **16** **17**: Bz = $C_6H_5CH_2$

Figure 3 shows that the ability of these ligands to extract alkaline earth cations depends strongly on the nature of the two remaining alkoxy groups which, in turn, determine the stereochemistry around the binding region.

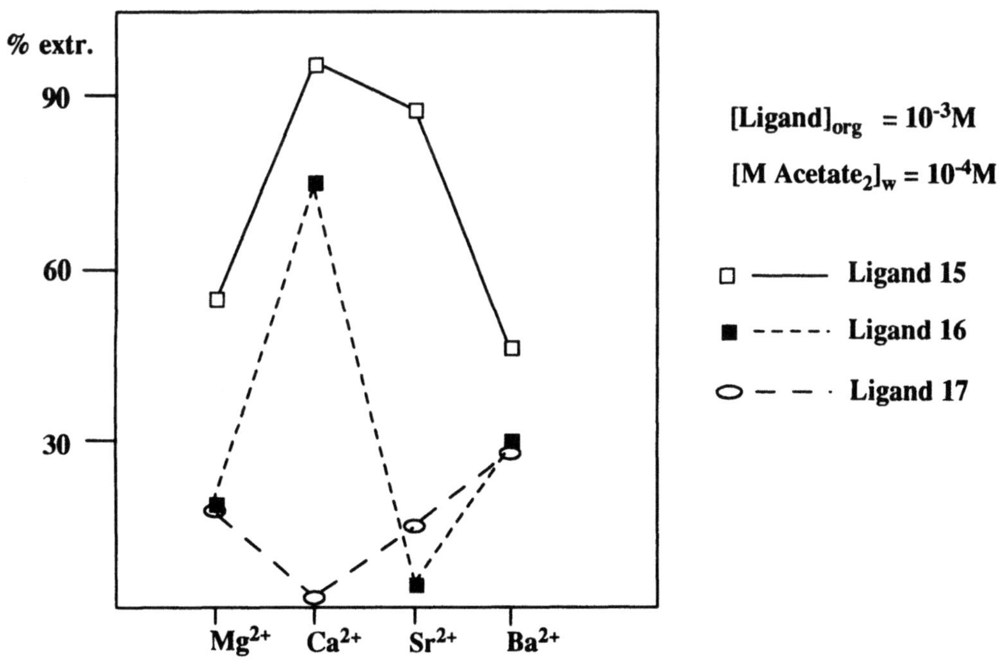

Fig. 3: Extraction of alkaline earth cations by ligands **15-17** from water to dichloromethane.

The 1,3-dimethoxy derivative **15** in which the OCH_3 groups can rotate inside the cavity, is more efficient in extracting alkaline earth cations but compound **16**, in the partial cone structure, is more selective toward Ca^{2+}, whereas the dibenzyloxy derivative **17** in the fixed cone structure is a very poor extractant [27].

4. Selective Functionalization of Calix[6]arenes

Fully O-alkylated calix[6]arenes, either lipophilic or water soluble, have been previously synthesized and studied for the complexation of ions and neutral molecules [28]. However, very little work has been done on the selective functionalization of calix[6]arenes both at the upper and the lower rim, to create new structures useful for Molecular Recognition. By treating p-*tert*-butylcalix[6]arene **2** with CH_3I in acetone, in the presence of K_2CO_3, the symmetrical 1,3,5-trimethyl derivative **18** was prepared in 27% yield [29,30]. This compound shows a very simple pattern in the ^1H and ^{13}C NMR

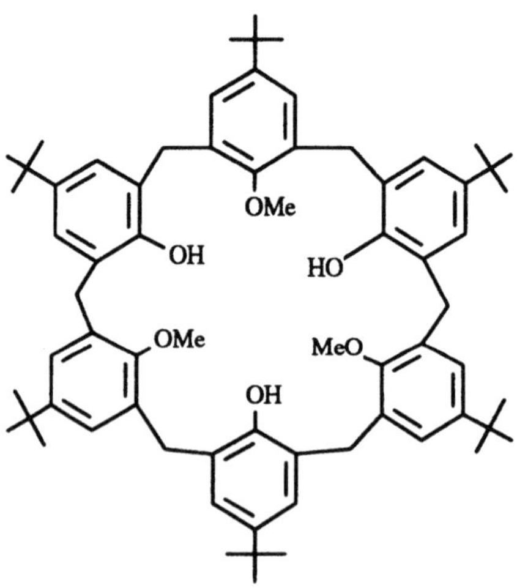

18

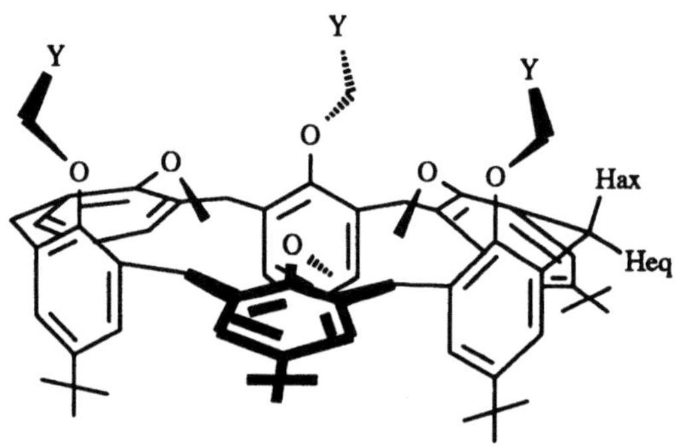

19: Y = $CON(C_2H_5)_2$

20: Y = $COO^tC_4H_9$

21: Y = C_6H_5

spectra indicating a very symmetrical and mobile structure. This contrasts with the partially methylated calix[4]arenes where rigidified cone structures have been observed both in solution [31] and in the solid state [32]. The weakening of the intramolecular hydrogen bonding system in the larger calix[6]arenes compared with calix[4]arenes [2, 9] is responsible for this different conformational behaviour. The remaining OH groups in compound **18** were then treated with bromoacetic acid derivatives or benzyl bromide to give fully alkylated products **19-21** in almost quantitative yields [29, 30].

The ^{1}H and ^{13}C NMR spectra of compounds **19-21** are very simple and do not change between -70 and +110°C, indicating the presence of a particularly rigid conformation which should be the cone because of the high symmetry of ^{1}H and ^{13}C NMR spectra and the presence of only one AB system for the methylene protons of the bridge (e.g. H_{ax} = 4.53 δ, H_{eq} = 3.43 δ in **19**). Moreover the absorption of methoxy groups at high field (δ ≅ 2.2 ppm) and the large Δδ between axial and equatorial protons (Δδ > 1 ppm) suggest that the OCH_3 are inside the calix cavity. These data have been confirmed in solution for compound **19** and **20** by 2D NMR experiments where NOE cross-peaks between equatorial and aromatic protons and between the two singlets of aromatic protons have been observed. The X-ray diffraction analysis of a single crystal of compound **19**, although characterized by high disorder of the amide-groups which does not allow to reach a satisfactory R value, clearly indicates a cone conformation with the OCH_3 groups bent inside the cavity ("flattened cone").

The behaviour of the trimethoxy derivatives **19-21** is even more surprising if we compare it with that of p-*tert*-butylcalix[6]arene hexamide (**22**). The latter at room

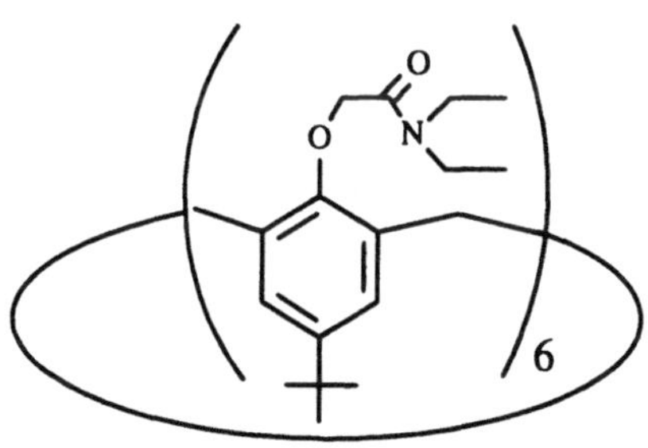

22

temperature is present as a mixture of conformations, one of which corresponds to the 1,2,3-alternate structure found in the solid state [10a], but at T > 100°C it is conformationally mobile [30].

Interestingly compound **22** is able to complex both Na^+ and K^+ picrates in $CDCl_3$ solution with a metal to ligand 2:1 stoichiometry. When the first cation is complexed a substantial rearrangement of the amide chains takes place. This sort of self-organization enhances the second complexation constant to such an extent that also at low values of metal/ligand molar ratio no appreciable amount (less than 5%) of the 1:1 complex can be detected [30].

The smaller sodium cations keep the ligand in a cone conformation, while with the larger potassium cations repulsive electrostatic interactions probably bring some of the nuclei in an anti orientation. Hexamide **22** and triamide **19** strongly bind guanidinium cation in $CDCl_3$ but ligand **19** is not able to complex sodium and potassium cations. Transport experiments of thiocyanates through a Supported Liquid Membrane reveal a great efficiency of hexamide **22** toward guanidinium, sodium and potassium. Triamide **19** although less efficient than hexamide in guanididium transport is more selective, since sodium and potassium are not carried across the polimeric membrane [30]. These data open a way for the selective complexation and transport of amino acids and other polar neutral molecules.

5. Inclusion of Neutral Molecules in the Apolar Cavity of Calixarenes

5.1. WATERS SOLUBLE CALIX[4]ARENES

With the aim of obtaining evidences for the inclusion of neutral molecules inside the apolar cavity of calixarenes we previously synthesized a tetracarboxylic acid derivative **23** of p-*tert*-butylcalix[4]arene in the fixed cone conformation [13a]. The alkali metal salts of this acid were water soluble up to 5 x 10^{-3} M, but they were unable to complex neutral molecules such as benzene, toluene, xylenes and methanol which were easily included in the apolar cavity of this class of macrocycles in the solid state showing a well defined stoichiometry [6].

Later Shinkai and coll. have reported several upper rim sulfonated calixarenes (e.g. **24**), which have a much higher water solubility than **23** and are able to complex organic cations and neutral molecules in water solution [28]. Therefore, we synthesized a tetracarboxylic acid tetrasulfonate **25** in the fixed cone structure in order to compare its binding properties with those of the flexible tetrasulfonate **24**, also studied by Atwood and coll. [33].

The protonation constants of ligand **24** and **25** were determined by Arena and coll. [34], who reevaluated the literature pKa values of compound **24** [35] because they were not in agreement with solid state evidences [33]. Only two titrable protons were found in the pH region 2.5-11.0 and the proton formation constants determined (Log K = 3.34 and 11.5) rule out the existence of a "super acidic proton" (pKa < 1) for compound **24** [35]. However, this proton is much more acidic (5.3 log units) than the monomeric

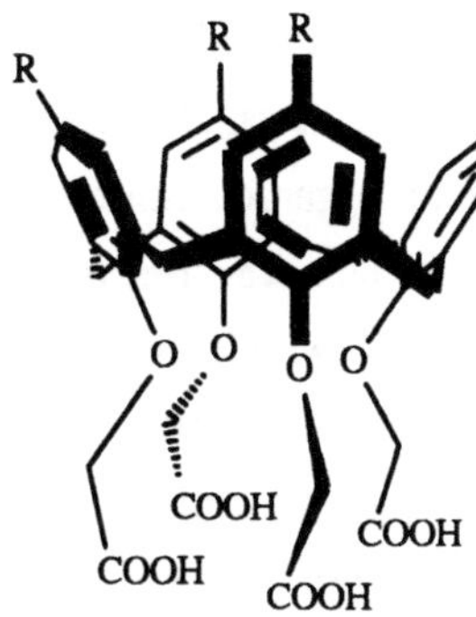

23: R = $^{t}C_4H_9$

25: R = SO_3Na

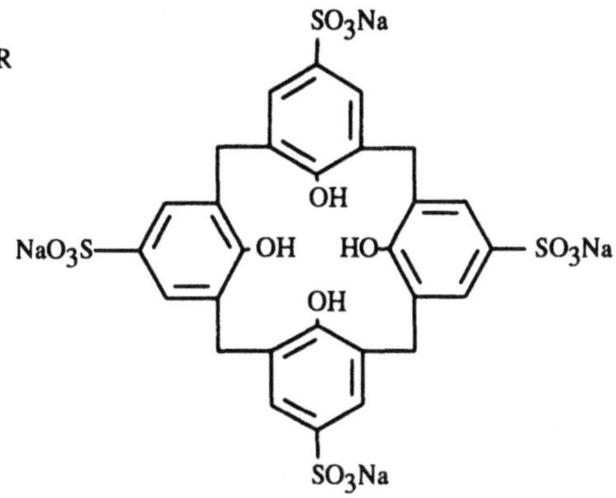

24

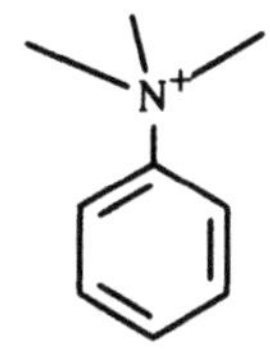

26

hydroxybenzene sulfonate, probably because the calix[4]arene-p-tetrasulfonate penta anion of **24**, which is in the cone structure [33,35], is stabilized by strong adjacent hydrogen bonds. The stability constants for the protonation of the tetraacid tetrasulfonate **25** (Log K = 3.19; 3.39; 4.19; 4.71) [34] show a normal behaviour being very close to the log K for the protonation of phenoxy acetate [36].

Compound **25** is able to include organic cations and neutral molecules in water solution at neutral pH. In order to compare the two ligands **24** and **25** trimethylanilinium chloride **26** and acetone were selected as guest molecules since they have been already studied with host **24** by ^1H NMR. At pH = 7.3 the aromatic protons of **26** are shifted upfield, particularly the para hydrogen which absorbs at 5.37 δ when the guest is 100% complexed, whereas the N^+CH_3 protons are only slightly affected by complexation. This behaviour resembles that observed by Shinkai and coll. with host **24** at pH 0.4 [37], thus suggesting that only one mode of binding is operating with compound **25**, namely that where the phenyl moiety is selectively bound to the calixarene cavity. The Japanese authors have in fact observed that at neutral pH both the ammonia-methyl and the phenyl moiety are unselectively bound to the cavity of calix[4]arene tetrasulfonate **24**.

An association constant K = 3100 M^{-1} for the 1:1 complex between **25** and trimethylanilinium chloride in water at pH 7.3 and 25°C was established by the computer assisted non linear least-square analysis of ^1H NMR titration experiments [38]. This value is lower than K_a = 5390 M^{-1} found in the same conditions for the inclusion of the guest **26** into the cavity of the flexible host **24** [37]. On the other hand the association constant of acetone with the preorganized host molecule **25** was found to be K_a = 1800 M^{-1} which is higher than the K_a = 1000 M^{-1} value established by Atwood and coll. [33] for the analogous complex with the flexible tetrasulfonate **24**. These data indicate that the driving forces for the inclusion of organic cations and neutral molecules inside the apolar cavity of the tetracarboxylate **25** are mainly hydrophobic and a little role is played by electrostatic interactions, which instead are responsible for the more efficient but non specific binding of trimethylanilinium cation **26** by calix[4]arene tetrasulfonate **24**. The selectivity in the recognition of ions and neutral molecules should be therefore higher with preorganized water soluble calixarenes such as **25** than with the flexible compound **24**.

5.2. CALIX[4]ARENE CAGE MOLECULES

In spite of several efforts of various laboratories little evidences have been obtained so far for the inclusion of neutral organic molecules inside the apolar cavity of calix[4]arenes in organic media [2,3].This can be due to the very low stability constant of such complexes which, in turn, depends on the extensive solvation of the host and to the nature of the *inter*molecular interactions involved, which are rather weak. Cram [39], Collet [40] and more recently Still [41], Diederich [42] and their coworkers have clearly shown a strong dependence of the stability of Host-Guest complexes on the nature (including polarity and size) of the competing solvent. From these and other studies [1,43] the tendency to synthesize new host molecules with enforced cavities of molecular dimension (cavitands) [44] has emerged. We were attracted by these studies and, inspired by the results of the X-ray analysis of calix[4]arene solid state inclusion complexes [3,6], we undertook a general project aimed at designing and synthesizing new cage molecules based on calix[4]arenes.

In order to have a rigid building block, the tetraethoxyethylcalix[4]arene **27** in the fixed cone conformation was first synthesized and then procedures for its selective functionalization at the upper rim were developed.

In this compound the para-position of the aromatic nuclei are equivalent and, to reach selectivity in the functionalization, only steric effects or complexation phenomena could be exploited.

Therefore we choose the Gross formylation, which is a mild electrophilic reaction easy to be controlled [45]. By playing around catalyst, solvent, temperature and molar ratio between substrate and reagent it was possible to obtain in good yield mono-, di-, tri- and tetra-formyl derivatives of tetraethoxyethylcalix[4]arene [46].

Particularly interesting and rather impressive is the regioselectivity observed in the bis-formylation: the 1,3-diformylcalix[4]arene **28** is practically the sole reaction product with less than 5% of the 1,2-isomer.

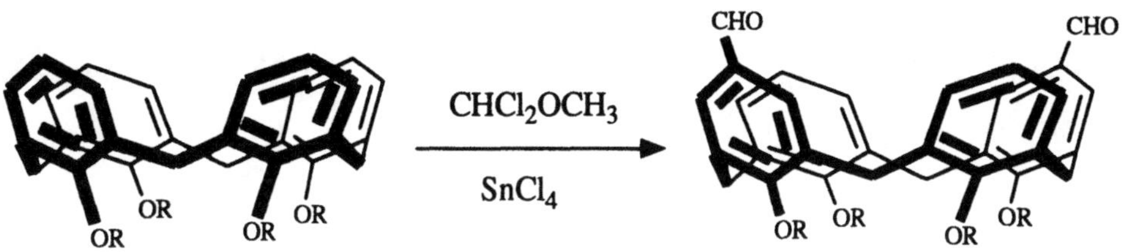

27 **28**

R = $CH_2CH_2OCH_2CH_3$

The 1,3-diformylated compound **28** has been the key intermediate for the synthesis of calix[4]arene cage molecules. In order to increase the van der Waals interactions and to control the solvation of the host cavity we synthesized new calix[4]arenes capped at the upper rim with bridges containing aromatic rings [46]. The 1,3-diformyl cavitand **28** was reduced to the corresponding dimethylol derivative **29** whose disodium salt was reacted with α,α'-dibromo-p-xylene under high dilution conditions to give in 30% yield the desired calix[4]arene capped at the upper rim. Variable temperature ^{1}H NMR spectra of compound **30** show that this bridge is flexible since the aromatic nucleus can rotate on the top of the calixarene cavity.

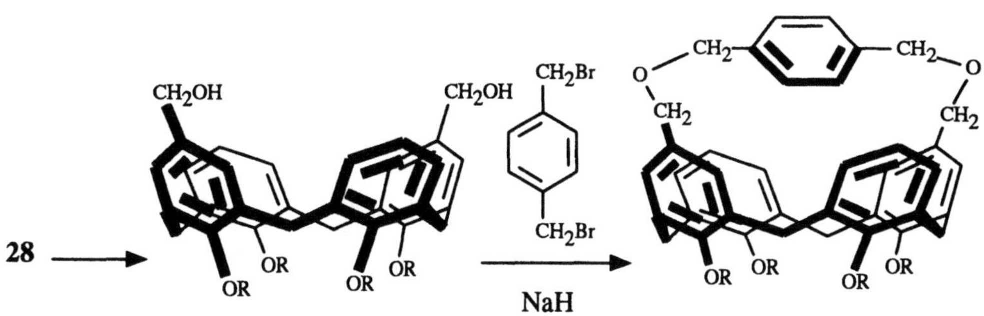

29 **30**

R = $CH_2CH_2OCH_2CH_3$

A more rigid compound **31** was obtained in 30% yield by reacting the dimethylol **29** with α-α'-dibromo-9,10-dimethylanthracene in the same conditions.

The observation that *p-tert*-butylcalix[4]arene **1** is able to form, in the solid state, a

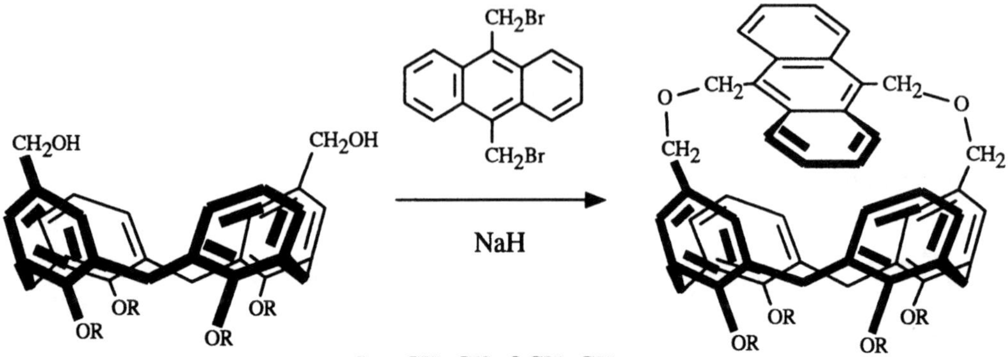

29 R = $CH_2CH_2OCH_2CH_3$ 31

2:1 complex with anisole, which is encapsulated in a tridimensional cage created by two macrocycles facing each other *via* the upper rim [47] has inspired us and others [48] to synthesize double calixarenes. Our approach has benefitted from the selective 1,3-formylation of tetraethoxyethyl calix[4]arene **27** discussed previously. In fact the double calixarene **33** was obtained by reaction of the sodium salt of dimethylol **29** and the bis-electrophile **32**.

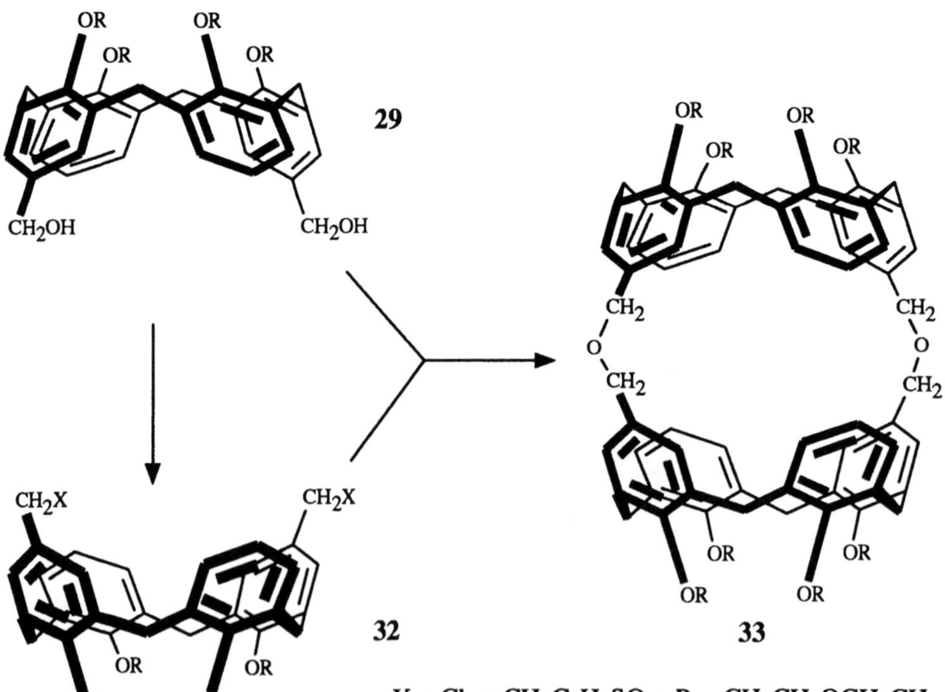

29

32

33

X = Cl or $CH_3C_6H_4SO_3$; R = $CH_2CH_2OCH_2CH_3$

Work is in progress to synthesize more elaborated double calix[4]arenes with different bridges and to study the complexing properties of these new cage molecules toward neutral organic molecules.

6. Acknowledgements

Part of this work has been performed in collaboration with Prof. Reinhoudt's group (Enschede), Prof. Arena and Prof. Rizzarelli (Catania), Prof. Balzani and Prof. Sabbatini (Bologna). Finantial support from the Italian Ministero Università e Ricerca Scientifica e Tecnologica (M.U.R.S.T.) and from EEC Science Program is gratefully acknowledged.

7. References

1 Diedrich, F. (1991) Cyclophanes, Monograph in Supramolecular Chemistry, Stoddart, J.F. (Ed), Royal Society of Chemistry, Cambridge.
2 Gutsche, C.D. (1989) Calixarenes, Monograph in Supramolecular Chemistry, Stoddart J.F. (ed.), Royal Society of Chemistry, Cambridge.
3 Vicens, J. and Böhmer, V. (eds.) (1991), Calixarenes. A Versatile Class of Macrocyclic Compounds, Kluwer Academic Publishers, Dordrecht.
4 Gutsche, C.D. (1987) 'Calixarenes and the Art of Molecular Basketmaking', in Izatt, R.M. and Christensen, J.J. (eds.), Progress in Macrocyclic Chemistry vol. 3, Wiley & Sons, New York, pp. 93-165.
5 Ungaro, R. and Pochini, A. (1991) 'Flexible and Preorganized Molecular Receptors Based on Calixarenes', in Frontiers in Supramolecular Organic Chemistry and Photochemistry, Schneider, H.-J. and Dürr, H. (eds.), VCH, Weinheim, pp. 57-81.
6 Andreetti, G.D., Ugozzoli, F., Pochini, A. and Ungaro, R. (1991) 'Inclusion of Ions and Neutral Molecules in Calixarenes', in Inclusion Compounds vol. 4, Atwood, J.L.; Davies, J.E.D. and Mac Nicol, D.D. (eds.), Oxford University Press, Oxford, pp. 64-125.
7 Gutsche, C.D. (1989) Org.Syntheses 68, White, J.D. (ed.), 233-246.
8 Andreetti, G.D., Ungaro, R., and Pochini, A. (1979) J. Chem. Soc. Chem.

Commun., 1005-1007.

9 Gutsche, C.D. and Bauer, L.J. (1985) J. Am. Chem. Soc. 107, 6052-6059.

10 a) Andreetti, G.D., Calestani, G., Ugozzoli, F., Arduini, A., Ghidini, E., Pochini, A., Ungaro, R. (1987) J. Inclusion Phenom. 5, 123-126; b) Andreetti, G.D., Ugozzoli, F., Casnati, A., Ghidini, E., Pochini, A. and Ungaro, R. (1989) Gazz. Chim. Ital. 119, 47-50.

11 Gutsche, C.D., Gutsche, A.E. and Karaulov, A.I. (1985) J. Inclusion Phenom. 3, 447-451.

12 Gutsche, C.D. and Bauer, L.J. (1981) Tetrahedron Lett. 22, 4763-4766.

13 a) Arduini, A., Pochini, A., Reverberi, S. and Ungaro, R. (1984) J. Chem. Soc. Chem. Commun., 981-982; b) Arduini, A., Pochini, A., Reverberi, S., Ungaro, R., Andreetti, G.D. and Ugozzoli, F. (1986) Tetrahedron 42, 2089-2100.

14 Calestani, G., Ugozzoli, F., Arduini, A., Ghidini, E. and Ungaro, R. (1987) J. Chem. Soc. Chem. Commun., 344-346; Arduini, A., Ghidini, E., Pochini, A.,Ungaro, R., Andreetti, G.D., Calestani, G. and Ugozzoli, F. (1988) J. Inclusion Phenom. 6, 119-134.

15 Groenen, L.C., Ruël, B.H.M., Casnati, A., Timmerman, P., Verboom, W., Harkema, S., Pochini, A., Ungaro, R. and Reinhoudt, D.N. (1991) Tetrahedron Lett. 32, 2675-2678.

16 Ungaro, R. and Pochini, A. (1991) 'Calixarene Based Cation Receptors and Carriers' in Ref. 3, pp. 127-147; Schwing, M.-J. and Mc Kervey M.A. (1991) 'Chemically Modified Calixarenes as New Selective Receptors for Monovalent Cations', ibidem, pp. 149-172.

17 Sabbatini, N., Guardigli, M., Mecati, A., Balzani, V., Ungaro, R., Ghidini, E., Casnati, A. and Pochini, A. (1990) J. Chem. Soc. Chem. Commun., 878-879.

18 a) Sabbatini, N., Perathoner, S., Balzani, V., Alpha, B. and Lehn, J.-M. (1987) 'Antenna Effect in Eu^{3+} and Tb^{3+} Cryptates', in Supramolecular Photochemistry, Balzani, V. (ed.), Reidel, Dordrecht, pp. 187-206; Alpha, B., Lehn, J.-M. and Mathis, G. (1987) Angew. Chem. Int. Ed. Engl. 26, 266-267; b) Alpha, B., Balzani, V., Lehn, J.-M., Perathoner, S. and Sabbatini, N. (1987) Angew. Chem. Int. Ed. Engl. 26,1266-1267.

19 Balzani, V., Lehn, J.-M., van de Loosdrecht, J., Mecati, A., Sabbatini, N. and Ziessel, R. (1991) Angew.Chem. Int.Ed.Eng. 30, 190-191.

20 Raymond, K.N., Miller, G. and Matzanke, B.F. (1984) Top.Curr.Chem. 123,

49-102; Lifson, S., Felder, C.E., Shanzer, A. and Libman, J. (1987) 'Biomimetic Macrocyclic Molecules: An Interactive Theoretical-Experimental Approach', in Progress in Macrocyclic Chemistry vol. 3, Izatt, R.M. and Christensen J.J. (eds.), Wiley & Sons, New York, pp. 241-307.

21 Raymond, K.N. and Garrett, T.M. (1988) Pure & Appl. Chem. 60, 1807-1816; Sun, Y. and Martell, A.E. (1989) J. Am. Chem. Soc. 111, 8023-8024; Ebmeyer, F. and Vögtle, F. (1991) 'New Hosts for the Molecular Recognition and Encapsulation of Guest Compounds', in Inclusion Compounds vol. 4, Atwood, J.L.; Davis, J.E.D. and Mc Nicol, D.D. (eds.), Oxford University Press, Oxford, pp. 263-282.

22 Casnati, A., Arduini, A., Ghidini, E., Pochini, A. and Ungaro, R. (1991) Tetrahedron 47, 2221-2228; Groenen, L.C., Ruël, B.H.M., Casnati, A., Verboom, W., Pochini, A., Ungaro, R. and Reinhoudt, D.N. (1991) Tetrahedron 47, 8379-8384.

23 Casnati, A., Arduini, A., Pochini, A. and Ungaro, R. (1990) 'Selective Dealkylation of Calix[4]arenes Ethers and Synthesis of New Hydroxamate Siderophores', VI Int. Symposium on Molecular Recognition and Inclusion, Berlin, Abst. F14.

24 Casnati, A., Pochini, A., Ungaro, R., Cacciapaglia, R. and Mandolini, L. (1991) J. Chem. Soc. Perkin Trans. 1, 2052-2054.

25 Ghidini, E., Ugozzoli, F., Ungaro, R., Harkema, S., El-Fadl, A.A. and Reinhoudt, D.N. (1990) J.Am.Chem.Soc. 112, 6979-6985.

26 Casnati, A., Ghidini, E., Pochini, A., Ugozzoli, F., Ungaro, R. and Reinhoudt D.N., unpublished results.

27 Casnati, A., Arduini, A., Mangia, A., Notari, M., Pochini, A. and Ungaro, R. (1989) 'New Shapes for the Selective Encapsulation of Divalent Metal Cations using p-t-Butylcalix[4]arene as a Template', Workshop on Supramolecular Organic Chemistry and Photochemistry, Saarbrücken, Abst. P14.

28 Shinkai, S. (1991) 'Functionalized Calixarenes: New Applications as Catalysts, Ligands and Host Molecules', in Ref. 3, pp. 173-198.

29 Casnati, A., Minari, P., Pochini, A. and Ungaro, R. (1991) J. Chem. Soc. Chem. Commun., 1413-1414.

30 Casnati, A., Minari, P., Pochini, A., Ungaro, R., Nijenhuis, W.F., de Jong, F. and Reinhoudt, D.N. (1992) Israel J. Chem., in press.

31 Alfieri, C., Dradi, E., Pochini, A., Ungaro, R. (1989) Gazz. Chim. Ital. 119,

335-338.

32 Grootenhuis, P.D.J., Kollman, P.A., Groenen, L.C., Reinhoudt, D.N., van Hummel, G.J., Ugozzoli, F. and Andreetti, G.D. (1990) J. Am. Chem. Soc. 112, 4165-4176.

33 Atwood, J.L. and Bott, S.G. (1991) 'Water Soluble Calixarene Salts. A Class of Compounds with Solid State Structures Resembling those of Clays', in Ref. 3, pp. 199-210.

34 Arena, G., Calì, R., Lombardo, G.G., Rizzarelli, E., Sciotto, D., Ungaro, R. and Casnati, A. (1992) J. Supramolecular Chem., in press.

35 Shinkai, S., Araki, K., Koreishi, H., Tsubaki, T. and Manabe, O. (1986) Chem.Lett., 1351-1354.

36 Christensen, J.J., Hansen, L.D., Izatt, R.M. (1976) 'Handbook of Proton Ionization Heats and Related Thermodynamic Quantities', Wiley & Sons, New York.

37 Shinkai, S., Araki, K., Matsuda, T. and Manabe, O. (1989) Bull. Chem. Soc. Jpn. 62, 3856-3862; Shinkai, S., Araki, K., Matsuda, T., Nishiyama, N., Ikeda, H., Takasu, I. and Iwamoto M. (1990) J. Am. Chem. Soc. 112, 9053-9058.

38 Casnati, A., Pochini, A. and Ungaro, R., unpublished results.

39 Cram, D.J., Stewart, K.D., Goldberg, I and Trueblood, K.N.(1985) J. Am. Chem. Soc. 107, 2574-2575; Briant, J.A., Ericson, J.L. and Cram, D.J.(1990) J. Am. Chem. Soc., 112, 1255-1256.

40 Collet, A. (1987) Tetrahedron 24, 5725-5759.

41 Chapman, K.T. and Still, W.C. (1989) J. Am. Chem. Soc. 111, 3075-3079.

42 Smithrud, D.B., Sanford, E.M., Chao, I., Ferguson, S.B., Carcanague, D.R., Evanseck, J.D., Houk, K.N. and Diedrich, F. (1990) Pure & Applied Chem. 62, 2227-2236.

43 Diederich, F.J. (1988) Angew. Chem. Int. Ed. Engl. 27, 362-386.

44 Cram, D.J. (1983) Science 219, 1177-1183.

45 Rieche, A., Gross, H. and Holt, E. (1960) Chem. Ber. 93, 88-94; Rieche, A., Gross, H., Hoft, E. (1967) Org. Synth. 47, 1-3.

46 Arduini, A.; Manfredi, G.; Pochini, A.; Sicuri, A.R. and Ungaro, R. (1991) J. Chem. Soc. Chem. Commun., 936-937.

47 Ungaro, R.; Pochini, A.; Andreetti, G.D. and Domiano, P. (1985) J. Chem. Soc. Perkin Trans. II, 197-201.

48 Böhmer, V., Goldmann, H., Vogt, W., Vicens, J. and Asfari, Z. (1989) Tetrahedron Lett., 1391-1394.

CALIX[4]ARENES, MOLECULAR PLATFORMS FOR SUPRAMOLECULAR STRUCTURES

L. C. GROENEN and D. N. REINHOUDT
Laboratory of Organic Chemistry
University of Twente
P.O. Box 217
7500 AE Enschede
The Netherlands

ABSTRACT. A new strategy for the synthesis of receptor molecules comprises the combination of medium-sized molecules to which functional groups for intermolecular interactions can be attached. Most efforts concern the functionalization of calix[4]arenes, but other building blocks like cyclodextrins and octols are used as well. Selective mono-, 1,2-di, 1,3-di-, and tetraalkylation of calix[4]arenes can be achieved by variation of the alkylation conditions. The calix[4]arenes can be (selectively) functionalized at the "upper rim" by a variety of methods including *ipso*-nitration. Combination of the calix[4]arenes with crown ethers to calixcrowns gave receptors with a high K^+/Na^+-selectivity and with terphenyls calixspherands were obtained that form kinetically stable complexes with Na^+, K^+ and Rb^+. Bridging of calix[4]arene with a salophene moiety gave a receptor for urea that can effectively transport urea through a supported liquid membrane. Calix[4]arenes can be combined covalently to double and triple calixarenes and with selectively functionalized octols. When substituted with hydrogen bond donors and acceptors like 2-pyridone, calixarenes undergo self-association to larger aggregrates.

1. Introduction

Nature constructs biological receptors by combining large numbers of simple building blocks, which carry different functional groups, into large molecules that form three-dimensional structures by specific intramolecular interactions between different parts of the linear structure. Amino acids are combined to proteins, nucleosides to DNA or RNA and monosaccharides to carbohydrates. The recognition sites are formed by a very precise stereochemistry and the recognition process is often very efficient.

V. Balzani and L. De Cola (eds.), Supramolecular Chemistry, 51–70.

However, this is achieved at the expense of a high molecular weight.

Synthetic receptors are obtained by de novo synthesis using modern synthetic methodologies, which allow almost unlimited variation. The strategy focusses on the complementarity of functional groups between receptor and guest species and aims for minimal reaction steps and molecular weights. The drawback is that for each individual guest a new synthetic pathway has to be developed; the learning experience is not efficiently accumulated.

We are currently exploring an approach that is a compromise of these two extremes in that we start from medium-sized molecules that we use as frameworks or platforms to which the functional groups for intermolecular interactions can be attached. In a later stage several of these platforms can be connected either by covalent or non-covalent bonds to build up larger structures. Examples of such molecular building blocks are cyclodextrins (**1**), the octols or Högberg compounds (**2**) and the calixarenes (**3**). Cyclodextrins (**1**) are being used by quite a few groups as starting structures for the synthesis of host molecules and enzyme models [1]. The octols **2** have mainly been used by Cram and coworkers as building blocks for cavitands [2]. A few years ago we have become interested in the calixarenes (**3**) as promising new building platforms [2,3], especially the calix[4]arenes (**3**, n = 4), and more recently also the calix[6]arenes (**3**, n = 6).

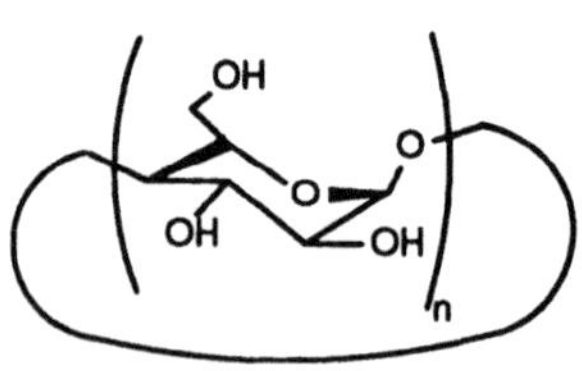

1

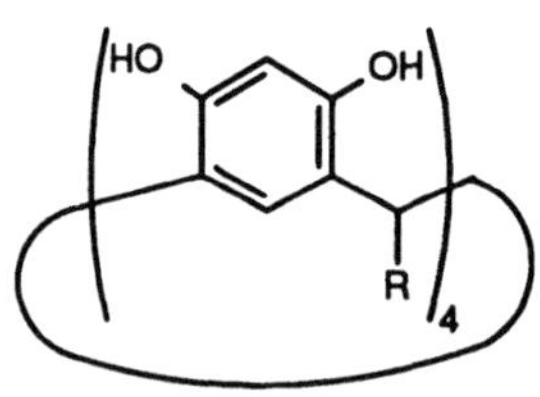

2

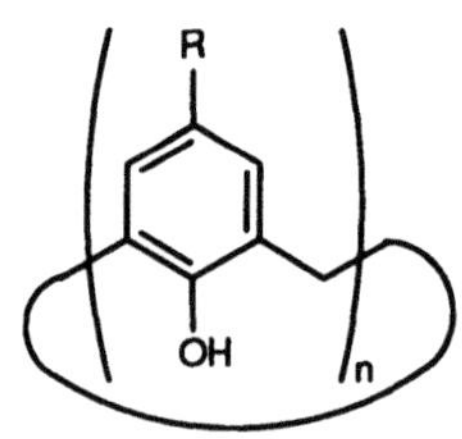

3

In this article we will outline our efforts towards the shaping of these molecular platforms to our needs. Although we have found some interesting new results for the selective functionalization of β-cyclodextrin (**1**, n = 7) and the octols **2**, our main research activity is dedicated to calix[4]arenes and our results in this field will be discussed in detail. Beside new methods for selective functionalization of calix[4]arenes, also ways for controlling the conformation of the calix[4]arene and thus the shape of the building block will be discussed. Finally, a few examples of the coupling of calix[4]arenes with other building blocks will be described.

2. Selective Functionalization of β-Cyclodextrin

Recently, the synthesis of the heptasilylated β-cyclodextrin **4** has been reported [4]. This compound, which is soluble in organic solvents, can be monotosylated at the secondary side in 30 % yield by reaction with one equivalent of NaH in tetrahydrofuran and treatment with *p*-tosyl chloride (Scheme 1). Subsequent epoxidation and ring-opening with LiN_3 afforded in good yield the monoazide **5**, which, after reduction, can be used in coupling reactions with other building blocks [5].

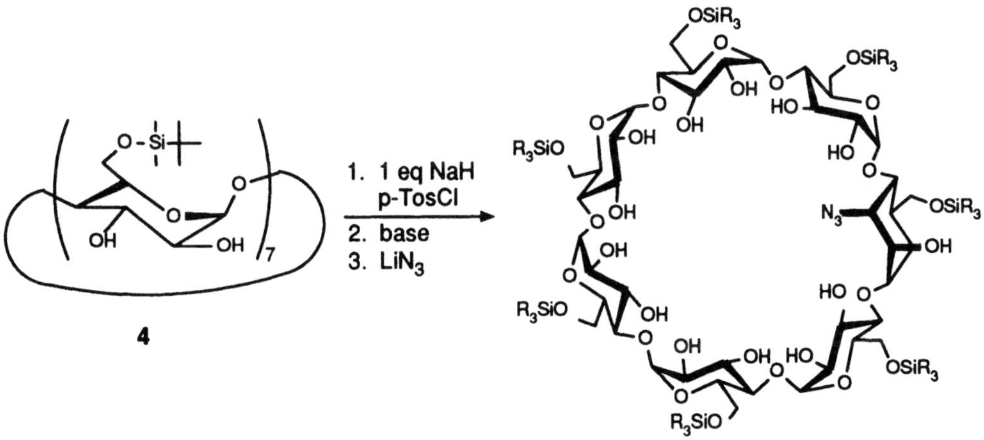

Scheme 1

3. Selective Functionalization of Octols

Two of the four bromo atoms of the tribridged octol **6** [6] can be selectively removed in 76 % yield by treatment with five equivalents of *n*-BuLi and quenching with acid after 15 seconds (Scheme 2). After bridging of the last two hydroxyl groups the remaining two bromo atoms can be replaced by electrophiles via the dilithio compound resulting in compounds **8**, which can also be used for coupling with other building blocks [7].

4. Selective Functionalization of Calix[4]arenes

In *p-tert*-butylcalix[4]arene **9b**, which exists in a cone conformation as drawn, two very different reactive sites can be distinguished. The four hydroxyl groups together form

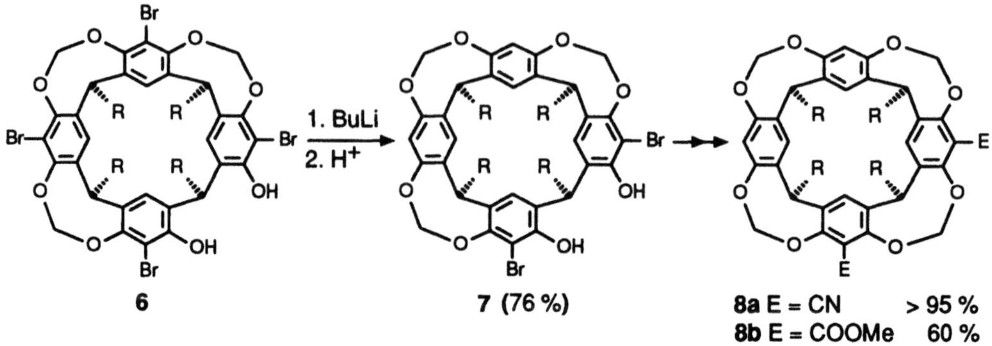

Scheme 2

the lower rim of the calix[4]arene and the para-positions of the four phenol rings form the upper rim. We have been able to design new methods for the selective functionalization of both the lower and the upper rim of calix[4]arenes.

4.1. SELECTIVE FUNCTIONALIZATION OF THE LOWER RIM

4.1.1. *Monoalkylation.* Monoalkylated calix[4]arenes have been obtained by selective dealkylation of the corresponding dialkyl or tetraalkyl ethers with one or three equivalents of Me_3SiI, respectively [8]. We recently found that direct alkylation of calix[4]arene with excess of alkylating agent in the presence of 1.2 equivalents of a very weak base, CsF in *N,N*-dimethylformamide (DMF), is an attractive alternative (see Scheme 3) [9]. The use of 0.6 equivalents of K_2CO_3 in acetonitrile also afforded reasonable amounts of monoalkylated products, but 1,3-disubstituted products were always formed as side-products. The alkylations with weak base proceed via the monoanion of the calix[4]arene. The oxyanion is stabilized by two hydrogen bonds from neighboring phenolic groups. However, also the monoanion of the monoalkylated calix[4]arene can be stabilized by two hydrogen bonds and consequently there exists

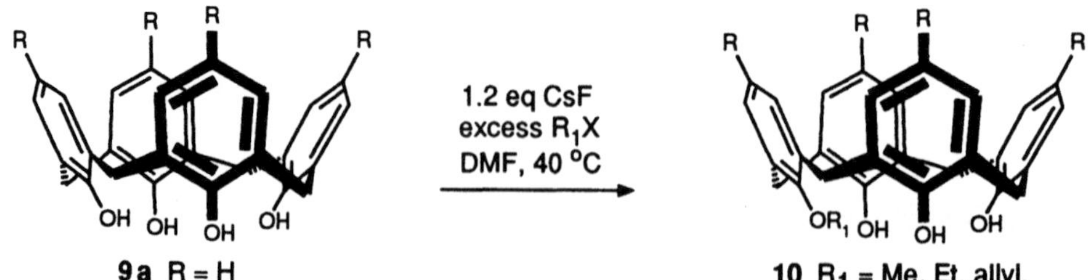

Scheme 3

only a small difference in basicity and reactivity of these two monoanions. The use of a very weak base is necessary to distinguish between the two anions. Apparently, K_2CO_3 and $KHCO_3$, which is formed during the reaction, are too strong bases to affect selective monoalkylation.

4.1.2. *1,3-Dialkylation.* Calix[4]arenes can be alkylated selectively on two diametrically located phenol rings by reaction with two equivalents of an alkylating agent in the presence of one equivalent of K_2CO_3 [10,11]. The yields are generally excellent and the products (**11**) exist in the cone conformation. This reaction is very selective because the monoanion of the 1,3-disubstituted calix[4]arene, from which further substitution would have to take place, cannot be stabilized by hydrogen bonds, or only a very weak one from the diametrically positioned phenol ring. Thus, this anion is not formed in any significant amounts under these reaction conditions with weak base.

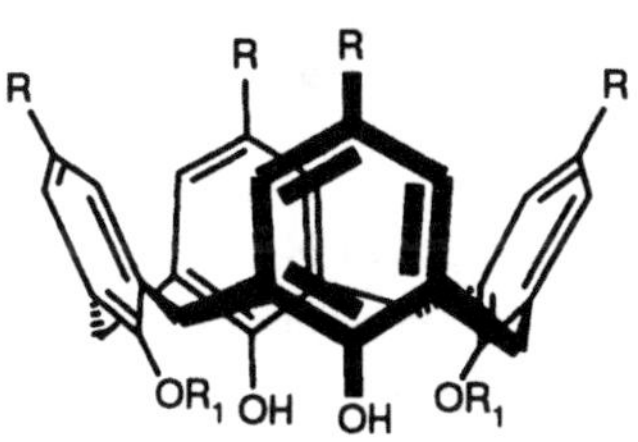

11 R_1 = Me, Et, Bn, allyl $CH_2C(O)R$

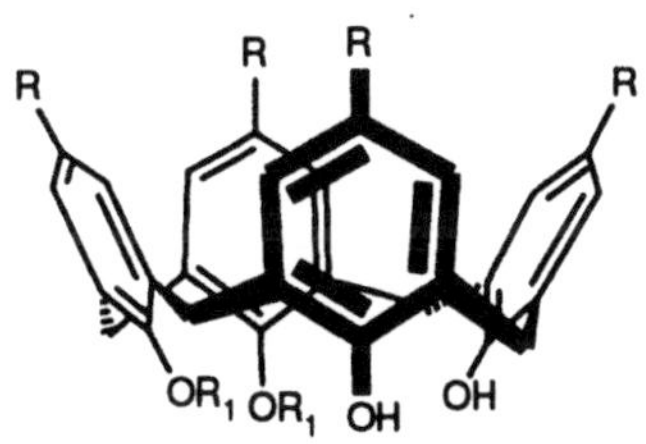

12 R_1 = Me, Et, Bn, allyl, CH_2CH_2OEt

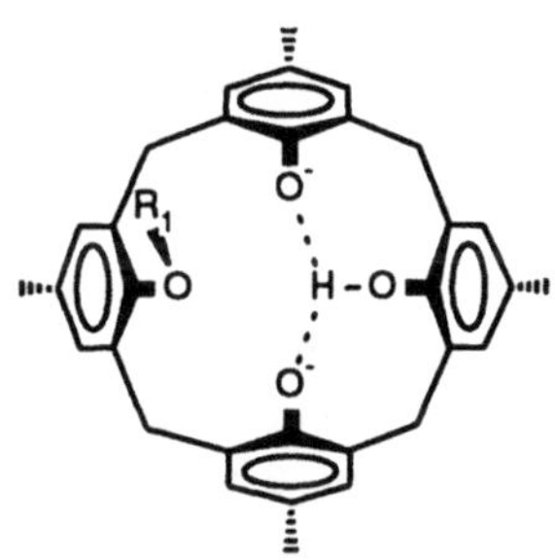

13

4.1.3. *1,2-Dialkylation.* The use of strong base instead of weak base completely alters the selectivity of the alkylation reaction. Reaction of unsubstituted calix[4]arenes with excess of NaH in *N,N*-dimethylformamide or acetonitrile and 2.2 equivalents of alkylating agent yields the 1,2-disubstituted calix[4]arenes **12** in the cone conformation in 15-55 % yield [12]. An important side-product is the tetrasubstituted calix[4]arene in the cone conformation, beside starting material. Under the strongly basic conditions of this reaction the calix[4]arene is multideprotonated. The formation of the 1,2-disubstituted calix[4]arenes can be rationalized by assuming that the dianion of the monoalkylated calix[4]arene (**13**) is an intermediate. This will have the structure as shown because of minimization of repulsion between the two oxyanions and the stabilizing effect of the hydrogen bond of the hydroxyl group in the middle. Apparently the calix[4]arene is not completely deprotonated under these conditions.

4.1.4. *Replacement of the Phenolic OH Groups*. Recently the synthesis of completely dehydroxylated calix[4]arene **14d** by reductive cleavage (K in liquid ammonia) of the tetrakis(diethyl phosphate) ester of *p-tert*-butylcalix[4]arene **9b** has been reported [13]. Starting from the 1,3-di- and tetraesters of **9b** we could isolate the partially dehydroxylated calix[4]arenes **14a-c** [14].

The hydroxyl groups can be replaced by thiol groups via the Newman-Kwart rearrangement of the corresponding *O*-dimethylthiocarbamates. The dithiols **15b** and **15c** were synthesized by reduction of the *S*-dimethylthiocarbamates that were obtained after the Newman-Kwart rearrangement of the *O*-dimethylthiocarbamates of **15a** and **14b** [14].

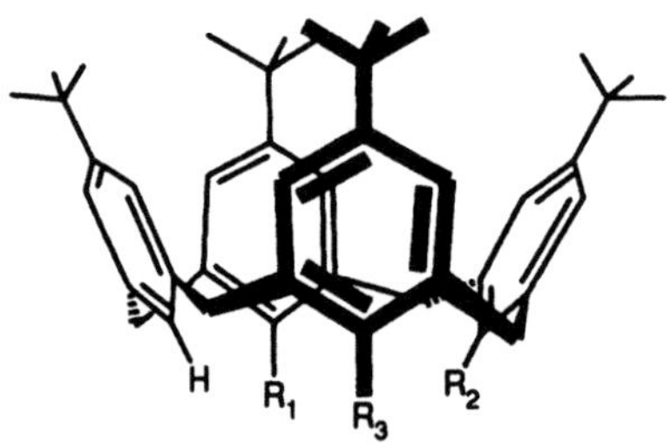

14a R_1=R_2=R_3=OH
14b R_1=R_3=OH, R_2=H
14c R_1=OH, R_2=R_3=H
14d R_1=R_2=R_3=H

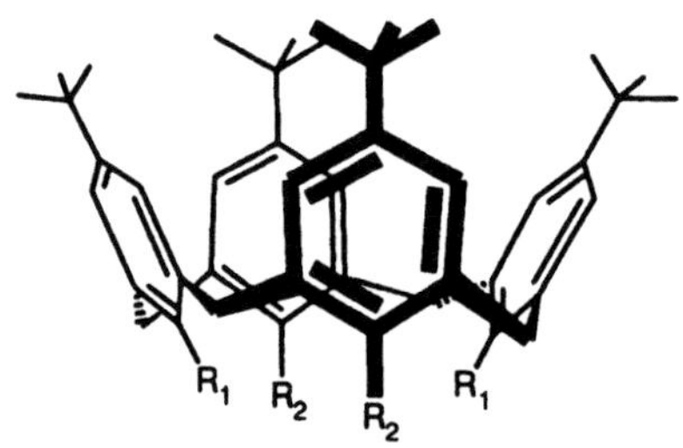

15a R_1=OMe, R_2=OH
15b R_1=OMe, R_2=SMe
15c R_1=H, R_2=SH

4.2. CONFORMATIONS OF TETRASUBSTITUTED CALIX[4]ARENES

The calix[4]arenes discussed so far all exist in the cone conformation because of the syn-orientation of the substituents and the presence of hydroxyl groups that stabilize the cone conformation by the formation of hydrogen bonds. However, beside the cone a calix[4]arene can adopt three other conformations, viz. the partial cone, the 1,2-alternate and the 1,3-alternate (see Scheme 4). In tetrasubstituted calix[4]arenes, in which no free hydroxyl groups are present, all these four conformations can in principal be formed.

4.2.1. *Flexible Calix[4]arenes*. The tetramethyl ether of *p-tert*-butylcalix[4]arene (**16**), which is a mobile compound, was first synthesized by Gutsche and coworkers [15]. They concluded from the low-temperature ^{1}H NMR spectrum that it consists as a mixture of two partial cone conformations, one in which the methyl group of the

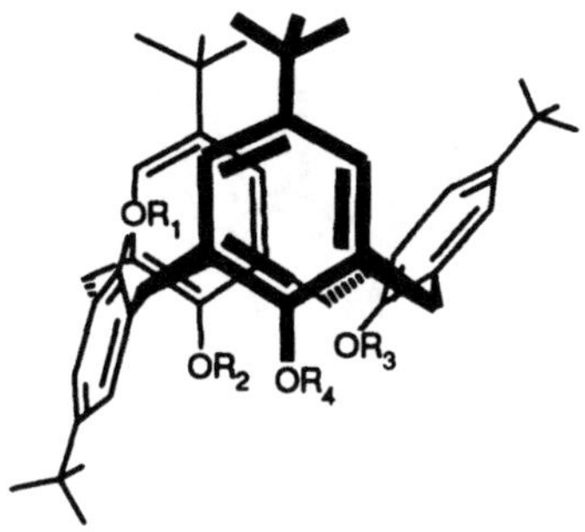

16 $R_1=R_2=R_3=R_4=Me$
17a $R_1=R_2=Me$, $R_3=R_4=Et$
17b $R_1=R_3=Me$, $R_2=R_4=Et$
17c $R_2=R_3=Me$, $R_1=R_4=Et$
17d $R_2=R_4=Me$, $R_1=R_3=Et$
18 $R_1=R_2=R_3=R_4=Et$

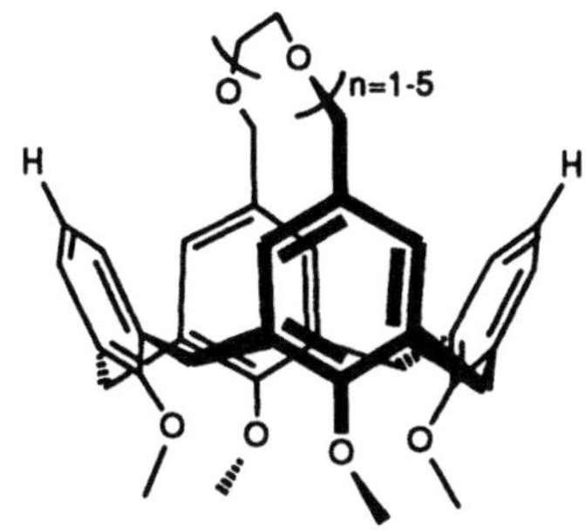

19

rotated aromatic ring points outwards and one in which this methyl group points into the cavity. Reinvestigation of the ^{1}H NMR spectrum showed that compound **16** exists as a mixture of all four conformations (see Table 1) [16]. In a 2D EXSY spectrum at the same low temperature cross peaks due to chemical exchange were present between signals of the partial cone and the cone, and of the partial cone and the 1,3-alternate, which indicates that these conformations interconvert. Strange enough no crosspeaks corresponding with signals of the 1,2-alternate were present. In the ^{1}H NMR spectrum at room temperature the signals of the 1,2-alternate conformation are still sharp, whereas the other signals have broadened, and only at higher temperature also the signals of the 1,2-alternate broaden and coalesce with the other signals. This indicates that the interconversion involving the 1,2-alternate is much slower than the interconversion between cone, partial cone and 1,3-alternate. *The 1,2-alternate is the kinetically most stable conformation.* The complete interconversion process is depicted in Scheme 4.

In order to obtain more information about the factors that influence the ratios of the conformations and the interconversion process we investigated the properties of partially fixed calix[4]arenes. Using different methods for selective functionalization, we synthesized a series of mixed diethyldimethyl ethers (**17**), that are partially rigidified because of the presence of the ethoxy groups [16]. The partial fixation precludes one or two of the conformations and in this way also certain pathways for interconversion are excluded. Table 1 shows the ratios of conformations of the four

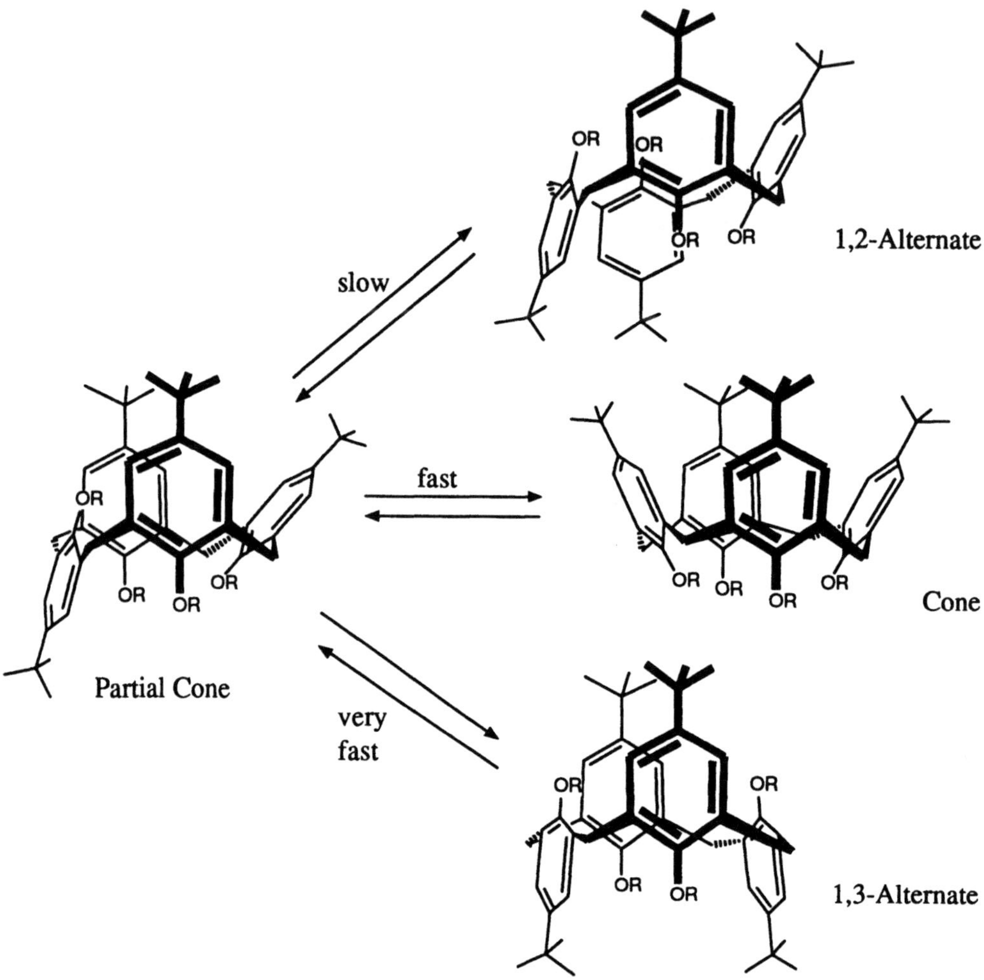

Scheme 4

different diethyldimethyl ethers. In compound **17b** the 1,2-alternate is precluded and the ^{1}H NMR spectrum at room temperature shows only broad peaks, indicating a fast interconversion between cone, partial cone and 1,3-alternate. For compounds **17a** and **17c** sharp peaks of the 1,2-alternate are visible beside broad signals of the other conformations, showing that the 1,2-alternate is also in these compounds a kinetically stable conformation. When both the cone and the 1,3-alternate are precluded as in compound **17d** only sharp peaks appear in the ^{1}H NMR spectrum because the two fast interconversion pathways are precluded. Interestingly, the 1,2-alternate is also thermo-

Table 1. Relative Equilibrium Concentrations of the Tetramethyl, Diethyldimethyl and Tetraethyl Ethers of *p-tert*-ButylCalix[4]arene. (Measured in $CDCl_3$ at -30 °C, in %.)

	cone	partial cone	1,2-alternate	1,3-alternate
16	4	85	8	3
17a	<5[a]	76	24	--[b]
17b	20	80	--[b]	<5[a]
17c	--[b]	64	25	11
17d[c]	--[b]	35	65	--[b]
18[d]	7	47	43	3

[a] This conformation could not be ruled out because of the low resolution of the ^1H NMR spectrum. [b] Precluded conformation for this compound. [c] Measured at room temperature. [d] Measured in $CDCl_2CDCl_2$ at 132 °C.

dynamically the most stable conformation for this compound.

Another way of rigidifying the tetramethyl ether is by bridging the upper rim of the calix[4]arene. We prepared compounds **19** from the selectively functionalized 1,3-*p*-diacid chloride of tetramethoxycalix[4]arene and the appropriate polyethylene glycol [17]. As for the comparable compound **17b** only the cone and the partial cone conformations could be observed, although the interconversion process involving the 1,3-alternate also takes place. A kinetic investigation of the interconversion process of compounds **19** (n=1 and n=2) by quantitative 2D EXSY NMR spectroscopy showed that the interconversion via the 1,3-alternate is more than a factor of 10 faster than via the cone.

Even the tetraethyl ether of *p-tert*-butylcalix[4]arene (**18**) is flexible, but only at temperatures higher than 100 °C. Shinkai and coworkers have reported that the partial cone conformation of this compound slowly isomerizes to the 1,2-alternate, resulting in a nearly 1 : 1 mixture of partial cone and 1,2-alternate [18]. With the above results in mind we reinvestigated this system and we could show that also this compound adopts all four conformations (see Table 1) [16]. A kinetic investigation of the conversion of the partial cone to the other three conformations confirmed the relative rates of the three processes.

4.2.2. *Conformationally Fixed Calix[4]arenes.* For calix[4]arenes with four substituents larger than the ethoxy group the interconversion is completely blocked [18]. Therefore, the conformation of these compounds is determined during their synthesis and depends on kinetic factors. Control of the conformation of the final tetrasubstituted calix[4]arenes would allow for the synthesis of four different building blocks, namely the four conformations.

We have investigated the influence of the strong base and the solvent on the conformational outcome of the tetraethylation of *p-tert*-butylcalix[4]arene **9b** (see Table 2) [12]. The use of NaH in *N,N*-dimethylformamide (DMF) or acetonitrile at room temperature exclusively leads to the tetraethylated calix[4]arene in the cone conformation in good yield. Also in reactions with other electrophiles exclusively the cone conformation is obtained [19]. The previously discussed 1,2-dialkylated calix[4]arenes are intermediates in these reactions [12]. When KH or potassium *tert*-butoxide (KOtBu) was used instead of NaH a mixture of all four conformations was obtained with the partial cone as the major product.

Shinkai and coworkers have reported that the tetra-*n*-propylation of **9b** with Cs_2CO_3 in DMF yields a mixture of partial cone, 1,2-alternate and 1,3-alternate conformations [18b]. Recently we found that alkylation of calix[4]arenes with other para-substituents under the same conditions yields mainly the 1,3-alternate conformation of the tetraalkylated calix[4]arenes [20].

The 1,2-alternate conformation cannot yet be obtained from direct alkylation reactions in attractive yields. We were able to show that ethylation of the *anti*-1,3-diethyl ether of **9b** with KOtBu in tetrahydrofuran afforded the 1,2-alternate conformation of the tetraethyl ether in 64 % yield [16]. Apparently, the *anti*-1,3-diethyl ether is not formed in large quantities in the reaction starting from *p-tert*-butylcalix[4]arene **9b** under the same conditions.

Table 2. Results of Tetraethylation of *p-tert*-Butylcalix[4]arene (EtI, RT, 24 h).

Reaction conditions		Relative yields of conformation (%)			
base	solvent	cone	partial cone	1,2-alternate	1,3-alternate
NaH	DMF	100	a	a	a
NaH	MeCN	100	a	a	a
NaH	THF/DMF 10:1	62	28	10	a
NaH	THF	b	b	b	b
KH	DMF	2	64	17	17
KH	THF	a	64	5	31
KOtBu	MeCN	5	84	2	9
KOtBu	THF	a	77	a	23

[a] Not detected in the ^1H NMR spectrum. [b] No product had been formed after 24 h.

4.3. SELECTIVE FUNCTIONALIZATION OF THE UPPER RIM

Except for two examples of mono-*p*-allylcalix[4]arenes [21] functionalization of the upper rim of calix[4]arenes has only been performed at all four para-positions simultaneously. We realized that the selectively *lower rim*-functionalized calix[4]arenes can be useful starting materials for the selective introduction of functional groups at the *upper rim*, and we have developed several strategies to prove this.

4.3.1. *Transfer of Functionality.* The selective functionalization of the lower rim allows for the synthesis of mixed alkylallyl ethers that can be subjected to a *p*-Claisen rearrangement [10]. In this way compounds of general structure **20a**, in which the selectivity has been transferred from the lower rim to the upper rim, were obtained in good yields. The allyl group can easily be transformed to other functional groups (CH_2X, CH_2=O, COOH).

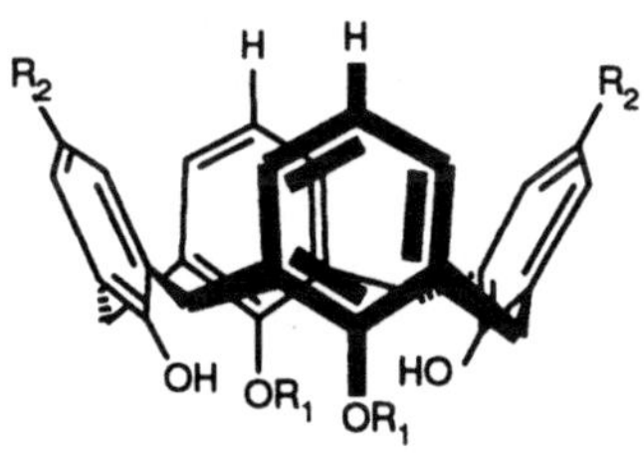

20a R_1 = alkyl, R_2 = allyl
20b R_1 = alkyl, R_2 = E

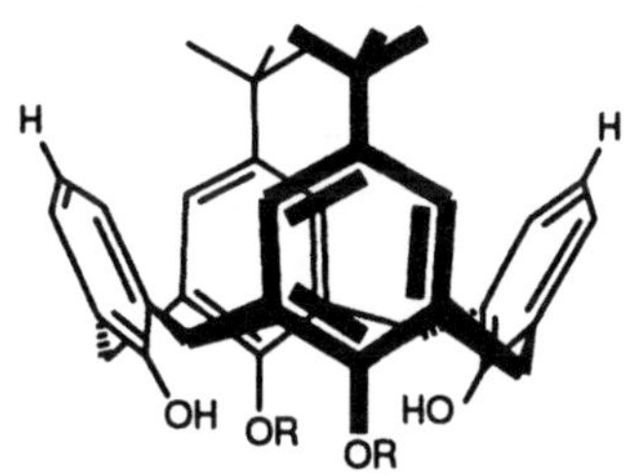

21 R = alkyl

4.3.2. *Direct Substitution.* A second method for the selective functionalization of the upper rim takes advantage of the fact that electrophilic aromatic substitutions are faster on phenols than on alkylated phenols. Starting with dialkylated calix[4]arenes a wide variety of electrophiles (Br, NO_2, CH_2=O, HgOTFA) can be introduced selectively on the para-positions of the phenol rings in yields > 80% (compounds **20b**) [10].

4.3.3. *Selective Removal of* tert-*Butyl Groups.* The difference in reactivity between phenols and alkylated phenols can also be exploited for the selective removal of two *tert*-butyl groups from dialkoxy-*p*-*tert*-butylcalix[4]arenes [10]. This leads to calix[4]arenes **21** in which the two alkoxy-substituted rings remain protected, while the two phenol rings can be further functionalized on the para-positions by the methods described above.

4.3.4. *Selective Nitration of Tetrasubstituted Calix[4]arenes.* The presence of a selectively functionalized lower rim is not an absolute prerequisite for selective functionalization of the upper rim. We found that the reaction of the tetra-*n*-propyl ether of de-*tert*-butylated calix[4]arene **9a** in the cone conformation with 15 equivalents of 65 % HNO_3 in CH_2Cl_2 for 30 minutes gave the mononitrated calix[4]arene **22a** in 55 % yield [22]. When the reaction time is prolonged to 2.5-5 hours and a little acetic acid is added a mixture of nitrated calix[4]arenes is obtained from which the 1,2-dinitro (**22c**) and the 1,3-dinitro compounds (**22b**) could be isolated in 10 and 30 %, respectively. The reaction of the tetrakis(ethoxyethyl) ether with acetyl nitrate in CH_2Cl_2 at room temperature afforded 30-40 % of the 1,3-dinitrated product together with 20 % of the mononitro compound [23]. The 1,2-dinitrated product was not observed in this reaction.

We recently discovered that the *tert*-butyl groups of tetrasubstituted calix[4]arenes can also be replaced directly by nitro groups (*ipso*-nitration) [19]. When this reaction is performed with 100 % HNO_3 in CH_2Cl_2/AcOH tetranitrocalix[4]arenes (e.g. **22d**) were obtained, but under milder conditions (65 % HNO_3) each of the four partially nitrated calix[4]arenes was formed. The tetranitration not only works for tetrasubstituted *p*-*tert*-butylcalix[4]arenes in the cone conformation, but also for the partial cone and the 1,3-alternate conformations (compounds **23** and **24**) [22].

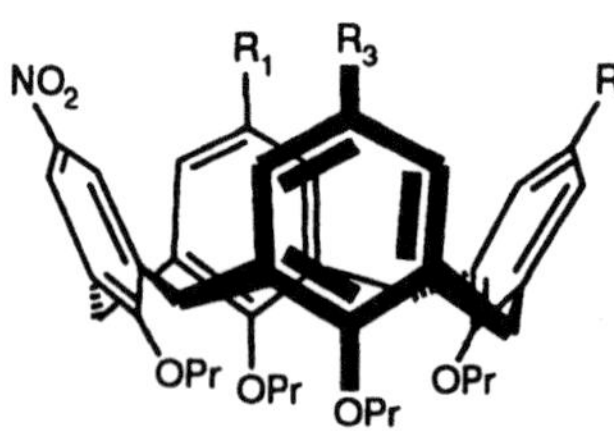

22a $R_1=R_2=R_3=H$
22b $R_1=R_3=H$, $R_2=NO_2$
22c $R_2=R_3=H$, $R_1=NO_2$
22d $R_1=R_2=R_3=NO_2$

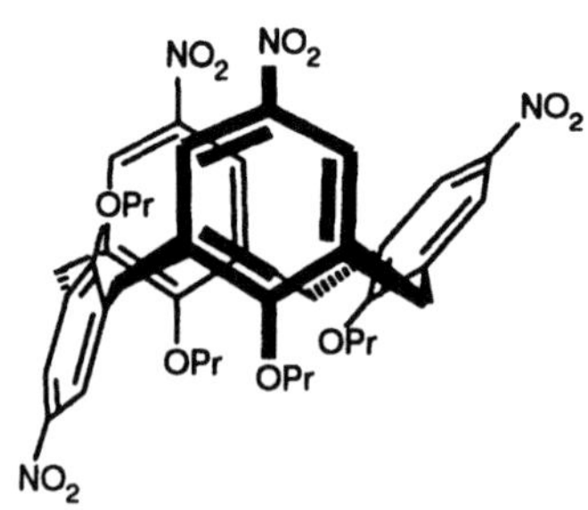

23

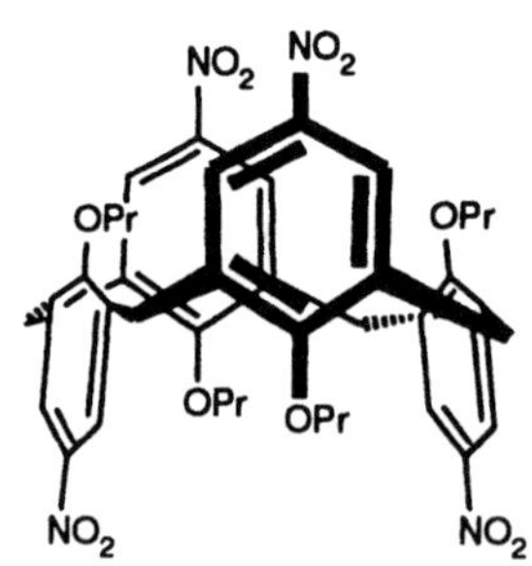

24

5. Combination of Calix[4]arenes with Other Building Blocks

Although the chemistry of calix[4]arenes is very interesting and (selectively) functionalized calix[4]arenes can have very useful properties, as was demonstrated above, the full potential of calix[4]arenes becomes apparent in their combinations with other building blocks. We will discuss a number of very successful compounds that are being used for different applications.

5.1 CALIXCROWN ETHERS

The reaction of calix[4]arene **9b** with tetraethylene glycol ditosylate gave the calixcrown ether **25** in 53 % yield [24]. Subsequent alkylation afforded the 2,4-di-alkylated calixcrowns **26** in mixtures of the cone, the partial cone and in some cases the 1,3-alternate conformations, except for the dimethyl ether **26a**, which is flexible and adopts the cone conformation. These alkylated calixcrown ethers form complexes with alkali metal ions and all of them are selective for K^+. The partial cone is not only the best conformation for complexation, which was nicely illustrated by compound **26a**, which changed from a cone to a partial cone conformation upon complexation of a cation, but also shows the highest selectivity for K^+. The dimethyl ether has been applied in a K^+-selective electrode based on ISFET-technology [25] and in supported liquid membranes that are used for the selective transport of K^+ [26].

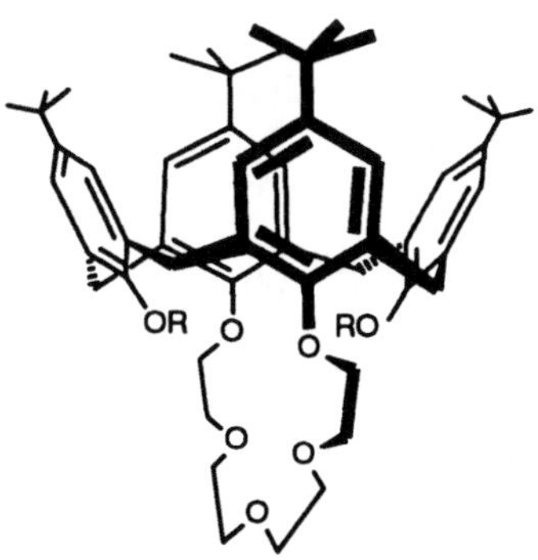

25 R = H
26a R = Me
26b R = Et
26c R = n-Pr
26d R = i-Pr

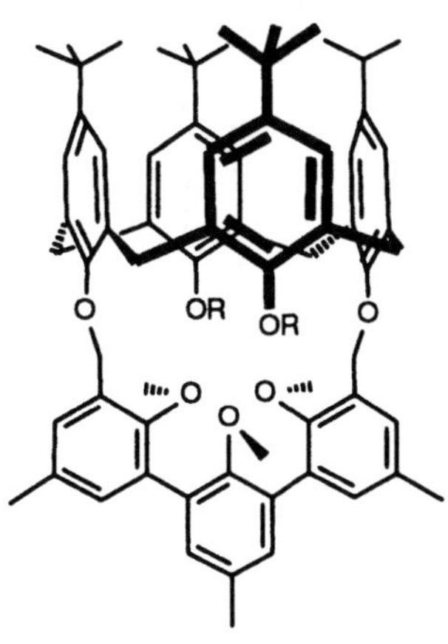

27 R = H
28 R = Me

5.2 CALIXSPHERANDS

A few years ago the synthesis of a new type of molecules, the spherands, has been reported that form complexes with Li^+ and Na^+ that are not only thermodynamically but also kinetically very stable [27]. Combination of the 1,3-dimethyl ether of *p-tert*-butylcalix[4]arene with half a spherand molecule afforded the calixspherand **28** as its Na^+ complex in 23 % yield [28]. The complexes of **28** with Na^+ and K^+ are kinetically very stable in $CDCl_3$ saturated with D_2O, and have half-lives of 3.7 and 2.2 years. The conformation of the calix[4]arene moiety in these complexes is a "flattened" partial cone as in the complexes of the calixcrown ether **26a**. Only recently we were able to prove that in the free ligand the calix[4]arene moiety adopts a cone conformation in complete analogy with compound **26a** [29].

Currently we are trying to increase the kinetic stability of the Rb^+ complex, which has a half-life of only 2.8 hours, by introducing more bulky groups on the 2- and 4-positions of the calix[4]arene, so that it can be used in organ imaging. Furthermore it is necessary to have a functional group in the calixspherand with which the calixspherand can be attached to an organ-specific compound (e.g. a peptide). For both purposes we employ a new strategy in which first the (functionalized) hemispherand

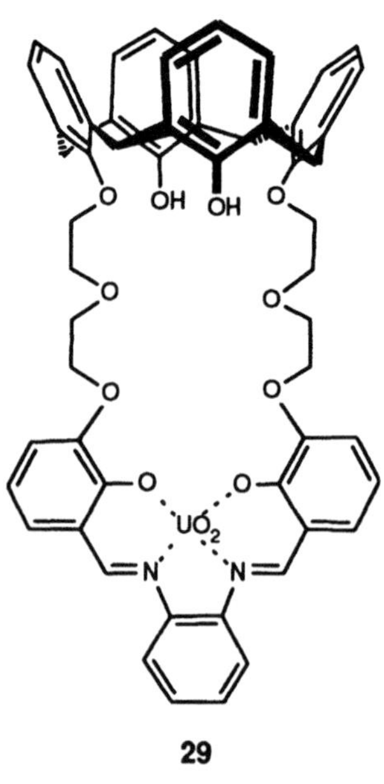

29

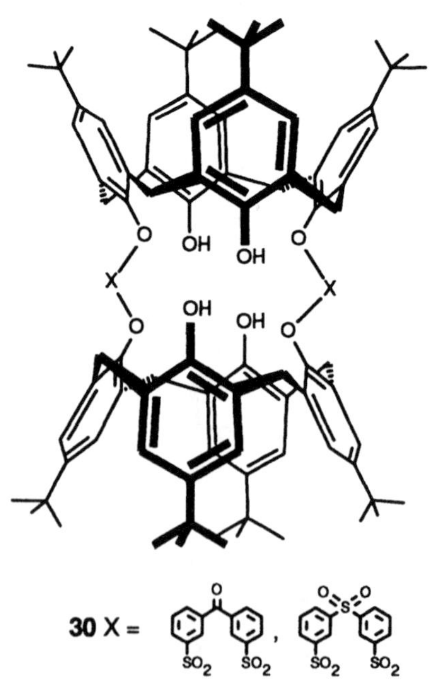

30 X =

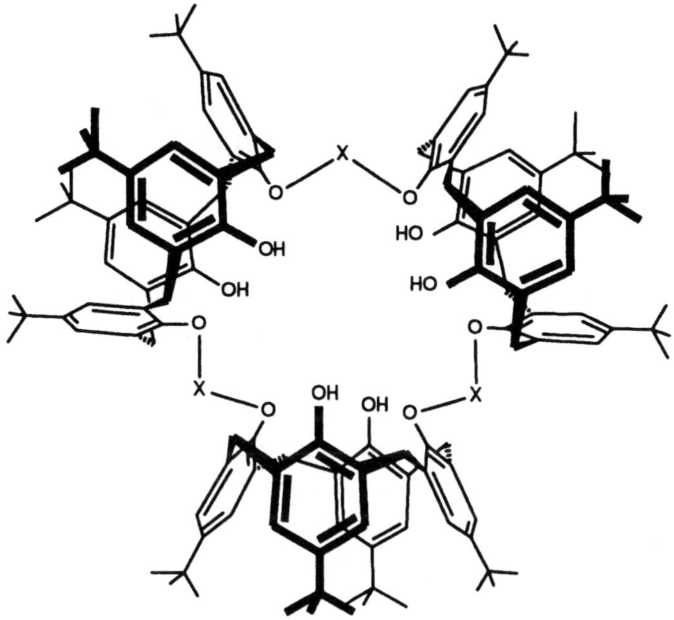

31 X = O_2S-biphenyl-SO_2

reacts with the unsubstituted *p-tert*-butylcalix[4]arene to yield the dihydroxycalix-spherand **27** in 75 % yield, and subsequently the calixarene moiety is further alkylated [30].

5.3. CALIX[4]ARENES COMBINED WITH SALOPHENE UNITS

Macrocycles that contain a salophene unit in which a uranyl cation is immobilized are very good complexing agents for urea and other neutral molecules containing a nucleophilic center [31]. Coupling of the salophene unit to calix[4]arene **9a** yielded compound **29**, which is very effective in the transport of urea through supported liquid membranes [32]. Further functionalization of the two remaining hydroxyl groups can lead to compounds that might be used as models for the enzyme urease [33].

5.4. DOUBLE AND TRIPLE CALIX[4]ARENES

In a study of alternative ways for the selective functionalization of the lower rim, by letting the unsubstituted calix[4]arene react with electrophiles with two reactive groups, we obtained not only the desired 1,2- and 1,3-bridged calix[4]arenes but in some cases also double calix[4]arenes that are coupled by two bridges (compounds **30**) [34]. The use of bridging agents that cannot bridge two hydroxyl groups of a calix[4]arene because of the too large distance between the two reactive centers, leads exclusively to double and even triple calix[4]arenes like compound **31** [35].

5.5 COMBINATION OF CALIX[4]ARENES AND OCTOLS

The coupling of the tribridged octol **6**, in which the two free hydroxyl groups have been transformed to oxyacetic acid chlorides, with a 1,3-*p*-diaminocalix[4]arene, which was obtained after reduction of the corresponding dinitrocompound, yielded compound **32** in 46 % yield [7]. This molecule can be regarded as a first impulse to the synthesis of large unsymmetrical cavities that cannot collapse. Molecules with a large dipole moment will be complexed in such cavities in one favorite orientation. Changing the dipole moment of the guest molecule by external influences will change the orientation of the guest molecule, so that a simple molecular switch would be obtained.

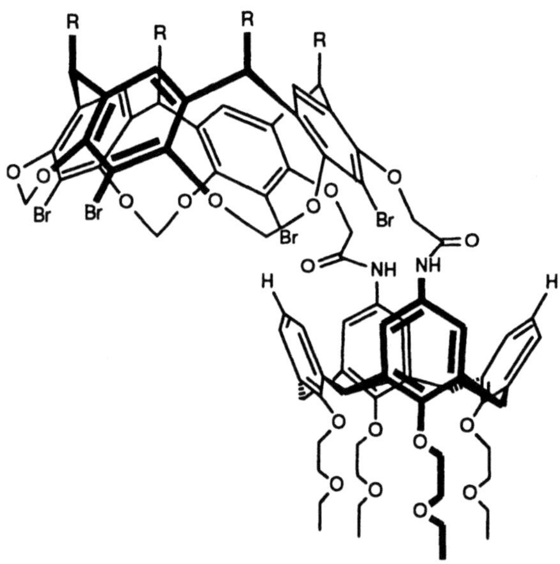

32 R = CH_2CH_2Ph

5.6 ASSOCIATION OF CALIX[4]ARENES BY NON-COVALENT INTERACTIONS

The calix[4]arene molecule cannot only be used as a building block for new receptor molecules in which it participates in the binding of the guest, but it can also be utilized as a molecular platform to which other molecules are attached that provide for the only binding sites. In compound **33**, which was synthesized from the already mentioned 1,3-*p*-diaminocalix[4]arene and 2,4-dichloro-6-methoxy-1,3,5-triazine and subsequent reaction with *n*-butyl amine, the calix[4]arene moiety is only used to position the two binding sites (the triazine rings) close to each other [23]. This molecule binds 5-ethyl-5-phenylbarbituric acid (**34**) with an association constant of $520 \pm 50\ M^{-1}$.

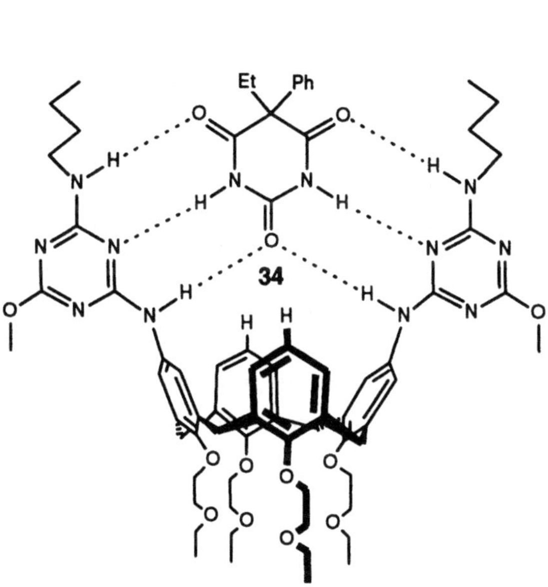

33

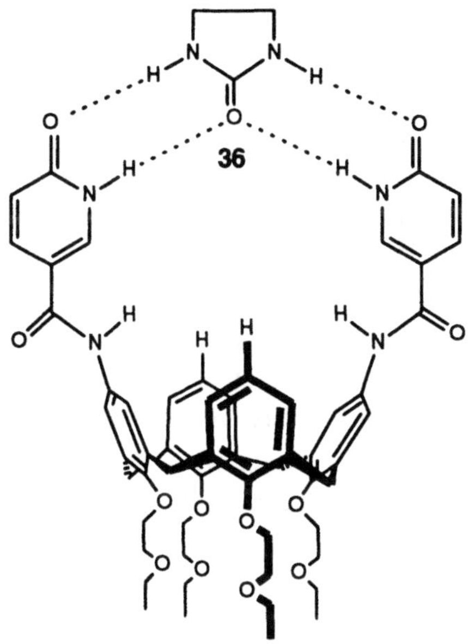

35

A more interesting example is compound **35**, which was synthesized from again the same 1,3-*p*-diaminocalix[4]arene and 6-hydroxynicotinic acid [36]. The four sites for hydrogen bonding in this molecule are slightly more divergent than in compound **33** and as a consequence it forms large associates in $CDCl_3$ solution. The associates are oligomers of calix[4]arene units, the compound can therefore be regarded as a (supramolecular) polymer. The number and weight average molecular weights of this polymer are 5100 ± 300 and 30000 ± 3000, respectively. The high ratio between the weight and number average molecular weights indicates that the polymer has a broad mass distribution, and that there is not a favored length of the polymer chains. The polymers are easily broken by adding molecules that can compete for the sites for hydrogen bonding on the calix[4]arenes. Addition of a cyclic urea derivative, 2-imidazolidone (**36**), leads to the formation of 1 : 1 and 1 : 2 complexes of calix[4]arene and guest molecule. The calix[4]arene can also solubilize urea itself in $CDCl_3$, up to 0.35 molecules of urea per calix[4]arene molecule.

Compound **35** opens the way to the development of molecular materials on the basis of intermolecular interactions. We are currently investigating how enough structural information can be incorporated into the monomeric calix[4]arene units so that they will form assemblies with a better defined geometry.

6. Conclusions

By the development of new methods for the selective functionalization of calix[4]arenes we have reached the point at which we can shape the calix[4]arene molecule in virtually any way that we would like. This is a prerequisite for the use of calix[4]arenes as building blocks in the synthesis of larger molecules with specific properties. The combination with several other small and well-known building blocks resulted in molecules with very useful complexation properties. We are currently investigating the selective functionalization of other building blocks (octols, cyclodextrins, calix[6]arenes) and the coupling of these building blocks with calix[4]arenes. Undoubtedly this will lead to new compounds with interesting properties. Finally, we are also exploring ways for building well-defined associates of calix[4]arenes on the basis of hydrogen bonds.

7. Acknowledgments

The work described in this paper is the result of the efforts of more than 20 coworkers whose names appear in the references of this paper. Our calixarene work is carried out in close collaboration with the University of Parma (Professors Ungaro and Pochini). Financial support from the Netherlands Organization for Scientific Research (NWO), the EEC Science Program and AKZO Research BV Arnhem, The Netherlands is gratefully acknowledged.

8. References

1 Breslow, R. (1990) Pure Appl. Chem. 62, 1859-1866.
2 Gutsche, C. D. (1989) 'Calixarenes', Royal Society of Chemistry, Cambridge.
3 Vicens, J. and Böhmer, V. (eds.) (1991) 'Calixarenes. A Versatile Class of Compounds', Kluwer Academic Publishers, Dordrecht.
4 Pregel, M. J. and Bruncel, E. (1991), Can. J. Chem. 69, 130-137.

5 van Dienst, E., Eissink, I. M., den Hertog, Jr., H. J. and Reinhoudt, D. N., unpublished work.

6 Cram, D. J., Karbach, S., Kim, H.-E., Knobler, C. B., Maverick, E. F., Ericson, J. L. and Helgeson, R. C. (1988) J. Am. Chem. Soc. 110, 2229-2237.

7 Timmerman, P., van Mook, M., Verboom, W. and Reinhoudt, D. N., unpublished results.

8 Casnati, A., Arduini, A., Ghidini, E., Pochini, A. and Ungaro, R. (1991) Tetrahedron 47, 2221-2228.

9 Groenen, L. C., Ruël, B. H. M., Casnati, A., Verboom, W., Pochini, A., Ungaro, R. and Reinhoudt, D. N. (1991) Tetrahedron 47, 8379-8384.

10 a) van Loon, J.-D., Arduini, A., Verboom, W., Ungaro, R., van Hummel, G. J., Harkema, S. and Reinhoudt, D. N. (1989) Tetrahedron Lett. 30, 2681-2684. b) van Loon, J.-D., Arduini, A., Coppi, L., Verboom, W., Pochini, A., Ungaro, R., Harkema, S. and Reinhoudt, D. N. (1990) J. Org. Chem. 55, 5639-5646.

11 Collins, E. M., McKervey, M. A. and Harris, S. J. (1989) J. Chem. Soc., Perkin Trans. 1, 372-374.

12 Groenen, L. C., Ruël, B. H. M., Casnati, A., Timmerman, P., Verboom, W., Harkema, S., Pochini, A., Ungaro, R. and Reinhoudt, D. N. (1991) Tetrahedron Lett. 32, 2675-2678.

13 Goren, Z. and Biali, S. E. (1990) J. Chem. Soc., Perkin Trans. 1, 1484-1487.

14 Ting, Y., Verboom, W., Groenen, L. C., van Loon, J.-D. and Reinhoudt, D. N. (1990) J. Chem. Soc., Chem. Commun., 1432-1433.

15 Gutsche, C. D., Dhawan, B., Levine, J. A., No, K. H. and Bauer, L. J. (1983) Tetrahedron 39, 409-426.

16 Groenen, L. C., van Loon, J. D., Verboom, W., Harkema, S., Casnati, A., Ungaro, R., Pochini, A., Ugozzoli, F. and Reinhoudt, D. N. (1991) J. Am. Chem. Soc. 113, 2385-2392.

17 van Loon, J. D., Groenen, L. C., Wijmenga, S. S., Verboom, W. and Reinhoudt, D. N. (1991) J. Am. Chem. Soc. 113, 2378-2384.

18 a) Araki, K., Iwamoto, K., Shinkai, S. and Matsuda, T. (1989) Chem. Lett., 1747-1750. b) Iwamoto, K., Araki, K. and Shinkai, S. (1991) J. Org. Chem. 56, 4955-4962.

19 Verboom, W., Durie, A., Egberink, R. J. M., Asfari, Z. and Reinhoudt, D. N., J. Org. Chem., in the press.

20 Datta, S., Janssen, R. G., van Loon, J.-D., Verboom, W. and Reinhoudt, D. N., unpublished results.

21 Gutsche, C. D. and Lin, L.-G. (1986) Tetrahedron 42, 1633-1640.

22 Kelderman, E., Derhaeg, L., Heesink, G., Verboom, W., Engbersen, J., van Hulst,

N., Persoons, A. and Reinhoudt, D. N., manuscript in preparation.
23 Heida, J. F., van Loon, J.-D., Verboom, W. and Reinhoudt, D. N., unpublished results.
24 Ghidini, E., Ugozzoli, F., Ungaro, R., Harkema, S., El-Fadl, A. A. and Reinhoudt, D. N. (1990) J. Am. Chem. Soc. 112, 6979-6985.
25 Sudhölter, E. J. R., van der Wal, P. D., Skowronska-Ptasinska, M., van den Berg, A., Bergveld, P. and Reinhoudt, D. N. (1990) Recl. Trav. Chim. Pays-Bas 109, 222-225.
26 Nijenhuis, W. F., Buitenhuis, E. G., de Jong, F., Sudhölter, E. J. R. and Reinhoudt, D. N. (1991) J. Am. Chem. Soc. 113, 7963-7968.
27 Cram, D. J. and Lein, G. M. (1985) J. Am. Chem. Soc. 107, 3657-3668.
28 a) Reinhoudt, D. N., Dijkstra, P. J., in 't Veld, P. J. A., Bugge, K.-E., Harkema, S., Ungaro, R. and Ghidini, E. (1987) J. Am. Chem. Soc. 109, 4761-4762. b) Dijkstra, P. J., Brunink, J. A. J., Bugge, K.-E., Reinhoudt, D. N., Harkema, S., Ungaro, R., Ugozzoli, F. and Ghidini, E. (1989) J. Am. Chem. Soc. 111, 7567-7575.
29 Groenen, L. C., Harkema, S. and Reinhoudt, D. N., manuscript in preparation.
30 Iwema Bakker, W. I., Ostaszewski, R., Khoo, C., Verboom, W., den Hertog, Jr., H. J. and Reinhoudt, D. N., unpublished results.
31 Van Doorn, A. R., Schaafstra, R., Bos, M., Harkema, S., van Eerden, J., Verboom, W. and Reinhoudt, D. N. (1991) J. Org. Chem. 56, 6083-6094.
32 Nijenhuis, W. F., van Doorn, A. R., Reichwein, A. M., de Jong, F. and Reinhoudt, D. N. (1991) J. Am. Chem. Soc. 113, 3607-3608.
33 Reichwein, A. M., Verboom, W. and Reinhoudt, D. N., unpublished results.
34 van Loon, J.-D., Kraft, D., Ankoné, M. J. K., Verboom, W., Harkema, S., Vogt, W., Böhmer, V. and Reinhoudt, D. N. (1990) J. Org. Chem. 55, 5176-5179.
35 Kraft, D., van Loon, J.-D., Owens, M., Verboom, W., Vogt, W., McKervey, M. A., Böhmer, V. and Reinhoudt, D. N. (1990) Tetrahedron Lett. 31, 4941-4944.
36 Janssen, R. G., van Loon, J.-D., Verboom, W. and Reinhoudt, D. N., unpublished results.

MOLECULES WITH LARGE CAVITIES - SELECTIVE COMPLEXATION OF INORGANIC AND ORGANIC GUESTS

R. BERSCHEID, I. LÜER, C. SEEL and F. VÖGTLE*
Institut für Organische Chemie und Biochemie der Universität
Gerhard-Domagk-Straße 1
5300 Bonn 1
Germany

ABSTRACT. The design of artificial complexing agents renders information about the steric and electronic requirements of molecular recognition, which will help in the construction of selective molecular catalysts. This article presents a concise overview on macrobi- and macrooligocyclic host compounds. Metal complexes of cagelike ligands have been used for photophysical investigations which might lead to the development of efficient sensitizers for solar energy harvesting. The selective complexation of neutral organic molecules has been underlined by extraction and transport experiments and can be of use in environmental technique and medical applications. Tailormade bicyclic host compounds indicate the inclusion of organic guests by changing their colour and might thus be used as sensors.

1. Introduction

The general acknowledgement of the significance of intermolecular interactions for life science as well as for materials science is one of the reasons for the recent boom of Supramolecular Chemistry. One of its major goals is the synthesis of molecular cavities, which are designed for the selective complexation of certain organic and inorganic guests [1]. The purpose is not only to study the phenomenon of molecular recognition - a basis of enzyme activity - but furthermore the properties of molecular aggregates which might be valuable for a variety of applications.

Macropolycyclic phane systems constitute an important class of host compounds which are able to encapsulate stereofunctionally complementary guests in the center of a spheric cavity. They can be synthesized according to a building block concept which allows to vary the size of the cavity and the properties of intraannular donor functionalities.

The very strategy, which should be applied in the synthesis of a cavity depends on its final shape and on steric requirements. Following path A usually renders higher yields than a synthesis according to path B, since fewer bonds have to be formed in the cyclization step. On the other hand, path B allows the use of easy accessible building blocks which might be an advantage over path A. The appropriate strategy is supported by the use of simple but efficient synthetic tools, such as the dilution principle [2], the cesium effect [3] or the template effect [4].

V. Balzani and L. De Cola (eds.), Supramolecular Chemistry, 71–86.

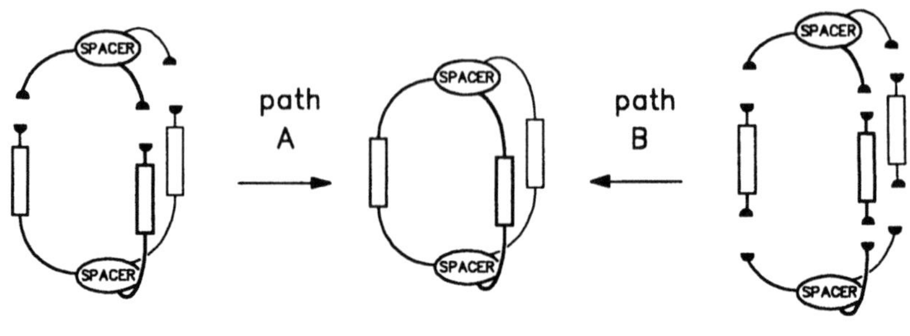

2. Siderands

Microorganisms produce siderophores (greek: iron carriers), hexadentate chelate ligands which are capable of binding extracellular iron even at physiological pH, thus making it available to the cell [5]. The prototype of a catechol siderophore is enterobactin (1), which is the strongest natural occuring iron ligand [6]. The association constant K_{ass} of the 'sideroplex' (Fe^{III} complex of 1) was determined to be around 10^{52} M^{-1} [7]. The synthesis of structural analogues of enterobactin is of interest for the study of natural occuring iron transport processes as well as for medical applications [8].

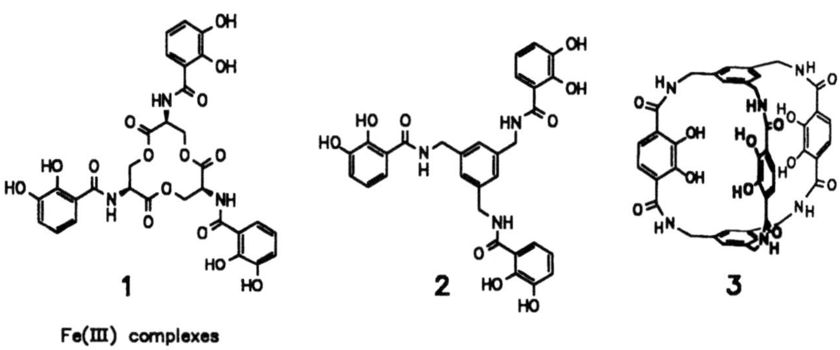

1 **2** **3**

Fe(III) complexes

$K_{ass} = 10^{52}$ M^{-1} $K_{ass} = 10^{46}$ M^{-1} $K_{ass} \approx 10^{59}$ M^{-1}

Such structural analogy is found in MECAM (2), yet this synthetic molecule is weaker than 1 as far as its iron binding properties are concerned [9]. In 1984, we synthesized the macrobicyclic siderand 3, which provides an optimal donor geometry: due to the cagelike molecular framework the catechol units are preorganized for complexation and are able to provide the demanded octahedral coordination sphere of the hexaanion [10]. This leads to strong metal binding properties of 3 which is very stable towards air, in contrast to e. g. enterobactin and other siderophores. The association constant for its iron(III) complex seems to be the highest one ever measured for any such complex (K_{ass} ≈ 10^{59} M^{-1}) [11]. Furthermore, 3 forms stable complexes with $Pb^{2\oplus}$ and $Bi^{3\oplus}$ which is of interest for a potential use of complexed isotopes in tumor imaging and therapy [12]. The expansion of the cavity of 3 by use of more extended spacers like 1,3,5-triphenylbenzene units leads to the hexamethoxy ligand 4 which binds $Cs^{\oplus}$

six times stronger ($pK_{ass} \approx -5.3 \pm 0.5$) than $K^{\oplus}$. In the case of metal complexation in the center, the methoxy groups of **4** are arranged in a convergent and cooperative manner[13] .

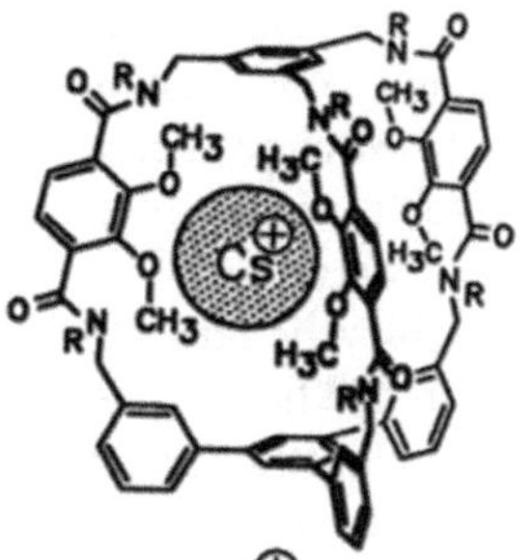

$[Cs \cdot 4]^{\oplus}$

3. Tris(bipyridine) Ligands

Due to its exceptional complexing abilities, 2,2'-bipyridine is predestinated for being incorporated in macrocyclic host systems. Thus, a series of macrobicyclic ligands containing three bipyridine units could be obtained. They exhibit unique properties with regard to selective molecular recognition, complex stability and photophysical processes.

Whereas the hexalactame ligand **5** was first synthesized in a stepwise procedure [14], the cage complex $[Ru \cdot 5]^{2\oplus}$ can only be obtained in a 'one-pot' synthesis from the ruthenium complex of 5,5'-bis(chlorocarbonyl)-2,2'-bipyridine since the complex is not formed by reaction of the cation with the cage **5** directly due to kinetic inhibition of the Ru(II) complex formation[15] .

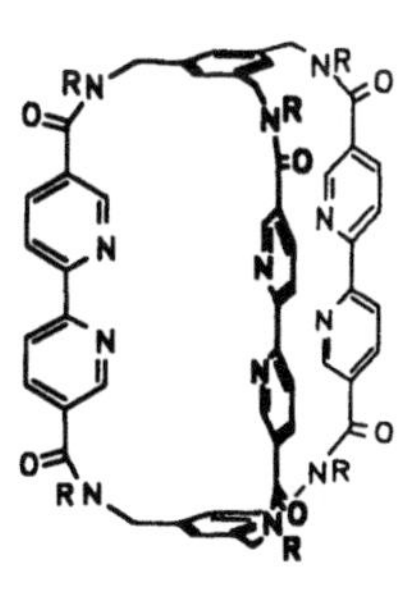

5

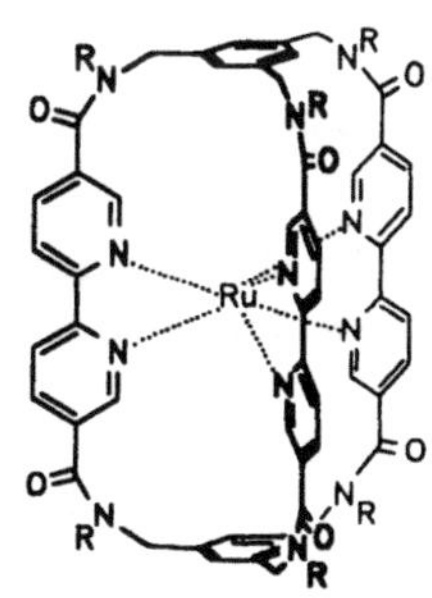

$[Ru \cdot 5]^{2\oplus}$

R = benzyl

The Fe(II) complex of **5** shows a remarkable stability towards strong oxidants such as H_2O_2. Its ruthenium(II) complex is of particular interest due to its photochemical properties. In general, ruthenium(II) polypyridine complexes like $[Ru(bipy)_3]^{2\oplus}$ are of importance as sensitizers in photochemical processes [16]. Severe limitations to that use are due to the quick radiationless decay of their lowest excited state, which determines their photochemical properties, and are due to their sensibility towards photodissociation [17]. One way of avoiding photodissociation is the incorporation of the bipyridine ligands in a cagelike molecular framework. As long as the first coordination sphere of the metal is not disturbed by steric effects, the photochemical properties of the cage complex are nearly the same as those of $[Ru(bipy)_3]^{2\oplus}$. The ligand **5** offers the desired features: while the clamping of the donor-bridges prevents photodissociation, the conformational flexibility of the cage allows complexation of Ru(II) in an ideal octahedral sphere (the drawings do not show this octahedral geometry). In photochemical investigations, an enlarged photostability and a remarkable higher lifetime of the luminescence state of $[Ru\cdot\mathbf{5}]^{2\oplus}$ compared to $[Ru(bipy)_3]^{2\oplus}$ was observed [18].

In a 'one-pot' reaction the tritopic macrobicyclic host **6** could be obtained without template assistance in a remarkable yield of 78%. In addition to the bipyridine units, the tris(2-aminoethyl)amine caps present two more metal coordinating centers. Consequently, $[Ag_3\cdot\mathbf{6}]^{3\oplus}$ could be synthesized, the first trinuclear silver cage complex ever known, as well as the dicopper complex $[Cu_2\cdot\mathbf{6}]^{2\oplus}$ [19].

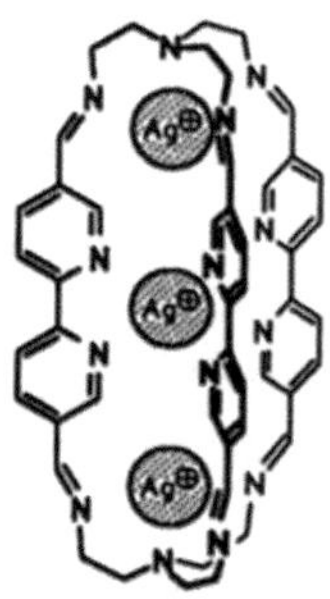

$[Ag_3\cdot\mathbf{6}]^{3\oplus}$

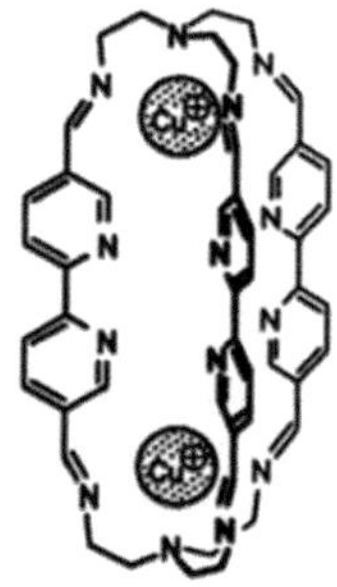

$[Cu_2\cdot\mathbf{6}]^{2\oplus}$

4. Complexation of Organic Guests

Using three crown ether building blocks as bridges, the macropentacyclic triscrowns **7-10** were obtained. They form 3:1 complexes with KSCN. In acidic aqueous solution ß-naphthol and

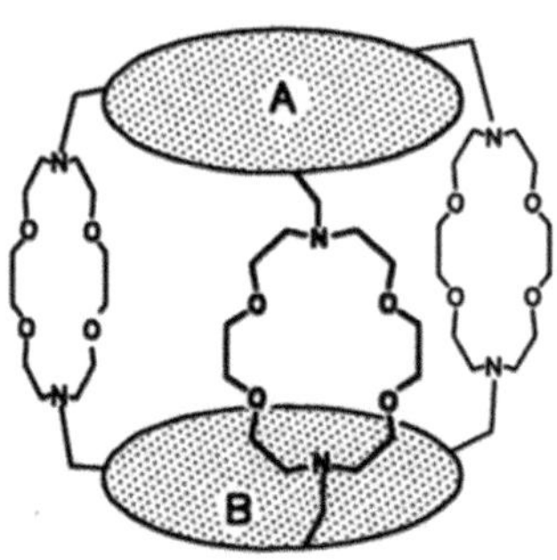

	7	8	9	10
(A)	I	I	III	II
(B)	I	III	III	II

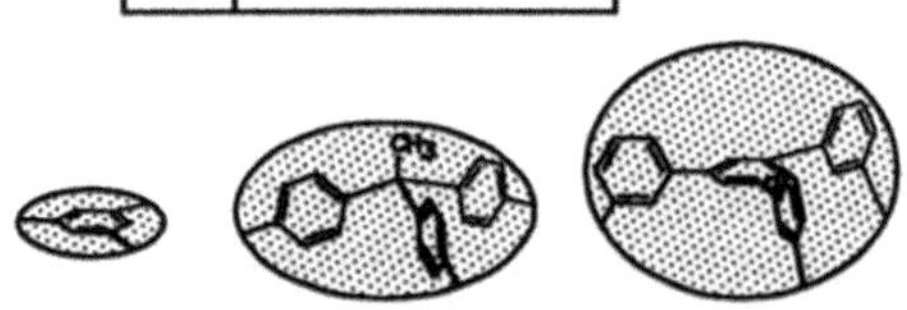

I II III

some naphthalenediols are complexed by **8** and **9** which is indicated by significant upfield shifts of the guest's protons in ^1H NMR spectra. Phenol, di- and trihydroxybenzenes as well as 1,3,5-benzenetris(methylammonium) salts are not appreciably complexed under these conditions [20].

The expansion of the framework of the tris(bipyridine) macrocyclus **5** could be achieved by the use of larger spacer groups. The successively enlarged cage compounds **11-13** present molecular cavities with donor functionalities which can enclose complementary guests by means of multiple hydrogen bonds [21]. As indicated by the table below, appropriately substituted trihydroxy benzenes like 1,3,5-trihydroxybenzene are well complexed, while isomers like 1,2,3-trihydroxybenzene and comparable phenols are discriminated. The guests are hardly soluble in CH_2Cl_2, but are solubilized by the bipyridine cage compounds **11-13**. The ^1H NMR spectra show the signals of both host and guest in 1:1 stoichiometry. The association constant of [phloroglucinol·**12**] was determined to be 11000 $\pm$2000 M^{-1}.

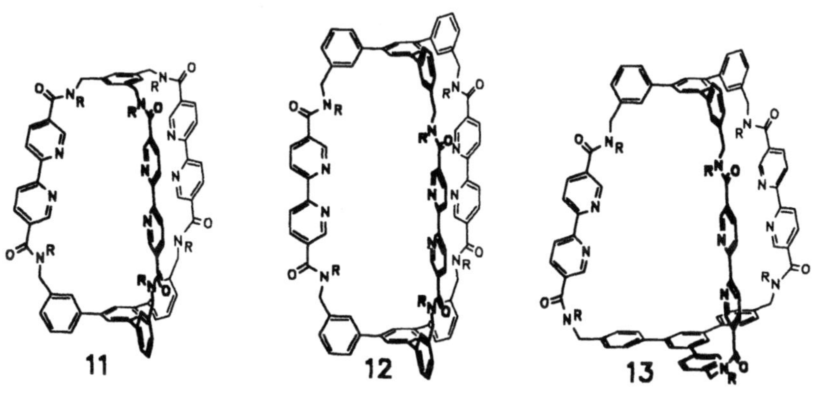

11 12 13

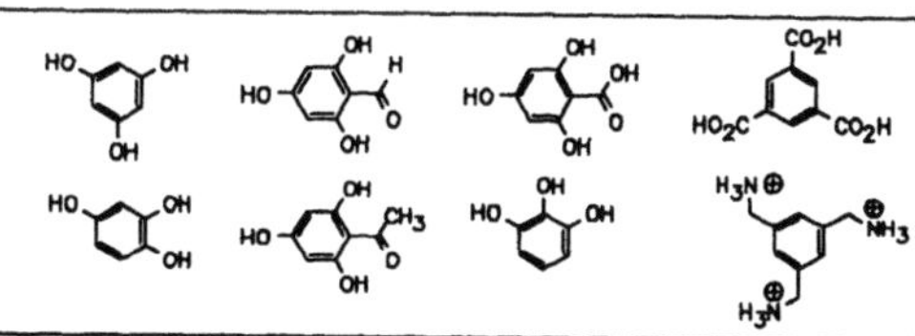

guests				
11	+	+	−	−
12	+	+	−	−
13	−	+	−	−

+ : complex formed − : no complexation

The complex binding abilities of the macrobicyclus **14** and the podand **15** could not be examined by NMR methods because the signals of possible guest molecules were hidden by the signals of the host protons. Therefore, the complexation of nucleobases by **14** was determined by solid-liquid extraction and by transport experiments [22].

The solubility of the host compounds **14** and **15** shows an inverse behaviour compared to the nucleobase guests: the latter are sparingly soluble in CH_2Cl_2 but soluble in acidic aqueous

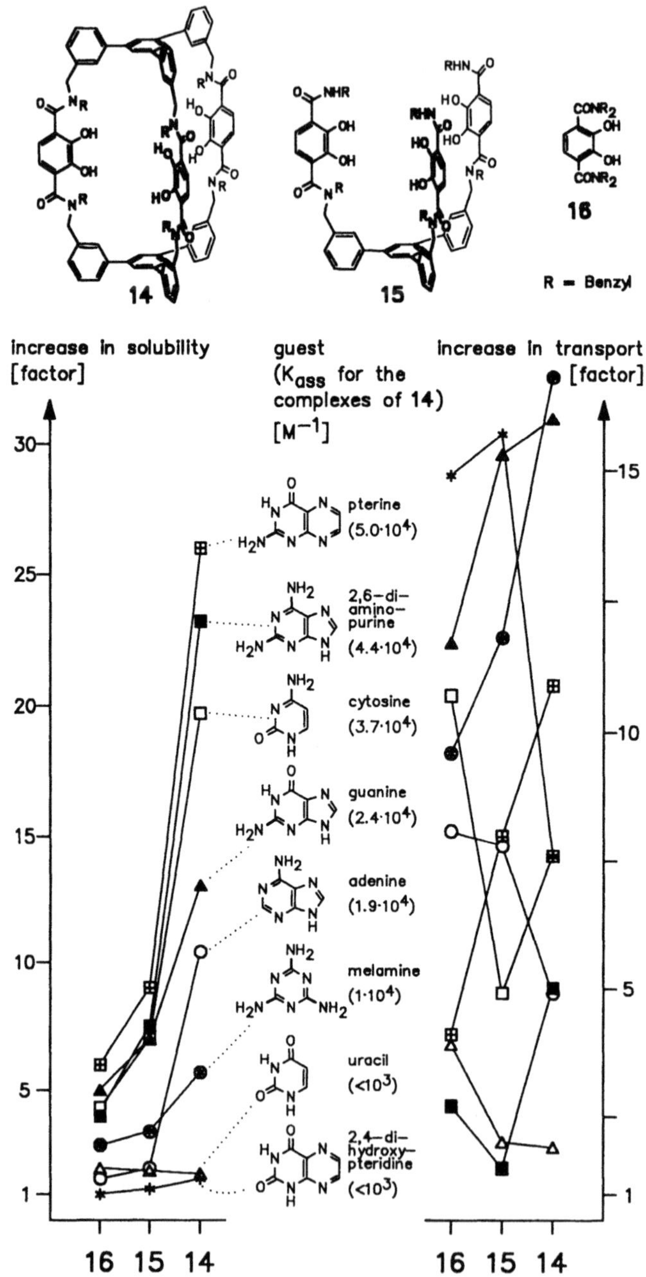

Fig. 1. Left: Enhancement factors of the solubility of the guests in CH_2Cl_2 at host concentrations of $5 \cdot 10^{-4}$ mol/l (in the case of **14** and **15**) and $1.5 \cdot 10^{-3}$ mol/l (in the case of **16**). Right: Enhancement factors of the transport of the guests (solid) through CH_2Cl_2 /*n*-hexane (3:1) in a 1 N hydrochloric acid solution at host concentrations of $1 \cdot 10^{-4}$ mol/l (in the case of **14** and **15**) and $3 \cdot 10^{-4}$ mol/l (in the case of **16**).

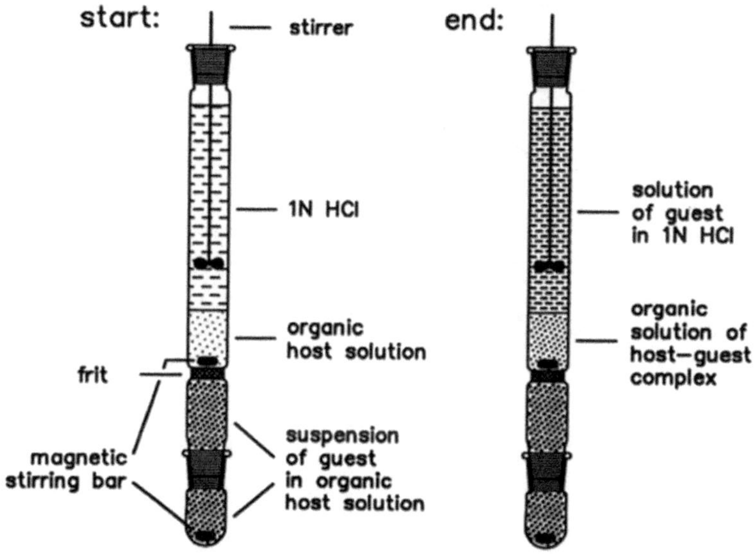
start:
stirrer
1N HCl
organic
host solution
frit
magnetic
stirring bar
suspension
of guest
in organic
host solution
end:
solution
of guest
in 1N HCl
organic
solution of
host–guest
complex

are directed rigidly towards the interior of the molecular cavity. Such convergent endopreorganization is realized in the macrobicyclic host **17**. The synthesis of **17** was accomplished in a one-step reaction of 2,9-bis(4-hydroxyphenyl)-1,10-phenanthroline with 1,3,5-tris(bromomethyl)benzene in 11% yield [23]. Experiments towards the complexation of 1,3,5-trihydroxybenzene remained unsuccessful, as well as efforts with 1,3,5-tris(hydroxyphenyl)benzene as guest, which should fit better into the cavity. However, an inclusion of three 2,9-disubstituted 1,10-phenanthrolines as their Cu(I) complexes could be achieved. Treatment of a solution of **17** in dimethylformamide with $[Cu(CH_3CN)_4]BF_4$ in acetonitrile, followed by addition of a solution of **18** or **19** in CH_2Cl_2 results in quantitative formation of the tricopper(I) complexes **21a** and **21b**, respectively.

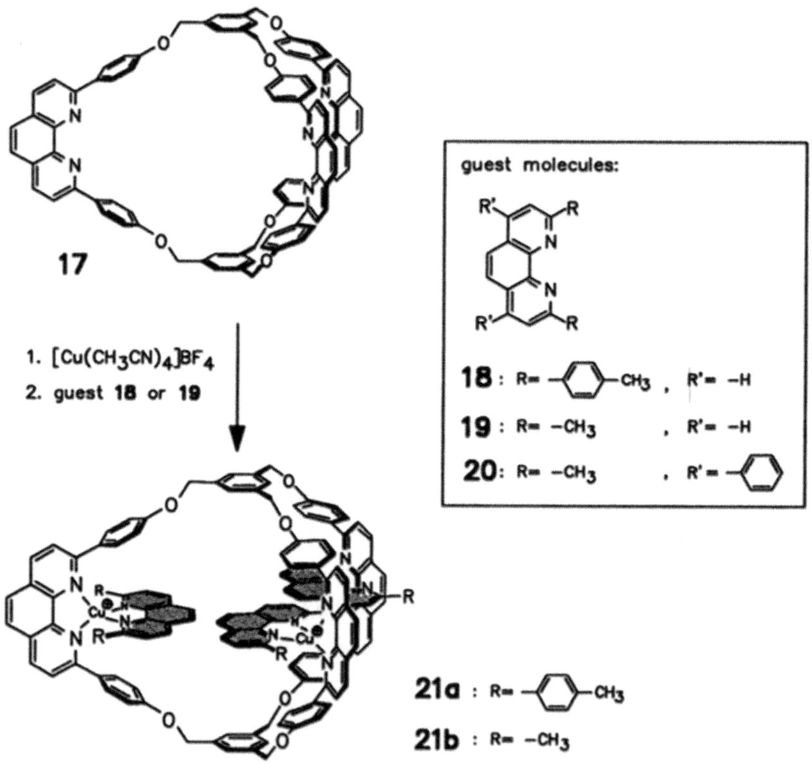

Elemental analysis, FAB mass spectra and ^{1}H NMR spectra confirm the suggested composition and structure. In the FAB MS, peaks for $[\mathbf{21a}(BF_4)_2]^{\oplus}$ and $[\mathbf{21b}(BF_4)_2]^{\oplus}$ can be detected, as well as further peaks which correspond to the successive loss of phenanthroline guests and Cu(I). The ^{1}H NMR spectrum gives evidence for the symmetric inclusion of three guests in the cavity: the signal of the OCH_2 host protons is shifted to higher field due to inclusion of the guests.

Macroscopically, the complex formation leads to an increased solubility: though the ligand **17** is hardly soluble in CH_2Cl_2, the complexes exhibit good solubility in that solvent. In experiments to enclose bathocuproine (**20**) as Cu(I) complex, the size limitation of the cavities becomes evident which is a further confirmation of endocomplexation: only the complex of **17** with two molecules of **20** can be detected by FAB mass spectrometry.

5. Complexation of Organic Guests by Non-bonding Interactions

The complexation of non-polar guests by a lipophilic cavity is especially favourable in aqeous medium due to the hydrophobic effect as driving force.

Suitable ligands are able to surround the guest, binding it by dispersive interactions, while peripheral hydrophilic groups effect their water solubility. The diphenyl methane system has been used in the construction of several effective hosts [24]. In a one-step reaction under dilution conditions of trimesic acid chloride with the matching diamine, followed by reduction of the carbonyl groups, we obtained the macrobicyclic host **22** [25]. **22** exhibits remarkable qualities concerning the complexation of complementary formed condensed aromatic hydrocarbons, like triphenylene, perylene or pyrene. For the latter, a binding constant of $K_{ass} \approx 10^{2.5}\ M^{-1}$ was determined.

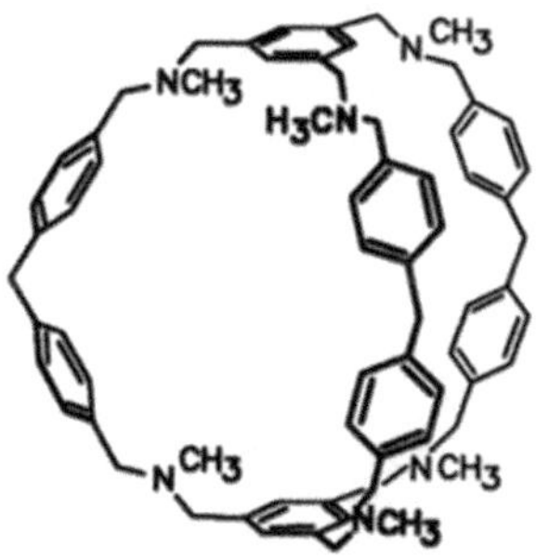

22

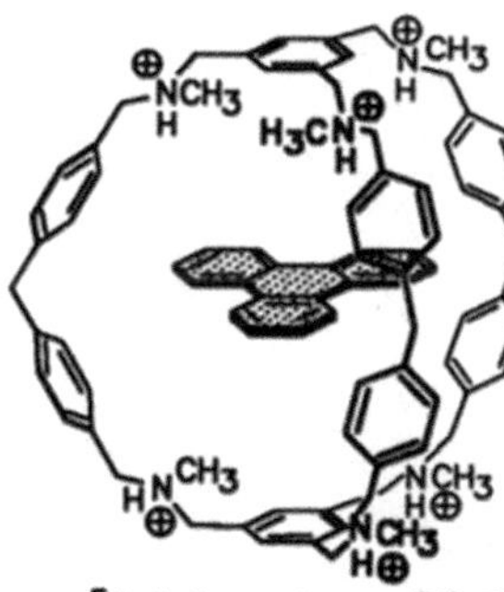

[triphenylene·**22**·6H]$^{6\oplus}$

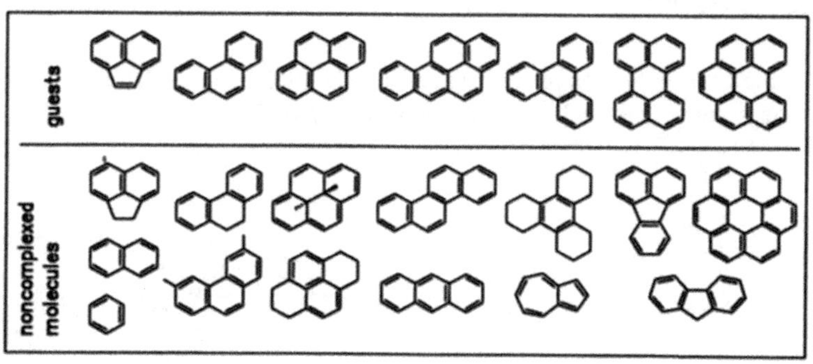

The complexation is proved by ^{1}H NMR spectroscopy. Signals of the guests experience strong shifts ($\Delta\delta$ = 1.5 ppm) to higher field; the otherwise broad signals of the host are sharpened and shifted. The inclusion of guest compounds hinders the conformational flexibility of the host and adapts its shape to the enclosed molecule ('induced fit'). Small compounds like benzene or naphthalene are not complexed, neither are large compounds like coronene. Partially hydrogenated derivatives of enclosed guests are not complexed. The binding by 'edge-to-face-stacking' is apparently obstructed by aliphatic CH_2- and CH_3-groups. For the first time, a discrimination of anthracene against phenanthrene was observed. The table above summarizes some inclusion experiments. Phase transfer experiments (liquid-liquid and solid-liquid) lead to the separation of pyrene from decahydropyrene and triphenylene from dodecahydrotriphenylene. Phenanthrene could be extracted from its mixture with naphthalene in *n*-hexane by application of the U-shaped tube method.

A special case of configurational isomerism is represented by the bicyclic host compounds **23** and **24**. The synthesis yields both compounds in a 1:1 ratio (total yield: 79%) [26]. Since **23** and **24** are in/out isomers and can not be interconverted by conformational changes, a pair of ligands with different cavity size is obtained. The difference in size influences their binding abilities: both molecules bind naphthalene as well as 2,6- and 2,7-dihydroxynaphthalene in aqueous solution, but only the more spacious out/out isomer **23** is able to complex adamantane and 1-adamantanethiol. To our knowledge this is the first complexation of a pure aliphatic hydrocarbon by any synthetic concave receptor.

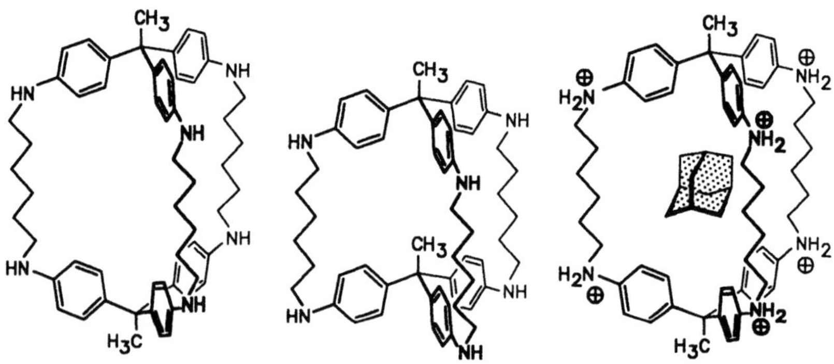

23: out/out **24** : in/out [adamantane·**23**·6H]$^{6\oplus}$

In contrary to the host molecules mentioned before, the molecular skeleton of **27** consists exclusively of carbon atoms. The cyclization is achieved by sixfold malonic ester alkylation. Expe-

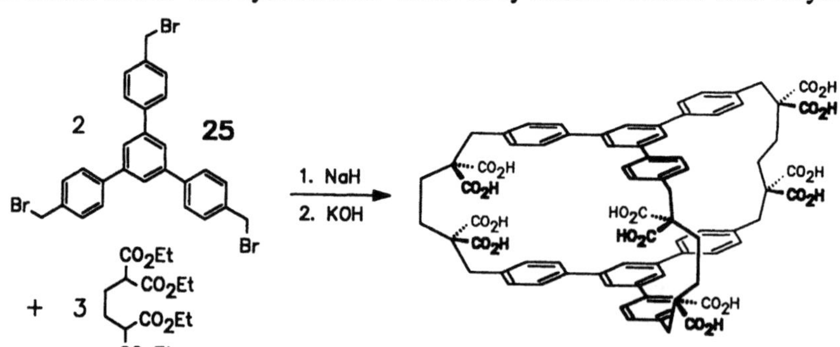

riments towards the complexation of benzene, toluene and mesitylene lead to shifts of the guest signals in the ^{1}H NMR spectra, indicating an inclusion of those guests inside the host's cavity. On the other hand, no complex formation is observed with 1,4-benzenebis(methylammonium)-diiodide in aqueous solution.

Another method to synthesize hosts by formation of carbon-carbon bonds is the well known *Eglinton* alkyne coupling. Cutting the desired cage compound retrosynthetically into two half molecules the application of this method leads to short synthetic pathways with good yields. Therefore the oxidative alkyne coupling has become more and more important for the design of supramolecular host structures [27,28] and led to the development of new variants of the *Eglinton* coupling by changing the reagent [27a] or solvent [29]. Following this strategy we have synthesized the macrobicyclic hosts **30a** [28b] and **30b** [28a] in a two step synthesis from the trisphenols **28a,b**.

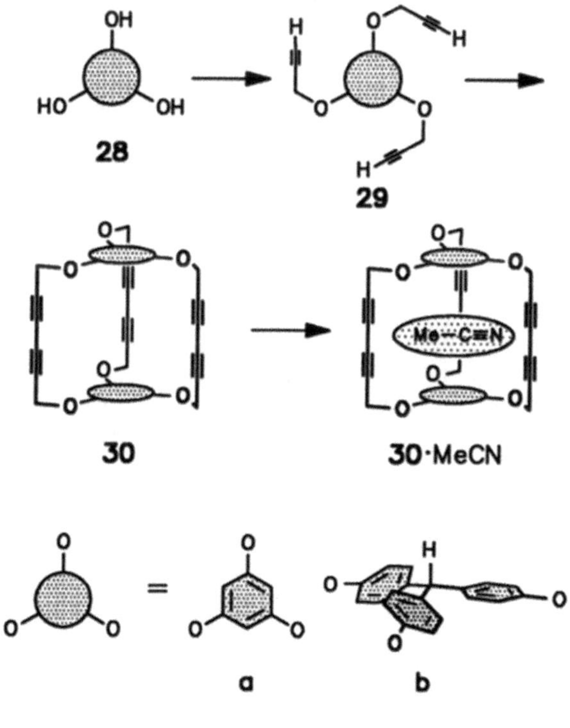

The host compounds **30a** and **30b** crystallize from acetonitrile as 1:1 inclusion complexes. The geometry of these complexes was established by X-ray structural analyses.

Both hosts enclose acetonitrile *inside* the cavity; empty intermolecular cavities are unoccupied. The complexes **30a**·MeCN and **30b**·MeCN are the first examples for the inclusion of acetonitrile in a molecular cavity [30] and show an interesting orientational selectivity relative to the enclosed acetonitrile. So in both structures the enclosed guest is found at the equatorial level perpendicular to the pseudo threefold axis. Whereas in **30b**·MeCN the *nitrile* group is surrounded by two host phenylene rings in **30a**·MeCN the *methyl* group most probably

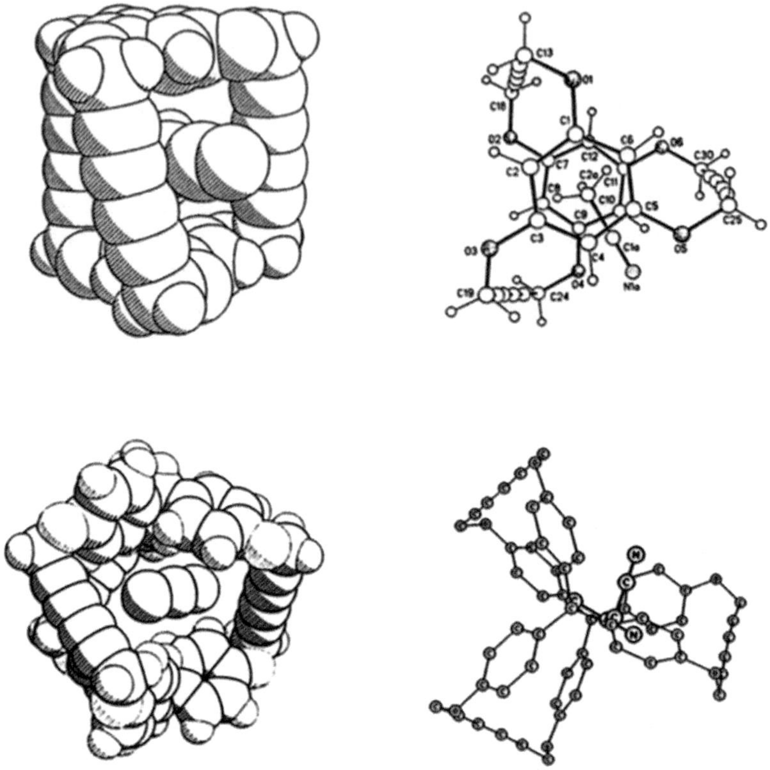

Fig. 3. Structures of **30a**·MeCN (above) and **30b**·MeCN (below) in the crystal. Left: space-filling drawings; for **30b**·MeCN only one the two split positions (1:1) for the enclosed acetonitrile is shown. Right: view of the complexes along their pseudo-C_3 axes; hydrogen atoms are omitted for clarity.

interacts with one phenoxy ring of host **30a** by an attractive σ-π-interaction between the acidic methyl proton of acetonitrile with the 'π-basic' 1,3,5-tris-alkoxybenzene donor of the host compound **30a**.

The IR spectra of both crystalline complexes in KBr matrix show a significant shift of the nitrile band of $\Delta\tilde{\nu}$ = 36 cm^{-1} towards smaller wavenumbers ($\tilde{\nu}$ = 2218 cm^{-1} resp. 2218.5 cm^{-1}) with nearly identical positions of the IR bands of the complexed host compared to the spectra of the guest-free hosts **30a** and **30b**. The nitrile function here is acting as a sensor for its structural environment detected by IR.

Crystallization of host **30b** from phenylacetonitrile affords the 1:3 inclusion/addition complex **30b**·$3C_6H_5CH_2CN$ [31]. The X-ray structure (see below) shows that one phenylacetonitrile molecule is enclosed with its aromatic ring inside the host **30b**; two phenylacetonitrile molecules are only attached to the macrobicyclus **30b**. The enclosed phenylacetonitrile is found again on the equatorial level; but the nitrile group here, however, is pointing outside the host cavity. This observation is in accordance with the IR spectrum of crystalline

30b·$3C_6H_5CH_2CN$: *no* shift of the nitrile band compared to free phenylacetonitrile is observed.

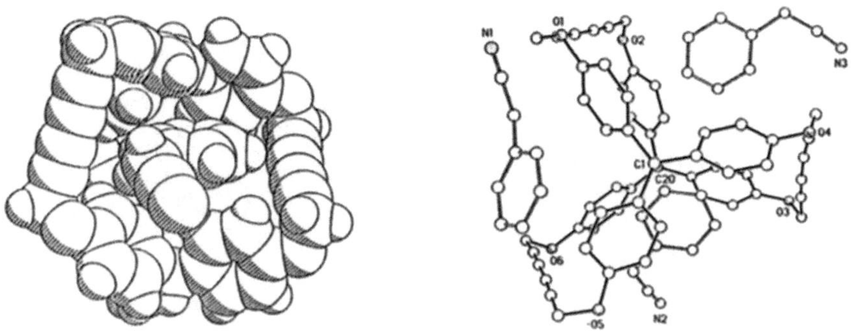

Fig. 4. Structure of **30b**·$3C_6H_5CH_2CN$ in the crystal. Left: space-filling drawing; only the enclosed phenylacetonitrile is shown. Right: view of complex **30b**·$3C_6H_5CH_2CN$ along its pseudo-C_3 axis; hydrogen atoms are omitted for clarity.

The successful inclusion of acetonitrile and phenylacetonitrile led us to the structurally related orange coloured triphenylmethyl dication **31** [1d,29]. It can be considered as a concave dye. **31** has a suitable cage size for the inclusion of anions and flat aromatic systems. Donor substituted aromatic compounds such as 1,3,5-trimethoxybenzene seem to be enclosed inside the molecular cavity. They cause bathochromic shifts up to 6 nm compared to open chained reference dyestuffs. Here the host is acting as a sensor: a bathochromic shift in the UV/Vis of the longest wavelength absorption maximum of host **31** indicates an inclusion of flat donor substituted guest molecules.

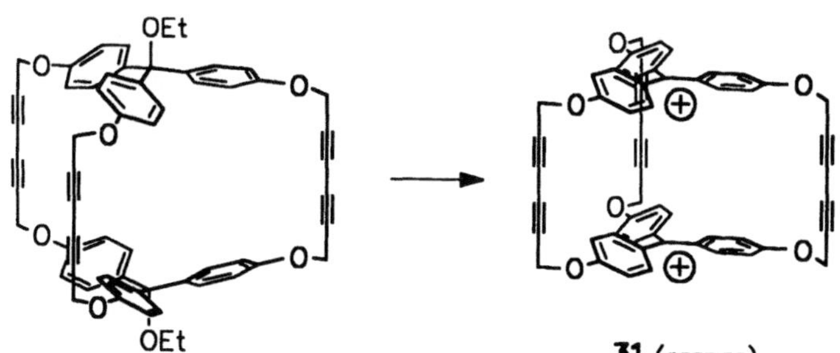

31 (orange)

For further concave dyestuffs bearing azobenzene chromophore units embracing the molecular cavity such as **32** and **33** we refer to literature [1d].

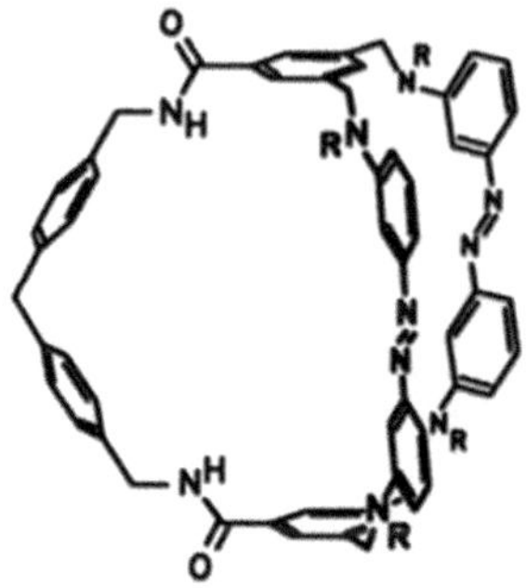

32

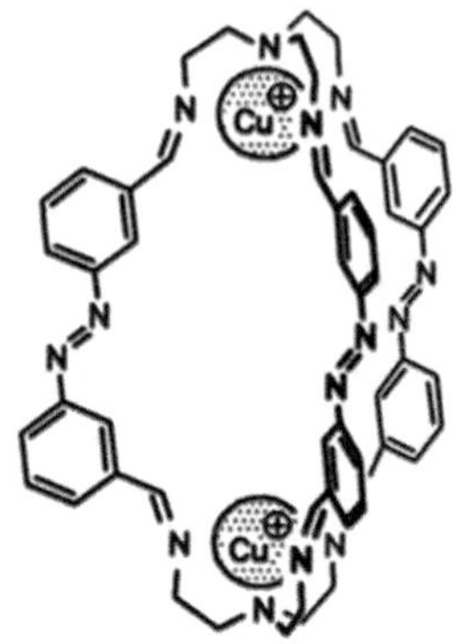

$[Cu_2 \cdot \mathbf{33}]^{2\oplus}$

6. Outlook

The synthesis of new molecular cavities is of interest for a variety of reasons. Investigations on the influence of cavity size and conformation on complexation abilities renders information on fundamental principles of molecular recognition which are basic for the life sciences. The ability to construct hosts for selective complexation is a prerequisite for the design of artificial catalysts for various reactions.

In this meaning also sensor properties of cage compounds are of general interest for the detection of host guest interactions in solution and can be used further for analytical applications.

Some metal complexing hosts are of interest for possible applications in medicine. They could enhance the possibilities of tumor imaging and therapy or work as metal decorporating agents.

Metal complexes of bipyridine and phenanthroline units are of interest for photophysical and photochemical investigations since they are suitable tools for the examination of photoinduced intramolecular energy transfer [32]. Deeper understanding of these processes will help in the development of efficient systems for the conversion of solar into chemical energy [33]. In this context we are trying to develop heterooligometallic complexes of the ligands **6** and **17**.

7. References

[1] Reviews: a) Ch. Seel, F. Vögtle, Angew. Chem., in press; b) F. Ebmeyer, F. Vögtle in *Bioorganic Chemistry Frontiers, Vol. 1*, Springer, Berlin, Heidelberg, **1990**, S. 143-159; c) F. Vögtle, Supramolecular Chemistry, an Introduction, Wiley, Chichester, **1991**; d) cf. also F. Vögtle, M. Bauer, C. Thilgen, P. Knops, *Chimia* **45** (1991) 319-321.

[2] a) L. Rossa, F. Vögtle, *Top. Curr. Chem.* **113** (1983) 1; b) P. Knops, N. Sendhoff, H.B. Mekelburger, F. Vögtle, *Top. Curr. Chem.*, in press.

[3] A. Ostrowicki, E. Koepp, F. Vögtle, *Top. Curr. Chem.*, in press.

[4] a) D.A. Laider, J.F. Stoddard in S. Patai (Ed.): 'The chemistry of functional groups', Suppl. E, part 1, Wiley, Chichester, **1980**; b) B.R. Bowsher, A.J. Rest, *J. Chem. Soc. Dalton Trans.* **1981**, 1157; c) S. Kulstad, L.A. Malmsten, *Tetrahedron* **36** (1980) 521; d)

A. Shanzer, N. Mayer-Shochet, *J. Chem. Soc. Chem. Commun.* **1980**, 176; e) C.K. Poon, C.M. Che, *J. Chem. Soc. Chem. Commun.* **1979**, 861; f) C.O. Dietrich-Buchecker, J. Guilhem, A.K. Khemiss, P. Kintzinger, C. Pascard, J.P. Sauvage, *Angew. Chem.* **99** (1987) 711; *Angew. Chem. Int. Ed. Engl.* **26** (1987) 661; g) C.O. Dietrich-Buchecker, A.K. Khemiss, J.P. Sauvage, *J. Chem. Soc. Chem. Commun.* **1986**, 1376; h) A. Shanzer, J. Libman, *J. Chem. Soc. Chem. Commun.* **1983**, 846; i) K. Saigo, R.J. Lin, M. Kubo, A. Youda, M. Hasegawa, *J. Am. Chem. Soc.* **108** (1986) 1996; j) H.L. Anderson, J.K.M. Sanders, *Angew. Chem.* **102** (1990) 1478; *Angew. Chem. Int. Ed. Engl.* **29** (1990) 1400.

[5] a) J.B. Neilands (Hrsg.): *Microbial Iron Transport: A Comprehensive Treatise*, Academic Press, New York, **1974**; b) K.N. Raymond, G. Müller, B.F. Matzanke, *Top. Curr. Chem.* **123** (1984) 49; c) *Struct. Bonding* **58** (1984); d) R.J. Bergeron, *Chem. Rev.* **84** (1984) 587; e) T.R. Swinburne (Hrsg.): *Iron, Siderophores and Plant Diseases*, Nato ASI Series, Series A, Life Science Vol. 117, Plenum, New York, **1986**; f) G. Winkelmann, D. Van der Helm, J.B. Neilands (Hrsg.): *Iron Transport in Microbes, Plants and Animals*, VCH, Weinheim, **1987**; g) M.J. Miller, *Chem. Rev.* **89** (1989) 1563; h) T.M. Loehr (Hrsg.): *Iron Carriers and Iron Proteins (Physical Bioorganic Chemistry*, Bd. 5), VCH, Weinheim, **1989**.

[6] a) J.R. Pollack, J.B. Neilands, *Biophys. Biochem. Res. Comm.* **38** (1970) 989; b) I.G. O'Brien, F. Gibson, *Biochem. Biophys. Acta* **215** (1970) 393.

[7] W.Harris, C.J. Carrano, S.R. Cooper, S.R. Sofen, A. Avdeef, J. McArdle, K.N. Raymond, *J. Am. Chem. Soc.* **101** (1979) 6097.

[8] Review: K.N. Raymond, G. Müller, G.F. Matzanke, *Top. Curr. Chem.* **123** (1984) 49.

[9] a) W. Harris, F.L. Weitl, K.N. Raymond, *J. Chem. Soc. Chem. Commun.* **1979** 177; b) M.C. Venuti, W.H. Rastetter, T.J. Erickson, J.B. Neilands, *J. Med. Chem.* **22** (1979) 123.

[10] a) W. Kiggen, F. Vögtle, *Angew. Chem.* **96** (1984) 712; *Angew. Chem. Int. Ed. Engl.* **23** (1984) 714; b) W. Kiggen, F. Vögtle, S. Franken, H. Puff, *Tetrahedron* **42** (1986) 1859; c) P. Stutte, W. Kiggen, F. Vögtle, *Tetrahedron* **43** (1987) 2065.

[11] For a critical discussion of this value in comparison with similar macrobicyclic siderands see: T.M. Garrett, T.J. McMurry, M.W. Husseini, Z.E. Reyes, F.E. Hahn, K.N. Raymond, *J. Am. Chem. Soc.* **109** (1991) 2965.

[12] M. Magerstädt, O.A. Gansow, L.K. Pannell, F. Vögtle, W. Kiggen, *Nucl. Med. Biol.* **17** (1990) 409.

[13] J. Peter-Katalinic, F. Ebmeyer, Ch. Seel, F. Vögtle, *Chem. Ber.* **122** (1989) 2391.

[14] S. Grammenudi, F. Vögtle, *Angew. Chem.* **98** (1986) 1119; *Angew. Chem. Int. Ed. Engl.* **25** (1986) 1122.

[15] P. Belser, L. De Cola, A. v. Zelewsky, *J. Chem. Soc. Chem. Commun.* **1988** 1057.

[16] V. Balzani, F. Barigeletti, L. De Cola, *Top. Curr. Chem.* **158** (1990) 30.

[17] B. Durham, J.V. Caspar, J.K. Nagle, T.J. Meyer, *J. Am. Chem. Soc.* **104** (1982) 4803.

[18] a) L. De Cola, F. Barigeletti, V. Balzani, P. Belser, A. v. Zelewsky, F. Vögtle, F. Ebmeyer, S. Grammenudi, *J. Am. Chem. Soc.* **110** (1988) 7210; b) F. Barigeletti, L. De Cola, V. Balzani, P. Belser, A. v. Zelewsky, F. Vögtle, F. Ebmeyer, S. Grammenudi, *J. Am. Chem. Soc.* **111** (1989) 4662; c) L. De Cola, P. Belser, F. Ebmeyer, F. Barigeletti, F. Vögtle, A. v. Zelewsky, V. Balzani, *Inorg. Chem.* **29** (1990) 495.

[19] J. de Mendoza, E. Mesa, J.C. Rodríguez-Ubis, P. Vásquez, F. Vögtle, P.M. Windscheif, K. Rissanen, J.M. Lehn, D. Lilienbaum, R. Ziesel, *Angew. Chem.* **103** (1991) 1365; *Angew. Chem. Int. Ed. Engl.* **30** (1991) 1331.

[20] A. Wallon, U. Werner, W.M. Müller, M. Nieger, F. Vögtle, *Chem. Ber.* **123** (1990) 859.

[21] F. Ebmeyer, F. Vögtle, *Angew. Chem.* **101** (1989) 95; *Angew. Chem. Int. Ed. Engl.* **28** (1989) 79.

[22] Ch. Seel, F. Vögtle, *Angew. Chem.* **103** (1991) 433; *Angew. Chem. Int. Ed. Engl.* **30** (1991) 442.

[23] F. Vögtle, I. Lüer, V. Balzani, N. Armaroli, *Angew. Chem.* **103** (1991) 1367; *Angew. Chem. Int. Ed. Engl.* **30** (1991) 1333.

[24] Review: F. Diederich, *Angew. Chem.* **100** (1988) 373; *Angew. Chem. Int. Ed. Engl.* **27** (1988) 362.

[25] F. Vögtle, W.M. Müller, U. Werner, H.W. Losensky, *Angew. Chem.* **99** (1987) 930; *Angew. Chem. Int. Ed. Engl.* **26** (1987) 901.

[26] a) J. Franke, F. Vögtle, *Angew. Chem.* **97** (1985) 224; *Angew. Chem. Int. Ed. Engl.* **24** (1985) 219; b) H. Schrage, M. Franke, F. Vögtle, E. Steckhan, *Angew. Chem.* **98** (1986) 335; *Angew. Chem. Int. Ed. Engl.* **25** (1986) 336.

[27] For the use of the *Eglinton* coupling in the syntheses of host molecules see: a) D. O'Krongly, S.R. Denmeade, M.Y. Chiang, R. Breslow, *J. Am. Chem. Soc.* **107** (1985) 5544. b) B.J. Whitlock, H.W. Whitlock, *J. Am. Chem. Soc.* **112** (1990) 3910 and refs. cited therein. c) A.D. Hamilton, D. Little, *J. Chem. Soc. Chem. Commun.* **1990**, 297. d) H.L. Anderson, J.K.M. Sanders, *Angew. Chem.* **102** (1990) 1478, *Angew. Chem. Int. Ed. Engl.* **29** (1990) 1400.

[28] a) F. Vögtle, R. Berscheid, W. Schnick, *J. Chem. Soc. Chem. Commun.* **1991**, 414. b) R. Berscheid, M. Nieger, F. Vögtle, *J. Chem. Soc. Chem. Commun.* **1991**, 1364.

[29] R. Berscheid, F. Vögtle, *Synthesis* **1992**, in press.

[30] Acetonitrile as guest molecule was published until recently only in complexes of crown ether compounds and as clathrate with hydroquinone. The acetonitrile therein is mostly bound by (weak) hydrogen bridges between its methyl group and the oxygen bridge atoms of the crown ether counterpart. Single crystal X-ray structural analyses of such complexes are cited in: a) C.L. Liotta, H.P. Harris, F.L. Cook, *J. Org. Chem.* **39** (1974) 2445; b) I. Goldberg, *Acta Cryst.* **B31** (1975) 754; c) T.C.W. Mak, K.-S. Lee, *Acta Cryst.* **B34** (1974) 3631; d) I. Tanaka, I. Tajima, Y. Hayakawa, M. Okada, M. Bitoh, T. Ashida, H. Sumitomo, *J. Am. Chem. Soc.* **102** (1980) 7873; e) I. Sakuragi, I. Tanaka, T. Ashida, I. Tajima, M. Okada, H. Sumitomo, *J. Am. Chem. Soc.* **104** (1982) 6035; f) T.-L. Chan, T.C.W. Mak, *J. Chem. Soc. Perkin Trans. 2*, **1983**, 777; g) C.J. Gilmore, D.D. Mac Nicol, A. Murphy, M.A. Russell, *Tetrahedron Lett.* **24** (1983) 3269; h) R.D. Rogers, L.K. Kurihara, P.D. Richards, *J. Chem. Soc. Chem. Commun.* **1987**, 604; i) E. Weber, S. Franken, J. Ahrendt, H. Puff, *J. Org. Chem.* **52** (1987) 5291; j) R.D. Rogers, P.D. Richards, E.J. Voss, *J. Incl. Phen.* **6** (1988) 65; k) R.L. Garrell, J.C. Smyth, F.R. Fronczek, R.D. Gandour, *J. Incl. Phen.* **6** (1988) 73; l) R.D. Rogers, *J. Incl. Phen.* **6** (1988) 629; m) P.D.J. Grootenhuis, P.A. Kollman, *J. Am. Chem. Soc.* **111** (1989) 4046; n) K. Panneerselvam, K.K. Chacko, E. Weber, H.-J. Köhler, *J. Incl. Phen.* **9** (1990) 337.

[31] R. Berscheid, M. Nieger, F. Vögtle, *Chem. Ber.* **1992**, in press.

[32] a) V. Balzani, F. Scandola, *Supramolecular Photochemistry*, Ellis Horwood, Chichester, **1991**; b) M.A. Fox, M. Chanon (Ed.): *Photoinduced Electron Transfer*, Elsevier, Amsterdam, **1988**.

[33] a) J.S. Connoly (Ed.): *Photochemical Conversion and Storage of Solar Energy*, Academic Press, New York, **1981**; b) T.J. Meyer, *Acc. Chem. Res.* **22** (1989) 163.

MULTI-ELECTRON REDOX ACTIVITY OF SUPRAMOLECULAR COORDINATION COMPOUNDS CONTAINING METALLOCYCLAM AND FERROCENE FRAGMENTS

A. M. De Blas,[1] G. De Santis, L. Fabbrizzi, M. Licchelli,
P. Pallavicini, and A. Poggi
Dipartimento di Chimica Generale,
Università di Pavia,
Via Taramelli, 12
27100 Pavia,
Italy

ABSTRACT. Multi-electron redox systems of supramolecular nature have been prepared by binding to a metal centre M two ligands L-X, which display independent redox activity. L-X conjugated systems have been obtained by linking a ligand L (e.g. pyridine) and a redox active fragment (ferrocene, Fc, active through the Fc/Fc^+ change, or metallocyclam, active through the the Ni^{II}/Ni^{III} couple). In particular, novel convenient template routes have been devised to append the metallocyclam subunit to the pyridine ring. $[M(L\text{-}X)_2]$ supercomplexes have been prepared, M = $>[Pt^{II}Cl_2]$ or $>[Ru^{II}(bipy)_2]$, and electrochemically characterized in their multi-electron redox behaviour. Whereas the Pt^{II} metal centre is redox inactive and plays a merely architectural rôle, in the ruthenium supercomplexes the metal centre displays further redox activity through the Ru^{II}/Ru^{III} couple. Finally, redox systems releasing two electrons at distinct potentials have been synthesized by linking a ferrocene fragment and a poly-aza macrocyclic subunit. Again, linking has been obtained by means of novel high yield template procedures. Appropriate tetra-aza rings were used to allow the Ni^{II}/Ni^{III} (cyclam) or the Cu^{II}/Cu^{III} (dioxocyclam) redox changes to take place.

1. Introduction

Metal containing systems, deriving from (i) coordination chemistry or (ii) organometallic chemistry, typically display one-electron redox activity. Two classical examples are given below: metallocyclam complexes, **1**, representatives of category (i), and ferrocene, **2**, representative of category (ii).

V. Balzani and L. De Cola (eds.), Supramolecular Chemistry, 87–103.

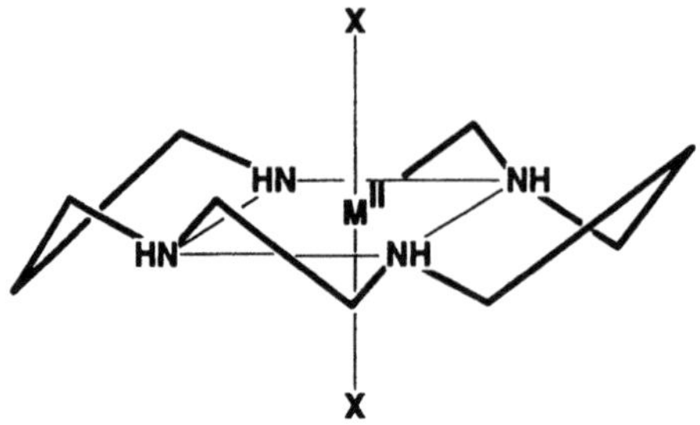

1

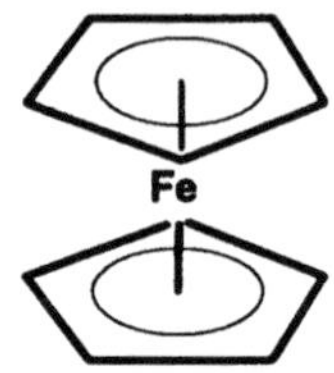

2

As far as type **1** compounds are concerned, it is well known that encircling of a *d* metal centre by the 14-membered ring of saturated tetra-aza macrocycles (1,4,8,11-tetra-aza-cyclotetradecane, *cyclam*), allows access to unusually high and otherwise unstable oxidation states (e.g. Ni^{III}, Cu^{III}, Ag^{II}, Ag^{III}) [2]. Moreover, the redox reactivity of the incorporated metal can be controlled by varying ligand's donor tendencies through structural modifications to the tetra-aza-macrocyclic framework. This opportunity has been extensively verified in the case of the Ni^{II}/Ni^{III} redox change [3]. On the other hand, ferrocene, Fc, undergoes a reversible one-electron oxidation to the stable ferrocenium ion, Fc^+ [4]. The Fc/Fc^+ redox change can be regulated through substituent effects on the cyclopentadienide ring [5]. Noticeably, in the case of complexes of type **1**, the released/uptaken electrons are conveyed from/to the metal along the *z* axis and the electron transfer takes place according to an *inner sphere* mechanism. On the other hand, type **2** derivatives typically exchange electrons through an *outer sphere* mode.

Two-electron redox systems can be generated by coupling two fragments of type **1** or **2**. The most obvious synthetic way to these systems involves the covalent linking of the two fragments. In the case of type **1** systems, this route has been explored through the synthesis of bicyclams, i.e. molecules in which two cyclam subunits have been linked together through a C-C bond [6] or by a $-(CH_2)_n-$ chain joining two nitrogen atoms of the two tetraaza rings [7]. A bicyclam molecule may incorporate two equivalent redox active metal centres (e.g. a pair of Ni^{II} ions) and the corresponding dimetallic complex can exchange a couple of electrons through two consecutive one-electron redox steps, whose electrode potentials are separated by the quantity ΔE. The value of ΔE results from the combination of a statistical contribution (a constant term, 36 mV at 25°C) and of an electrostatic contribution, whose magnitude decreases with the increasing distance between the two tetra-aza rings [7]. Biferrocenes, i.e. molecules in which two ferrocene subunits are covalently linked, are well known two-electron redox systems [8]. In this case, the bridging segment may allow electron communication between the two subunits, inducing a ΔE separation much larger than that expected on the basis of the statistical and electrostatic effects.

We are now exploring a different approach to the synthesis of multicentre redox systems, in which the redox fragments are held together not by *covalent* bonds, but through *coordinative* interactions. Assembled systems in which more subunits are linked together through non-covalent interactions have been defined as *supramolecular* systems [9]. In this sense, coordinatively linked multicentre systems belong to the realm of supramolecular chemistry and, in particular, they should be defined as *supramolecular coordination compounds*.

The coordinative approach to the design of multicentre systems displaying multi-electron activity is described by the following equations.

$$L + X = L\text{-}X \qquad (1)$$

$$M + 2\ L\text{-}X = [M(L\text{-}X)_2 \qquad (2)$$

First, according to reaction (1), the redox active fragment X is appended to a molecule L, possessing distinctive coordinating tendencies. Then, according to reaction (2), two (or more) molecules of the conjugate system L-X are coordinatively bound to a metal centre M, to form the *supercomplex* $[M(L\text{-}X)_2]$.

The metal centre M may be redox inactive. Thus, it plays a purely architectural role and the redox activity is located on the appended X fragments. However, the metal centre may also display its own redox activity. In this case, the $[M(L\text{-}X)_2]$ supercomplex will present a richer multi-electron redox activity located on both X and M. It is important that M-L coordinative bonds are stable from both a thermodynamic and a kinetic point of view. Moreover, their stability should not be altered by the redox changes taking place on the proximate subunits X and, in particular, by a change of the oxidation state of M, if occurring.

The present contribution reports some supramolecular redox systems recently prepared in our laboratory, in which metallocyclam or ferrocene subunits have been used as X fragments. Moreover, most of the examples described below use pyridine as the ligand L. Pyridine, one of the very classical ligands of coordination chemistry, forms stable complexes with most transition metal ions. We have appended ferrocene or metallocyclam subunits to the pyridine ring, by using the standard procedures of synthetic organic chemistry or some novel metal template reactions recently developed in our laboratory.

2. Redox active supramolecular coordination compounds containing the ferrocene fragment

Ferrocene can be appended at the backbone of the pyridine ligand through the synthetic route described in the Scheme below. The reaction involves the coupling of ferrocene and the diazoderivative of 3-aminopyridine: 3-ferrocenylpyridine (**3**, Fcpy) is obtained in satisfactory yield.

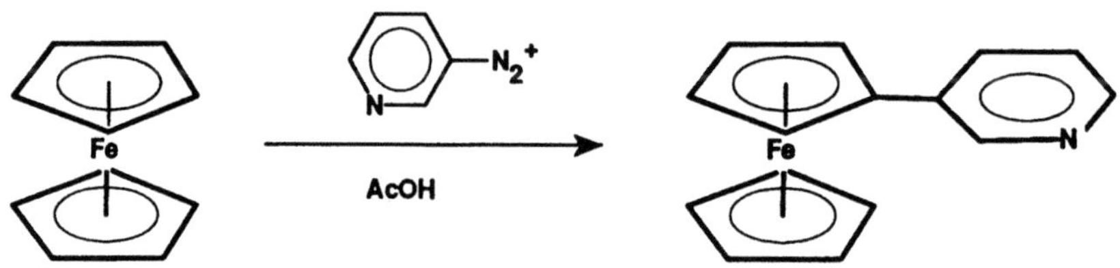

3

Reaction of Fcpy with $[Pt^{II}Cl_4]^{2-}$ in a 2:1 molar ratio gives the $[Pt^{II}(Fcpy)_2Cl_2]$ complex. The *trans*-effect should drive the stepwise substitution of Cl^- ions to give the *cis*-derivative. Formation of the *cis*-$[Pt^{II}(Fcpy)_2Cl_2]$ complex has been confirmed by single crystal x-ray diffraction analysis [10]. The structural arrangement of the complex is sketched below.

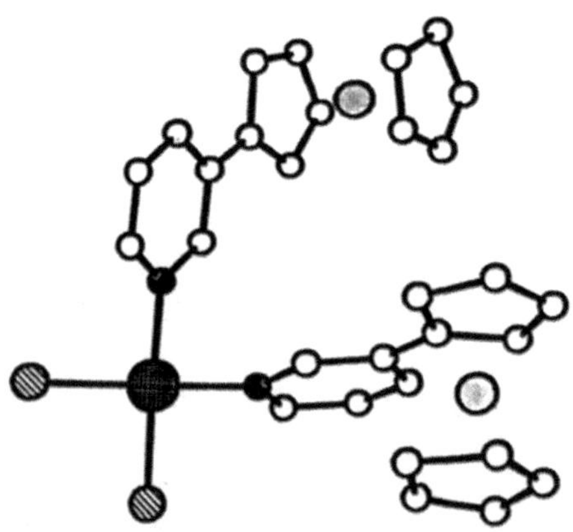

Figure 1. Structure of *cis*-$[Pt^{II}(FcPy)_2Cl_2]$ supercomplex

Cyclic voltammetry and controlled potential coulometry investigations on a MeCN solution, made 0.1 M in Bu_4NClO_4, have indicated that *cis*-$[Pt^{II}(FcPy)_2Cl_2]$ undergoes a two-electron oxidation process at the platinum electrode, according to the two consecutive one-electron steps (4) and (5):

$$cis\text{-}[Pt^{II}(Fcpy)_2Cl_2] = cis\text{-}[Pt^{II}(Fcpy)(Fcpy^+)Cl_2]^+ + e^- \quad (4)$$

$$cis\text{-}[Pt^{II}(Fcpy)(Fcpy^+)Cl_2]^+ = cis\text{-}[Pt^{II}(Fcpy^+)_2Cl_2]^{2+} + e^- \quad (5)$$

whose potentials are separated by 36 mV, the statistical term. This means that the comproportionation equilibrium (6):

$$cis\text{-}[Pt^{II}(Fcpy)_2Cl_2] + cis\text{-}[Pt^{II}(Fcpy^+)_2Cl_2]^{2+} = 2\ cis\text{-}[Pt^{II}(Fcpy)(Fcpy^+)Cl_2]^+ \quad (6)$$

which results from the difference of equations (4) and (5), is characterized by the statistical equilibrium constant K = 4. This demonstrates that the mixed valence species *cis*-$[Pt^{II}(Fcpy)(Fcpy^+)Cl_2]^+$ is not stabilized by any kind of electron delocalization, indicating that the Pt^{II} metal center prevents electron communication from a ferrocene moiety to the other one. Moreover, no repulsive electrostatic effects operate between the two redox sites. In principle, a mixed valence species of general formula A^{n+} is electrostatically stabilized with respect to the $A^{(n-1)+}$ and $A^{(n+1)+}$ forms only if n>1 [11]. Thus, if one considers the ferrocene fragment as definitely neutral (n=1), equilibrium (6) should not profit at all from the electrostatic term. Moreover, the electrostatic advantage decreases with the increasing distance between the redox sites: thus, for an Fe-Fe distance similar to that observed in the x-ray analysis of the *cis*-$[Pt^{II}(Fcpy)_2Cl_2]$ complex (7.42 Å), the electrostatic effect, if any, would disappear. Therefore, the *cis*-$[Pt^{II}(Fcpy)_2Cl_2]$ supercomplex is a novel reducing agent able to release reversibly a couple of electrons *in a single shot*. In this sense, it behaves like a diferrocene molecule, in which the two subunits are linked by a reasonably long aliphatic chain.

The ferrocene fragment has also been appended to a well known chelating agent: the salicylate ion. In particular, the 5-ferrocenyl-2-hydroxy-benzoic acid (**4**, $FcsalH_2$) has been obtained through the diazo-coupling of ferrocene and salicylic acid [12].

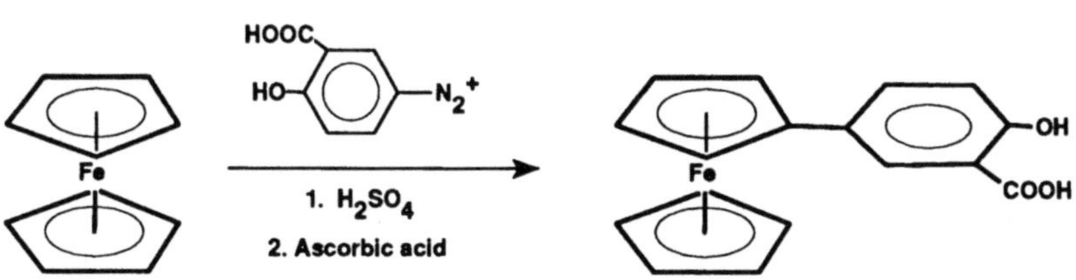

4

As the salicylate ion is known to form stable complexes in solution with the Fe^{III} ion, the complexation of Fe^{III} by $FcsalH_2$ in solution has been investigated. In particular, titration with standard NaOH of a 50/50 water/dioxane solution containing 1 equivalent of Fe^{III} and 3 equivalents of $FcsalH_2$ indicated that, at pH = 7, 100% of the $[Fe^{III}(Fcsal^{2-})_3]^{3-}$ is formed. By analogy with the corresponding salicylate complex, the $[Fe^{III}(Fcsal^{2-})_3]^{3-}$ ion should present an octahedral stereochemistry, as illustrated below.

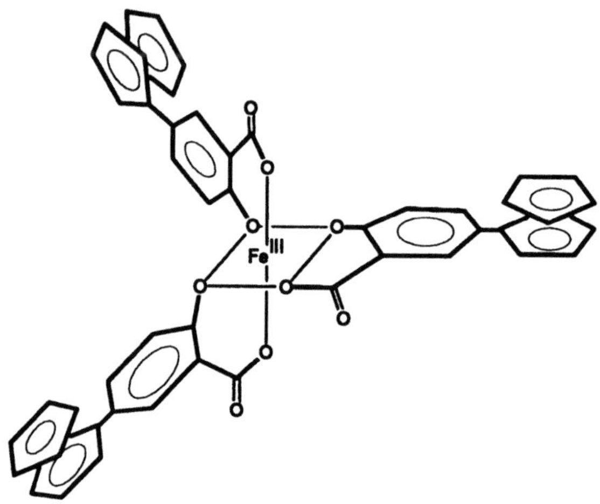

The controlled potential coulometry experiment on a 50/50 water/dioxane solution containing $[Fe^{III}(Fcsal^{2-})]^{3-}$ indicated that the complex undergoes a three-electron oxidation process. As the oxidation process involves the three ferrocene subunits appended on the back of each salicylate moiety, the neutral species $[Fe^{III}(Fc^{+}sal^{2-})_3]$ forms. The cyclic voltammetry investigation on the same solution disclosed a single reversible wave, indicating that the release of the three electrons is only controlled by statistical effects. In particular, the $[Fe^{III}(Fcsal^{2-})_3]^{3-}$ complex releases three electrons from each ferrocene subunit, according to three one-electron steps, whose potentials are separated by 28 mV (= {RT/F}ln3, at 25°C).

The above examples demonstrate that metal centres can conveniently be used to link redox active fragments in order to generate multi-electron redox active systems. If a stable system is desired, a metal ion that forms inert complexes should be chosen. This is the case of the *cis*-$[Pt^{II}(Fcpy)_2Cl_2]$ system. On the contrary, the $[Fe^{III}(Fcsal^{2-})_3]^{3-}$ species possesses a labile metal centre and the redox active ligand can be removed, for instance on addition of a strong acid. It should be noted that the two considered metal centres (Pt^{II}, Fe^{III}) do not present individual redox activity in the coordinative environments examined and play a purely architectural role.

3. Redox active supramolecular coordination compounds containing the metallocyclam subunit

3.1 A NOVEL TEMPLATE ROUTE TO FUNCTIONALIZED CYCLAM-LIKE MACROCYCLES.

Functionalization of cyclam (for instance obtained by appending on it a pyridine fragment) is generally being carried out through the nucleophilic attack of one of the four secondary amine nitrogen atoms of cyclam on the bromo-derivative, RBr, of the functional group of interest, R. If monosubstituted cyclam is desired, (i) all but one nitrogen atoms should be protected [13]; alternatively, (ii) cyclam has to be present in a large excess compared to RBr [14]. Both sinthetic pathways are rather tedious and time consuming, due to the protection/deprotection procedure, in case (i), or to the separation of the wanted product from the unreacted cyclam, present in a overwhelming amount, in case (ii).

We have now devised a template synthetic methodology that allows us to obtain the metal complex of the appropriately functionalized cyclam-like macrocycle in a satisfactory yield, through a single step reaction, which involves inexpensive starting materials and does not present any serious purification problem. The reaction takes inspiration from the Ni^{II} assisted synthesis of poly-aza macrocycles from polyamines and formaldehyde, using a secondary amine, RNH_2, as a locking fragment, as reported by Suh *et al.* [15]. We have recently applied this procedure to the synthesis of the penta-aza macrocycle **5** (methylazacyclam) from the open-chain tetramine 1,4,8,11-tetra-aza-undecane (2.3.2-tet), as illustrated in the following Scheme.

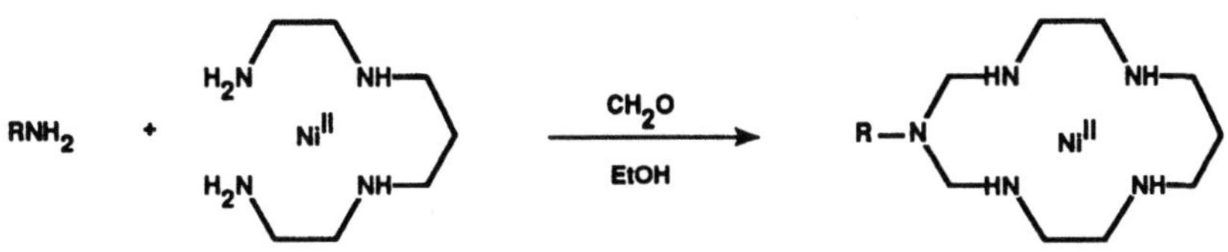

R = alkyl, CH_3: **5**

X-ray investigation of the low-spin $[Ni^{II}(azacyclam)](ClO_4)_2$ complex produced through the above procedure has shown that the tertiary nitrogen atom of the penta-aza macrocycle is not coordinated and that azacyclam displays coordinating tendencies (in terms of bond distances and angles) strictly similar to those of cyclam [16]. Moreover, the tertiary amine group appears as distinctly flattened, thus presenting a pronounced sp^2, rather than sp^3, character. This should drastically reduce its basicity. As a matter of fact, the tertiary amine group presents extremely low basic tendencies and is protonated only in a strong acid solution of concentration 1 M or higher. This demonstrates that the fifth nitrogen atom introduced by template syntheses of the type described above exerts a purely structural function and does not alter the cyclam-like donor properties of the macrocycle.

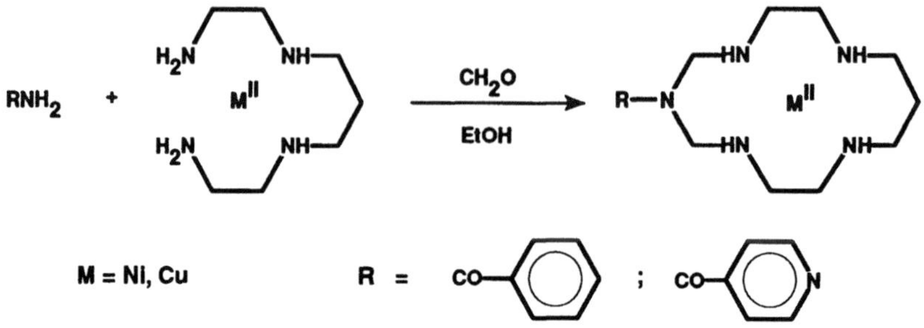

6 7

We have recently observed that arylamides, e.g. benzamide, $C_6H_5CONH_2$, can efficiently replace primary amines, RNH_2, as a locking fragment in the template synthesis of the type described above: the novel functionalized cyclam-like macrocycle **6** is obtained. Cu^{II} works well as a template, but satisfactory yields are also obtained with Ni^{II}. The corresponding complexes exhibit the typical macrocyclic inertness towards demetallation, persisting in strongly acidic solution. Using the 4-pyridinamide derivative (isonicotinamide, see the Scheme above), novel cyclam-like metal complexes are obtained, in which the macrocycle carries an appended pyridine subunit (**7**): $[M^{II}(cyclamCOpy)]^{2+}$ (M = Ni, Cu) [17]. In particular, the Ni^{II} complex undergoes in MeCN or DMSO solution the typical one-electron oxidation process to form the stable trivalent species. Therefore, the $[Ni^{II}(cyclamCOpy)]^{2+}$ complex represents a novel example of the class of L-X systems, i.e. ligands possessing an appended redox active subunit, and, by analogy with that observed with the ferrocene-pyridine conjugate system **3**, described in the previous Section, it can be used as a building block to redox active supramolecular coordination compounds.

3.2 Appending two metallocyclam subunits at a redox inactive metal centre: Pt^{II}.

Reaction of 2 equivalents of $[Ni^{II}(cyclamCOpy)](ClO_4)_2$ with 1 equivalent of $K_2[Pt^{II}Cl_4]$ gave the supercomplex $[\{Ni^{II}(cyclamCOpy)\}_2Pt^{II}Cl_2](ClO_4)_4$. No crystals suitable for x-ray investigation were obtained. However, experimental conditions should drive the chloride substitution of the tetrachloroplatinate(II) complex to the *cis*-derivative. An idealized structural arrangement of the supercomplex is sketched below.

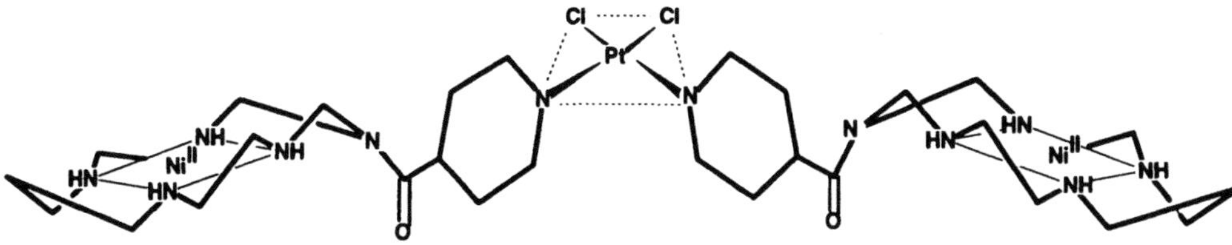

Cyclic voltammetry investigation on a DMSO solution of the supercomplex, abbreviated as $[\{Ni^{II}(Lpy)\}_2Pt^{II}Cl_2]^{4+}$, disclosed just one reversible wave, with $E_{1/2}$ = 0.023 V vs Fc^+/Fc. This potential is very close to that observed for the oxidation of the reference system $[Ni^{II}(\mathbf{7})](ClO_4)_2$, under the same conditions: 0.032 V vs Fc^+/Fc. Moreover, the coulometry experiment on a solution of $[\{Ni^{II}(Lpy)\}_2Pt^{II}Cl_2]^{4+}$ at a potential 200 mV more positive than $E_{1/2}$ showed the consumption of 2 electrons. This indicates that the $[\{Ni^{II}(Lpy)\}_2Pt^{II}Cl_2]^{4+}$ supercomplex undergoes a two-electron oxidation process according to the following two one-electron reversible steps:

$$[\{Ni^{II}(Lpy)\}_2Pt^{II}Cl_2]^{4+} = [\{Ni^{III}(Lpy)\}\{Ni^{II}(Lpy)\}Pt^{II}Cl_2]^{5+} + e^- \qquad E_1$$

$$[\{Ni^{III}(Lpy)\}\{Ni^{II}(Lpy)\}Pt^{II}Cl_2]^{5+} = [\{Ni^{III}(Lpy)\}_2Pt^{II}Cl_2]^{6+} + e^- \qquad E_2$$

Careful examination of the parameters associated to the voltammetric profile indicated that the stepwise two-electron oxidation process is merely controlled by statistical effects: thus, E_2 - E_1 = 36 mV, E_1 = 5 mV and E_2 = 41 mV. Therefore, by analogy with that observed for the corresponding Pt^{II} supercomplex containing ferrocene subunits, described in the previous Section, no communication exists between the two metallocyclam redox centres. In particular, in view of the higher electrical charge of the redox active metal centres, Ni^{II} and Ni^{III}, one would expect more intense electrostatic repulsive effects than for a system based on the Fc/Fc^+ change. However, due to the large distance, and to the shielding effect exerted by the aliphatic backbone of the macrocycle, the intermetal electrostatic repulsion is slight or nil and does not affect the redox behaviour.

3.3 APPENDING TWO METALLOCYCLAM SUBUNITS AT A REDOX ACTIVE METAL CENTRE: Ru^{II}.

It is well known that in the *cis*-$[Ru^{II}(bipy)_2Cl_2]$ complex (bipy = 2,2'-bipyridine) the two chloride ions can easily be replaced by nitrogeneous ligands, to give a very stable species from both a thermodynamic and a kinetic point of view. Thus, the *cis*-$[Ru^{II}(bipy)_2Cl_2]$ complex has been reacted with two equivalents of the functionalized ligand $[Ni^{II}(cyclamCOpy)]^{2+}$ to form the

$[Ru^{II}(bipy)_2\{Ni^{II}(cyclamCOpy)\}_2]^{6+}$ supercomplex, whose structural arrangement is sketched below.

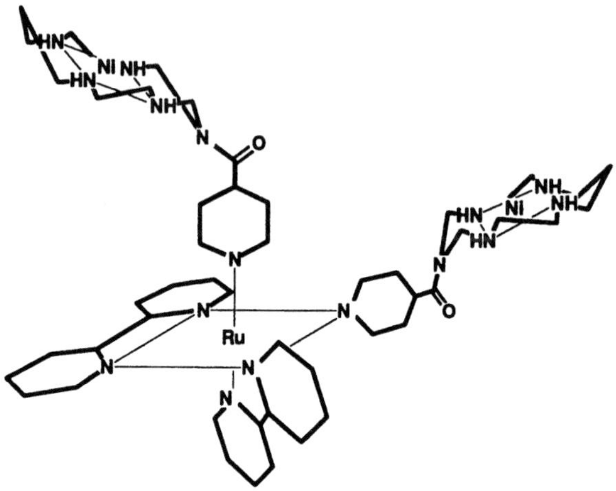

The $[Ru^{II}(bipy)_2\{Ni^{II}(cyclamCOpy)\}_2]^{6+}$ species is very similar to the previously described $[\{Ni^{II}(cyclamCOpy)\}_2Pt^{II}Cl_2]^{4+}$ supercomplex, in that it contains two metallocyclam subunits bound, through a pyridine fragment, at the adjacent corners of a square. However, as the $Ru^{II}N_6$ moiety is typically redox active, through the Ru^{II}/Ru^{III} change, the metal centre does not exert in this case a simply structural role, but makes richer the multi-electron redox activity of the supercomplex.

As a matter of fact, the electrochemical investigation (cyclic voltammetry and coulometry) of a MeCN solution 0.1 M in Bu_4NClO_4 and 10^{-3} M in $[Ru^{II}(bipy)_2\{Ni^{II}(cyclamCOpy)\}_2]^{6+}$ showed first a one-electron oxidation process, to be ascribed to the Ru^{II}/Ru^{III} change, followed, at a more anodic potential, by a two-electron oxidation process, to be assigned to the statistically controlled oxidation of the two metallocyclam subunits. The corresponding potential values are reported in the unidimensional diagram in Figure 2.

Replacing Bu_4NClO_4 with Bu_4NCl as a background electrolyte induces a drastic change in the voltammetric behaviour of the $[Ru^{II}(bipy)_2\{Ni^{II}(cyclamCOpy)\}_2]^{6+}$ supercomplex. In particular, the two-electron wave becomes less anodic than the one-electron wave (see the potential values in the lower part of the diagram in Fig. 2). This effect seems to be connected with the different stabilizing effect exerted by Cl^- ions on the Ru^{III} and Ni^{III} centres. As a matter of fact, oxidation of the Ni^{II} tetra-aza-macrocyclic fragment involves the formation of a *trans*-octahedral complex (d^7, low spin), in which the two axial positions are occupied by solvent molecules or by anions. Whereas the poorly coordinating ClO_4^- ions can hardly compete with the solvent molecules for the axial coordination sites, the strongly ligating Cl^-

ions definitively occupy the apices of the elongated octahedron and stabilize the trivalent state. Therefore, replacing of ClO_4^- by Cl^- as an anion of the supporting electrolyte causes a dramatic decrease of the $E_{1/2}$ value: more than 0.6 V! Such a stabilizing effect is not expected for the coordinatively saturated $Ru^{II}N_6$ subunit. Indeed, a very moderate potential decrease is observed

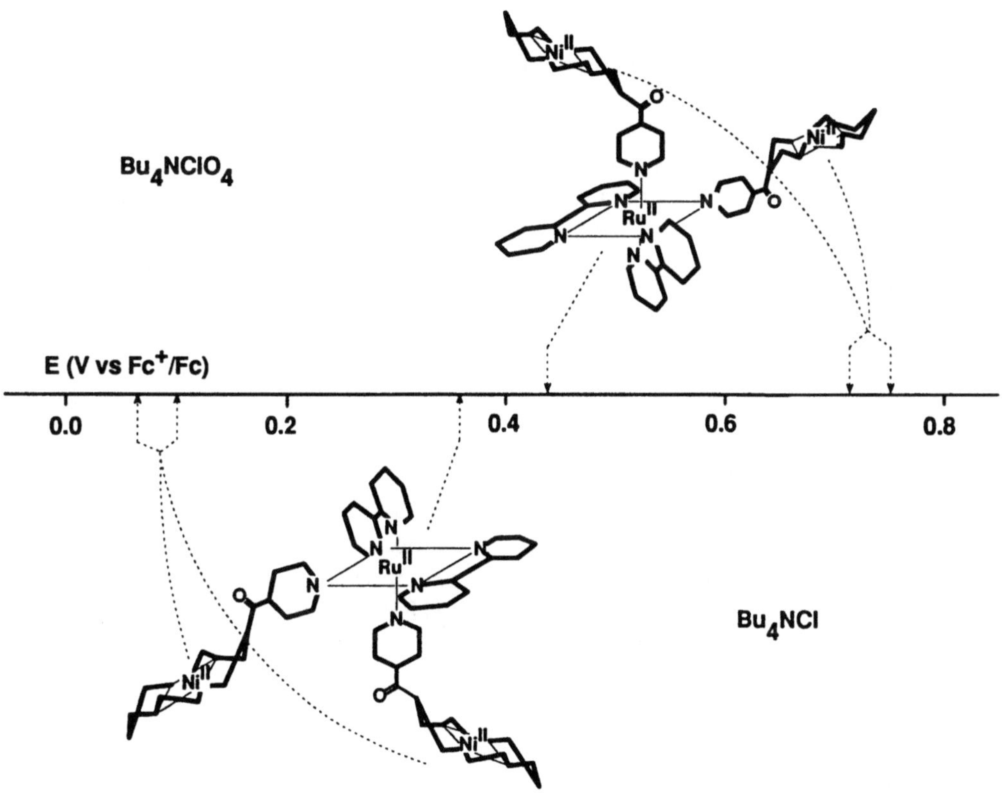

Figure 2. Electrode potentials measured for MeCN solutions of $[Ru^{II}(bipy)_2\{Ni^{II}(cyclamCOpy)\}_2]^{6+}$. Upper part refers to a solution made 0.1 M in Bu_4NClO_4, lower part refers to a solution made 0.1 M in Bu_4NCl.

also in the case of the Ru^{II}/Ru^{III} redox change (0.1 V). It should be noted that the sp^2 nitrogen atoms of 2,2'-bipyridine and of the functionalized pyridine molecules are so firmly bound to the Ru centre that they cannot be replaced by any anion, in the conditions of the electrochemical investigation. However, it should be considered that anions present in solution of a not too polar solvent as MeCN can interact electrostatically with the RuN_6 moiety, in the form of an ion pair. Thus, the moderate stabilization of the Ru^{III} state may reflect the formation of tighter ion pairs with Cl^- ions, compared to ClO_4^- ions, which present a much lower charge density.

In any case, the $[Ru^{II}(bipy)_2\{Ni^{II}(cyclamCOpy)\}_2]^{6+}$ species represents an interesting example of a coordination compound displaying three-electron redox activity, in which the mode of the electron release, (1 + 2) or (2 + 1), can be switched by varying the nature of the background electrolyte.

4. Design of ferrocene/metallocyclam conjugates.

If, for some reasons, a two-electron redox system is desired, in which the electrons are exchanged at distinctly different potentials, two different fragments X and Y, each one active through a one-electron process, should be linked. At this stage, the reader does not need to be convinced that ferrocene and $[Ni^{II}(cyclam)]^{2+}$ could be reasonable candidates to play the roles of X and Y fragments.

Among the many possible synthetic routes to ferrocene/metallocyclam assemblies, we have first considered a template procedure quite analogous to that used for the synthesis of functionalized cyclam-like complexes of type **6** and **7**. In fact, we have observed that arylsulphonamides are even more efficient than arylamides as locking fragments for the Ni^{II} and Cu^{II} assisted closure of open-chain tetramines [18]. In particular, ferrocenesulphonamide reacts with $[Ni^{II}(2.3.2\text{-tet})](ClO_4)_2$, in an ethanolic solution containing excess formaldehyde, to give a silky diamagnetic precipitate, orange in colour, of the ferrocene/metallocyclam conjugate **8**, $FcSO_2[Ni^{II}(cyclam)](ClO_4)_2$.

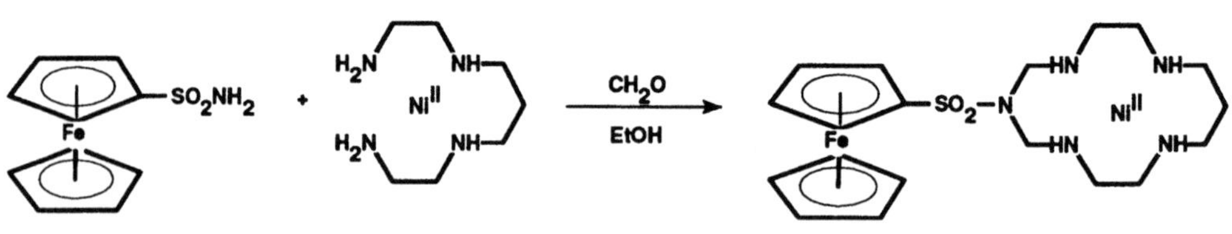

8

The metallocyclam moiety of the conjugate system **8** displays typical macrocyclic properties: for instance, in strongly acidic solutions the Ni^{II} ion is not extruded from the tetra-aza ring and the system remains intact for weeks. Moreover, the conjugate molecule **8** discloses the expected two-electron redox activity, as documented by voltammetric investigations. Figure 3 reports the Differential Pulse Voltammetry profile of a MeCN solution 0.1 M in Bu_4NClO_4 and 5×10^{-4} M in **8**.

Two one-electron peaks are observed, separated by 440 mV. The more anodic peak has an $E_{1/2}$ value (0.80 V vs Fc^{+}/Fc) slightly more positive than that observed with the Ni^{II} complex obtained by the template cyclization procedure in which 4-methyl-sulphonamide has been used as a locking fragment: $ArSO_2[Ni^{II}(cyclam)]^{2+}$. Thus, the peak is assigned to the Ni^{II}/Ni^{III} redox change taking place inside the cyclam subunit. As a consequence, the less anodic peak has to be assigned to the Fc/Fc^{+} oxidation process, taking place in the organometallic moiety of **8**.

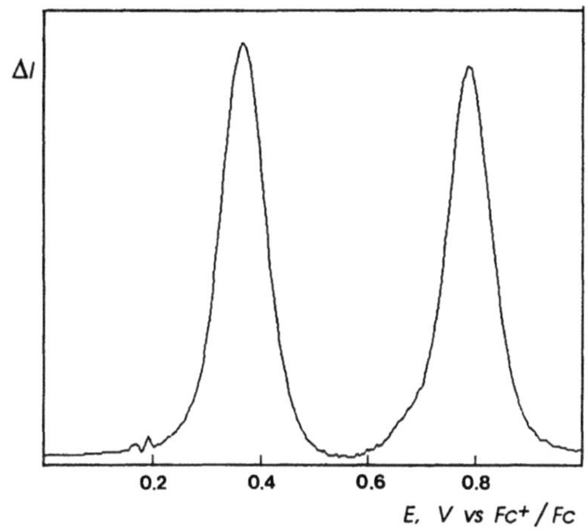

Figure 3. Differential Pulse Voltammetry profile obtained at a platinum microsphere for a MeCN solution of the conjugate system $FcSO_2[Ni^{II}(cyclam)](ClO_4)_2$, **8**. The less anodic peak refers to the one-electron oxidation of the ferrocene moiety, the more anodic one corresponds to the Ni^{II} to Ni^{III} oxidation process taking place in the metallocyclam fragment.

It should be noted that the potential associated with the reversible oxidation of the ferrocene subunit of **8** (0.37 V vs Fc^{+}/Fc) is about 170 mV more positive than that observed for the oxidation of N,N-diethyl-ferrocene-sulphonamide, which has been taken as a reference system ($E_{1/2}$ values vs Fc^{+}/Fc for all the mentioned systems are displayed in the potential diagram in Fig. 4). This potential increase should be ascribed to the electrostatic repulsive effects exerted by the adjacent Ni^{II} cation on the ferrocene subunit, which makes the Fc/Fc^{+} oxidation process remarkably more difficult. On the contrary, very small or nil electrostatic repulsive effects are observed in the oxidation process involving the cyclam fragment of **8**, whose $E_{1/2}$ potential is only slightly more positive than that observed for the oxidation of the reference system $ArSO_2[Ni^{II}(cyclam)]^{2+}$ (see Fig. 4). In this connection, it should be considered that the increase of the positive electrical charge of the redox active metal centre (from 2+ to 3+) takes place under a much more moderate repulsive field, i.e. that exerted by the adjacent monopositive ferrocenium moiety.

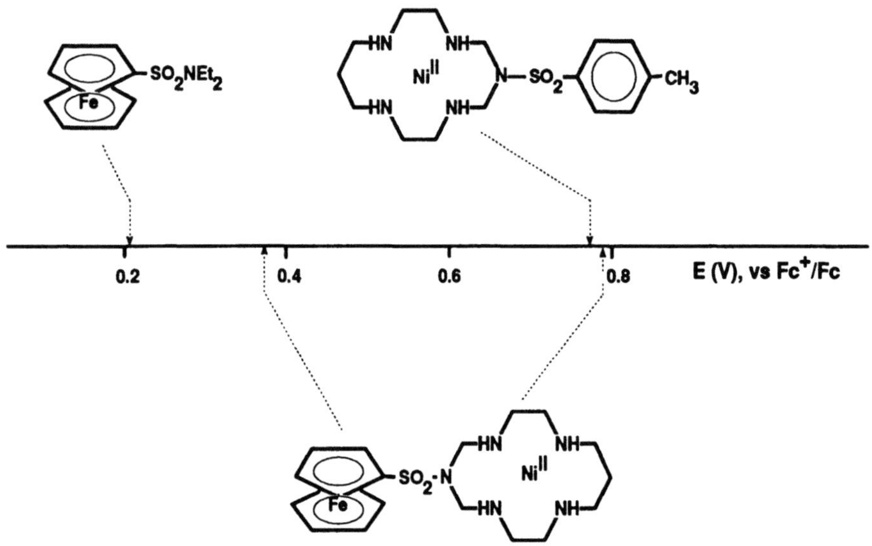

Figure 4. Electrode potentials measured for MeCN solutions of the ferrocene/metallocyclam conjugate system **8** and for related fragments.

In a different approach to the design of a two-electron redox system, we appended the ferrocene fragment to a tetra-aza macrocycle (a further example of an L-X system): the functionalized ligand should incorporate a *d* metal centre and should promote its redox activity. The tetra-aza macrocycle we have chosen is dioxocyclam, **9**. This ligand, which can be prepared through reaction of diethylmalonate with 2.3.2-tet, on addition of 2 equivalents of a strong base incorporates in aqueous solution a Cu^{II} ion, with simultaneous extrusion of two protons from the two amido groups [19].

In the special coordinative environment of the dioxocyclamato(2-) ligand, Cu^{II} can be oxidized at a moderate potential to the Cu^{III} state [20]. The dioxocyclamato(2-) coordinating framework represents a rather unique ligand able to stabilize in aqueous solution the otherwise elusive Cu^{III} cation. This behaviour seems to be ascribed to the fact that the redox change from the Cu^{II} complex (d^9, octahedral) to the Cu^{III} complex (low-spin d^8, square) profits from an extremely large Ligand Field Stabilization Energy contribution and that the dioxocyclamato(2-) system, due to the presence of two negatively charged nitrogen donor atoms (the two deprotonated amido groups), exerts particularly strong coordinative interactions [21].

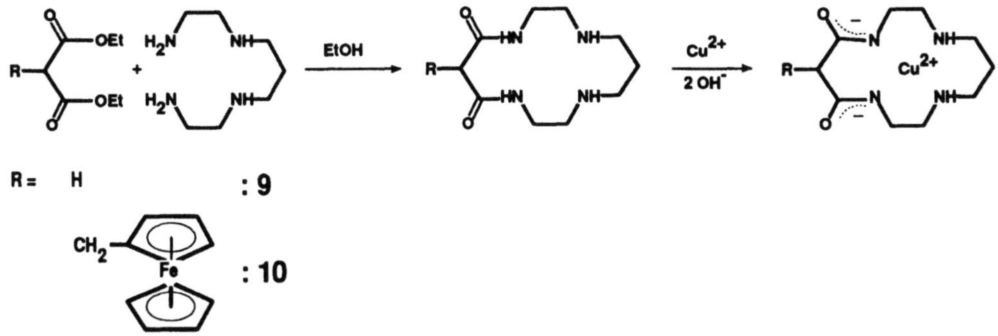

In the present sinthetic approach, the ferrocenemethyl group was first appended at the α-carbon atom of diethylmalonate. Then, ferrocenemethyldiethylmalonate was reacted with the 2.3.2-tet tetramine in refluxing ethanol to give ferrocenemethyldioxocyclam, **10**, H_2L-Fc, the diamine-diamide macrocycle with an appended ferrocenemethyl subunit, as sketched in the Scheme above [22]. Spectrophotometric and pH-metric titrations in an 80/20 v/v MeCN/water solution indicated that after the addition of 2 equivalents of strong base H_2L-Fc extrudes two protons and simultaneously incorporates a Cu^{II} ion, to give the neutral complex [Cu^{II}{L(2-)-Fc}] [22]. This species undergoes a two-electron oxidation process at the platinum electrode, consisting of two consecutive one-electron reversible steps, which are separated by 510 mV. The first step is ascribed to the oxidation of the ferrocene moiety. The second step involves the oxidation of the encircled metal centre. Corresponding $E_{1/2}$ values are indicated in the unidimensional potential diagram in Figure 5.

Interestingly, the potential associated with the Fc/Fc^+ oxidation process in the uncomplexed ferrocenemethyldioxocyclam system, H_2L-Fc, is almost coincident with that observed for the Fc/Fc^+ oxidation process which takes place in the conjugate complex. Thus, dropping a dipositive metal in the adjacent tetra-aza ring does not seem to influence, through electrostatic repulsive effects, the oxidation process involving the ferrocene fragment. This behaviour contrasts with that observed for the ferrocene/metallocyclam conjugate **8**, in which the oxidation of the organometallic moiety was found to be substantially more difficult when a Ni^{II} cation was present in the adjacent macrocyclic ring. However, it should be noted that in the [Cu^{II}{L(2-)-Fc}] complex, the metal centre is merged in a doubly negatively charged coordinative framework, which shields its doubly positive electrical charge and minimizes repulsive electrostatic effects on the close redox change. It should also be noted that the potential value associated with the Cu^{II} oxidation in the [Cu^{II}{L(2-)-Fc^+}]$^+$ complex is substantially more positive (by 100 mV) than that observed for the oxidation of the plain dioxocyclamato(2-) complex,

$[Cu^{II}\{L(2\text{-})\}]$. Thus, whereas the 0 to +1 charge increase taking place on the organometallic moiety is not influenced by the adjacent neutral metallomacrocycle subunit, the 0 to +1 charge increase on the dioxocyclamato fragment is sensitive to the repulsive effect exerted by the proximate monopositive organometallic moiety.

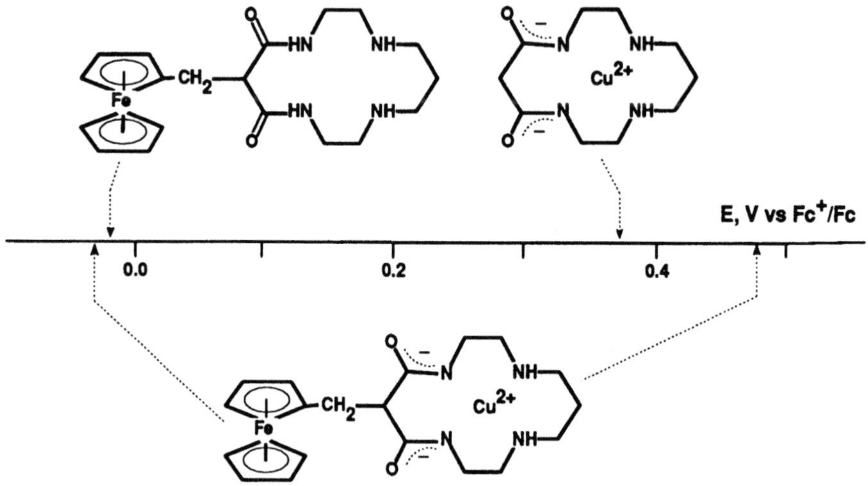

Figure 5. Electrode potentials measured for MeCN solutions of the ferrocene/metallocyclam conjugate system **10** and for related fragments.

5. Conclusions

One hundred years of coordination chemistry have produced many fascinating metal centred systems, redox active through a one-electron redox change, and have taught chemists the secrets of modulating the corresponding electrode potential through synthetic modifications to the coordination framework. Supramolecular science now allows synthetic chemists to build much more complex systems able to exchange a *given* number of electrons, according to a *predetermined* sequence, at *desired* values of potential. Moreover, the use of metal centres as structural elements allows us to control quite easily the topology of the multielectron redox systems, permitting us to place the redox sites in chosen positions, at predetermined distances. For instance, in the cases discussed before, the employed structural metal centres, $>Pt^{II}Cl_2$ and $>Ru^{II}(bpy)_2$, allow us to place two equivalent redox active fragments in the adjacent corners of a square. The use of further metal centres may allow the design of multisite redox systems of varying topologies, which are related to the type of coordination polyhedron which has been chosen. In this sense, the *coordinative* approach seems much more versatile and promising than the *covalent* one, as far as the design of multi-electron redox systems is concerned.

References and Notes

[1] present address: Departamento de Quimica Pura y Aplicada, Universidad de Vigo, Spain.
[2] L. Fabbrizzi, *Comments Inorg. Chem.*, **1985**, *4*, 33.
[3] A. Buttafava, L. Fabbrizzi, A. Perotti, A. Poggi, G. Poli, and B. Seghi, *Inorg. Chem.,* **1986**, *25*, 1456.
[4] G. Wilkinson, M. Rosenblum, M. C. Whiting, and R. D. Woodward, *J. Am. Chem. Soc.*, **1952**, *74*, 2125.
[5] S. P. Gubin, S. A. Smirnova, L. I. Denisovich, and A. A. Lubovich, *J. Organometal. Chem.*, **1971**, *30*, 243.
[6] L. Fabbrizzi, L. Montagna, A. Poggi, T.A. Kaden, and L. Siegfried, *J. Chem. Soc., Dalton Trans.,* **1987**, 2631.
[7] M. Ciampolini, L. Fabbrizzi, A. Perotti, A. Poggi, B. Seghi, and F. Zanobini, *Inorg. Chem.,* **1987**, *26*, 3527.
[8] J. C. Kotz, Topics in Organic Electrochemistry, A. J. Fry and W. E. Britton Eds, Plenum Press, New York 1986.
[9] J.-M. Lehn, *Angew. Chem.Int. Ed. Engl.*, **1988**, *27*, 89.
[10] O. Carugo, G. De Santis, L. Fabbrizzi. M. Licchelli, A. Monichino, and P. Pallavicini, *Inorg. Chem.,* in the press.
[11] D. E. Richardson and H. Taube, *Inorg. Chem.*, **1981**, *20*, 1278.
[12] G. De Santis, L. Fabbrizzi, M. Licchelli, A. Monichino, P. Pallavicini, A. Perotti, and A. Poggi, *Inorg. Chim. Acta*, **1991**, *188*, 1.
[13] M. Ciampolini, L. Fabbrizzi, M. Licchelli, A. Perotti, F. Pezzini, and A. Poggi, *Inorg. Chem.,* **1986**, *25* 4131.
[14] M. Studer and T. A. Kaden, *Helv. Chim. Acta*, **1986**, *69*, 2081.
[15] M. P. Suh and S.-G. Kang, *Inorg. Chem.*, **1988**, *27*, 2544.
[16] L. Fabbrizzi, A. M. Manotti Lanfredi, P. Pallavicini, A. Perotti, A. Taglietti, and F. Ugozzoli, *J. Chem. Soc., Dalton Trans.,* in the press.
[17] A. De Blas, G. De Santis, L. Fabbrizzi, M. Licchelli, and P. Pallavicini, manuscript in preparation.
[18] A. De Blas, G. De Santis, L. Fabbrizzi, M. Licchelli, C. Mangano, and P. Pallavicini, submitted for publication.
[19] L. Fabbrizzi, A. Poggi, and B. Seghi, *Inorg. Synth.,* **1985**, Vol. XXIII, 82.
[20] L. Fabbrizzi and A. Poggi, *J. Chem. Soc., Chem. Comm.,* **1980**, 646.
[21] L. Fabbrizzi, A Perotti, and A. Poggi, *Inorg. Chem.*, **1983**, *22*, 1411.
[22] G. De Santis, L. Fabbrizzi, M. Licchelli, P. Pallavicini, and A. Perotti, to be published.

NEW APPROACHES TO ANION COORDINATION CHEMISTRY - ACYCLIC QUATERNARY POLYBIPYRIDINIUM RECEPTORS DESIGNED TO BIND ANIONIC GUEST SPECIES

P. D. BEER[*a,b], J. W. WHEELER[b] and C. MOORE[c]
[a]Inorganic Chemistry Laboratory
University of Oxford
South Parks Road
Oxford OX1 3QR U.K.

[b]School of Chemistry
University of Birmingham
P.O. Box 363
Birmingham B15 2TT U.K.

[c]Kodak Limited
Headstone Drive
Harrow
Middlesex HA1 4TY U.K.

ABSTRACT. New acyclic redox-active quaternary polybipyridinium receptors L^1-L^3 containing the 5,5'-disubstituted-N,N'-dimethyl-2,2'-bipyridinium moiety have been prepared and isolated as their hexafluorophosphate salts. ^1H and ^{13}C n.m.r. titration investigations in d_6-dimethyl sulphoxide solutions reveal L^1-L^3 to co-ordinate chloride and bromide anionic guest species. Under identical experimental conditions model compounds, N,N'-5,5'-tetramethyl-2,2'-bipyridinium bis(hexafluorophosphate) (9) and N,N'-dimethyl-4,4'-bipyridinium bis(hexafluorophosphate) (10) do not complex these anions. Preliminary electrochemical anion complexation studies show the redox sensitivity of the acyclic polybipyridinium receptors L^1-L^3 towards the chloride anion guest is low, a maximum electrostatic perturbation of the respective receptor's reduction waves of $\Delta E \leq 10mV$ is observed.

Introduction

The molecular recognition of anionic guest species by positively charged or electron-deficient neutral abiotic organic receptor molecules is a relatively new area of chemical investigation [1]. This is somewhat surprising in view of the ubiquitous role anionic substrates play in chemical and biochemical processes. Only a few classes of anion receptor have been reported including Lewis acid - containing ligands [2], ammonium quaternary salts [3], protonated polyamines [4] and guanidines [5]. We have recently reported the first redox-responsive class of anion receptor based on the organometallic redox-active cobalticinium moiety[6] and have shown that simple acyclic derivatives containing the positively charged cobalticinium unit in combination with an amide N-H group are the essential components for the molecular and electrochemical recognition of

V. Balzani and L. De Cola (eds.), Supramolecular Chemistry, 105–118.

anionic guest species [7]. We describe here the syntheses, anion co-ordination and electrochemical studies of a new class of acyclic anion receptor containing the redox-active, pH-independent positively charged quaternary bipyridinium group.

Results and Discussion

Interest in the chemistry of 2,2'-bipyridine and related polypyridine derivatives has grown rapidly in recent years due to their numerous applications in a variety of fields. Examples include effective ligands for coordinating a large diversity of metal ions [8] and the versatile photoactivity of the corresponding metal complexes in electron- and energy-transfer processes [9]. One application that has yet to be exploited is utilising the positively charged redox-active diquaternary 2,2'-bipyridinium moieties [10] as potential *anion* binding sites and redox responsive centres.

SYNTHESIS OF RECEPTOR MOLECULES

The synthesis of a molecule containing two 5,5'-disubstituted-2,2'-bipyridyl units was achieved by the reaction of two equivalents of 5-bromomethyl-5'-methyl 2,2'-bipyridine (1) [11] with N,N'-dimethylethylene diamine (2) in acetonitrile using potassium carbonate as base. Purification by column chromatography on alumina, using CH_2Cl_2/MeOH (99:1) as eluent, and recrystallisation from acetonitrile afforded (3) in 43% yield (Scheme 1). Reaction of (3) with dimethyl sulphate at 75°C for 7 days gave a mixture of products that were not fully methylated. Following conversion to the hexafluorophosphate salts, this mixture was reacted further with methyl iodide in acetonitrile at reflux for 18 days. After this time the product was isolated and converted to the hexa-hexafluorophosphate salt to give L^1 in 11% overall yield (Scheme 1).

The need for the two stage methylation is most likely because the partially quaternised products are not sufficiently soluble in dimethyl sulphate to remain in solution and undergo the final quaternisation reactions. In converting the partially methylated products to the hexafluorophosphate salts solubility in polar solvents is conferred upon them and hence they may then undergo completion of the quaternisation procedure.

Reaction of 5,5'-bis(bromomethyl)-2,2'-bipyridine (4) [11] with a large excess of 4,4'-bipyridine (5) in acetonitrile, followed by conversion of the resultant precipitate to the hexafluorophosphate salt, gave the dicationic compound (6) in 88% yield (Scheme 2). The reaction of (6) with methyl iodide in nitromethane at reflux for 24h produced an orange precipitate which was collected, dissolved in water and converted to the hexafluorophosphate salt to afford the tetra-cationic compound (7) in 89% yield (Scheme 2). Alkylation of (7) with dimethyl sulphate in acetonitrile at reflux for 48h gave initially a white precipitate which was converted to the hexa-cationic hexafluorophosphate acyclic receptor molecule L^2 in 71% yield (Scheme 2).

4,4-Bipyridine (5) was reacted with two equivalents of 5-bromomethyl-5'-methyl-2,2'-bipyridine (1) to produce, on addition of ammonium hexafluorophosphate, the dicationic compound (8) in 78% yield (Scheme 3). Exhaustive methylation of (8) was achieved via alkylation reactions with methyl iodide and subsequently dimethyl sulphate followed finally by conversion to the hexafluorophosphate salt to give the desired hexa-cationic receptor L^3 in 44% overall yield. (Scheme 3). All these new acyclic receptors gave spectroscopic and analytical data in accordance with assigned structures.

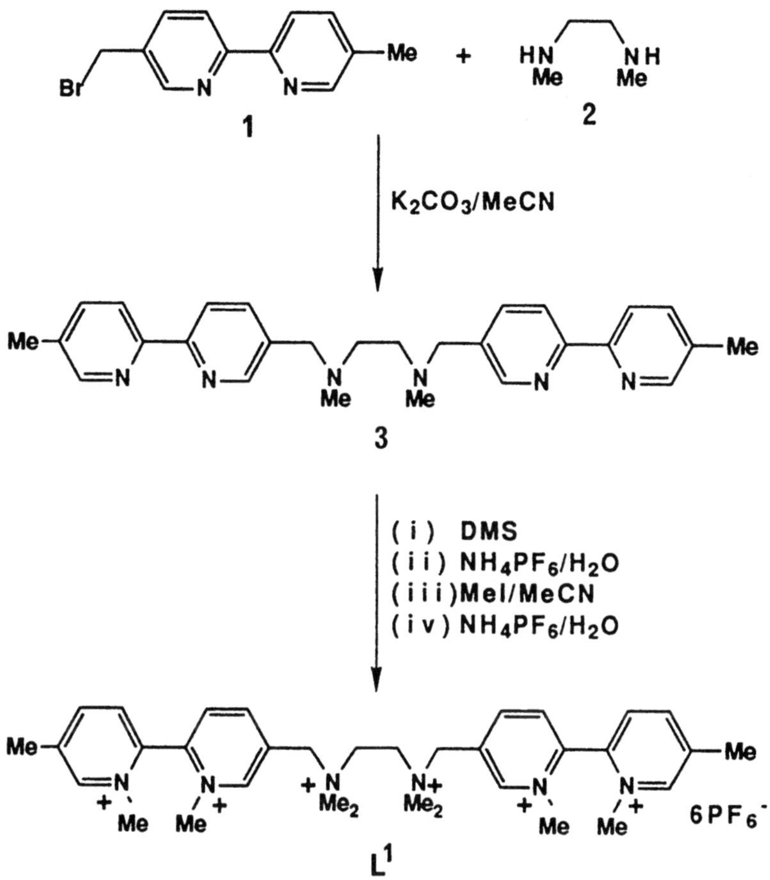
Br
Me
+
HN
NH
Me
Me
1
2
K2CO3/MeCN
Me
N
N
N
N
Me
Me
N
N
Me
3
(i) DMS
(ii) NH4PF6/H2O
(iii) MeI/MeCN
(iv) NH4PF6/H2O
L1
6PF6-

Scheme 1.

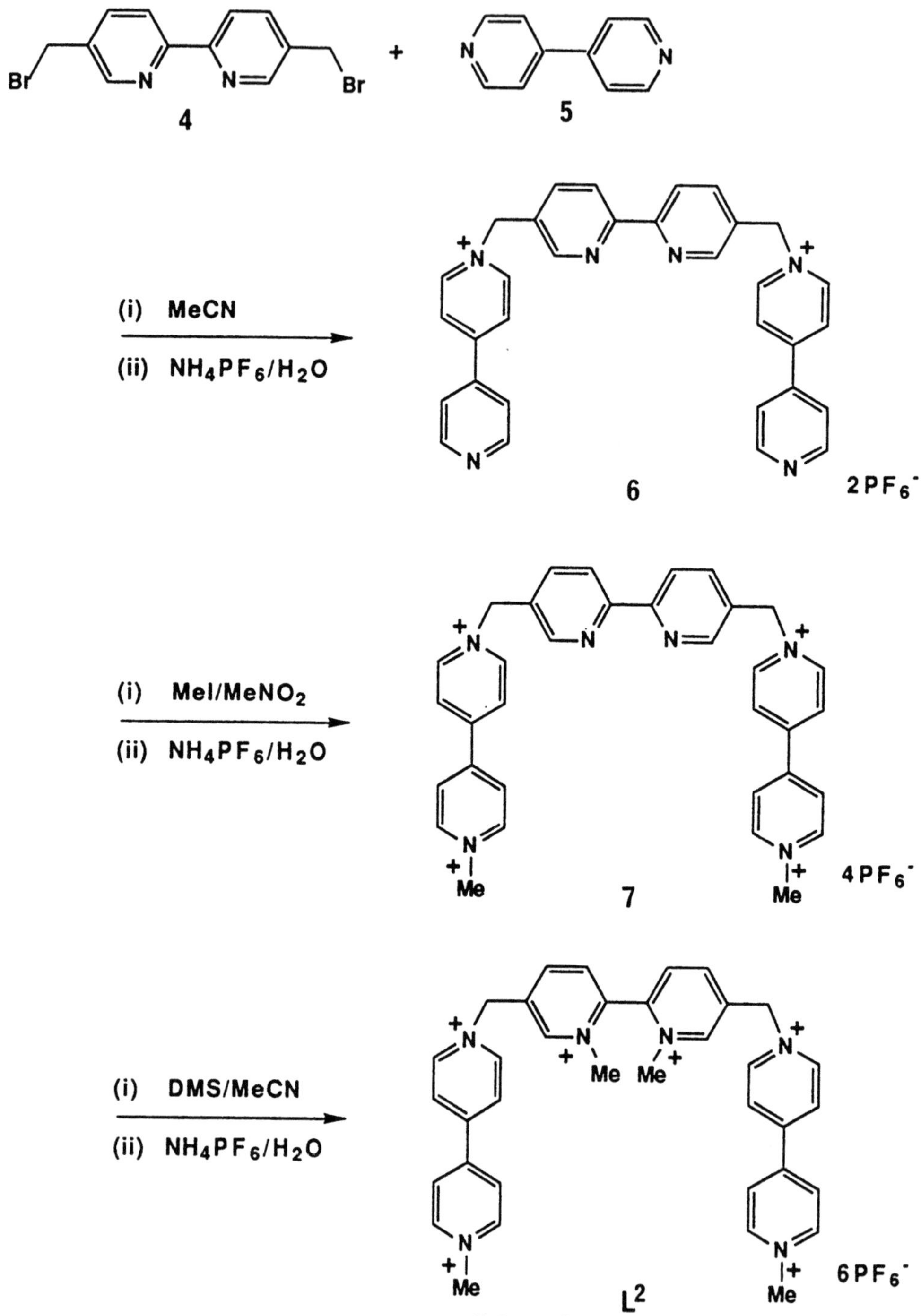
Br
N
N
Br
4
+
N
N
5
(i) MeCN
(ii) NH4PF6/H2O
N+
N
N
N+
N
N
6
2PF6-
(i) MeI/MeNO2
(ii) NH4PF6/H2O
7
4PF6-
Me
(i) DMS/MeCN
(ii) NH4PF6/H2O
L2
6PF6-

Scheme 2.

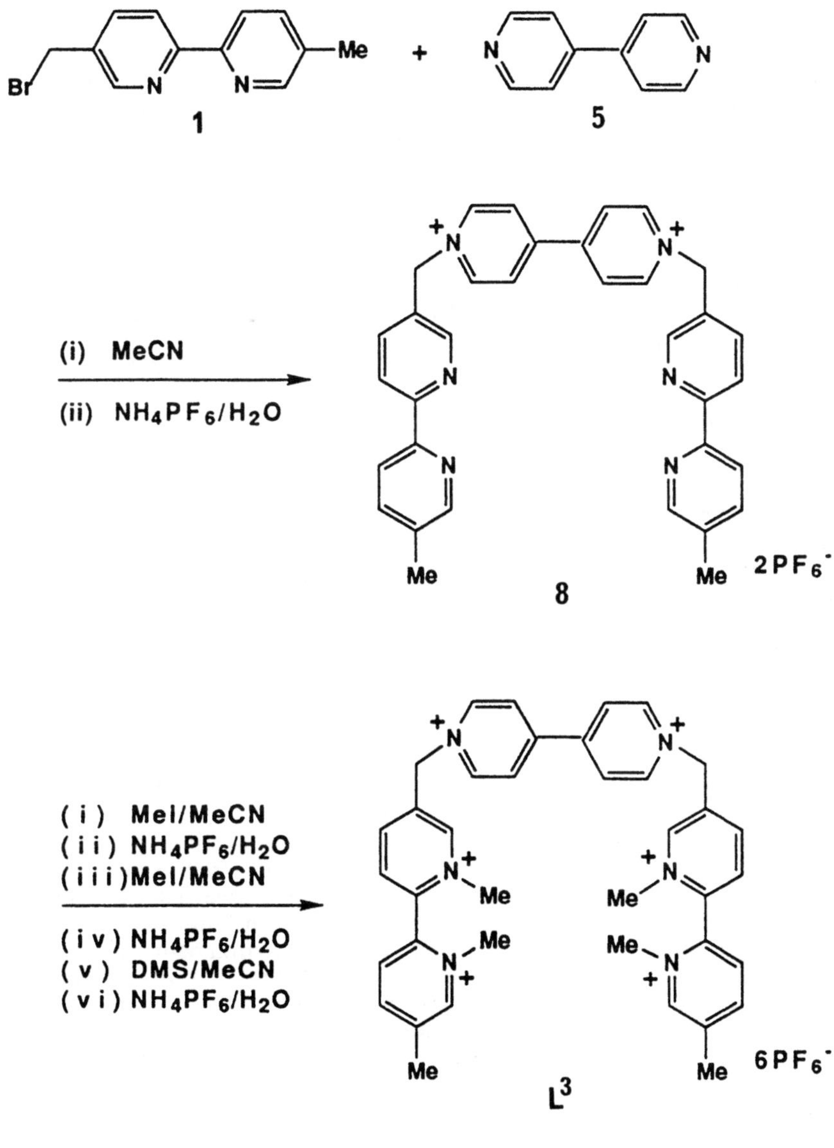
Br
N
N
Me
1
+
N
N
5
(i) MeCN
(ii) NH4PF6/H2O
+N
N+
N
N
N
N
Me
Me
2PF6-
8
(i) MeI/MeCN
(i i) NH4PF6/H2O
(i i i)MeI/MeCN
(i v) NH4PF6/H2O
(v) DMS/MeCN
(v i) NH4PF6/H2O
+N
N+
N+ Me
+N Me
N+ Me
Me N+
Me
Me
6PF6-
L3

Scheme 3.

ANION BINDING STUDIES

The combination of ^{1}H and ^{13}C n.m.r. spectroscopy together with electrochemical cyclic voltammetric experiments were used to investigate the preliminary anion coordination chemistry of the new acyclic polybipyridinium receptors L^1, L^2, L^3 and simple model bipyridinium compounds (9), (10), with halide anions.

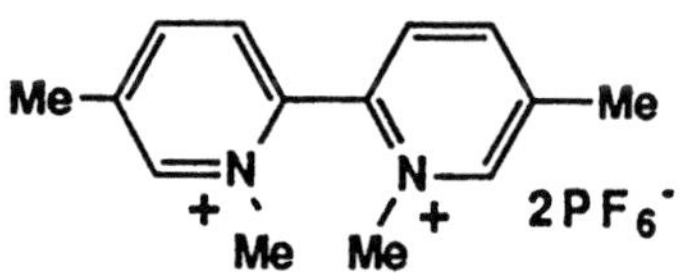

9

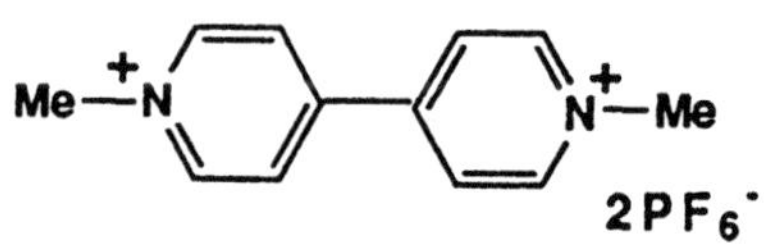

10

In a typical ^{1}H n.m.r. titration experiment the addition of one equivalent of tetra-butyl ammonium chloride to a d_6-dimethyl sulphoxide solution of L^1 led to substantial shifts of the receptor's proton signals (Figure 1). The values of the proton resonances are given in Table 1 together with the observed shifts.

The largest shifts are seen for protons H_3, H_6, $H_{4''}$, H_7 and $H_{10,11}$. As no hydrogen-bonding interactions are expected for the quaternary bipyridinium receptors, these shifts may be attributed to the proximity of the anionic guest perturbing the electrostatic environment of the receptor and also causing alterations to its solution conformation.

Under identical experimental conditions no significant shifts ($\Delta\delta \leq 0.01$ppm) in the respective ^{1}H n.m.r. spectra of model compounds (9) and (10) were observed on addition of chloride anions, implying no anion complexation takes place and that simple anion exchange is not responsible for these $\Delta\delta$ppm observations with L^1 and chloride anion.

A noteworthy feature of the ^{1}H n.m.r. spectrum of the complex formed between L^1 and the chloride anion is that protons H_7, which appear as a sharp singlet in the spectrum of the free receptor (Fig. 1(a)) are observed as an AB system, (see inset of Fig. 1(b)). Clearly this implies some conformational restrictions have been imposed upon the host, causing the two H_7 protons to exist in differing chemical environments, and is further evidence for the formation of a host-guest specie. Figure 2 shows a possible structure of the chloride complex of L^1 with the anionic guest in close proximity to the protons of the host that undergo the largest shifts in the ^{1}H n.m.r. spectrum. (Table 1).

Chloride anion addition to ^{1}H n.m.r. solutions of receptor L^2 results in significant shifts of some of the host's protons (Table 2). Interestingly there are no shifts seen for the signals corresponding to the protons H_{13} and H_{14}, and the shifts for the protons H_9 and H_{12} are very small, suggesting the interaction with the guest anion is taking place primarily close to the central N,N'-dimethyl-2,2'-bipyridinium unit. This is to be expected as in this region the positive charge density is highest, especially if the molecule can arrange itself in a U-shaped conformation as depicted in Figure 3.

The binding studies of the hexa-cationic receptor L^3 were undertaken with both chloride and bromide anionic guests. The ^{1}H n.m.r. data and the resulting shifts observed on addition of each of these anions to a solution of L^3 is given in Table 3. It is noteworthy that the magnitude of the shifts on complexation of chloride anion are approximately three times the size of those produced on interaction with bromide anion. This is a result of the chloride anion having a higher negative charge density than the bromide anion and consequently a greater polarising electrostatic influence on L^3.

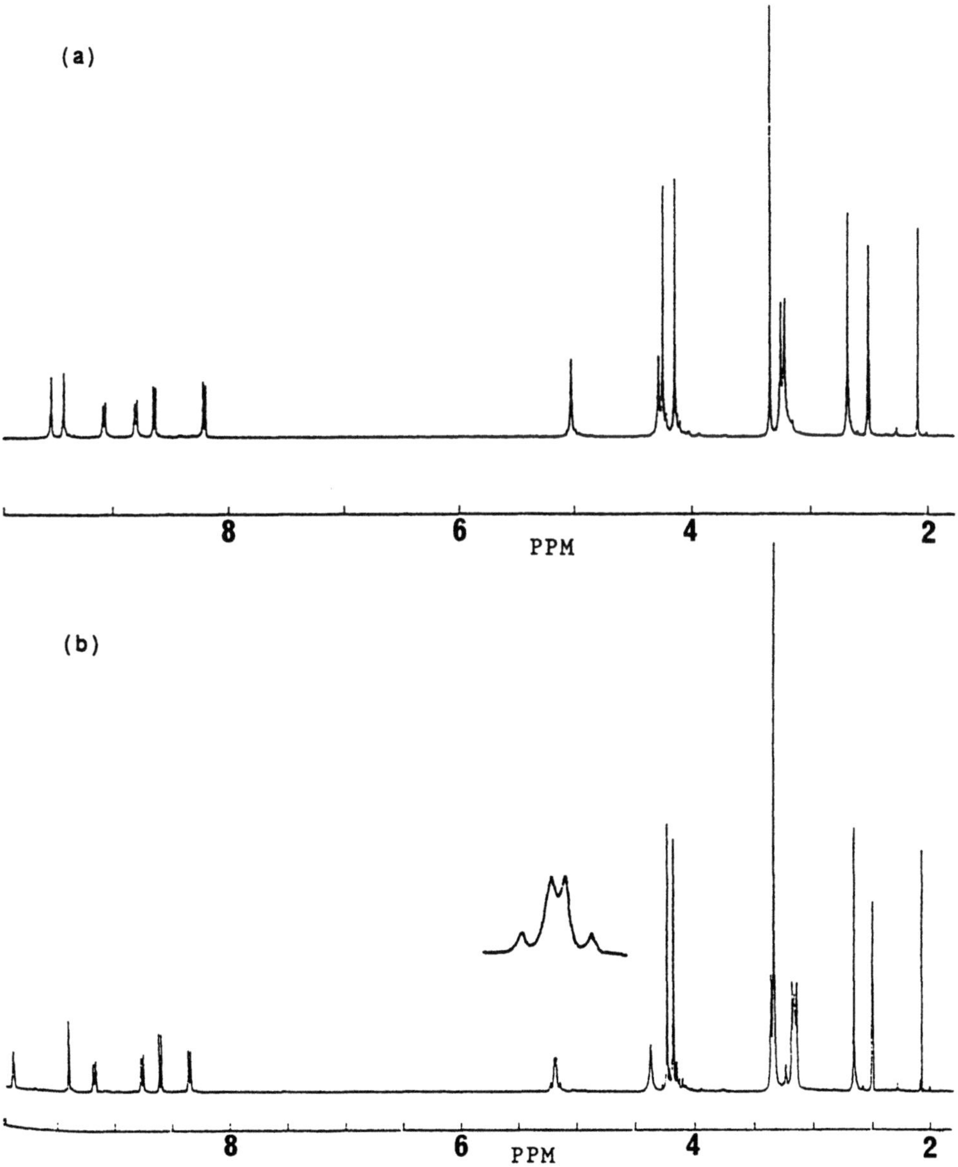

Figure 1 ^{1}H NMR of (a) the free receptor **L^{1}** and (b), on addition of one equivalent of tetra-butyl ammonium chloride (d_6-DMSO, 400MHz)

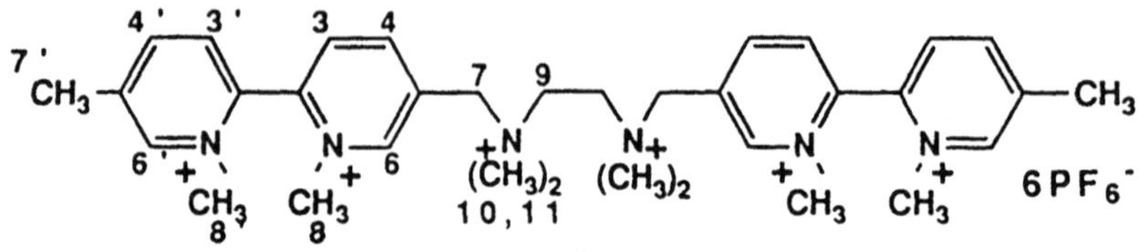

L^1

The numbering scheme for the hexa-cationic receptor

Table 1 ^{1}H NMR spectrum peak values and assignments for the free receptor and on addition of one equivalent of tetra-butyl ammonium chloride (d_6-DMSO, 400MHz)

Proton	**Receptor** (ppm)	**+ 1 equiv Bu_4NCl** (ppm)	$\Delta\partial^a$ (ppm)
H_3	9.07	9.17	+0.10
H_4	8.63	8.60	-0.03
H_6	9.53	9.91	+0.38
$H_{3'}$	8.79	8.76	-0.03
$H_{4'}$	8.19	8.34	+0.15
$H_{6'}$	9.42	9.39	-0.03
H_7	5.02	5.21/5.16	(+0.16)
$H_{7'}$	2.68	2.66	-0.02
H_8	4.24	4.23	-0.01
$H_{8'}$	4.14	4.18	+0.04
H_9	4.27	4.37	+0.10
$H_{10,11}$	3.24/3.21	3.36/3.33	+0.12/+0.12

a - The value for $\Delta\partial$ is the diference between the resonance position of the proton in the free receptor and on addition of one equivalent of tetra-butyl ammonium chloride. A positive number indicates a downfield shift and a negative number an upfield shift.

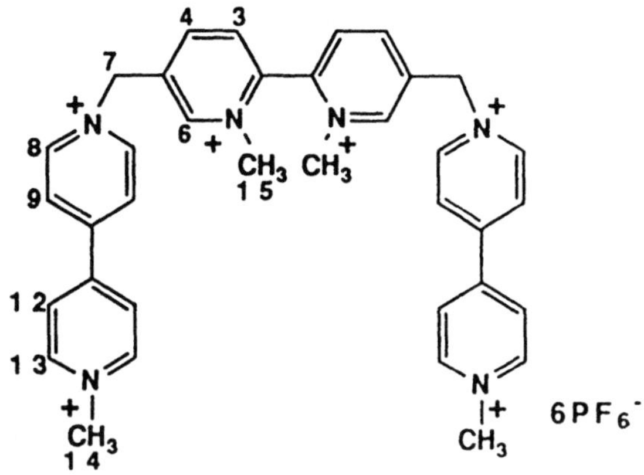

The numbering scheme for the hexa-cationic receptor L^2

Table 2 ^{1}H NMR spectrum peak values and assignments for L^2 and on addition of one equivalent of chloride anions (d_6-DMSO, 400MHz)

Proton	Receptor (ppm)	+ 1 equiv Bu_4NCl (ppm)	$\Delta\partial^a$ (ppm)
H_3	9.07	9.15	+0.08
H_4	8.47	8.57	+0.10
H_6	9.67	9.95	+0.28
H_7	6.30	6.39	+0.09
H_8	9.49	9.69	+0.20
H_9	8.91	8.92	+0.01
H_{12}	8.75	8.79	+0.04
H_{13}	9.31	9.31	-
H_{14}	4.45	4.45	-
H_{15}	4.13	4.19	+0.06

a - The value for $\Delta\partial$ is the diference between the resonance position of the proton in the free receptor and on addition of one equivalent of tetra-butyl ammonium chloride. A positive number indicates a downfield shift and a negative number an upfield shift.

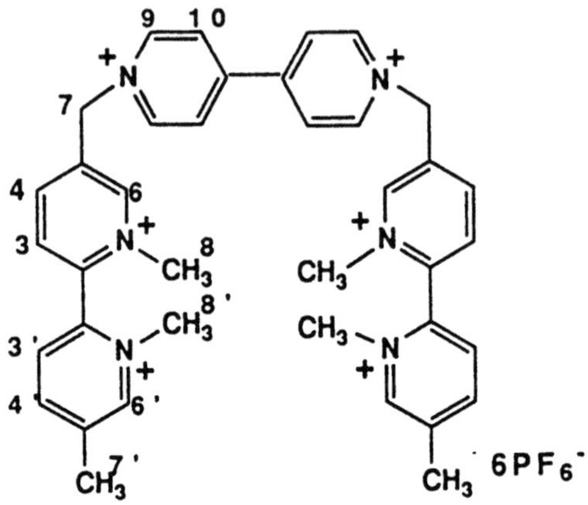

The numbering scheme for the hexa-cationic receptor L^3

Table 3 ^1H NMR spectrum peak values and assignments for L^3 and on addition of one equivalent of chloride anions and one equivalent of bromide ions(d_6-DMSO, 400MHz)

Proton	Receptor (ppm)	+ 1 equiv Bu_4NCl (ppm)	$\Delta\partial^a$ (ppm)	+ 1 equiv BuN_4Br (ppm)	$\Delta\partial^a$ (ppm)
H_3	9.05	9.16	+0.11	9.08	+0.03
H_4	8.52	8.55	+0.03	8.54	+0.02
H_6	9.66	9.98	+0.32	9.76	+0.10
H_7	6.29	6.39	+0.10	6.33	+0.04
$H_{3'}$	8.73	8.72	-0.01	8.72	-0.01
$H_{4'}$	8.25	8.35	+0.10	8.31	+0.06
$H_{6'}$	9.35	9.39	+0.04	9.35	-
$H_{7'}$	2.64	2.63	-0.01	2.63	-0.01
H_8	4.16	4.18	+0.02	4.16	-
$H_{8'}$	4.05	4.10	+0.05	4.08	+0.03
H_9	9.51	9.73	+0.22	9.58	+0.07
H_{10}	8.87	8.96	+0.09	8.91	+0.04

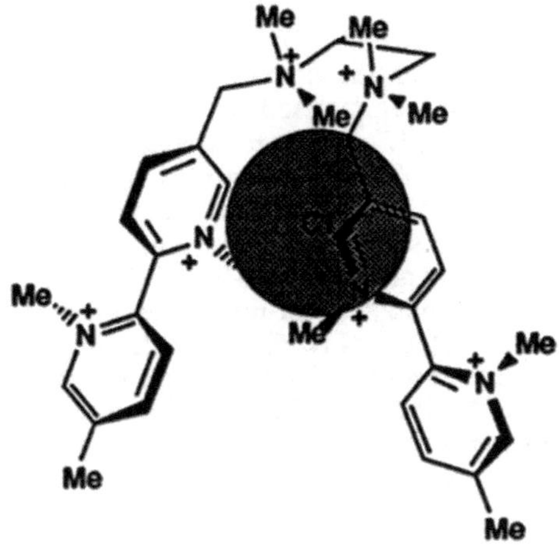

Figure 2 A possible form for the chloride complex of the receptor L^1

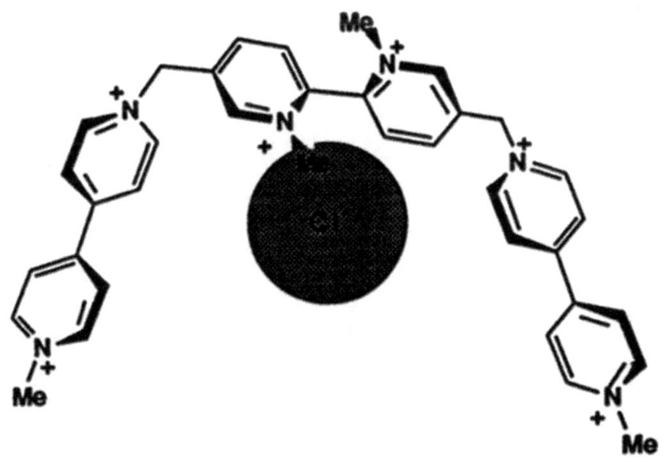

Figure 3 A possible form for the chloride complex of the receptor L^2

Analogous anion complexation results were observed with ^{13}C n.m.r. titration studies. Significant shifts of the respective carbon resonances of L^1, L^2, L^3, of up to 1.11 ppm were seen on addition of chloride anions to d_6-DMSO solutions of all three receptors. No shifts were observed with model compounds (9), (10) in agreement with ^1H n.m.r. titration investigations described previously.

The electrochemical properties of (7) L^1, L^2, L^3 were investigated using cyclic voltammetry in dimethyl sulphoxide containing 0.2M [$^nBu_4N^+$][PF_6^-] as supporting electrolyte and the results are summarised in Table 4. It is noteworthy that although compound (7) and receptors L^2, L^3 exhibit reversible electrochemical reduction processes by virtue of the redox-active 4,4'-bipyridinium moiety, the redox chemistry of L^1 containing only N,N'-dimethyl-2,2'-bipyridinium groups is not generally well behaved. The cyclic voltammogram of L^2 is shown in Fig.4. The waves for the quasi-reversible 4,4'-bipyridinium group occur at potentials of -0.38V and -0.64V.

At more cathodic potentials another quasi-reversible wave (-0.82V) that may be attributed to a one-electron reduction of the N,N'-dimethyl-2,2'-bipyridinium moiety is observed together with an irreversible second reduction redox process at -1.29V. The irreversible redox behaviour of simple N,N'-dialkyl-2,2'-bipyridinium molecules has previously been attributed to the reduced species possessing a non-planar conformation which can undergo an intramolecular chemical reaction [10].

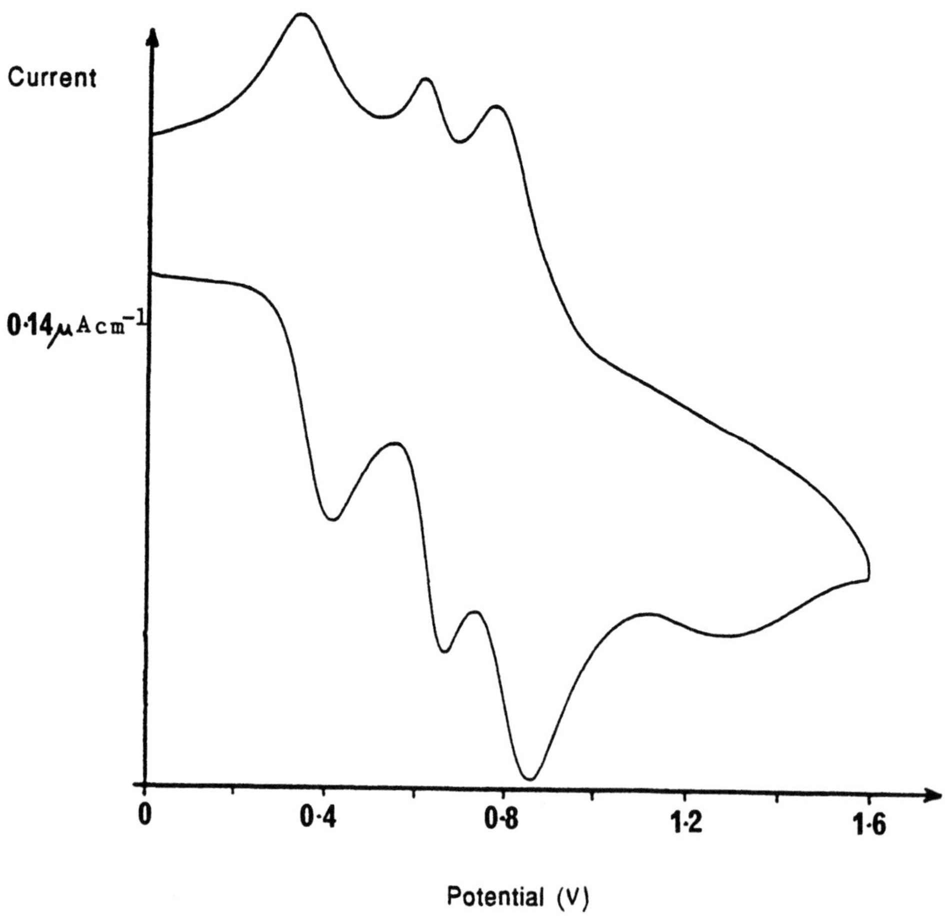

Figure 4 Cyclic voltammogram of L^2

TABLE 4. Electrochemical data for L^1-L^3

Compound	E_1(V) (ΔEp_1(mV)[a])	E_2(V) (ΔEp_2(mV)[a])	E_3(V) (ΔEp_3(mV)[a])	E_4(V) (ΔEp_4(mV)[a])
L^1 [b]	-0.69	-0.85	-1.225	-
(7)[c]	-0.43	-0.79	-	-
L^2 [d]	-0.38 (75)	-0.64 (55)	-0.82 (90)	-1.29
L^3 [e]	-0.29 (100)	-0.62 (100)	-0.82	-1.31

[a] Separation between the cathodic and anodic peak potentials.
[b] Single-wave two-electron process followed by two two-electron irreversible reduction processes. [c] Two-wave four-electron quasi-reversible reduction process. [d] Two-wave four-electron quasi-reversible reduction process, single wave one-electron quasi-reversible reduction process and a one electron irreversible reduction. [e] Two-wave four-electron quasi-reversible reduction process and two one-electron irreversible reduction processes.

The addition of chloride anions to electrochemical DMSO solutions of L^1, L^2, L^3 disappointingly had only small effects on the respective reduction waves. Perturbations of only $\Delta E \leq 10$mV were found suggesting under these experimental conditions the redox-active hosts are not sensitive enough to electrochemically recognise the chloride anionic guest.

In conclusion the novel acyclic polybipyridinium receptors containing the 5,5'-disubstituted-N,N'-dimethyl-2,2'-bipyridinium moiety have been shown from ^{1}H and ^{13}C n.m.r. titration studies to complex chloride and bromide guest anions. Thus they represent a new class of anion coordinating ligand. Although preliminary electrochemical anion complexation studies reveal the redox sensitivity of this type of acyclic receptor towards the chloride anion guest in DMSO is low, future work is being directed at constructing macrocyclic polybipyridinium analogues in an effort to enhance both the host-guest anion interaction and redox-responsive behaviour.

Acknowledgements

We thank the SERC for a CASE studentship to J.W.W., Kodak Limited for additional financial support, and the SERC for use of the Mass Spectrometry Service of University College Swansea and the high field n.m.r. service at the University of Warwick.

References

[1] For reviews of anion coordination chemistry see (a) Pierre J.-L. and Baret, P. (1983) Bull. Soc. Chim. Fr. 367; (b) Dietrich B. in Inclusion Compounds, vol. 2. Atwood, J.L. Davies J.E.D. and MacNicol D.D. (1984) eds. Academic Press, New York, 337; (c) Schmidtchen, F.P. (1988) Nachr. Chem. Tech. Lab., 36, 8.

[2] (a) Katz, H.E. (1987) Organometallics 6, 1134; (b) Wuest, J.D. and Zacharie B. (1987) J. Am. Chem. Soc. 109, 4714; (c) Newcomb, M. Horner, J.H. and Blandon, M.T. (1987) *ibid* 109, 7878; (d) Jung M.E. and Xia, H. (1988) Tetrahedron Lett 29, 297.

[3] (a) Schmidtchen, F.P. (1977) Angew Chem. In Edit 16, 720; (b) Schmidtchen, F.P. Gleich, A. and Schummer, A. (1989) Pure and Applied Chem. 61, 1535 and references therein.

[4] (a) Park, C.H. and Simmons, H.E. (1968) J. Am. Chem. Soc., 90, 2431; (b) Dietrich, B. Hosseini, M.W. and Lehn, J.-M. (1981) *ibid.* 103, 1282; (c) Lehn, J.-M. and Heyer D. (1986) Tetrahedron Lett. 27, 5869; (e) Hosseini, M.W. and Lehn, J.-M. (1986) Helv. Chim. Acta 69, 587.

[5] (a) Dietrich, B. Fyles, T.M. Lehn, J.-M. Pease, L.G. and Fyles, D.L. (1978) J. Chem. Soc. Chem. Commun. 934; (b) Schmidtchen, F.P. (1989) Tetrahedron Lett. 30, 4493; (c) Echararren, A. Gatan, A. Lehn, J.-M. and de Mendoza, J. (1989) J. Am. Chem. Soc., 111, 4994.

[6] Beer, P.D. and Keefe, A.D. (1989) J. Organometal. Chem. 375, C40.

[7] Beer, P.D. Hesek, D. Hodacova, J. and Stokes, S.E. (1992) J. Chem. Soc. Chem. Commun. in press.

[8] Creutz, C. (1982) Comments Inorg. Chem. 1, 293.

[9] Juris, A. Barigelletti, F. Campagna, S. Valzani, V. Belser, P. von Zelewsky, A. (1988) Coord. Chem. Rev. 84, 85

[10] Summers, L.A. (1980) 'The Bipyridinium Herbicides' Academic Press, New York.

[11] Rodriguez-Ubi J.-C. Alpha B. Plancherel D. and Lehn, J.-M. (1984) Helv. Chim. Acta 64, 2264.

SUPRAMOLECULAR CYCLOPHANE CHEMISTRY

FRANCOIS DIEDERICH
Department of Chemistry and Biochemistry
University of California at Los Angeles
Los Angeles, California 90024-1569, U.S.A.

ABSTRACT. This account illustrates the rich supramolecular chemistry of cyclophanes with preorganized apolar cavity binding sites. These artificial receptors, which are constructed via relatively short synthetic routes, form structurally well defined inclusion complexes with a diversity of organic substrates. Among the molecules that form stable liquid phase complexes are flat arenes, paracyclophanes, steroids, and nucleosides. Dispersion and aromatic-aromatic interactions in addition to solvophobic forces provide the major driving forces for the binding of neutral solutes. Comprehensive studies have provided unique insight on the individual molecular level into solvent effects on tight apolar complexation. For enantioselective binding in liquid phase, 1,1'-binaphthyl units and unnatural alkaloids are incorporated as chiral spacers into optically active cyclophanes. The functionalization of the receptors with coenzyme derivatives yields efficient supramolecular catalysts. Porphyrin-cyclophanes show cytochrome P-450 monooxygenase activity and epoxidize reactive aromatic hydrocarbons in protic solvents. In a synthetically useful process, aromatic methyl esters are prepared efficiently and selectively over aliphatic methyl esters by electrochemical oxidation of aldehydes mediated by biscoenzyme - flavin and thiazolium-cyclophane - catalysis at an extraordinarily low potential in methanol.

1. Introduction

Cyclophanes ('bridged aromatic compounds') represent a central class of synthetic receptors in molecular recognition studies [1]. All types of substrates, from inorganic and organic cations and anions to neutral molecules, have been complexed by tailor-made cyclophanes. Following the landmark paper by *Koga* et al. in 1980 [2], who reported the first unambiguous evidence for stoichiometric inclusion of an apolar guest in the cavity of a synthetic cyclophane host in both aqueous solution and the solid state, we started a broad program to develop and comprehensively explore the supramolecular chemistry of these receptors. Much of our work over the past ten years has been the subject of previous reviews which were written with a particular focus on the thermodynamics of complexation [3-5]. In a different approach, this account illustrates the main synthetic routes to the various types of macrocyclic receptors and catalysts that have been developed in our laboratories. In addition, molecular structure - supramolecular activity correlations should reveal advantages and disadvantages of the different systems and show the perspectives for relevant new developments.

V. Balzani and L. De Cola (eds.), Supramolecular Chemistry, 119–136.

2. Basic Tetraoxa[n.1.n.1]paracyclophane Receptors: Efficient Binders of Benzene and Naphthalene Substrates in Protic Solvents.

Some of the simplest cyclophanes, e.g. **1** - **6**, which we prepared in our laboratory [6-10] (Scheme 1) have also been among the most efficient ones in molecular recognition studies.

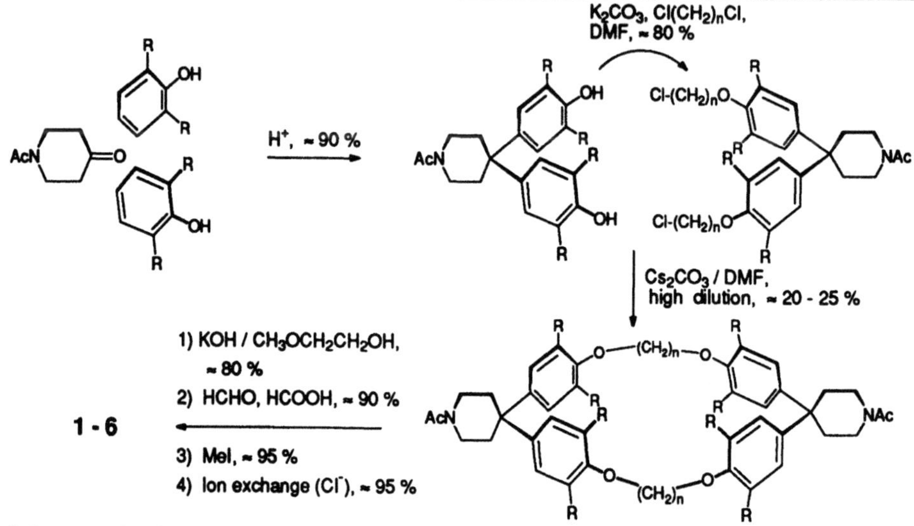

Scheme 1: Synthesis of basic tetraoxa[n.1.n.1]paracyclophane receptors.

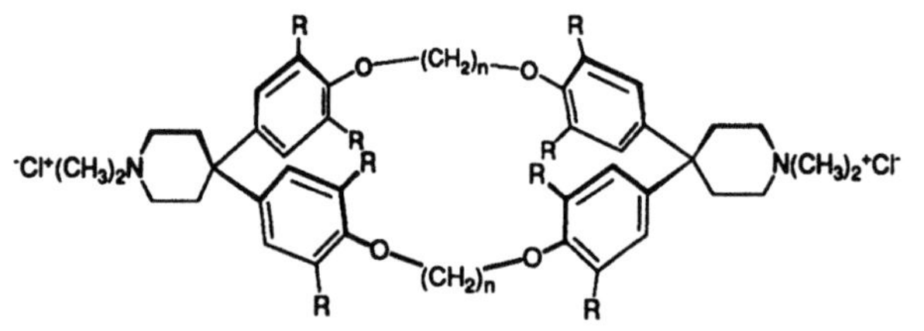

	n	R
1	2	H
2	3	H
3	4	H
4	4	CH_3
5	3	OCH_3
6	4	OCH_3

In the productive binding conformation, cyclophanes **1** - **6** adopt the shape of a rectangular cavity with four electron-rich aromatic rings as its walls. This is shown in the X-ray crystal structure of a benzene complex formed by the bis(*N*-methylpiperidine)-precursor to **3** (Figure 1) [10] and the computer-generated structure of the **5**·*p*-xylene complex (Figure 2) [5]. ^1H NMR binding studies showed that, contrary to expectations based on CPK molecular model examinations, the size of the cavity in **1** is too small to incorporate a benzene derivative [11]. The other receptors **2** - **6** all form stable inclusion complexes with benzene substrates, and the larger macrocycles **3**, **4**, and **6** also bind naphthalene derivatives.

Meaningful ^1H NMR titrations [12] to determine free energies for stoichiometric host-guest complexation in water need to be executed below the critical aggregation concentration (cac) of the binding partners [13]. Therefore, it is desirable that the hosts show high cac's, i.e. molecular-dispersed solubility in concentration ranges suitable for NMR spectroscopy. It was gratifying to find that the methoxy groups in **5** and **6** dramatically reduce the aggregation tendency of [n.1.n.1]paracyclophanes [7]. The cac's of these macrorings ($\geq 1 \times 10^{-2}$ mol L^{-1}) are much higher than those of derivatives **3** (1.6

x 10^{-4} mol L^{-1}) [11] and **4** (< 2 x 10^{-5} mol L^{-1}), which allowed to perform comprehensive ^{1}H NMR complexation studies with the methoxy-substituted derivatives in pure D_2O.

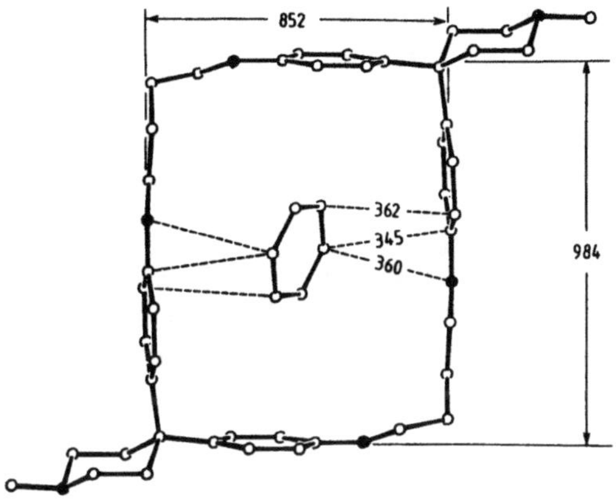

Figure 1. X-ray crystal structure of a benzene complex with the bis(*N*-methylpiperidine)-precursor to **3**

Figure 2. Computed lowest energy geometry of the *p*-xylene complex of cyclophane **5**.

All aromatic guests adopt an axial-type inclusion geometry in the cavities of **2** - **6**. This geometry allows highly solvated guest substituents to orient into the solution, which minimizes any unfavorable desolvation of these functional groups in the complexes (Figure 3). ^{1}H NMR studies strongly suggest that the geometries of benzene and naphthalene complexes with **2** - **6** in the liquid phase resemble the geometry of the *p*-xylene complex in Figure 2. This orientation allows the aromatic guests to undergo both attractive π-π stacking and dipolar edge-to-face interactions with the four cavity walls. Since the aromatic rings of the host are electron-rich, both interactions should become more energetically favorable with increasing electron affinity of the guest.

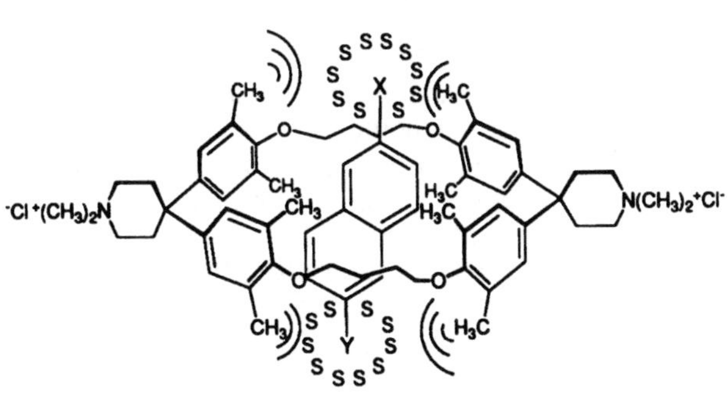

Figure 3. The polar functional groups X and Y of axially included arene guests are favorably oriented into the solution but their solvation can be sterically hindered by substituents at the aromatic rings of the host.

The methoxy-cyclophanes **5** and **6** form stable complexes with neutral benzene derivatives in water (association constants $K_a \approx 10^2 - 10^5$ L mol^{-1} at 293 K) [7,8]. Calorimetric investigations showed that these tight inclusion processes are enthalpically driven (negative ΔH^o) and that the enthalpic gain is partially compensated by a loss in entropy (negative $T\Delta S^o$) [9]. For example, the incorporation of *p*-dinitrobenzene into the apolar cavity of **6** is characterized by the thermodynamic quantities $K_a = 7.8 \times 10^3$ L mol^{-1}, $\Delta G^o = -5.22$ kcal mol^{-1}, $\Delta H^o = -9.8 \pm 0.2$ kcal mol^{-1}, $T\Delta S^o = -4.6$ kcal mol^{-1}, and $\Delta C_p{}^o = -40$ cal K^{-1} mol^{-1}. Thermodynamic data of similar sign have also been measured for other biotic and abiotic tight binding processes in water which involve small apolar substrates [5].

It was initially expected that electron-accepting guests would form the more stable complexes with the electron-rich cyclophane hosts **4** - **6** [14]. However in water, we did not find a good correlation between the binding free energy and the electronic character of the guests. For example, *p*-dimethoxybenzene ($\Delta G^o = -5.38$ kcal mol^{-1}) and *p*-dicyanobenzene ($\Delta G^o = -5.23$ kcal mol^{-1}) form complexes of similar stability with **6**. Specific substituent solvation effects in the complexes, as illustrated in Figure 3, apparently are particularly strong in water and mask the contributions from the aromatic π-π and edge-to-face interactions to the overall complex stability [8]. In contrast, electron donor-acceptor (EDA) interactions generally control the relative stability of cyclophane-arene inclusion complexes in CD_3OD and $(CD_3)_2SO$, where solvation effects are less important than in the more polar solvent water. In the organic environments, electron-deficient benzene and naphthalene guests form the most stable complexes with the electron-rich cyclophanes. For example, the complex of 2,6-dicyanonaphthalene with **4** in CD_3OD is 1.0 kcal mol^{-1} more stable that the complex of 2,6-dimethoxynaphthalene in the same solvent.

EDA interactions also control the relative stability of arene inclusion complexes in water if complexation does not induce steric hindrance to the solvation of guest substituents. The relative stability of complexes between host **2** and benzene derivatives in water follows the sequence expected on the basis of an EDA model [8]. In contrast to the methyl and methoxy groups in **4** - **6**, the small hydrogen atoms on the aromatic rings of **2** do not prevent substituents of axially included guests from regaining their full solvation in the complexes.

3. Complexation, Solubilization, and Transport of Polycyclic Aromatic Hydrocarbons in Water.

In 1983, we reported the synthesis of cyclophane **7** via a macrocyclization leading to the formation of four aryl ether bonds (Scheme 2) [15,16]. With its four quaternary nitrogens, this system shows high monomeric water solubility (cac $\approx 7.5 \times 10^{-3}$ mol L^{-1}) which allowed detailed investigations of the complexation of polycyclic aromatic hydrocarbons (PAH's) in aqueous solution [13,17]. These substrates are extremely insoluble in water but, upon host-guest complexation, the amount of PAH that can be solubilized is dramatically increased. As an example, the solubility of pyrene in water at 293 K is only 8×10^{-7} mol L^{-1}. In contrast, with a 5.5×10^{-3} molar solution of **7**, a complex solution can be prepared by solid-liquid extraction which contains a total pyrene concentration of 2.8×10^{-3} mol L^{-1}. The complexes formed between **7** and PAH's of geometric complementarity are highly stable: binding free energies for the perylene and pyrene complexes were measured as $\Delta G^o = -9.6$ kcal mol^{-1} ($K_a = 1.6 \times 10^7$ L mol^{-1}) and -8.4 kcal mol^{-1} ($K_a = 1.8 \times 10^6$ L mol^{-1}), respectively. The large PAH's like pyrene and fluoranthene are preferentially located in the cavity of **7** within a specific plane which is perpendicular to the mean plane of the host and passes through the two spiro carbons of the two diphenylmethane units (Figure 4) [18].

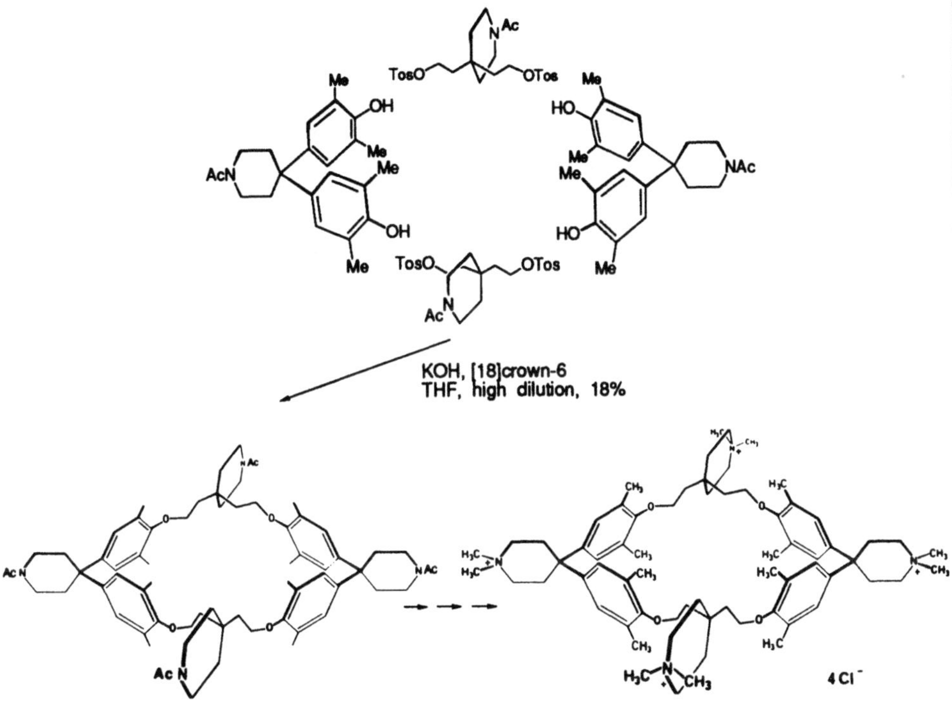

7

Scheme 2: Synthesis of cyclophane **7**, a receptor for polycyclic aromatic hydrocarbons.

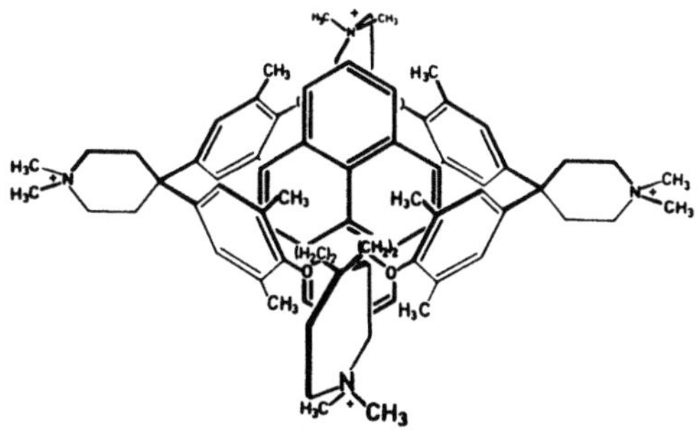

Figure 4. Schematic drawing of the highly favored geometry of the **7**·pyrene complex in aqueous solution.

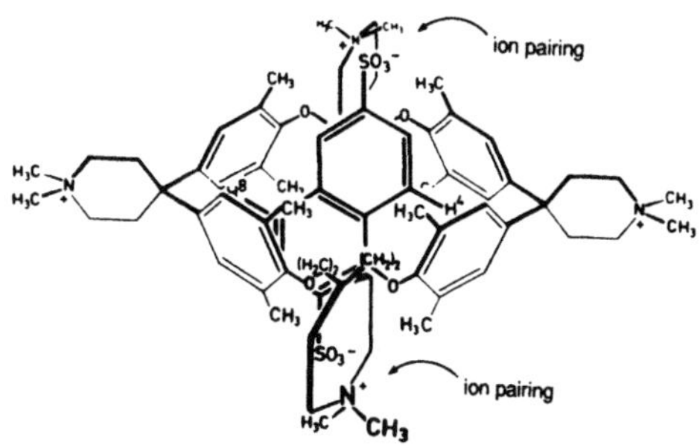

Figure 5. Schematic drawing of the geometry of the **7**·2,6-naphthalene-disulfonate complex.

A remarkable guest selectivity was observed with naphthalene derivatives bearing ionic functional groups [13,16]. The complexes of naphthalenemono- and disulfonates are stabilized by apolar binding interactions in the cavity along with attractive Coulombic interactions between the anionic residues of the guests and the quaternary nitrogen atoms of the piperidinium rings attached to the aliphatic bridges of the host. This is shown in Figure 5 for the **7**·2,6-naphthalenedisulfonate complex ($K_a > 10^6$ L mol^{-1}, $\Delta G^o < -8.0$ kcal mol^{-1} at 293 K). Accordingly, Coulombic interactions severely destabilize the complexes of **7** and cationic naphthalene derivatives. Protonated 1,5-bis(dimethylammonium)naphthalene no longer forms a stable complex ($K_a < 10$ L mol^{-1}, $\Delta G^o > -1.3$ kcal mol^{-1}) which shows that, for a series of structurally very similar guests, additional charge-charge interactions lead to differences in binding free energy $\Delta(\Delta G^o)$ of more than 6.5 kcal mol^{-1}.

Cyclophane **7** was found to be an excellent carrier for the transport of lipophilic arenes through an aqueous phase along a concentration gradient [13,17]. In U-type cell experiments, in which the relative rate of transport of pyrene from one hexane phase (source) into a second hexane phase (receiving) through an aqueous phase was measured, the transport of the arene was considerably accelerated over the rate for passage through pure water. For example, the rate of transport of pyrene (0.01 M in source phase) through a 5 x 10^{-3} molar solution of **7** showed a 2100-fold acceleration over the transport through pure water. High selectivity was observed in the acceleration of the arene transport mediated by **7**. The highest acceleration factors were measured for arenes like pyrene and fluoranthene which form the most stable complexes with **7** in water and have very low distribution constants $K_{dist} = [\text{Guest}]_{H2O}/[\text{Guest}]_{hexane}$.

4. Solvent Dependency of Tight Apolar Complexation Processes.

With **8**, prepared via two macrocyclizations (Scheme 3), we obtained a spherical receptor with the remarkable property of being soluble in solvents of all polarity [19,20]. This unique property allowed us to evaluate in a quantitative way the nature of solvent effects on apolar molecular recognition. When we studied the stability of the pyrene complex **9** in solvents that span the entire polarity range, we found a dramatic solvent dependency [21]. Upon changing from water, the most polar, to carbon disulfide, the least polar among 18 solvents, binding free energies decreased from $-\Delta G^o = 9.4$ kcal mol^{-1} to $-\Delta G^o = 1.3$ kcal mol^{-1}. Apolar binding strength increases regularly from apolar to dipolar aprotic solvents, to polar protic solvents, and to water. The empirical solvent polarity parameter $E_T(30)$ is useful for predicting the strength of apolar host-guest complexation in different solvents. A strong linear free energy relationship exists between the free energy for formation of complex **9** and the solvent polarity parameter $E_T(30)$ of the various solvents including water [21,22].

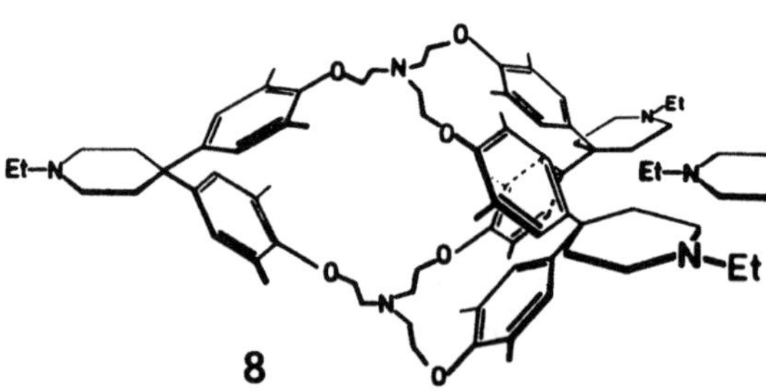

8

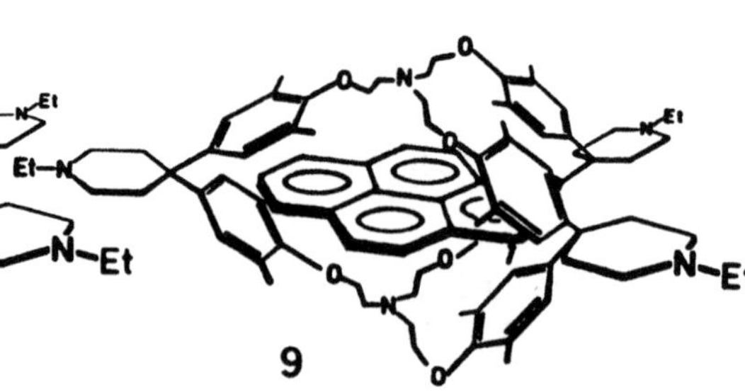

9

A calorimetric study revealed the thermodynamic characteristics for the formation of complex **9** in the various solvents [9,23]. The formation of **9** is enthalpically driven in all solvents, and complexation entropies are mostly unfavorable. Complexation in protic solvents exhibits the largest exothermicity and, in general, the enthalpic driving force decreases from polar protic, to dipolar aprotic, and to apolar solvents. Correspondingly, the complexation entropy becomes increasingly less favorable as the exothermicity increases, resulting in a strong isoequilibrium relationship.

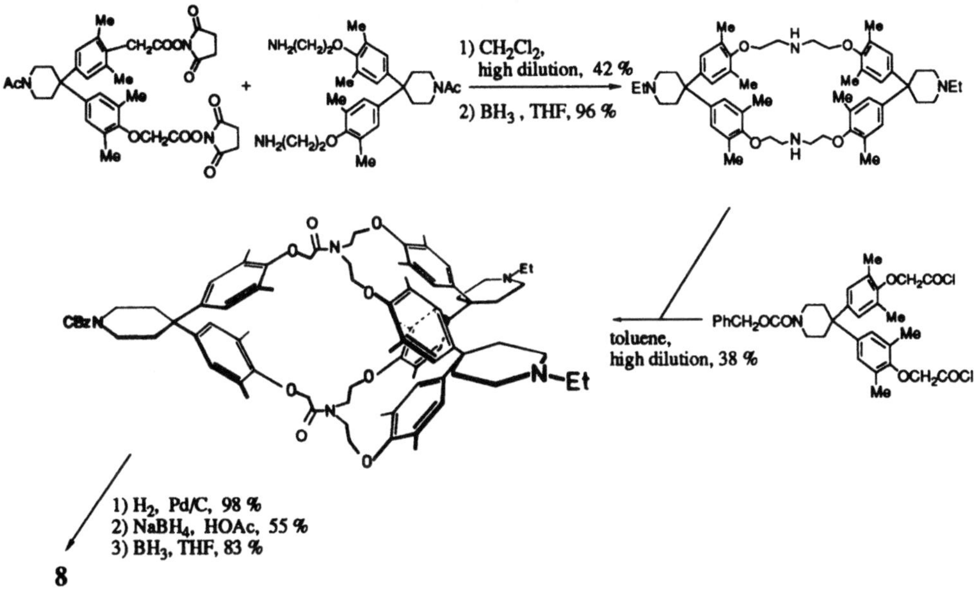

Scheme 3: Synthesis of the macrobicyclic receptor **8** which forms complexes with extended polycyclic aromatic hydrocarbons in solvents of all polarity.

5. Steroid Inclusion Complexation in Aqueous Solution

Cyclophane inclusion complexation is not limited to aromatic substrates. Following the synthetic pathway outlined in Scheme 4, we prepared macrocycle **10** capable of incorporating steroids into its large, wide open cavity [24,25]. ^1H NMR complexation studies in D_2O/CD_3OD (1:1, v/v) showed that bile acids, corticoids, and testosterone all prefer an axial inclusion geometry (Figure 6). In the series of structurally similar bile acid derivatives, high binding selectivity was observed. At 293 K, the complex of lithocholic acid ($K_a = 7075$ L mol^{-1}, $\Delta G^o = -5.18$ kcal mol^{-1}) is much more stable than the complex of desoxycholic acid ($K_a = 250$ L mol^{-1}, $\Delta G^o = -3.21$ kcal mol^{-1}) which has an additional hydroxy group at C(12α). According to the ^1H NMR analysis, the C and D rings of lithocholic acid are preferentially encapsulated by **10** (Figure 6). This generates a large number of favorable contacts between the apolar surfaces of the rings and the cavity walls. A similar orientation of desoxycholic acid in the cavity would require considerable desolvation of the hydroxy group at C(12α) since it would be located deeply inside the apolar cavity. Apparently this is too costly, and inclusion occurs in a different geometry to minimize the energetically unfavorable desolvation of the hydroxy group. Desoxycholic acid is preferentially encapsulated with ring B which positions the hydroxy group at C(12α) more outside the cavity. The solvation-induced changes in inclusion geometry therefore destabilize the complex of desoxycholic acid by ≈ 2 kcal mol^{-1} as compared to the complex of lithocholic acid.

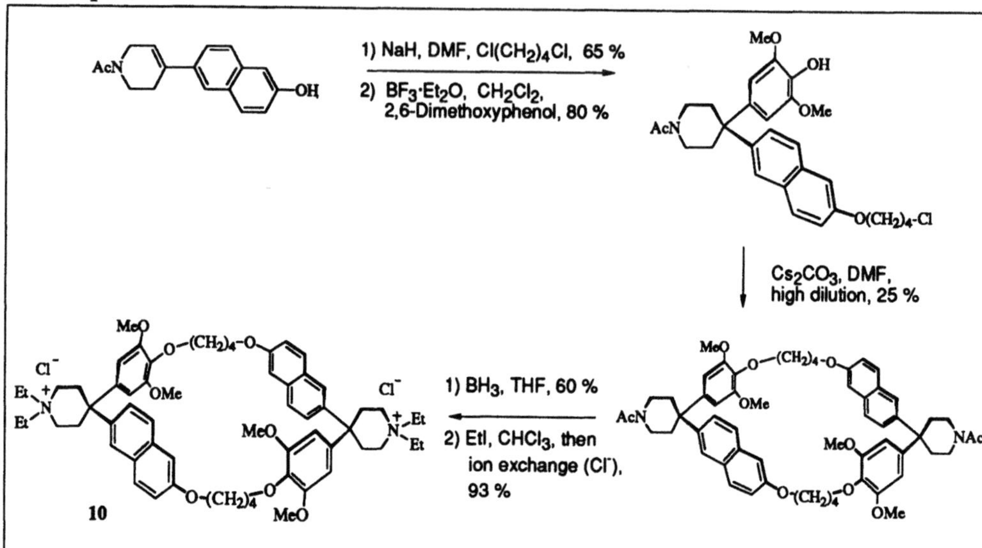

Scheme 4: Synthesis of the steroid receptor **10**.

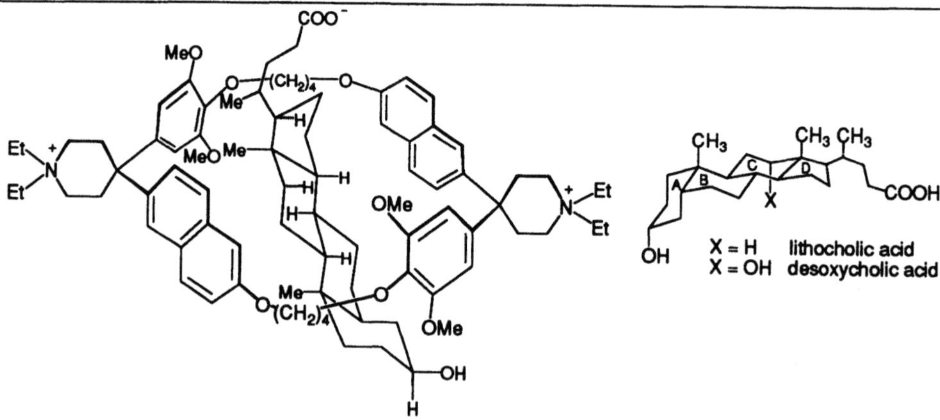

Figure 6. Preferred inclusion geometry of lithocholic acid in D_2O/CD_3OD (1:1, v/v).

In the preceding example, selectivity is a result of differential solvation effects upon binding in an apolar, non-functionalized cavity. In biological systems, binding selectivity is more often a result of hydrogen bonding interactions between receptor and substrate [26]. We now have initiated a program to develop synthetic complexes in aqueous solution that are stabilized mainly through strong apolar and solvophobic forces whereas host-guest selectivity is established through oriented hydrogen bonding [27,28]. Scheme 5 shows the recent synthesis of a new tritopic receptor (**11**) featuring one neutral molecule and two cation binding sites [29]. Two carbonyl residues converge in a precise geometrical array into the apolar binding site of **11**. Its remarkable X-ray crystal structure analysis is shown in Figure 7. It features a large rectangular cavity with dimensions of roughly 9 x 14 Å and a spacing of 9.7 Å between the two carbonyl oxygen atoms. This spacious cavity should be suitably sized to incorporate large functionalized molecules such as steroids. The complexation properties of **11** and related derivatives are currently being evaluated.

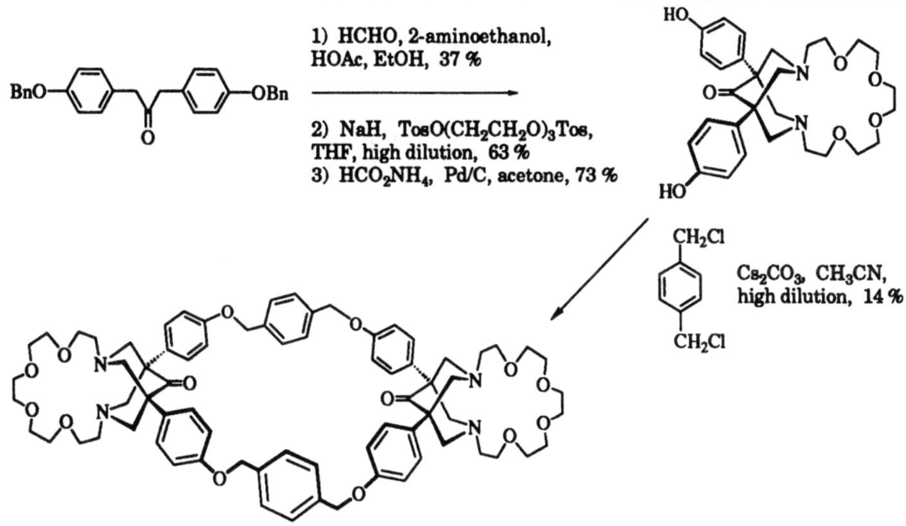

11

Scheme 5: Synthesis of the tritopic receptor **11**.

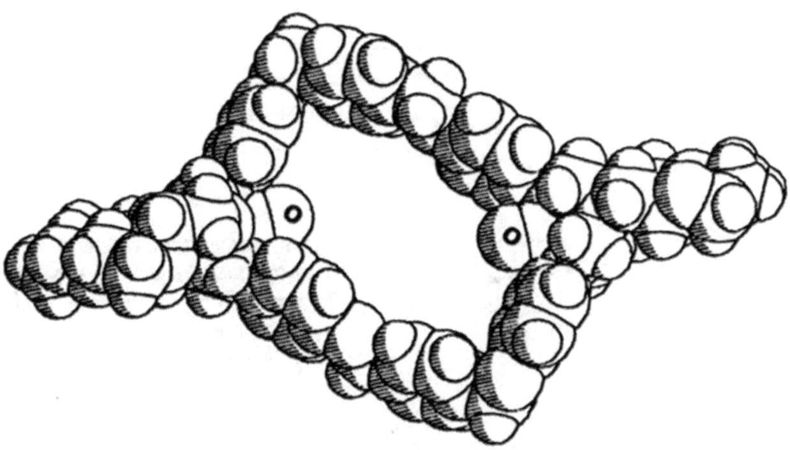

Figure 7. X-ray crystal structure of the tritopic receptor **11**.

6. Cyclophanes for Chiral Molecular Recognition Studies

In efforts to generate receptors for the enantioselective complexation of naproxen derivatives, e.g. **12a/b**, in aqueous solution, we prepared a series of optically active, water-soluble cyclophanes (**13** - **17**) with apolar cavities shaped by the major grooves of the 2,2',6,6'- and 2,2',7,7'-tetraoxa-1,1'-binaphthyl units [30,31]. The synthesis of macrocycle **15** is shown in Scheme 6.

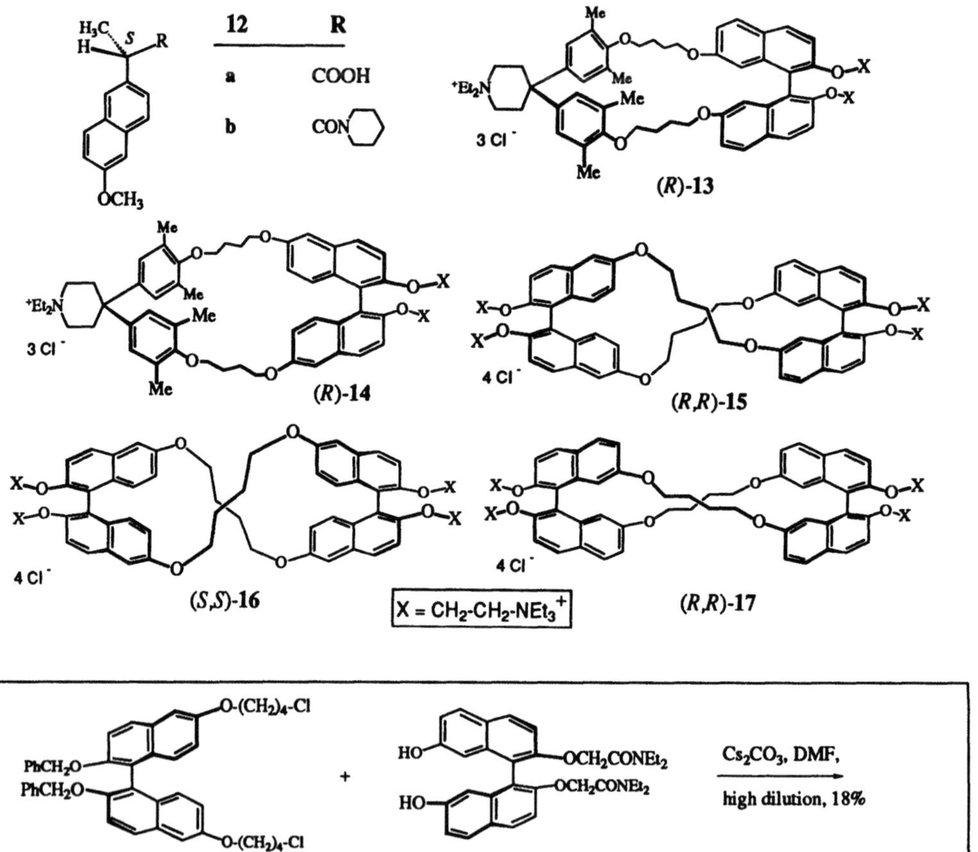

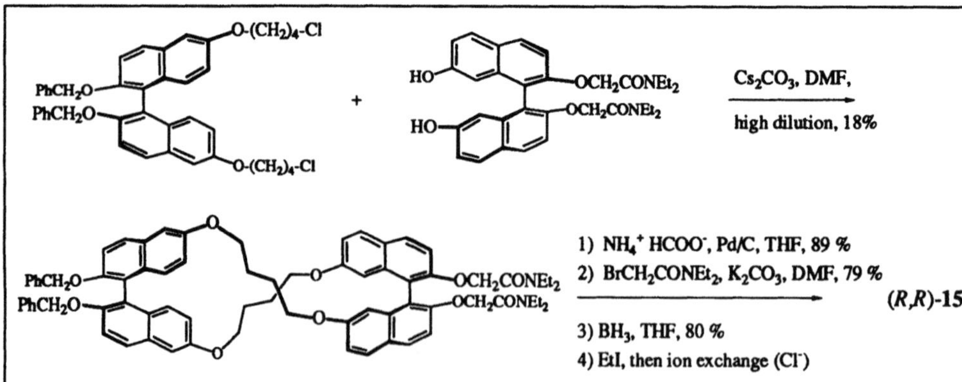

Scheme 6: Synthesis of the bis(binaphthyl)-cyclophane 15.

We had hoped that efficient chiral recognition in diastereomeric complexes of **13** - **17** and naproxen derivatives would result from a combination of (a) differential apolar interactions between the host cavity and the axially incorporated naphthalene part of the guest, (b) differential π-π interactions between the binaphthyl units and the carboxy residues of the guests, and (c) differential steric interactions between the chiral barrier in the host and the substituents at the naproxen stereocenter. Complexation studies in D_2O/CD_3OD (60:40, v/v, 293 K) revealed unexpectedly large differences in the overall binding performance of the various receptors as well as only a modest degree of enantioselectivity in binding [31]. Cyclophane **13**, in which a 2,2',7,7'-tetraoxa-1,1'-binaphthyl (7,7'-BN) unit is bridged by a diphenylmethane (DPM) unit, forms the most stable complex of all five receptors. In addition, this macrocycle gives the highest degree of chiral recognition. For example, the complex of (*R*)-**13** with (*R*)-**12b** (K_a = 2490 L mol^{-1}, ΔG^o = - 4.55 kcal mol^{-1}) is 0.33 kcal mol^{-1} more stable than the complex between (*R*)-**13** and (*S*)-**12b** (K_a = 1405 L mol^{-1}, ΔG^o = - 4.22 kcal mol^{-1}). Naproxen itself (**12a**) and other derivatives with smaller carboxy residues were bound less enantioselectively than the bulky piperidine derivative **12b**, which led to the conclusion that differential steric interactions are relevant for the enantioselectivity observed in the binding of **12b**.

All four other macrocycles performed much poorer than **13**, and a significant degree in chiral recognition ($\Delta\Delta G^o$ = 0.2 kcal mol^{-1}) in the binding of **12b** was only observed with host **14** in which a 2,2',6,6'-tetraoxa-1,1'-binaphthyl (6,6'-BN) is bridged by a DPM unit.

Perhaps the greatest surprise in the studies with **13** - **17** was the complete absence of binding power in the bis(7,7'-BN)-macroring **17**. In contrast, the cyclophanes **14** - **16** with one or two 6,6'-BN units all show good binding capability in the binary solvent mixture, forming complexes of K_a around 400 L mol^{-1} with various naphthalene derivatives. The lack of binding by **17** is due to the conformation of the binaphthyl units. In the lowest energy conformations of **17**, the binaphthyl units prefer to adopt a small dihedral angle (θ = 70 - 75°) about their chirality axis which narrows the cavity width which is crucial for binding [32]. The O···O distances at the major grooves of the two spacers are reduced to ≈ 5.5 - 6.0 Å which leads to the closing of any potential binding site. In contrast, bridging the chiral spacer in the best binder **13** with the wide open DPM unit (O···O distance ≈ 8.5 - 9.0 Å) enforces a larger dihedral angle in the binaphthyl unit (θ = 81°, O···O distance at the major groove = 6.4 Å in our calculated lowest energy conformation of **13**), and a more preorganized binding site is obtained. The O···O distance at the major groove of the 6,6'-BN unit (10.0 Å at θ = 90°) is significantly larger than in the 7,7'-BN unit (7.1 Å at θ = 90°), and, therefore, all 6,6'-BN cyclophanes have more open, productive binding sites.

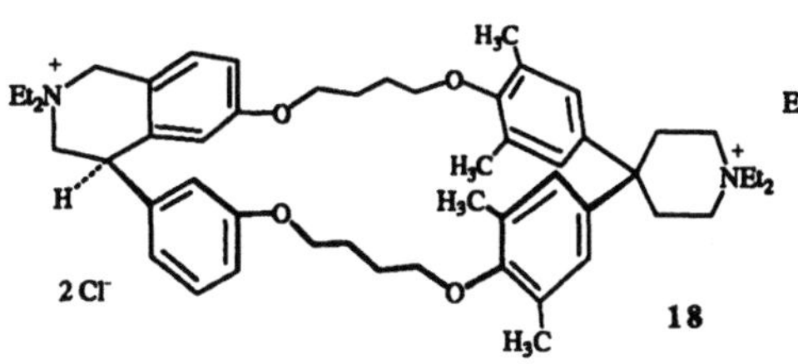

18

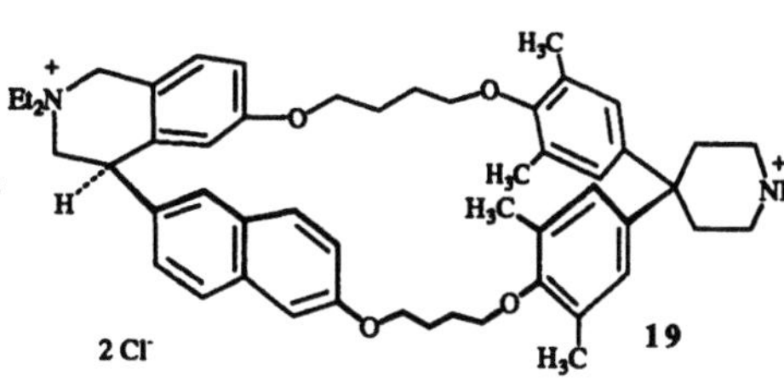

19

From these comprehensive studies as well as those performed with the chiral receptors **18** [32] and **19** [33], which incorporate unnatural alkaloid spacers, we draw the following general conclusions: (1) We are not yet able to predict and control in a significant way the conformational preferences of large flexible macrocycles with molecular weights around 1000 Daltons or their apolar inclusion complexes in protic environments. (2) The macrorings possess considerable conformational heterogeneity in the unbound as well as in the complexed state. This allows the guest enantiomers to form diastereomeric complexes that are geometrically and, hence, also energetically rather similar. Improving enantiomer differentiation properties of future chiral cyclophanes [34,35] will require a better control of macroring conformations. (3) Finally, we tend to conclude that it will be difficult to achieve a high degree of enantioselectivity in complexation processes based solely on differential non-directional interactions such as apolar, π-π, and steric interactions. Efficient chiral recognition seems to require the introduction of oriented bonding interactions, e.g. hydrogen bonding, as an additional discriminating force in diastereomeric complexes. Indeed, the highest degree of chiral recognition has been measured for diastereomeric complexes in which hydrogen bonding interactions represent a major binding force [30,36,37].

7. Supramolecular Catalysis by Coenzyme-Cyclophanes

Parallel to the fundamental binding studies with cyclophane receptors, we have developed over the past years a series of coenzyme-cyclophanes which show interesting supramolecular function [38] and activity. The binding properties of the flavinophane **20** can be switched reversibly in a redox process: only the reduced macrocycle is capable of inclusion complexation [39]. The thiazolium-cyclophane **21** is an excellent catalyst for the benzoin condensation [40]. Its thiazolium ion shows a remarkably enhanced C(2)-H acidity as a result of the microenvironment provided by the apolar macrocyclic cavity. The iron(III) porphyrin-cyclophane **22** is capable of binding arenes in protic solvents and, in the presence of iodosylbenzene as oxygen source, acts as efficient catalyst for the epoxidation of reactive aromatic C-C bonds in productively bound substrates [41,42]. For example, stirring a homogeneous solution with [**22**] = 0.54 mM, [acenaphthylene] = 11 mM, and [PhIO] = 15 mM in 2,2,2-trifluoroethanol for 30 min in the absence of oxygen affords acenaphthen-1-one with a turnover number (mols of substrate converted per mol of catalyst) of ≈ 6. This oxidation occurs via the epoxide which forms as the primary product but rapidly reacts under the reaction conditions to the ketone (Equation 1).

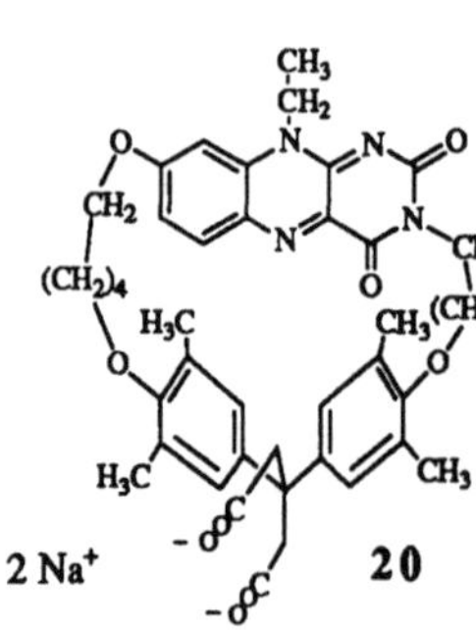

20

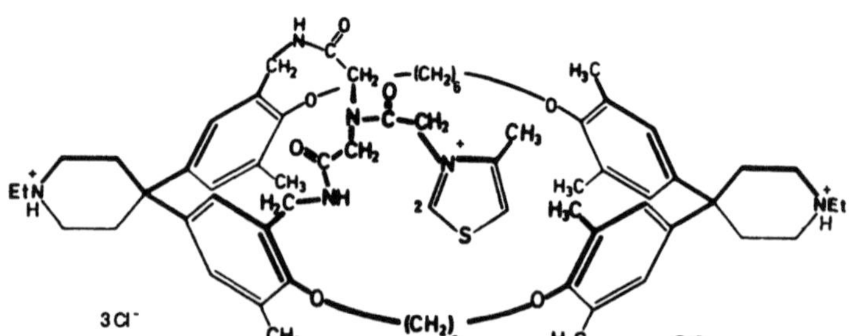

21

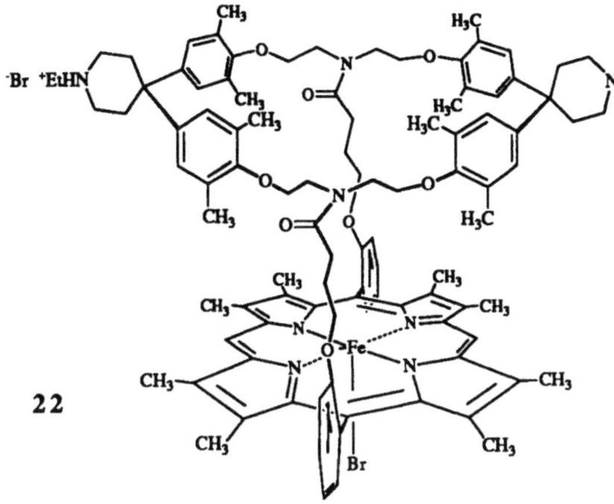

22

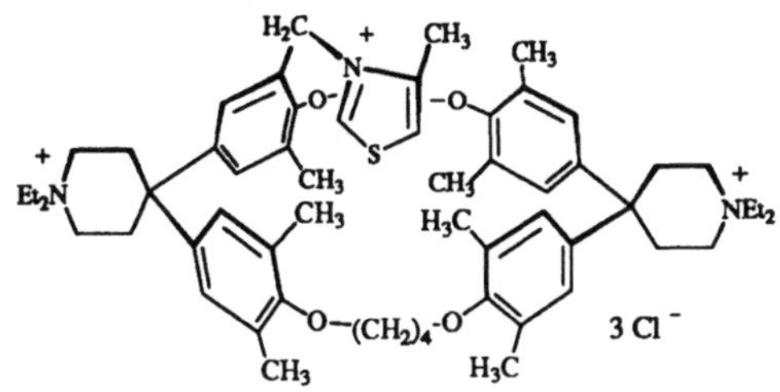

23

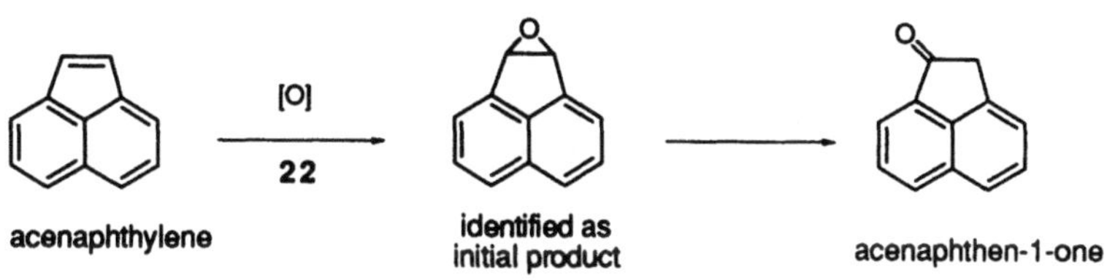

(1)

A particularly successful example for supramolecular cyclophane catalysis is provided by the efficient one-pot synthesis of aromatic methyl esters by electrochemical oxidation of aldehydes mediated by biscoenzyme - flavin and thiazolium-cyclophane (**23**) - catalysis [43,44]. The double coenzyme regeneration accomplished in this process in illustrated in Scheme 7. Equation 2 shows the conversion which takes place at an extraordinarily low oxidation potential (- 0.3 V vs Ag/AgCl) in methanol with the thiazolium-cyclophane **23** as the supramolecular catalyst. Under the conditions of equation 2, the chemical yields as well as the current efficiencies for the formation of aromatic methyl esters are very high. Starting from benzaldehyde, methyl benzoate is isolated after 10 hours in 74 % yield (7.4 turnovers and 88 % current efficiency). Starting from 4-cyanobenzaldehyde, the corresponding ester is obtained in 95 % isolated yield (9.5 turnovers) and 90 % current efficiency. Although these conversions can also be undertaken with simple thiazolium ions, e.g. the benzylthiazolium ion, the supramolecular catalysis by **23** has distinct advantages: the macrocyclic catalyst provides both enhanced reaction rates and increased yields of the desired aromatic esters.

$$\underset{(150\ \text{mM},\ 30\ \text{eq})}{\text{R-CHO}} \xrightarrow[\substack{35\,^{\circ}\text{C, Pt (- 300 mV vs. Ag/AgCl), methanol} \\ \text{in undivided electrochemical cell}}]{\textbf{23}\ (1\ \text{eq}),\ \text{MeFl}\ (3\ \text{eq}),\ \text{NEt}_3\ (30\ \text{eq}),\ \text{NEt}_4\text{Br}\ (9\ \text{eq})} \text{R-COOMe} \qquad (2)$$

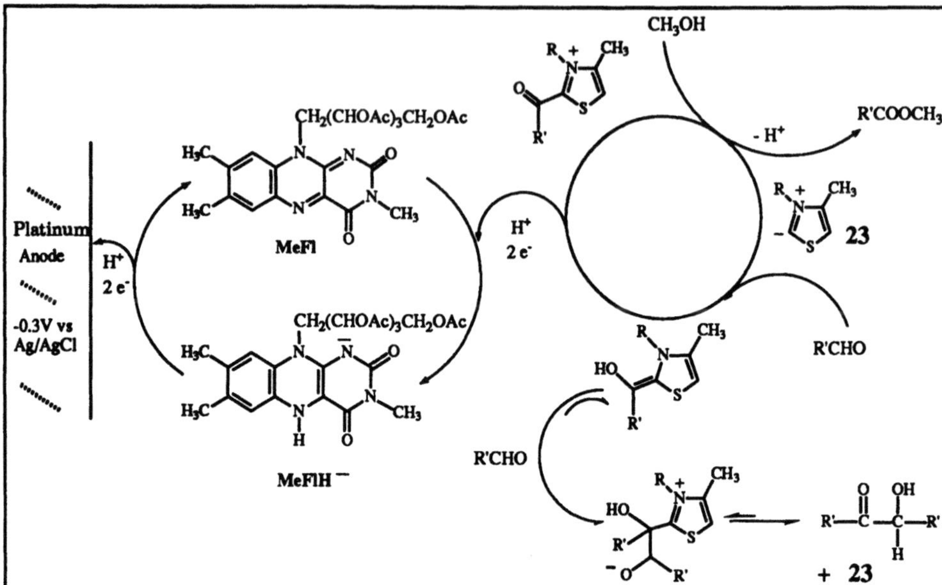

Scheme 7: Electrochemical regeneration of two coenzymes in the catalytic oxidation of aromatic aldehydes to methyl esters.

With its preformed cavity, the cyclophane catalyst **23** forms tight inclusion complexes with benzene and naphthalene substrates in methanol [43]. The increased rates probably are a result of (i) entropically favorable orientation and proximity between the reacting functional groups in the supramolecular complex and (ii) microenvironment effects in the apolar macrocyclic cavity. Since the binding site of **23** is complementary in size to only one benzene or naphthalene guest, the competitive benzoin condensation side reaction is suppressed which explains the enhanced yields in the reaction catalyzed by the thiazolium-cyclophane. The investigation of the full scope of the supramolecular electrochemical process and its selectivity in reactions with multifunctional substrates represent interesting future endeavors.

8. Conclusions

Cyclophanes provide powerful binding sites for a great diversity of organic molecules. Their studies have generated unprecedented individual molecular level understanding of solvent effects on apolar molecular recognition. Their selective functionalization has generated interesting supramolecular catalysts. In the hands of *J. F. Stoddart* and coworkers, cyclophanes have been converted into building blocks for unique molecular devices [45-47]. Cyclophane complexation studies in the group of *D. A. Dougherty* [48] have revealed the previously unknown magnitude of onium ion-aromatic π-system interactions, an essential force in the binding of the neurotransmitter acetylcholine by the enzyme acetylcholine esterase [49] and presumably a relevant force also in other important biological recognition events. There is no doubt that cyclophanes will continue to be a major class of receptors in future supramolecular chemistry [50].

9. Acknowledgement: We are very grateful to the Office of Naval Research (ONR), National Science Foundation (NSF), and the National Institutes of Health (NIH) for the continuing support of our supramolecular cyclophane chemistry program.

10. References

[1] Diederich, F. (1991) Cyclophanes. Monographs in Supramolecular Chemistry, Vol. 2, Stoddart, J. F. (ed.), Royal Society of Chemistry, Cambridge.

[2] Odashima, K., Itai, A., Iitaka, Y., and Koga, K. (1980) 'Host-Guest Complex Formation between a Water-Soluble Polyparacyclophane and a Hydrophobic Guest Molecule', J. Am. Chem. Soc. 102, 2504-2505.

[3] Diederich, F. (1988) 'Complexation of Neutral Molecules by Cyclophane Hosts', Angew. Chem. 100, 372-396; Angew. Chem. Int. Ed. Engl. 27, 362-386.

[4] Diederich, F. (1990) 'Molecular Recognition in Aqueous Solution', J. Chem. Ed. 67, 813-820.

[5] Smithrud, D. B., Sanford, E. M., Chao, I., Ferguson, S. B., Carcanague, D. R., Evanseck, J. D., Houk, K. N., and Diederich, F. (1990) 'Solvent Effects in Molecular Recognition', Pure & Appl. Chem. 62, 2227-2236.

[6] Diederich, F. and Dick, K. (1982) 'New Water-Soluble Macrocycles of the Paracyclophane Type: Aggregation Behaviour and Host-Guest-Interaction with Hydrophobic Substrates', Tetrahedron Lett. 23, 3167-3170.

[7] Ferguson, S. B., Seward, E. M., Diederich, F., Sanford, E. M., Chou, A., Inocencio-Szweda, P., and Knobler, C. B. (1988) 'Strong Enthalpically Driven Complexation of Neutral Benzene Guests in Aqueous Solution', J. Org. Chem. 53, 5593-5595.

[8] Ferguson, S. B., Sanford, E. M., Seward, E. M., and Diederich, F. (1991) 'Cyclophane-Arene Inclusion Complexation in Protic Solvents: Solvent Effects versus Electron Donor-Acceptor Interactions', J. Am. Chem. Soc. 113, 5410-5419.

[9] Smithrud, D. B., Wyman, T., and Diederich, F. (1991) 'Enthalpically Driven Cyclophane-Arene Inclusion Complexation: Solvent-Dependent Calorimetric Studies', J. Am. Chem. Soc. 113, 5420-5426.

[10] Krieger, C. and Diederich, F. (1985) 'Structure of Host-Guest Complexes of 1',1"-Dimethyldispiro[1,6,20,25-tetraoxa[6.1.6.1] paracyclophane-13,4':32,4"-bispiperidine] with Benzene and *p*-Xylene', Chem. Ber. 118, 3620-3631.

[11] Diederich, F., Dick, K., and Griebel, D. (1985) 'Water-Soluble Tetraoxa[n.1.n.1]paracyclophanes: Synthesis and Host-Guest Interactions in Aqueous Solution' Chem. Ber. 118, 3588-3619.

[12] Wilcox, C. S. (1991) 'Design, Synthesis, and Evaluation of an Efficacious Functional Group Dyad. Methods and Limitations in the Use of NMR for Measuring Host-Guest Interactions', in H.-J. Schneider and H. Dürr (eds.), Frontiers in Supramolecular and Organic Chemistry and Photochemistry, VCH, Weinheim, 1991, pp. 123-143.

[13] Diederich, F. and Dick, K. (1984) 'A New Water-Soluble Macrocyclic Host of the Cyclophane Type: Host-Guest Complexation with Aromatic Guests in Aqueous Solution and Acceleration of the Transport of Arenes Through an Aqueous Phase', J. Am. Chem. Soc. 106, 8024-8036.

[14] Ferguson, S. B. and Diederich, F. (1986) 'Electron Donor-Acceptor Interactions in Host-Guest Complexes in Organic Solutions', Angew. Chem. 98, 1127-1129; Angew. Chem. Int. Ed. Engl. 25, 1127-1129.

[15] Diederich, F. and Dick, K. (1983) 'Inclusion Complexes Between a Macrocyclic Host Molecule and Aromatic Hydrocarbons in Aqueous Solution', Angew. Chem. 95, 730-732; Angew. Chem. Int. Ed. Engl. 22, 715-716.

[16] Diederich, F. and Dick, K. (1985) 'A Water-Soluble Tetraoxa[7.1.7.1] paracyclophane: Synthesis and Host-Guest Interactions with Alicyclic and Cationic Guest Molecules in Aqueous Solution', Chem. Ber. 118, 3817- 3829.

[17] Vögtle, F., Müller, W. M., Werner, U., and Losensky, H.-W. (1987) 'Selective Molecular Recognition and Separation of Isomeric and Partially Hydrogenated Arenes', Angew. Chem. 99, 930-932; Angew. Chem. Int. Ed. Engl. 26, 901-903.

[18] Diederich, F. and Griebel, D. (1984) '^{1}H NMR Investigations of Host-Guest Complexation Between a Macrocyclic Host of the Cyclophane Type and Aromatic Guests in Aqueous Solution', J. Am. Chem. Soc. 106, 8037-8046.

[19] Diederich, F. and Dick, K. (1984) 'Spherical Host Molecules For the Complexation of Aromatic Hydrocarbons in Aqueous Solution', Angew. Chem. 96, 789-790; Angew. Chem. Int. Ed. Engl. 23, 810-812.

[20] Diederich, F., Dick, K, Griebel, D. (1986) 'Complexation of Arenes by Macrocyclic Hosts in Aqueous and Organic Solutions', J. Am. Chem. Soc. 108 2273-2286.

[21] Smithrud, D. B. and Diederich, F. (1990) 'Strength of Molecular Complexation of Apolar Solutes in Water and in Organic Solvents Is Predictable by Linear Free Energy Relationships: A General Model for Solvation Effects on Apolar Binding', J. Am. Chem. Soc. 112, 339-343.

[22] Schneider, H.-J., Kramer, R., Simova, S., and Schneider, U. (1988) 'Solvent and Salt Effects on Binding Constants of Organic Substrates in Macrocyclic Host Compounds. A General Equation Measuring Hydrophobic Binding Contributions', J. Am. Chem. Soc. 110, 6442-6448.

[23] Canceill, J., Cesario, M., Collet, A., Guilhem, J., Lacombe, L., Lozach, B., and Pascard, C. (1989) 'Structure and Properties of the Cryptophane-E/$CHCl_3$ Complex, a Stable van der Waals Molecule', Angew. Chem. 101, 1249-1251; Angew. Chem. Int. Ed. Engl. 28, 1246-1248.

[24] Carcanague, D. R. and Diederich, F. (1990) 'A Spacious Cyclophane Host for Inclusion Complexation of Steroids and [m.n]Paracyclophanes', Angew. Chem. 102, 836-838; Angew. Chem. Int. Ed. Engl. 29, 769-771.

[25] Kawakami, H., Yoshino, O., Odashima, K., and Koga, K. (1985) 'Design, Synthesis, and Properties of Novel Water-Soluble Cyclophanes Having Naphthylphenylmethane Units as Hosts for Aliphatic and Aromatic Guests' , Chem. Pharm. Bull. 33, 5610-5613.

[26] Jeffrey, G. A. and Saenger, W. (1991) 'Hydrogen Bonding in Biological Structures', Springer-Verlag, Berlin.

[27] Rebek, J., Jr. (1990) 'Molecular Recognition with Model Systems', Angew. Chem. 102, 261-272; Angew. Chem. Int. Ed. Engl. 29, 245-255.

[28] Hamilton, A. D. (1990) ' Molecular Recognition by Macrocyclic Receptors' in J.. L. Atwood (ed.), Inclusion Phenomena and Molecular Recognition, Plenum Press, New York, pp. 57-64.

[29] Carcanague, D. R., Knobler, C. B., and Diederich, F. 'Water Soluble Cyclophane Receptors with Convergent Functional Groups', J. Am. Chem. Soc., submitted for publication.

[30] Castro, P. P., Georgiadis, T. M., and Diederich, F. (1989) 'Chiral Recognition in Clefts and Cyclophane Cavities Shaped by the 1,1'-Binaphthyl Major Groove', J. Org. Chem. 54, 5835-5838.

[31] Castro, P. P. and Diederich, F. (1991) 'Optically Active Mono and Bis(1,1'-Binaphthyl)-Cyclophanes: Large Differences in the Complexation of Naproxen Derivatives in Aqueous Solution', Tetrahedron Lett., in press.
[32] Dharanipragada, R., Ferguson, S. B., and Diederich, F. (1988) 'A Novel Optically Active Host: Design, Computer Graphics, Synthesis, and Diastereomeric Complex Formation in Aqueous Solution', J. Am. Chem. Soc. 110, 1679-1690.
[33] Georgiadis, T. M., Georgiadis, M. M., and Diederich, F. (1991) 'Synthesis and Complexation Properties of a Water-Soluble Optically Active Cyclophane Incorporating a 4-Naphthyl-1,2,3,4-tetrahydroisoquinoline Unit as a Chiral Spacer', J. Org. Chem. 56, 3362-3369.
[34] Petti, M. A., Shepodd, T. J., Barrans, R. E., Jr., and Dougherty, D. A. (1988) 'Hydrophobic Binding of Water-Soluble Guests by High-Symmetry, Chiral Hosts. An Electron-Rich Receptor Site with a General Affinity for Quaternary Ammonium Compounds and Electron-Deficient π-Systems', J. Am. Chem. Soc. 110, 6825-6840.
[35] Webb, T. H., Suh, H., and Wilcox, C. S. (1991) 'Enantioselective and Diastereoselective Molecular Recognition of Alicyclic Substrates in Aqueous Media by a Chiral, Resolved Synthetic Receptor', J. Am. Chem. Soc. 113, 8554-8555.
[36] Jeong, K.-S., Muehldorf, A. V., and Rebek, J., Jr. (1990) 'Molecular Recognition. Asymmetric Complexation of Diketopiperazines', J. Am. Chem. Soc. 112, 6144-6145.
[37] Hong, J.-I., Namgoong, S. K., Bernardi, A., and Still, W. C. (1991) 'Highly Selective Binding of Simple Peptides by a C_3 Macrotricyclic Receptor', J. Am. Chem. Soc. 113, 5111-5112.
[38] Lehn, J.-M. (1988) 'Supramolecular Chemistry - Molecules, Supermolecules, and Molecular Devices (Nobel Lecture)', Angew. Chem. 100, 91-116; Angew. Chem. Int. Ed. Engl. 27, 89-112.
[39] Seward, E. M., Hopkins, R. B., Sauerer, W., Tam, S.-W., and Diederich, F. (1990) 'Redox-Dependent Binding Ability of a Flavin Cyclophane in Aqueous Solution: Hydrophobic Stacking versus Cavity-Inclusion Complexation', J. Am. Chem. Soc. 112, 1783-1790.
[40] Diederich, F. and Lutter, H.-D. (1989) 'Catalytic Cyclophanes. 4. Supramolecular Catalysis of Benzoin Condensations by a Thiazolium Cyclophane', J. Am. Chem. Soc. 111, 8438-8446.
[41] Benson, D. R., Valentekovich, R., and Diederich, F. (1990) 'Catalytic Cyclophanes: A Porphyrin-Bridged Cyclophane as a Model for Cytochrome P-450 Enzymes', Angew. Chem. 102, 213-216; Angew. Chem. Int. Ed. Engl. 29, 191-193.
[42] Benson, D. R., Valentekovich, R., Knobler, C. B., and Diederich, F. (1991) 'Porphyrin-Cyclophanes: Inclusion Complexation and X-ray Crystal Structure of a Zinc Octamethyldiphenylporphyrin', Tetrahedron Lett. 47, 2401-2422.
[43] Jimenez, L. and Diederich, F. (1989) 'Catalytic Cyclophanes: A Highly Efficient Model for Pyruvate Oxidase', Tetrahedron Lett. 30, 2759-2762.
[44] Tam, S.-W., Jimenez, L., and Diederich, F. 'Efficient One Pot Synthesis of Aromatic Methyl Esters by Electrochemical Oxidation of Aldehydes Mediated by Biscoenzyme Catalysis', J. Am. Chem. Soc., submitted for publication.
[45] Ashton, P. R., Goodnow, T. T., Kaifer, A. E., Reddington, M. V., Slawin, A. M. Z., Spencer, N., Stoddart, J. F., Vicent, C., and Williams, D. J. (1989) 'A [2]Catenane Made to Order', Angew. Chem. 101, 1404-1407; Angew. Chem. Int. Ed. Engl. 28, 1396-1399.
[46] Anelli, P. L., Spencer, N., Stoddart, J. F. (1991) 'A Molecular Shuttle', J. Am.

Chem. Soc. 113, 5131-5133.

[47] Ashton, P. R., Brown, C. L., Chrystal, E. J. T., Parry, K. P., Pietraszkiewicz, M., Spencer, N., and Stoddart, J. F. (1991) 'Molecular Trains: The Self-Assembly and Dynamic Properties of Two New Catenanes', Angew. Chem. 103, 1058-1061; Angew. Chem. Int. Ed. Engl. 30, 1042-1045.

[48] Dougherty, D. A. and Stauffer, D. A. (1990) 'Acetylcholine Binding by a Synthetic Receptor: Implications for Biological Recognition', Science (Washington, D.C.) 250, 1558-1560.

[49] Sussman, J. L., Harel, M., Frolow, F., Oefner, C., Goldman, A., Toker, L., Silman, I. (1991) 'Atomic Structure of Acetylcholinesterase from *Torpedo californica*: A Prototypic Acetylcholine-Binding Protein', Science (Washington, D.C.) 253, 872-879.

[50] For other recent highlights in supramolecular cyclophane chemistry, see: (a) Cram, D. J., Tanner, M. E., and Thomas, R. (1991) 'The Taming of Cyclobutadiene', Angew. Chem. 103, 1048-1051; Angew. Chem. Int. Ed. Engl. 30, 1024-1027. (b) Bonar-Law, R. P., Davis, A. P., and Murray, B. A. (1990) 'Artificial Receptors for Carbohydrate Derivatives', Angew. Chem. 102, 1497-1499; Angew. Chem. Int. Ed. Engl. 29, 1407-1408. (c) Anderson, H. L., Sanders, J. K. M. (1990), 'Amine-Template-Directed Synthesis of Cyclic Porphyrin Oligomers', Angew. Chem. 102, 1478-1480; Angew. Chem. Int. Ed. Engl. 29, 1400-1403. (d) Vögtle, F., Lüer, I., Balzani, V., and Armaroli, N. (1991) 'Endoreceptors with Convergent Phenanthroline Units; A Molecular Cavity for Six Guest Molecules', Angew. Chem. 103, 1367-1370; Angew. Chem. Int. Ed. Eng. 30, 1333-1336. (e) Neder, K. M. and Whitlock, H. W., Jr., (1990) 'Importance of Hydrogen Bond Acceptor Ability in Design of Host Molecules Capable of Molecular Recognition', J. Am. Chem. Soc. 112, 9412-9414. (f) Murakami, Y., Kikuchi, J., Ohno, T., Hirayama, T., Hisaeda, Y., Nishimura, H., Snyder, J. P., Steliou, K. (1991) 'Synthesis of Macrocyclic Enzyme Models. 8. Conformational Mobility and Molecular Recognition by the Internal Cage of Kyuphane' *J. Am. Chem. Soc.* 113, 8229-8242.

NEW SYNTHETIC RECEPTORS FOR COMPLEXATION AND CATALYSIS

ANDREW D. HAMILTON
Department of Chemistry
University of Pittsburgh
Pittsburgh, PA 15260
USA

ABSTRACT. Molecules with multiple directed hydrogen bonding or π-stacking interactions are shown to form stong and selective complexes with complementary substrates.

1. Introduction

An important goal in modern bioorganic chemistry concerns the design of synthetic molecules that mimic various aspects of enzyme chemistry[1]. Detailed study of such models can lead not only to insights into the nature of enzyme action but also to new chemical species with some of the specificity and speed normally associated only with enzymes. Central to the success of this endeavor is the development of molecular architecture in which different regions or functional groups are positioned in a well-defined array to provide a specific chemical microenvironment. The principal focus of our research has been in the area of molecular recognition. We have constructed a number of artificial receptors that show strong and selective binding to small organic substrates. These have been based on a complementary relationship between the host and its guest. This complementarity must include both the shape and the chemical characteristics of the substrate. Thus, for effective molecular recognition the receptor must contain a cavity or cleft in which spatially separated hydrogen bonding, hydrophobic and charged residues are directed to interact with complementary regions on a substrate. In recent years we have made significant progress towards the artificial recognition of such key biomolecules as peptides, nucleotides and clinically-used drugs. Artificial receptors of this type may lead to the development of novel pharmaceutical strategies, drug delivery systems or chemical sensor designs. These early studies are directed at the ground state structures of the target substrates. Extension of the molecular recognition approach to receptors that are complementary to the proposed transition state structures of a reaction will lead to "artificial enzymes" that accelerate the reaction by stabilization of the transition state[2].

2. Directed Hydrogen Bonding Interactions

A crucial binding interaction in biological recognition involves hydrogen bonding between the protein and its substrate. A synthetic cavity containing several directed hydrogen bonding groups should lead to binding and potential orientation of a substrate with complementary groups. We have prepared a series of receptors (e.g. **1**) for the

V. Balzani and L. De Cola (eds.), Supramolecular Chemistry, 137–144.

barbiturate family of drugs. These important seditive and anti-convulsant targets form hexahydrogen bonded complexes with macrocyclic receptors containing two 2,6-diamidopyridine derivatives (as in **2**). We have recently confirmed the nature and precise

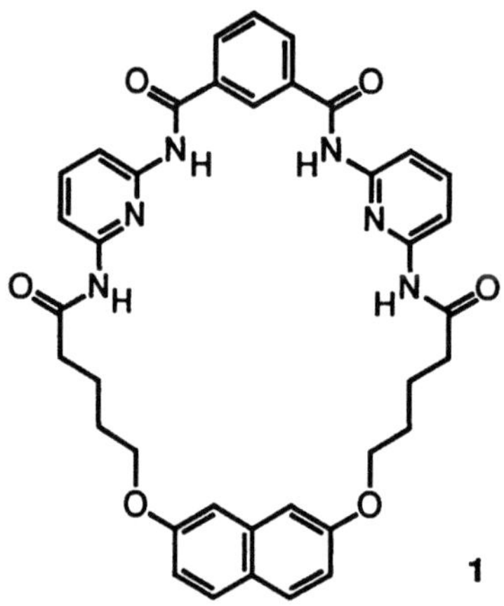

1

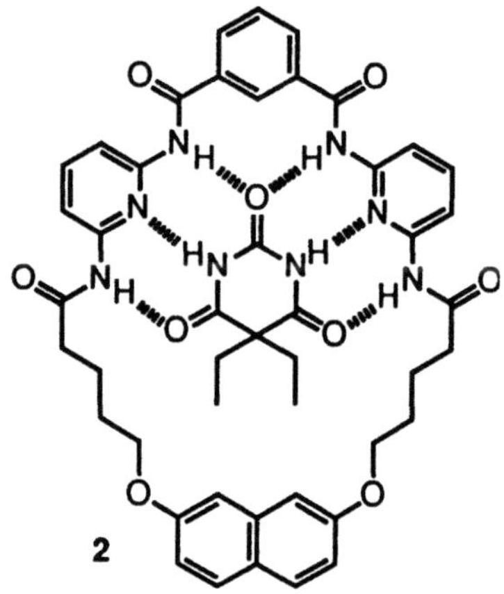

2

details of this interaction by X-ray crystallographic analysis of **1** and **2** (figure 1A and B, respectively). The strength and selectivity of binding is reflected in the large association constant for **2** ($1x10^5$ M^{-1}) which drops dramatically ($K_a=7x10^2$ M^{-1}) on alkylation of one of the imide nitrogen atoms[3].

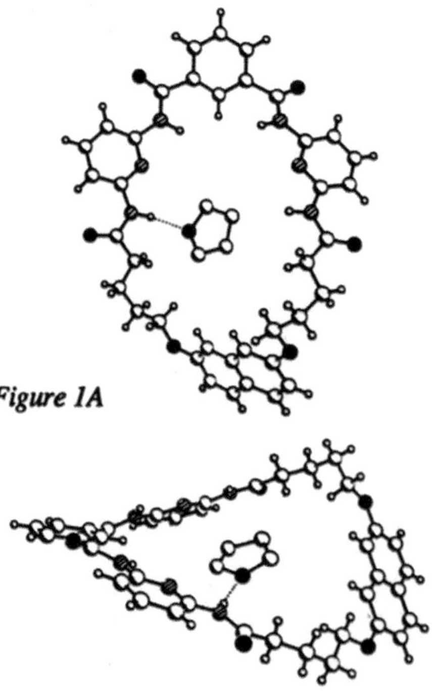

Figure 1A

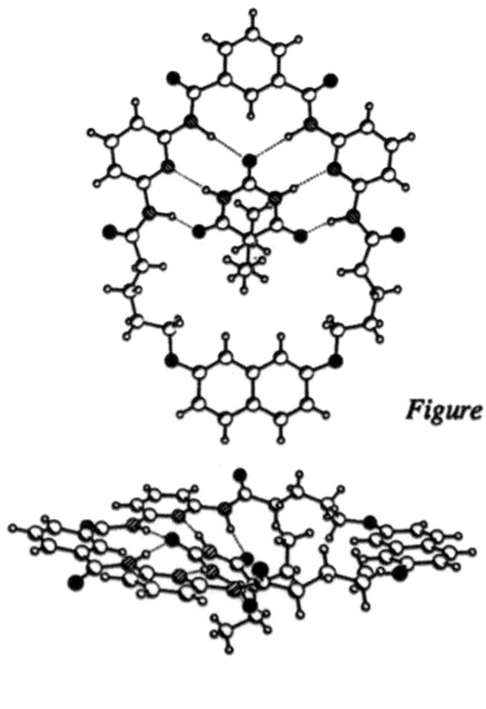

Figure 1B

In a very similar way a simple receptor for dicarboxylic acids can be constructed from the reaction of 2-aminopicoline with terephthaloyl dichloride. Diamide **3** now possesses two bidentate binding regions at a distance apart that coresponds to the spacing of the carboxylic acid groups in adipic or glutaric acid. Thus **3** would be expected to bind to appropriately lengthed dicarboxylic acids via four hydrogen bonds as in **4**.

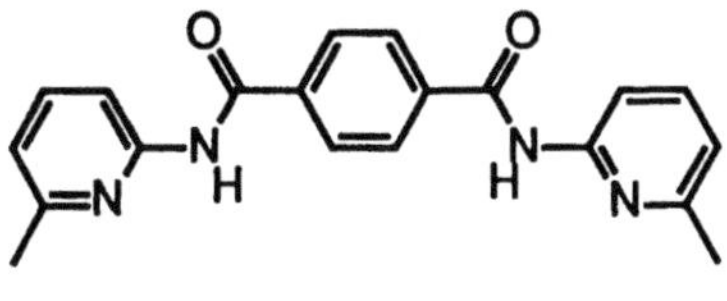

3

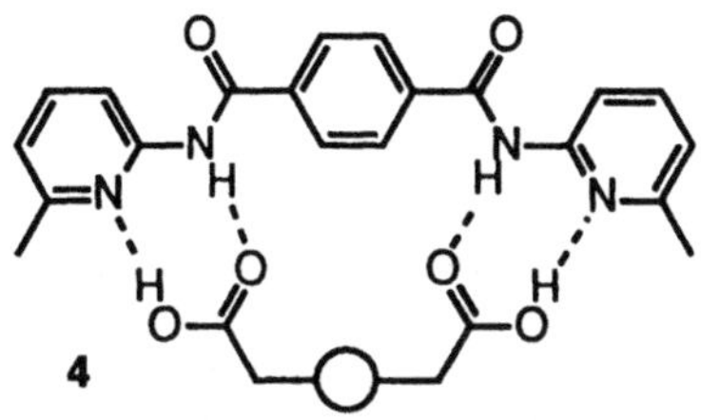

4

In $CDCl_3$ solution **3** binds very strongly to dicarboxylic acids containing 3 to 4 methylene groups. Monitoring the large downfield shifts of the amide resonances in **3** that occur on hydrogen bonding gave a binding isotherm from which K_a values of 10^4- 10^5 M^{-1} were calculated for glutaric and adipic acids[4]. An X-ray structure of the adipic acid complex with **3** confirmed the tetrahydrogen bonding character of the complex as well as the position of the polymethylene chain across the face of the terephthaloyl group(figure 2).

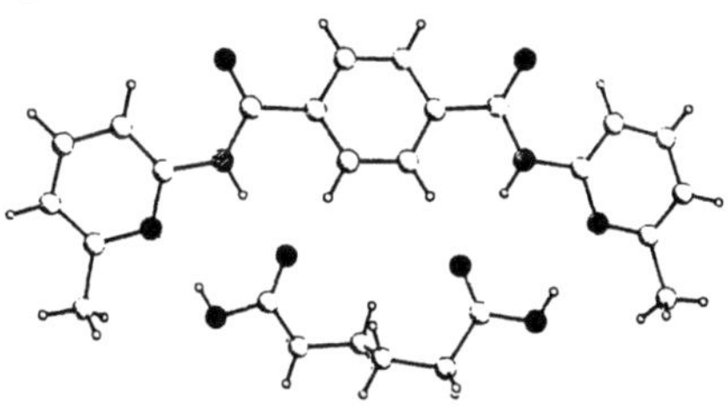

Figure 2. X-ray Stucture of 3: Adipic Acid Complex

3. Directed π-Stacking Interactions

The simultaneous influence of both hydrogen bonding and π-stacking interactions offers a powerful approach for the recognition of planar heterocyclic substrates such as the nucleotide bases. By exploiting this two-site binding strategy, we have developed a series of receptors for the nucleotide bases that involves the perpendicular convergence of hydrogen bonding and aromatic stacking interactions. Our strategy was to link within a macrocyclic framework, a group capable of stacking with the nucleotide base to one complementary to its hydrogen bonding periphery (figure 3). The geometry of the π-stacking groups is of particular interest since in a survey of protein crystal structures Petsko[5] has identified two important orientations for aromatic-aromatic interactions; namely face-to-face and edge-to-face (figure 4). We have used the synthetic receptor approach to investigate the importance of the electronic characteristics of the stacking group on its orientation in the complex. In particular, we have synthesized receptor **5**, containing a diaminopyridine hydrogen bonding site and a bis-ethoxycarbonyl substituted naphthalene as a π-stacking unit. In $CDCl_3$, **5** forms complexes with alkylthymine derivatives (K_a= 5

Figure 3

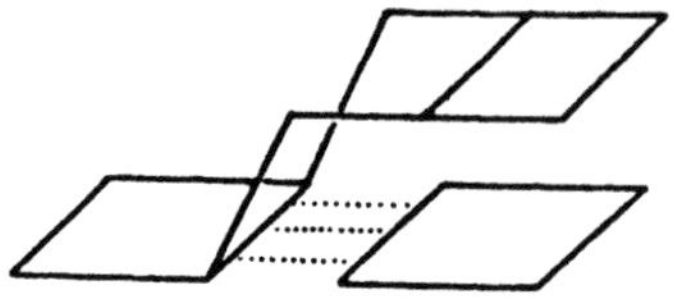

Figure 4

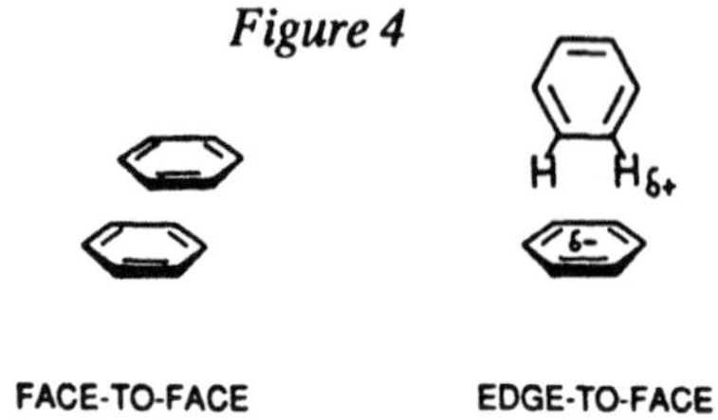

x 10^2 M^{-1}) that are characterized by downfield shifts of the H-bonding protons and upfield shifts of the protons on the periphery of the thymine. These ^{1}H NMR changes are consistent with a complex involving both hydrogen bonding and π-stacking interactions as in **6**. This was confirmed by an X-ray structure of **6** which showed in addition to the three hydrogen bonds a parallel or face-to-face positioning of the thymine and naphthalene rings (figure 5). A consideration of the atomic charge distribution around the thymine and

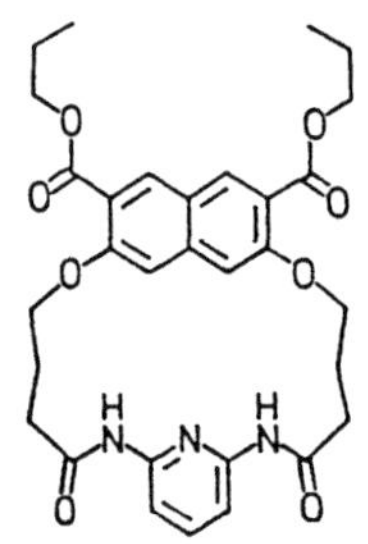

5

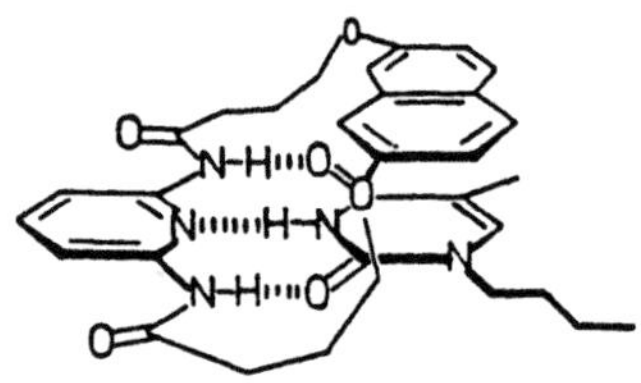

6

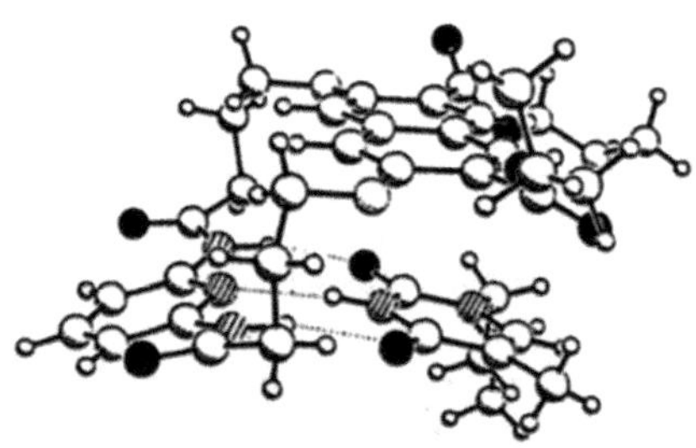

Figure 5. X-Ray Structure of 6.

naphthalene rings shows a remarkable electrostatic complementarity where five points of +ve charge on the naphthalene are positioned directly above five regions of -ve charge on the thymine(figure 6A). Changing the electron withdrawing ester groups to electron

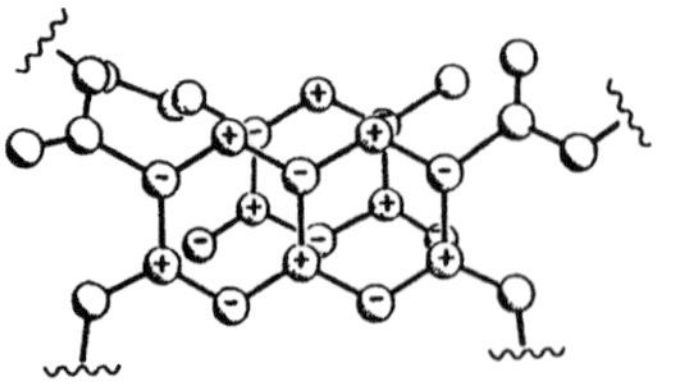

Figure 6A

Figure 6B

donating alkylethers will cause a change in the charge distribution around the naphthalene ring such that a face-to-face interaction with the thymine (shown hypothetically in figure 6B) with lead to repulsive interactions. We have prepared bis-butyloxy-substituted naphthalene **7** and shown that it binds more weakly to alkylthymines (K_a= 130 M^{-1}) with none of the upfield shifts in the ^1H NMR spectrum that would be expected for face-to-face π-stacking. An X-ray structure of the complex between **7** and thymine (figure 7)

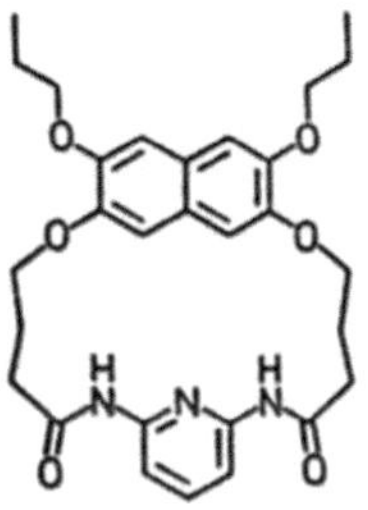

7

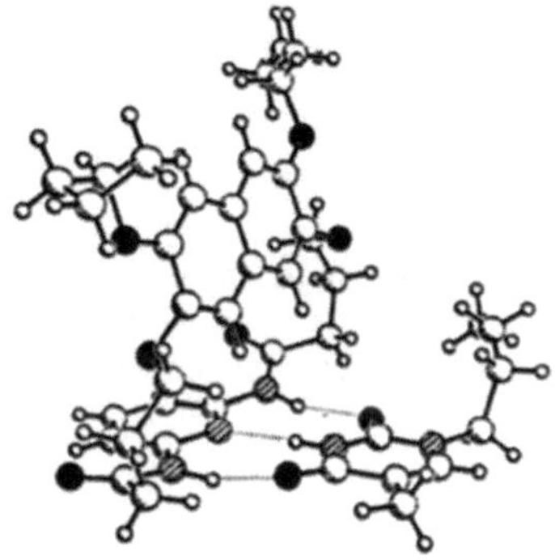

Figure 7

confirms that an alternative edge-to-face geometry has been adopted by the interacting rings [6]. Thus, using simple models, we have shown that the strength and geometry of π-stacking between highly polarized π-systems depends on a complementarity of charge distribution in the two interacting rings.

In the above studies the nature of the π-stacking component was changed. It is also possible to modify the hydrogen bonding region and in doing so alter the substrate selectivity of the receptor. We have recently used this strategy to prepare hydrogen bonding and π-stacking receptors for adenine[7] and guanine[8].

4. Directed Electrostatic Interactions.

We were interested in extending the directed bidentate hydrogen-bonding between a carboxylic acid and amidopyridines (**8**) to anionic substrates such as phosphate and

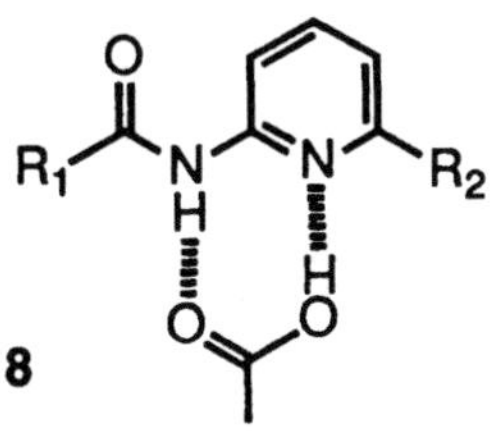

8

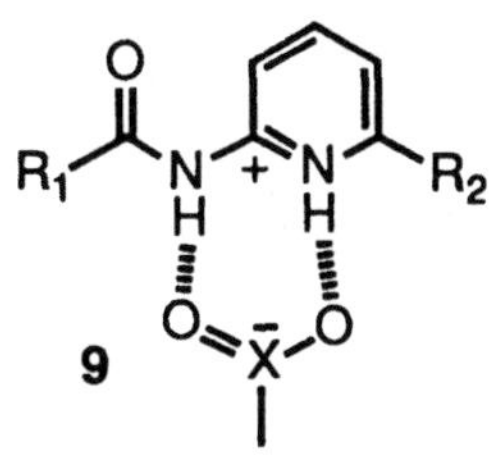

9

carboxylate. A straightforward strategy would involve protonating the pyridine to form the bidentate ion pair **9**. However, crystals grown from a 1: 1 mixture of 2,6-dibutyramidopyridine **10** and diphenylphosphoric acid show , in the solid state (figure 8), that while proton transfer occurs the cyclic bidentate complex does not. Instead the N-pyridine bonds rotate by 180° to form two intramolecular hydrogen bonds between the pyridiniumH and amideCO and two intermolecular hydrogen bonds between the

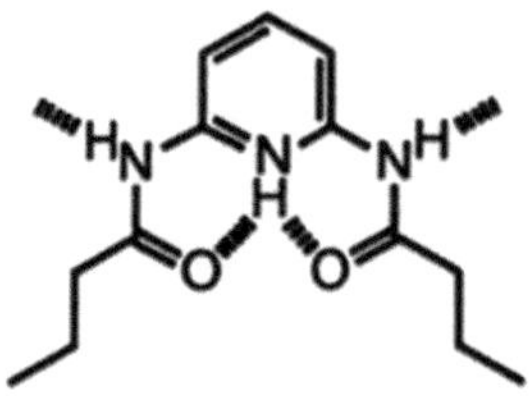

10

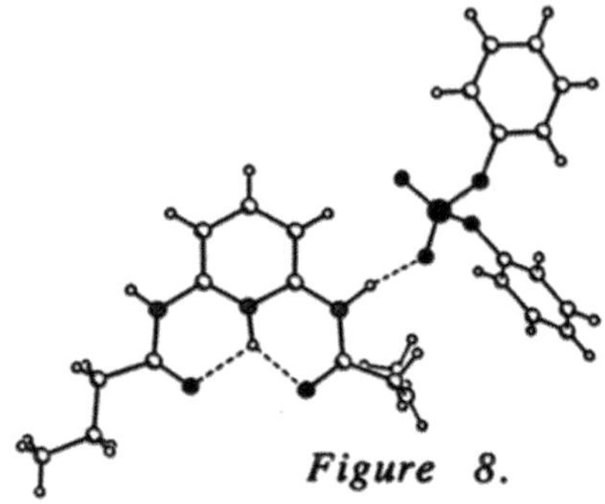

Figure 8.

amideNHs and two phosphate units.

An intramolecular hydrogen bond of this type might be exploited as a rigidifying element in receptor design if an additional binding group were positioned at a site corresponding to

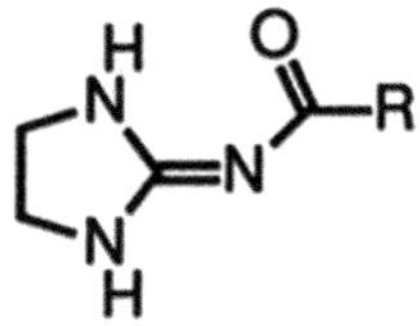

11

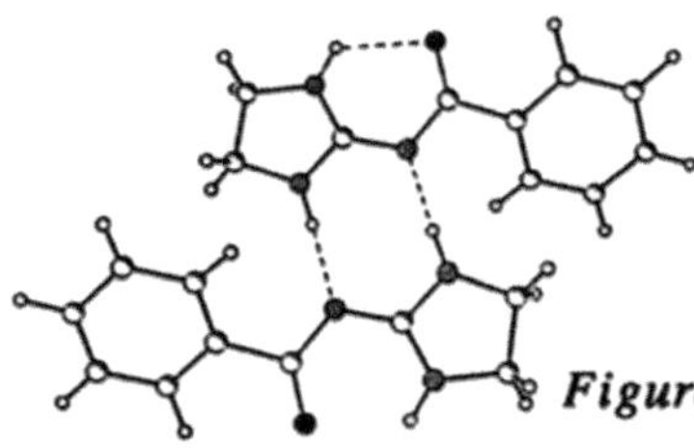

Figure 9

the pyridine 3-carbon in **10**. This arrangement exists in 2-acylaminoimidazolines, **11**, and a crystal structure of 2-benzoylaminoimidazoline (figure 9) showed that an intramolecular hydrogen bond between one ringNH and the benzoylCO was present. The other NH and the acylN are positioned to form a cyclic bidentate interaction with a second molecule. Based on our earlier studies of phosphorodiamidate recognition [9], we reasoned that linking two protonated aminoimidazolines through an isophthalic acid spacer should lead to a simple receptor for phosphate ester anions.

Reaction of dimethyl isophthalate with 2-aminoimidazolinium p-toluenesulfonate in MeOH and NaOMe gave the bis-guanidinne receptor in 16% yield. The basicity of acylaminoimidazolines (pKa = 7.09) is reduced relative to aminoimidazolines (pKa = 13.58); however, they can be readily protonated and treatment with picric acid gave dicationic receptor **12.** A simpler receptor **13** containing two acylguanidinium groups can be formed from the reaction of guanidinium hydrochloride and dimethylisophthalate. Both **12** and **13** could be converted into their tetraphenylborate (TBP) salts by treating the corresponding bis-hydrochloride with sodium tetraphenylborate. The ^{1}H NMR spectrum of **13** in CD_3CN shows three broad signals due to guanidinium-hydrogens at 7.4, 8.2 and 11.2 ppm, respectively. The large downfield shifted resonance reflects the intramolecular hydrogen bond to each of the isophthaloyl carbonyl groups, as depicted in **12** and **13.** By analogy to the intramolecular hydrogen bonds in protein α-helices or β-sheets, these provide additional rigidification to the synthetic receptors and ensure that the only flexibility is in the phenyl-carbonyl bonds.

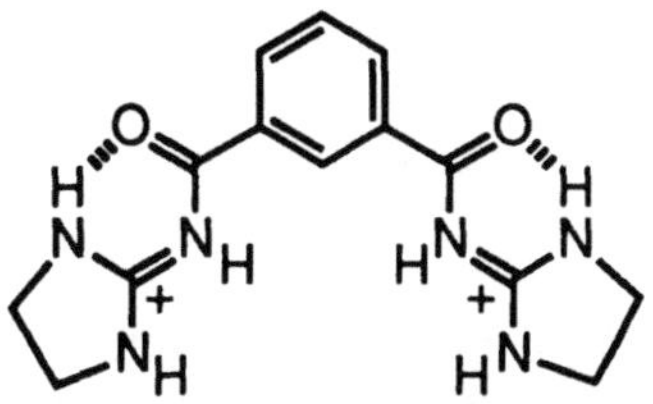

12 (picrate)$_2$

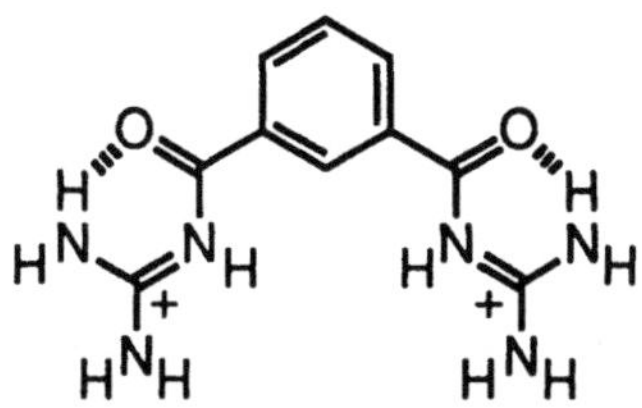

13 (picrate)$_2$

Addition of one equivalent of tetrabutylammmonium diphenylphosphate (TDPP) to a CD_3CN solution of **13**-TPB_2 leads to downfield shifts of the protons on the guanidinium directed into the cavity. There is almost no shift, however, in the position of the outwardly-directed protons. These results are consistent with the formation of a complex of type **14** in which both guanidiniums converge on the central cavity to form four hydrogen bonds to a single phosphate substrate.

A second series of ^1H NMR changes occur on further addition of TDPP (from one to approximately three equivalents) to the CD_3CN solution. Now the outwardly-directed guanidinium protons shift dramatically downfield , while the inwardly directed protons barely move. These NMR changes indicate the formation of a 3 : 1 complex, as in **15** in which the outwardly directed protons act as two additional binding sites for phosphodiester anions. Non-linear regression analysis of a binding isotherm generated by a 1: 1 dilution

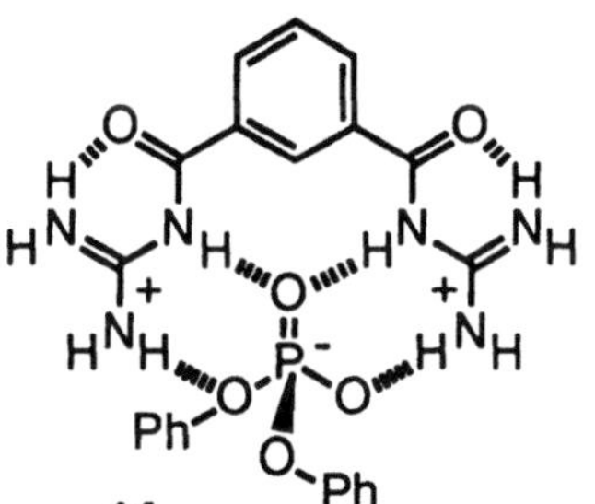

14

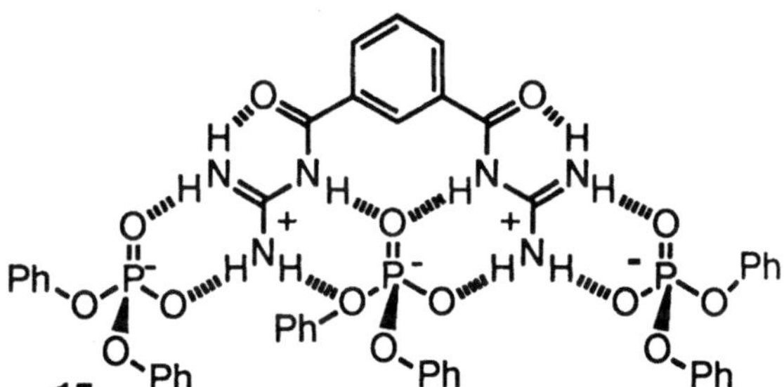

15

experiment gave a Ka for the complex between diphenyl phosphate and **13** of $4.6 \pm 1.7 \times 10^4$ M^{-1}.

Receptors **12** and **13** are well-suited for the catalysis of phosphodiester cleavage. They possess a dicationic trigonal binding cavity that should be complementary both in terms of shape and electrostatics to the dianionic trigonal bipyramidal intermediate for nucleophilic attack on a phosphodiester.

References.

1. For a recent review see Dugas, H. (1989) Bioorganic Chemistry, Springer Verlag, New York, 2nd Edition.

2. Kraut, J. (1988) Science (Washington, D.C.), ' How Enzymes Work' 242, 533.

3. Chang, S. K., Van Engen, D., Fan, E.and Hamilton, A. D. (1991) ' Molecular Recognition and Hydrogen Bonding: Synthetic, Complexation and Structural Studies on Barbiturate Binding to a Synthetic Receptor'J. Am. Chem. Soc. 113, 7640.

4. Garcia-Tellado, F., Goswami, S., Chang, S. K., Geib, S. and Hamilton, A. D. (1990)' Molecular Recognition: A Remarkably Simple Receptor for Dicarboxylic Acids J. Am. Chem. Soc. 112, 7393

5. Burley, S. K. and Petsko, G. A. (1985) ' Aromatic-Aromatic Interactions' Science (Washington, D.C.), 229, 23.

6. Muehldorf, A. V., Van Engen, D., Warner, J. C. and Hamilton, A. D. (1988) 'Aromatic-Aromatic Interactions in Molecular Recognition' J. Am. Chem. Soc. 110, 6561.

7. Goswami, S., Van Engen, D. and Hamilton, A. D. (1988) ' Nucleotide Base Recognition: A Macrocyclic Receptor for Adenine Employing Hydrogen Bonding and Aromatic Stacking Interactions' J. Am. Chem. Soc. 111, 3425.

8. Hamilton, A. D.and Pant, N. (1988) ' Ditopic Binding of Guanine to a Macrocyclic Receptor' J. Chem. Soc. Chem. Commun. 765.

9. Tecilla, P., Chang, S. K. and Hamilton, A. D. (1990) ' Transition State Stabilization and Molecular Recognition' J. Am. Chem. Soc.112, 9586.

Unimolecular Micelles.

George R. Newkome

Center for Molecular Design & Recognition, Department of Chemistry,

University of South Florida, Tampa, Florida 33620 USA

Over the past 25 years, diverse cavities within a macrocyclic framework have been created to probe all sorts of host-guest combinations with the futuristic goal of molecular recognition. These cavities have been designed with arms & legs, inner & outer functionalized coatings, small & large cavities with two- & three-dimensional frameworks, flexible & rigid superstructures, et cetera. Is the cavity necessary or would a functionalized surface or crevice offer easy ingress for reagent and egress for products? The creation of an unimolecular micelle has the advantage that the inner region is lipophilic and the surface is hydrophilic giving rise to enhanced water-solubility as well as porous enough to permit the intrusion of lipophilic reagents. Further if the inner 'void' regions or crevices have anchor points for guests or metal ions, chemical transformations may be possible within a suitable microenvironment.

The initial synthetic model was based on the architectural design for trees, thus the term 'arborol', in which a 'cascade' approach, put forth by Professor Vögtle[1] (Scheme 1), would be based on a simple mathematical progression [1→3→9→27→81→243→....]. This is best demonstrated in our initial published[2] example in 1985 (Scheme 2). The limiting feature is that a spacer moiety had to be utilized to circumvent the effects of the inherent neopentyl branching locus.

V. Balzani and L. De Cola (eds.), Supramolecular Chemistry, 145–155.

Scheme 1

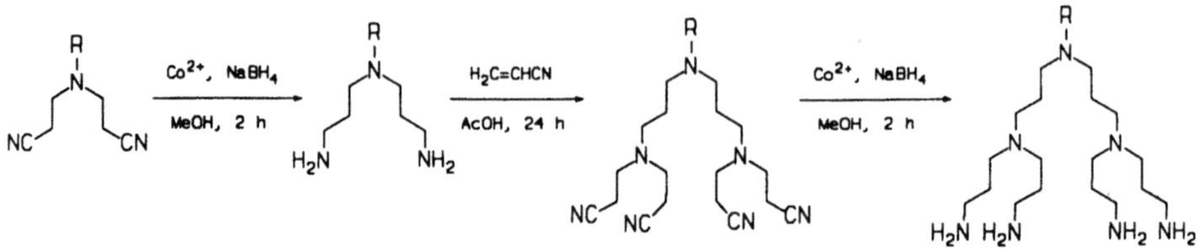

The effect of this quaternary center is further reflected in the retarded reactivity of **1**[3] with triethyl sodiomethanetricarboxylate. It was only when sufficient distance between the branching locus and the leaving group (in general > 3 atoms) was incorporated into the building blocks and cores the real progress occurred. To demonstrate this (Scheme 3), **2** was smoothly transformed to dodecaester **3**,[4] whereas pentaerythritol and **1** afforded either recovered starting materials or spiranes under drastic reaction conditions.

Scheme 3

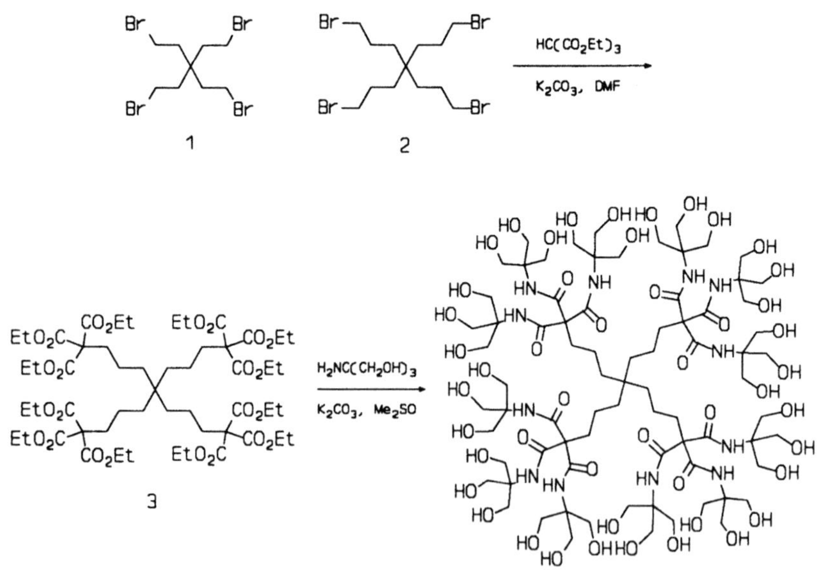

'Tris' also possesses this undesired neopenyl handicap and although added transformations to incorporate a spacer group proved successful,[2] they added several steps to the overall iterative

Scheme 2

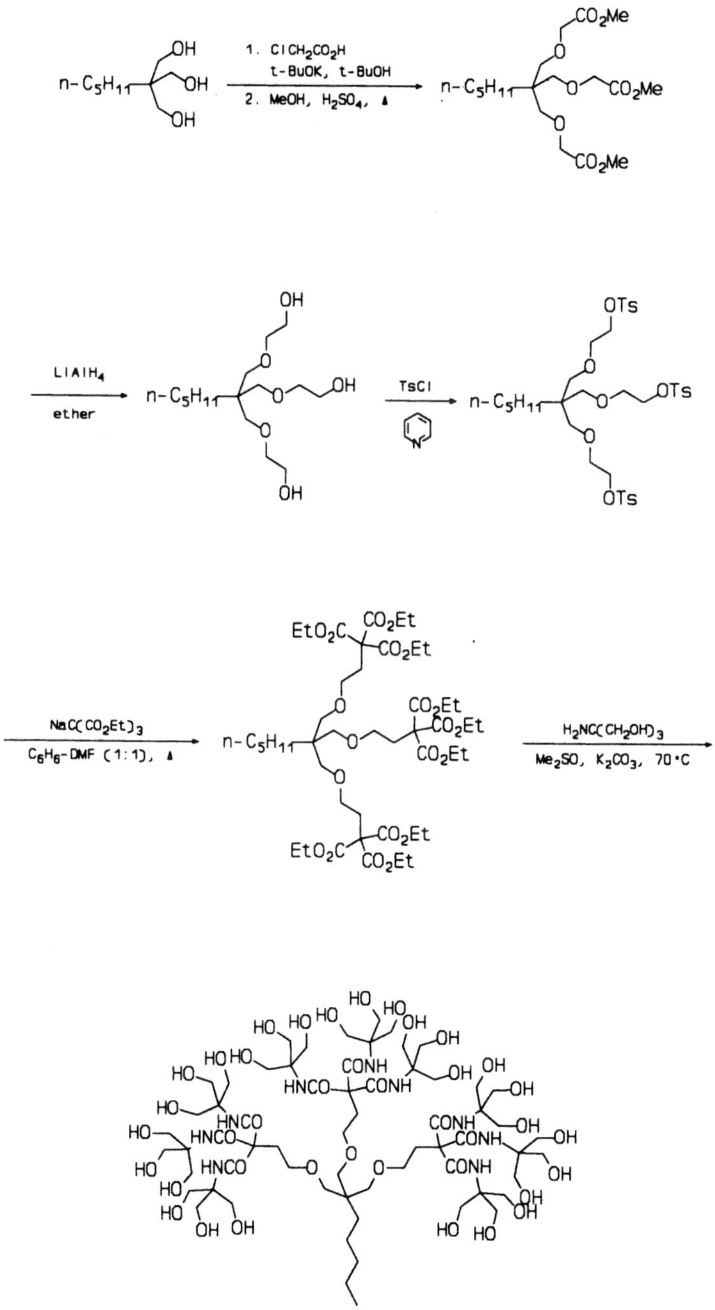
n-C5H11
OH
1. ClCH2CO2H
t-BuOK, t-BuOH
2. MeOH, H2SO4, Δ
CO2Me
O
LiAlH4
ether
TsCl
OTs
NaC(CO2Et)3
C6H6-DMF (1:1), Δ
EtO2C
CO2Et
H2NC(CH2OH)3
Me2SO, K2CO3, 70 °C
HO
CONH
HNCO

process. We thus created a series of new reagent[5] which directly incorporated the necessary 3-atom spacer linkage (Scheme 4). The addition of acroylnitrile to nitromethane afforded the trinitrile **4**, which was easily hydrolyzed to the corresponding triacid **5**. Reduction of **5** with borane quantitatively afforded triol **6**; further reduction with T1 Raney Nickel yielded the desired 'Bishomotris' **7**.

Scheme 4

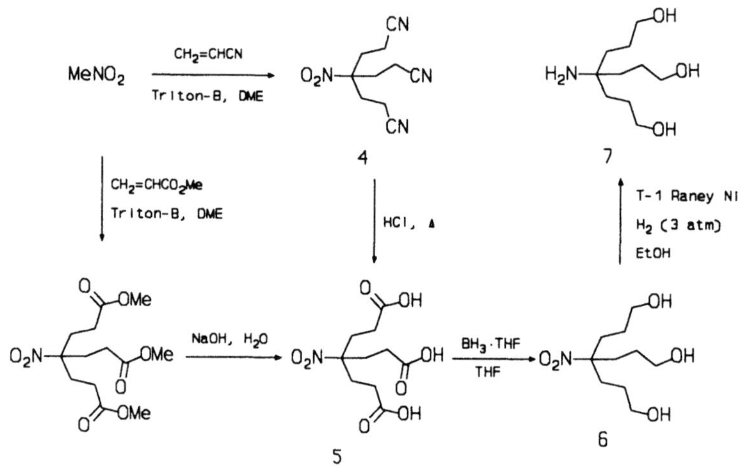

Simple modification of this sequence utilizing *tert*-butyl acrylate afforded the corresponding *tert*-butyl triester, which when reduced with Raney nickel gave the desired amine **8**[6] (Scheme 5). Interestingly, other unhindered alkyl ester under these reductive conditions gave rise to the worthless lactam. Treatment of a model tetraacyl chloride core (**9**) with amine **8** gave the desired dodecaester **10**, which when treated with formic acid quantitatively gave the corresponding dodecaacid **11**.[7] Thus the aminotriester **8**, shortens the cascade sequence to two simple steps (Scheme 6); amide formation (via either the acyl halide or classical peptide coupling procedures), hydrolysis, and repetition of the sequence with **8**.

Scheme 5

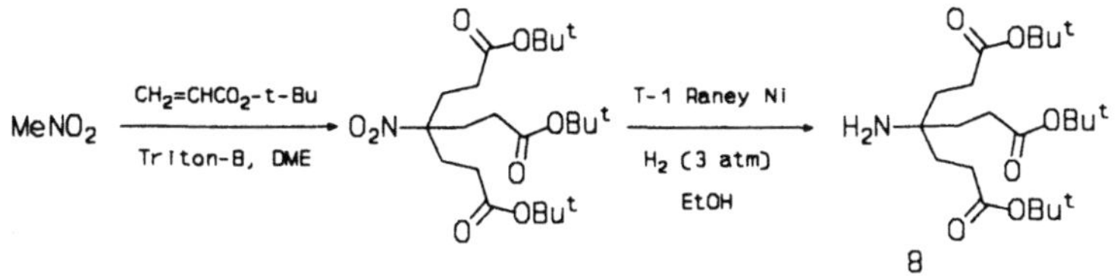

The pivotal reagent proved to be nitrotriol **6**, which can be readily acetylated with acetic anhydride or benzylated (**12**) with benzyl chloride. A family of key monomers and cores was derived from ether **12** by replacement of the nitro moiety by denitration-cyanoethylation (Scheme 7), via the method of Ono[8] generated the mononitrile **13**; reduction with borane gave amine **14**, whereas catalytic reduction generated the 'trishomobishomotris' (**15**).

Scheme 7

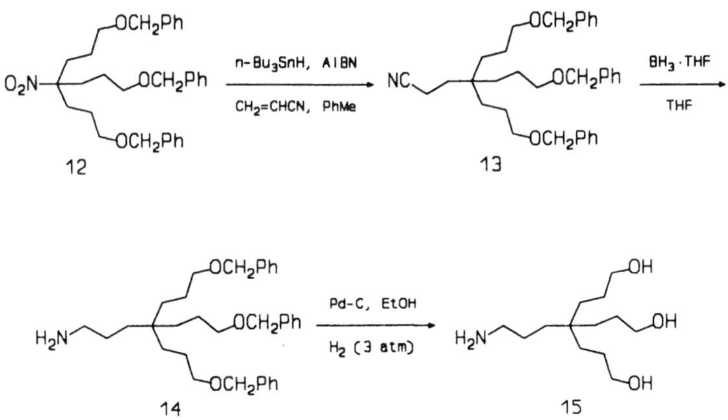

Nitrile **13** was readily converted to aldehyde **16**, via reduction with DIBAL, which was treated with 1,3-propanedithiol gave the dithiane **17** (Scheme 8). The carbanion of **17** afforded the first reagent that can give entrance to an all *C*-unimolecular micelle. Due to the bulky environment of this dithiane, it was rationalized that a cleaner procedure would utilize a carbanion with less steric congestion. Thus, when aldehyde **16** is reduced with sodium borohydride, alcohol **18** was isolated in excellent yield; alternatively, nitrile **13** can be hydrolyzed

Scheme 6

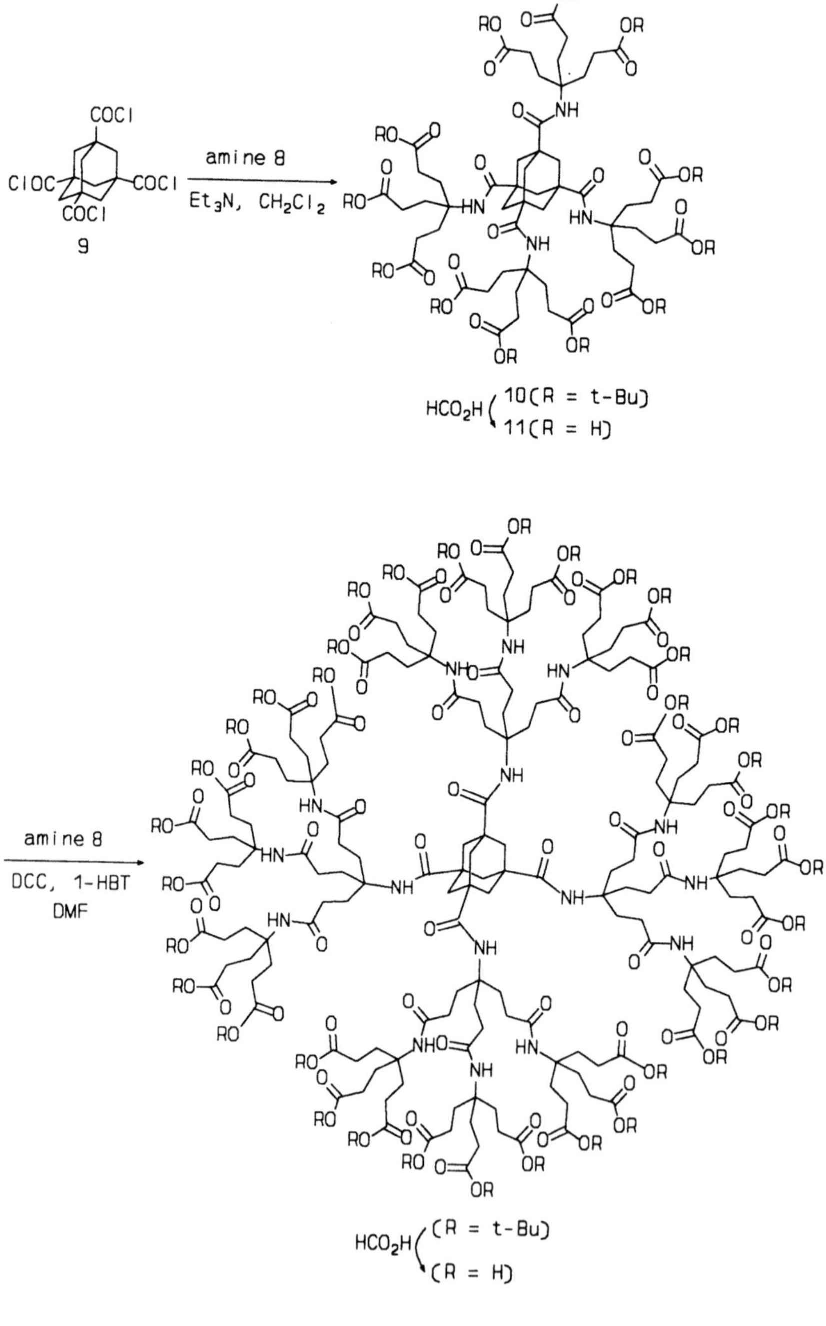

COCl
ClOC
COCl
COCl
9
amine 8
Et3N, CH2Cl2
HCO2H
10(R = t-Bu)
11(R = H)
amine 8
DCC, 1-HBT
DMF
HCO2H
(R = t-Bu)
(R = H)

to the corresponding acid, which can reduced with diborane. Treatment of alcohol **18** with thionyl chloride afforded chloride **19**, which when treated with lithium acetylide gave the desired alkyne **20**, an ideal monomer for the construction of the *C*-unimolecular micelle.[9]

Scheme 8

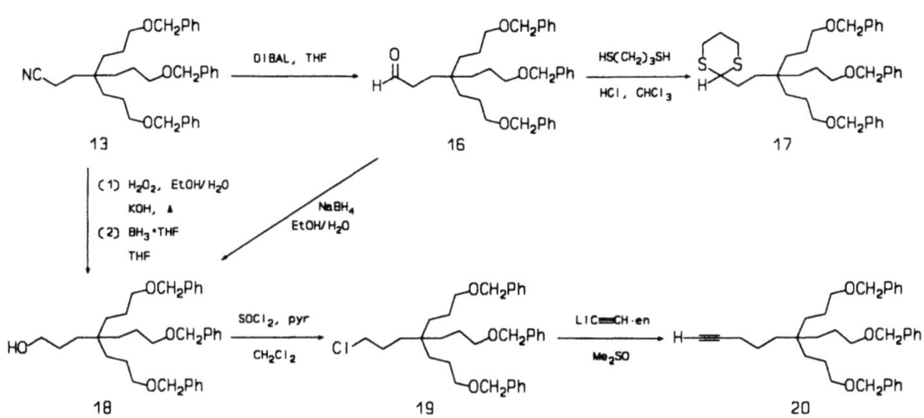

With the availability of diverse, easily prepared monomer building blocks, combinations of these reagents permit the generation of macromolecules possessing varied internal functionality. The surface can similarly be converted to whatever functionality is desired. The preparation of a simple four-directional core, however, was limited; it was deemed necessary to devise a scheme to a core possessing a neopentyl center, three carbon arms, and terminated with a good leaving group (Scheme 9). The hydrogenolysis of the hydroxybenzylether **18** gave tetraol **21**, which with HBr gave excellent overall yields of the tetrabromide **1**. Direct treatment of **18** with HBr afforded the identical tetrabromide.

Generation of the alkynide anion of **20**, followed by addition bromide **1** gave rise to the alkyne-ether protected first tier (**22**). Subsequent catalytic reduction of **22** afforded in one-step the dodecaol **23** in excellent yield. Treatment of **23** with HBr afforded the dodecabromide, which

Scheme 9

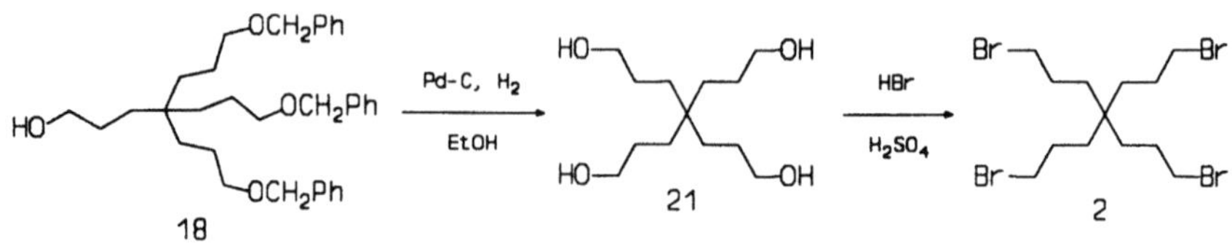

upon treatment with a slight excess of the same acetylide reagent gave the next tier (**24**). We have used the term Micellane^{TM} to denote the structural similarity to the traditionally envisioned spherical micelle. Repetition of the synthetic sequence will afford access to the larger members in this series.

These polyols (e.g. **24**) can readily be oxidized with $RuO_2/NaIO_4$ to afford the [8^2•3]Micellanoic acid^{TM} (**25**). Such structures are totally water soluble and possess a topological resemblance to globular micelles. They also exhibit the typical characteristics of a micelle such an ability to incorporate lipophilic reagents, are monodispersed, and show similar spectral properties with probe molecules to that of a spherical micelle.

The micellar characteristics[10] of the polytetramethylammonium salt of [8^2•3]Micellanoate^{TM} were investigated via UV analysis of guest molecules such as pinacyanol chloride, phenol blue, naphthalene, and diphenylhexatriene, as molecular probes. Such studies confirm the availability of 'void' regions within the Micellanoate^{TM} structure. These regions are variable depending on the solvent, similar to normal polymers, and pH changes. The expanded and contracted structures are shown in Figure 1. It is interesting to note the dimensions of the Micellanoate^{TM} structures when compared to that of buckminsterfullerene[11] - there is sufficient room for four (4) Bucky balls within these 'void' regions. With this knowledge, can one conduct chemistry within these regions. In order to probe this region, we treated the internal acetylenic groups with a metal(0) reagent, bis(triphenylphosphine)platinum(0), known to form stable complexes.[12] The resultant creation of platinacycles within these regions suggest that internal

Scheme 10

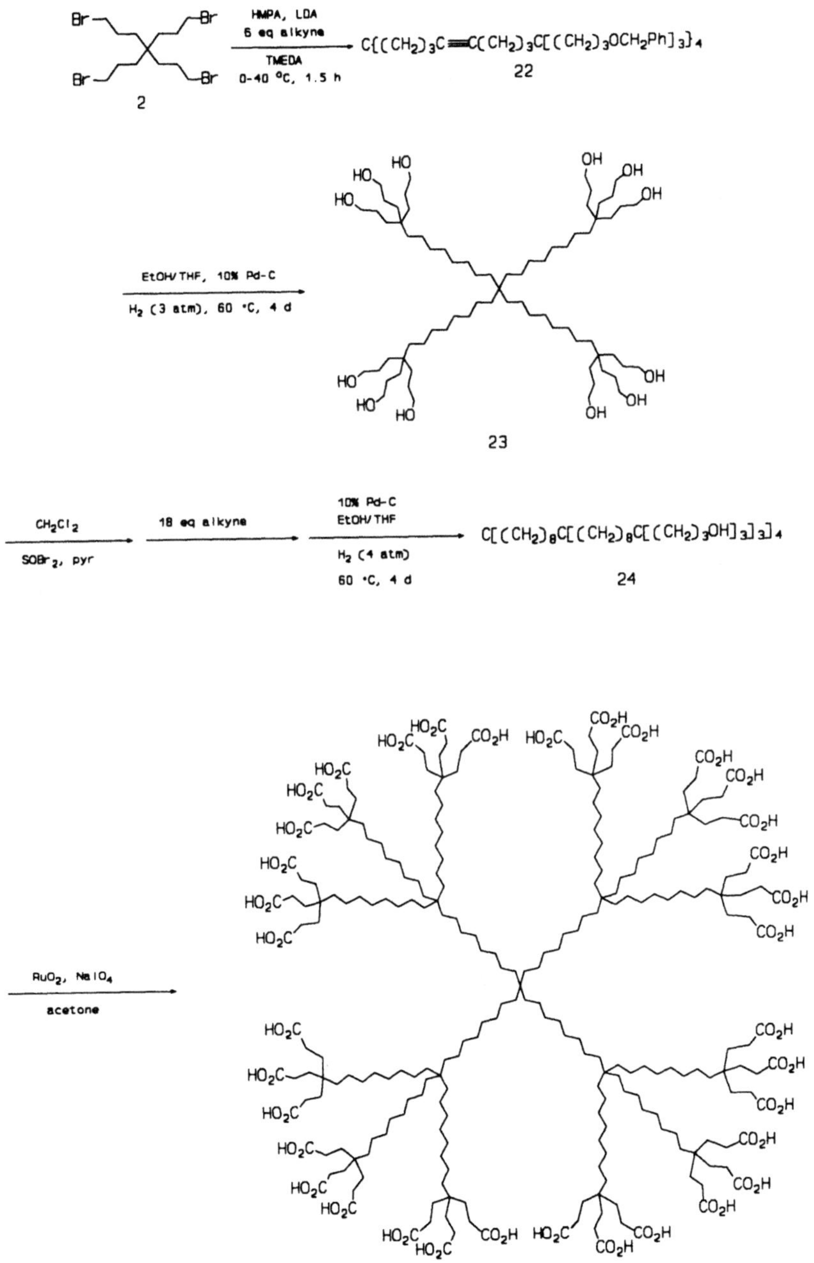

HMPA, LDA
6 eq alkyne
TMEDA
0-40 °C, 1.5 h
C{(CH2)3C≡C(CH2)3C[(CH2)3OCH2Ph]3}4
2
22
EtOH/THF, 10% Pd-C
H2 (3 atm), 60 °C, 4 d
23
CH2Cl2
SOBr2, pyr
18 eq alkyne
10% Pd-C
EtOH/THF
H2 (4 atm)
60 °C, 4 d
C[(CH2)8C[(CH2)8C[(CH2)3OH]3]3]4
24
RuO2, NaIO4
acetone
25

coating of the regions with metal ions is possible. The availability to introduce diverse metal at different depths within the micellar structure, control the oxidation state, and conduct catalytic processes within these porous structures suggest the creation of flexible organic zeolites.

Figure 1

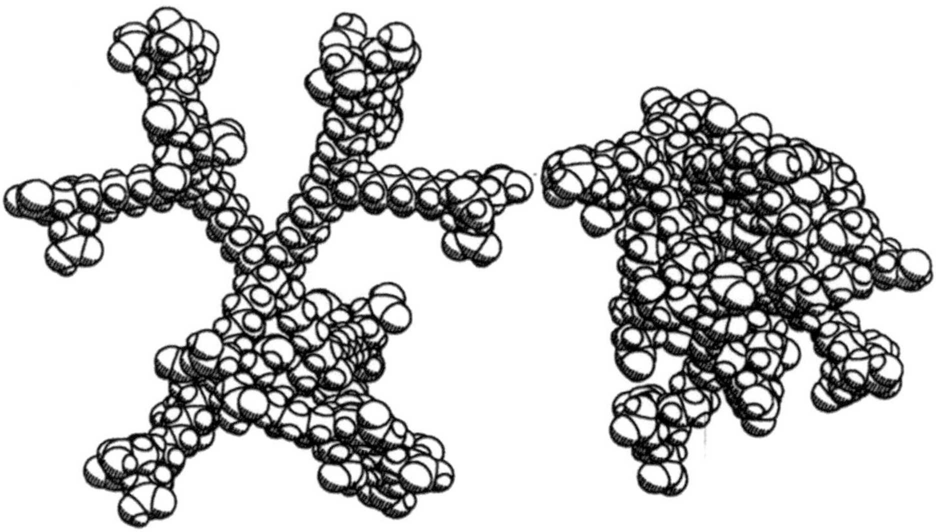

References

1. Buhleier, E.; Wehner, W.; Vögtle, F. *Synthesis* **1978**, 155.
2. Newkome, G. R.; Yao, Z.-q.; Baker, G. R.; Gupta, V. K. *J. Org. Chem.* **1985**, *50*, 2003.
3. Newkome, G. R.; Gupta, V. K.; Griffin, R. W.; Arai, S. *J. Org. Chem.* **1987**, *52*, 5480.
4. Newkome, G. R. American Chemical Society Meeting, Spring, 1987.
5. Newkome, G. R.; Moorefield, C. N.; Theriot, K. J. *J. Org. Chem.* **1988**, *53*, 5552.
6. Newkome, G. R.; Behera, R. K.; Moorefield, C. N.; Baker, G. R. *J. Org. Chem.* **1991**, *56*, 7162-7169.
7. Newkome, G. R.; Nayak, A.; Behera, R. K.; Moorefield, C. N.; Baker, G. R. *J. Org. Chem.* **1991**, in press.

8. Ono, N.; Miyake, H.; Kamimura, A.; Hamamoto, I.; Tamura, R.; Kaji, A. *Tetrahedron* **1985**, *41*, 4031.

9. Newkome, G. R.; Moorefield, C. N.; Baker, G. R.; Johnson, A. L.; Behera, R. K. *Angew. Chem. Int. Ed. Engl.* **1991**, *30*, 1176.

10. Newkome, G. R.; Moorefield, C. N.; Baker, G. R.; Saunders, M. J.; Grossman, S. H. *Angew. Chem. Int. Ed. Engl.* **1991**, *30*, 1178.

11. Krätschmer, W.; Lamb, L. D.; Fostiropoulos, K.; Huffman, D. R. *Nature* **1990**, *347*, 354.

12. Newkome, G. R.; McClure, G. L. *J. Am. Chem. Soc.* **1974**, *96*, 617.

RU(II) AND/OR OS(II) TRIMETALLIC COMPLEXES OF TRIS(BIPYRIDINE) BRIDGING LIGANDS. ABSORPTION SPECTRA, LUMINESCENCE PROPERTIES, ELECTROCHEMICAL BEHAVIOR, AND INTERCOMPONENT ENERGY TRANSFER.

Luisa De Cola, Francesco Barigelletti, and Vincenzo Balzani
Dipartimento di Chimica "G. Ciamician" dell'Università and Istituto FRAE-CNR, 40126 Bologna, Italy

Peter Belser and Alex von Zelewsky
Institut für Anorganische Chemie, Universität Freiburg, Freiburg, Switzerland

Christian Seel, Michael Frank, and Fritz Vögtle
Institut für Organische Chemie und Biochemie, Universität Bonn, Bonn, Germany

ABSTRACT. The tris(bipyridine) tripod ligands 1,3,5-tris[4-(((2,2'-bipyridyl-5-yl)carbonyl)benzylamino)methyl]benzene, **1**, and 1,3,5,tris-[4-(((2,2'-bipyridyl-5-yl)carbonyl)benzylaminomethyl)phenyl] benzene, **2**, have been synthesized and their complexes **1**$[Ru(bpy)_2]^{2+}$, **1**$[Ru(bpy)_2]_2{}^{4+}$, **1**$[Ru(bpy)_2]_3{}^{6+}$, **1**$[Os(bpy)_2]_3{}^{6+}$, **1**$[Ru(bpy)_2]_2[Os(bpy)_2]^{6+}$, **2**$[Ru(bpy)_2]_3{}^{6+}$, **2**$[Os(bpy)_2]_3{}^{6+}$, and **2**$[Ru(bpy)_2]_2[Os(bpy)_2]^{6+}$ have been prepared. All the complexes display very intense, ligand centered absorption bands in the UV region and moderately intense metal-to-ligand charge transfer bands in the visible. Electrochemical oxidation of each Ru(II) or Os(II) metal center occurs always at the same potential (+1.30 V for Ru(II), +0.87 V for Os(II)), regardless of the number and type of metal-based units that are present in the supramolecular structure. All the homometallic Ru(II) complexes exhibit the same luminescence properties, and this also happens for the homometallic Os(II) complexes. The luminescence data obtained for the mixed-metal complexes show that electronic energy transfer takes place from the Ru-based to the Os-based

V. Balzani and L. De Cola (eds.), Supramolecular Chemistry, 157–180.

components. The efficiency of energy transfer decreases in going from **1**$[Ru(bpy)_2]_2[Os(bpy)_2]^{6+}$ to **2**$[Ru(bpy)_2]_2[Os(bpy)_2]^{6+}$, i.e. as the size of the spacer which links the three arms of the bridging ligand increases. The mechanism of the energy transfer process is discussed.

1. Introduction

Light absorption by a molecular species causes electronic excitation. The transfer of the electronic excitation energy from a molecular species to another is a process of considerable basic and applied interest. In the natural photosynthetic process light absorption and electronic energy transfer occur in arrays of pigments incorporated into a protein matrix, and the transferred electronic energy is used for chemical synthesis [1]. In artificial photochemical molecular devices, light absorption and electronic energy transfer may be used to obtain important functions such as spectral sensitization, antenna effect, remote photosensitization, and light-energy up-conversion [2].
In an attempt to elucidate the role played by the various factors which control the occurrence of electronic energy transfer, much attention is currently devoted to the design and synthesis of supramolecular species [2,3] where energy transfer takes place between covalently linked components [4-11].

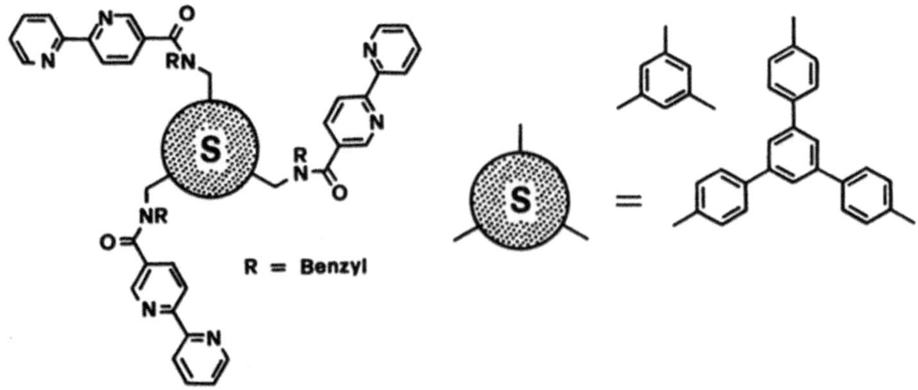

Fig. 1 - Structural formulas of the bridging ligands **1** and **2**.

Along this line, we have designed and synthesized the tris(bipyridine) ligands **1** and **2** (Fig. 1) which can coordinate three (equivalent or different) metal-containing building blocks. As shown in Fig. 1, ligand **2** differs from ligand **1** only for the presence of an additional phenylene group in each arm of the spacer S which bridges the three coordinating bpy-type sites. Coordination of $Ru(bpy)_2^{2+}$ and/or $Os(bpy)_2^{2+}$ units to such sites yields supramolecular species which contain three $M(bpy)_3^{2+}$-type complexes (M=Ru or Os) linked by a spacer (Fig. 2). On the basis of previous investigations on mononuclear [12,13] and oligonuclear [4-7,9-11] Ru(II) and Os(II) polypyridine complexes, it can be expected that the supramolecular species schematized in Fig. 2 exhibit

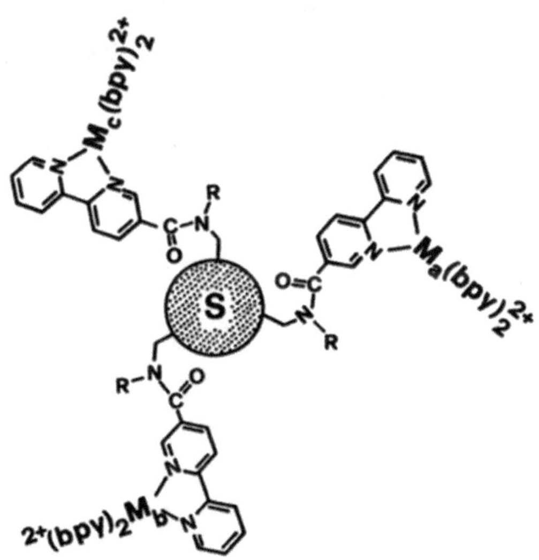

Fig. 2 - Structure of the trimetallic complexes. For the spacer S, see Fig. 1.

several interesting properties: (i) they should show intense absorption bands in the UV and visible spectral regions; (ii) they should display luminescence both in rigid matrix at 77 K and in fluid solution at room temperature; (iii) they should undergo reversible mono- and multi-electron redox processes; (iv) energy transfer from the Ru-based to the Os-based units should take place in the Ru(II)-Os(II) mixed-metal species; (v) photoinduced electron transfer processes should occur when the metal ions exhibit different oxidation states; (vi) it should be possible to study the interaction between excited chromophoric units upon laser excitation.

In this paper we describe: (a) the synthesis of the ligands **1** and **2**; (b) the absorption spectra, luminescence properties, and electrochemical behavior of the following complexes (hereafter indicated by the abbreviations given in bold into parenthesis): **1**$[Ru(bpy)_2]^{2+}$ (**1.Ru**), **1**$[Ru(bpy)_2]_2^{4+}$ (**1.Ru$_2$**), **1**$[Ru(bpy)_2]_3^{6+}$ (**1.Ru$_3$**), **1**$[Os(bpy)_2]_3^{6+}$ (**1.Os$_3$**), **1**$[Ru(bpy)_2]_2[Os(bpy)_2]^{6+}$ (**1.Ru$_2$Os**), **2**$[Ru(bpy)_2]_3^{6+}$ (**2.Ru$_3$**), **2**$[Os(bpy)_2]_3^{6+}$ (**2.Os$_3$**), and **2**$[Ru(bpy)_2]_2$ $[Os(bpy)_3]^{6+}$ (**2.Ru$_2$Os**); (c) the intercomponent energy transfer in the **1.Ru$_2$Os** and **2.Ru$_2$Os** species. Some results obtained for the complexes of **1** have been reported in a preliminary communication [14]. A detailed description of the synthesis and characterization of the above listed complexes and the results of investigations concerning intercomponent electron transfer in mixed-valent species and interactions between excited states will be reported elsewhere.

2. Experimental Section

Commercial 2-acetylpyridine, N,N-dimethylacetamide, triethylamine, 2,2'-bipyridine, $RuCl_3.3H_2O$ and $(NH_4)_2OsCl_6$ were used as received. $Ru(bpy)_2Cl_2.2H_2O$ and $Os(bpy)_2Cl_2$ were prepared according to literature procedures[15,16]. The solvents and reactants used were of the highest purity commercially available and were used as received.

2.1 Preparation of the ligands

The ligand **1** was synthesized in a four-step reaction as illustrated in scheme 1. The pyridinium-salt[17] of 2-acetylpyridine was transformed to 5-methyl-2,2'-bipyridine via the Kröhnke reaction[18]. The methyl group was then changed into a carboxylic group by oxidation with potassium permanganate[19]. The 2,2'-bipyridine-5-carboxylic acid was boiled in thionyl chloride to obtain the corresponding acid chloride. The condensation of three molecules of 2,2'-bipyridine-5-carboxylic chloride with one molecule of 1,3,5-tris[N-benzyl(aminomethyl)]benzene[20] (spacer) gave the ligand **1**. In the last step

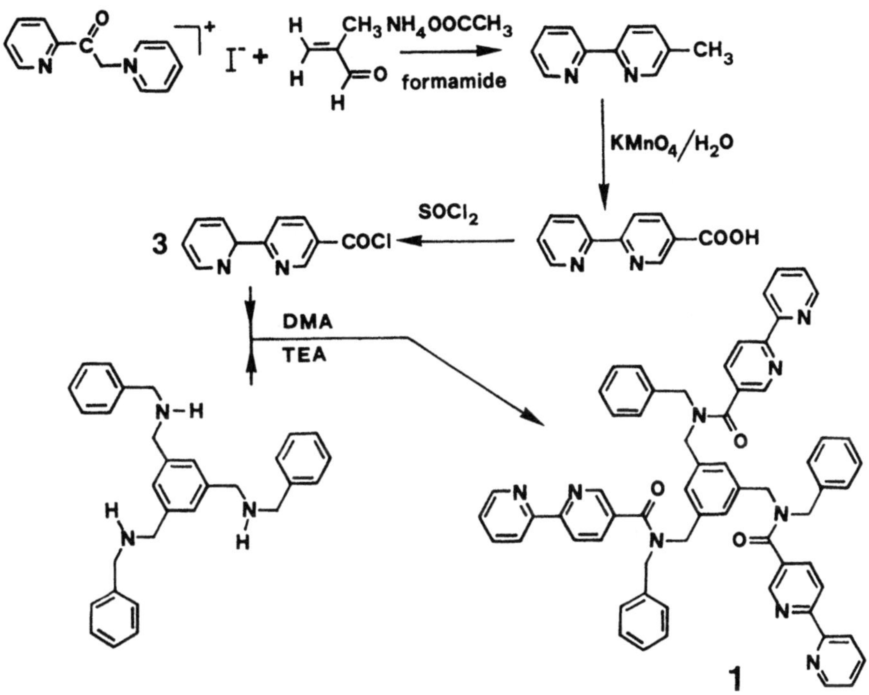

Scheme 1

of this synthesis, 2.5 g (5.74 mmol) of the spacer dissolved in 37 ml N,N-dimethylacetamide (DMA) and 3.76 g (17.21 mmol) of 2,2'-bipyridine-5-carboxylic chloride dissolved in 84 ml DMA were mixed together. Nine ml of triethylamine were then added and the mixture heated for 3 minutes at 120 °C. The solution was stirred overnight at room temperature. The salt of the triethylamine hydrochloride was filtered off and the solvent was distilled at 100 °C under reduced pressure. The remaining brown crystalline mass was taken up in ethyl acetate, treated with activated charcoal and dried with magnesium sulfate. The solution was filtered and evaporated to dryness. The foamy solid was then redissolved in dichloromethane (50 ml) and the solution saturated with n-hexane. Both solvents were evaporated slowly at low temperature. The last procedure was repeated and the remaining yellowish, foamy solid was vacuum dried at 70 °C (4.27 g, 75.7%). *UV (ε) [nm]*: 287 (54700), 248 sh (32900), 241 (36700). *FAB*: m/z = 983 (100%) M^+-peak, m/z = 843 (17%), m/z = 799 (30%), m/z = 512 (14%). M.P.: 100.7 °C. *1H NMR (360 MHz, $CDCl_3$)*: δ =

4.40 (s, 2 H broad, aliphatic protons); 4.68 (s, 2 H broad, aliphatic protons); 6.97 - 7.33 (m, 7 H); 7.72 (dxd, 1H); 7.86 (d, 1 H); 8.31 (t, 2 H broad); 8.58 (s, 1 H); 8.75 (s, 1 H).

The ligand **2** was prepared with the procedure shown in scheme 2. A mixture of 1,3,5-tris[4-((benzyl)aminomethyl)phenyl]benzene [21] (2.0 g, 3.0 mmol) and triethylamine (1.01 g, 10 mmol) in 10 ml dimethylformamide (10 ml) was added to a stirred mixture of 2,2'-bipyridine-5-acid-chloride (2.0 g, 10 mmol) in dimethylformamide (25 ml). Stirring was continued overnight and after evaporation of the solvent the residue was dissolved in chloroform. The chloroform layer was concentrated in vacuo and chromatographed on alumina

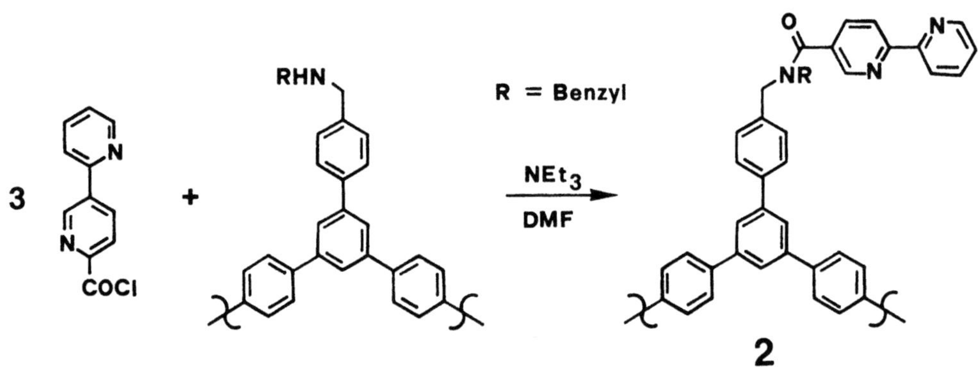

Scheme 2

with chloroform to afford nearly pure **2**. For final purification the compound was recrystallized in an ethylacetate/cyclohexane mixture (1.9 g, 55%). FAB: *m/z* = 1210.4 (MH^+, 100%; calc.:1210.5), 1119.4 (MH^+-benzyl, 2%), 1026.4 (M^+-bpyCO, 29%). M.P. 123-127 °C. *^{1}HNMR (200 MHz $CDCl_3$):* δ = 4.53 (6 H, Ar-CH_2), 4.82 (6 H, Ar-CH_2), 7.15 -7.45 (m, 24 H, 21 Ar-H, 3 bpy-H), 7.68 - 7.76 (m, 6 H, Ar-H), 7.79 - 7.84 (m, 3 H, Ar-H) 7.82 (td, 3 bpy-H, ^3J = 7.8 Hz, ^4J = 1.8 Hz), 7.97 (dd, 3 bpy-H, ^3J = 8.2 Hz ^4J = 2.2 Hz), 8.38 (dt, 3 bpy-C-H, ^3J = 7.8 Hz, ^4J = 1.0 Hz), 8.44 (dd, 3 bpy-H, ^3J = 8.2 Hz, ^4J = 0.7 Hz), 8.67 (ddd, 3 bpy-H, ^3J = 4.8 Hz, ^4J = 1.8 Hz, ^4J = 1.0 Hz), 8.86 (dd, 3 bpy-H, ^4J = 2.2 Hz, ^4J = 0.7 Hz). *^{13}C-NMR (22.62 MHz, $CDCl_3$):* δ = 47.42 (CH_2), 51.40 (CH_2), 120.70, 121.38 (CH), 124.23 (CH), 125.20 (CH), 127 - 129 (8 CH), 131.64 (C), 135.62 (CH), 136.46 (CH), 137.01 (CH), 140.57 (C), 141.90 (C), 147.18 (CH),

149.35 (CH), 155.20 (C), 157.21 (C), 169.90 (CO). *IR*: 710 (m), 770 (s), 820 (w), 1005 (m) , 1165 (m), 1260 - 1320 (m), 1435 (s), 1460 (s), 1600 (s), 1650 (vs), 2980 - 3020 (w).

2.2 Preparation of the complexes

The monometallic complexes were prepared according to the reaction

$$x[Ru(bpy)_2Cl_2] + \mathbf{1} \text{ or } \mathbf{2} \xrightarrow[2\ \ 2x\,NH_4PF_6]{1\ \uparrow\downarrow\ \text{Ethylene glycol}} (\mathbf{1} \text{ or } \mathbf{2})[Ru(bpy)_2]_x(PF_6)_{2x}$$

$$x[Os(bpy)_2Cl_2] + \mathbf{1} \text{ or } \mathbf{2} \xrightarrow[2\ \ 2x\,NH_4PF_6]{1\ \uparrow\downarrow\ \text{Ethylene glycol}} (\mathbf{1} \text{ or } \mathbf{2})[Os(bpy)_2]_x(PF_6)_{2x}$$

x = 1, 2, 3

Scheme 3

scheme 3, and the mixed-metal complexes according to the reaction scheme 4. The compounds were purified by chromatography on partially deactivated alluminium oxide, chromatography on silica gel, preparative thin layer plates on silica gel, and recrystallization from an acetonitrile/diethyl ether mixture (vapour diffusion method). Characterization was performed with UV/VIS, luminescence, ^{1}HNMR, IR, FAB MS, cyclic voltammetry, thin layer chromatography, and elemental analysis. Details on the procedures used for synthesis, purification, and characterization of the novel complexes will be reported elsewhere.

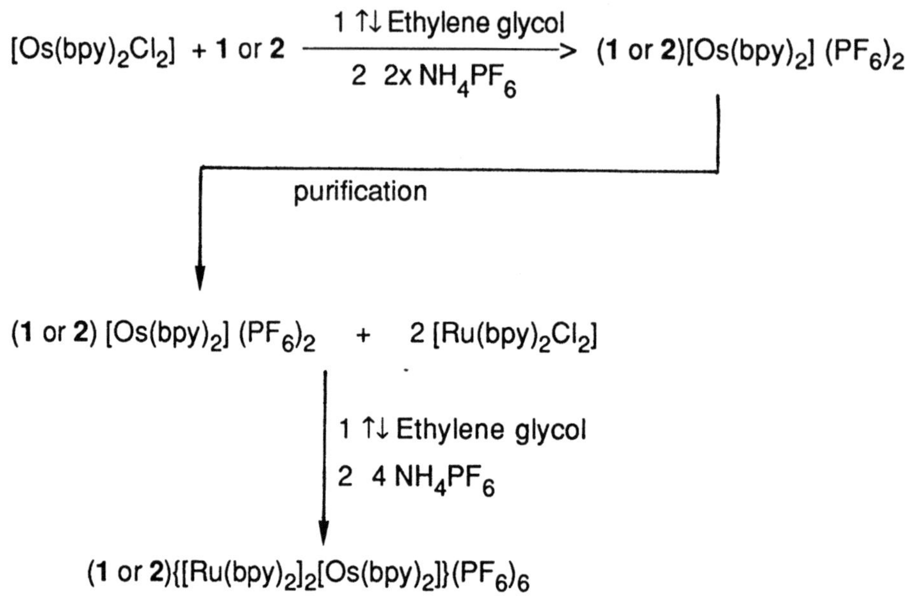

Scheme 4

2.3 Equipment and methods

Electrochemical measurements were carried out at room temperature (~ 25 ^{o}C) by using a Metrohm E/506 Polarecord, a Metrohm E/612 VA scanner, and a Hewlett-Packard 7044 x-y recorder. Cyclic voltammograms were obtained in acetonitrile solution by using a microcell equipped with a stationary platinum disk electrode, a platinum disk counter electrode, and a SCE reference electrode with tetrabutylammonium hexafluorophosphate as supporting electrolyte. In all cases $[Ru(bpy)_3](PF_6)_2$ was used as a standard, taking its oxidation potential equal to +1260 mV vs SCE [22, 23] The electrochemical window examined was between +2.0 and -2.0 V. Scanning speed was 200 mV s^{-1}. All the reported values are *vs* SCE. Half-wave potentials were calculated as an average of the cathodic and anodic peaks.

Luminescence experiments were performed in acetonitrile at room temperature and in a 4:5 v/v propionitrile-butyronitrile mixture at 77 K. The absorption spectra, emission spectra, luminescence decays, and electrochemical potentials were obtained as previously described [24]. Interference filters, cut-off filters, or monochromators were used to select appropriate spectral regions for lifetime measurements. The procedure used to obtain values for the quenching of the luminescence of the Ru-based components and the sensitization of the luminescence of the Os-based component in the mixed-metal compounds was as follows (for the sake of simplicity, we will describe the case of **1.Ru_2Os**; the same procedure was used for **2.Ru_2Os**). First, we have recorded the absorption spectra of equimolar (1.0×10^{-5} M) solutions of **1.Ru_2Os** and of its **1.Ru_3** and **1.Os_3** "parent" compounds and we have found that they exhibit an isosbestic point at 461 nm (Fig. 3). Then, solutions of **1.Ru_2Os** and of a 2:1 mixture of **1.Ru_3**

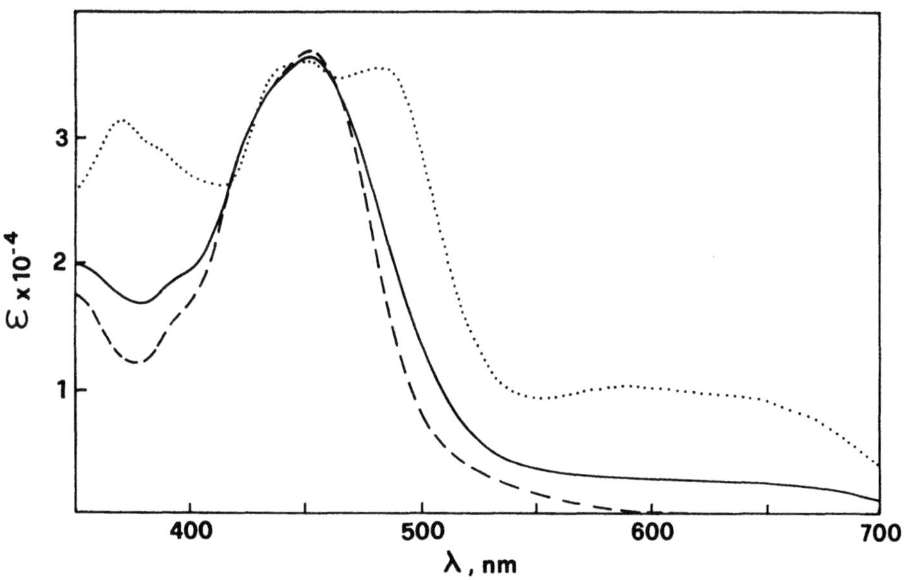

Fig. 3 - Absorption spectra of **1.Ru_3** (---), **1.Os_3** (...), and **1.Ru_2Os** (—) in acetonitrile solution at room temperature

and **1.Os_3** having the same concentrations were prepared and were found to exhibit identical absorption spectra. The luminescence spectra of such solutions were recorded with excitation in the isosbestic point at 461 nm under identical instrumental conditions. The quenching of the Ru-based luminescence was obtained by comparing the heights of the emission bands at 640 nm (Fig. 4) where the luminescence of the Os-based unit is negligible.

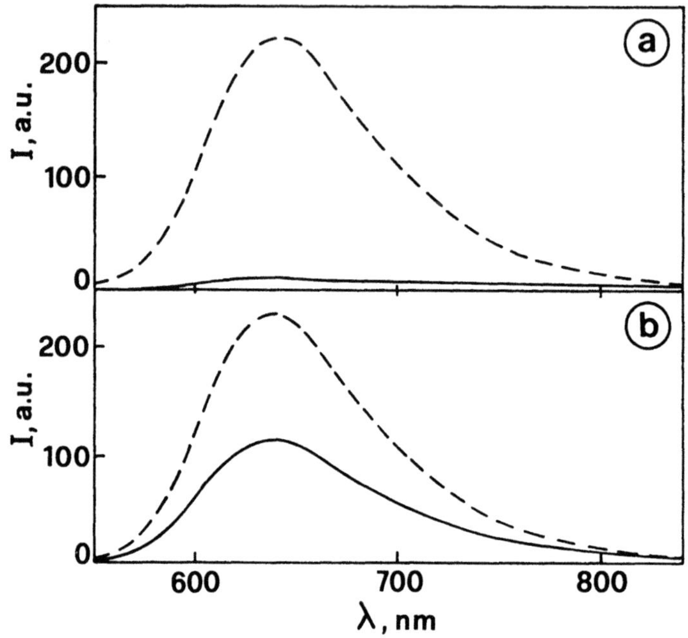

Fig. 4 - Luminescence spectra of isoabsorptive acetonitrile solutions (298 K) of: (a) **1.Ru_2Os** (—), a 2:1 mixture of **1.Ru_3** and **1.Os_3** (---); (b) **2.Ru_2Os** (—), a 2:1 mixture of **2.Ru_3** and **2.Os_3** (---). Excitation wavelength 461 nm.

In order to measure the sensitization of the luminescence of the Os-based unit, a more complex procedure had to be adopted because the tail of the (residual) luminescence of the Ru-based units strongly interferes with the measurement of the luminescence intensity of the Os-based units even at their emission maximum (780 nm). The luminescence spectra of the **1.Ru_2Os** and **1.Os_3** compounds were recorded with excitation in their isosbestic point at 461 nm. From the luminescence band of **1.Ru_2Os**, the contribution coming from the (unquenched) luminescence of the Ru-based units was subtracted by using a normalized (at 640 nm) spectrum of **1.Ru_3** (Fig. 5; for the analogous complexes of **2**, see Fig. 6). The difference in the heights of the **1.Ru_2Os** and **1.Ru_3** bands at 780 nm thus measures the contribution of the Os-based unit to the **1.Ru_2Os** luminescence at that wavelength and can be directly compared with the height of the **1.Os_3** band.

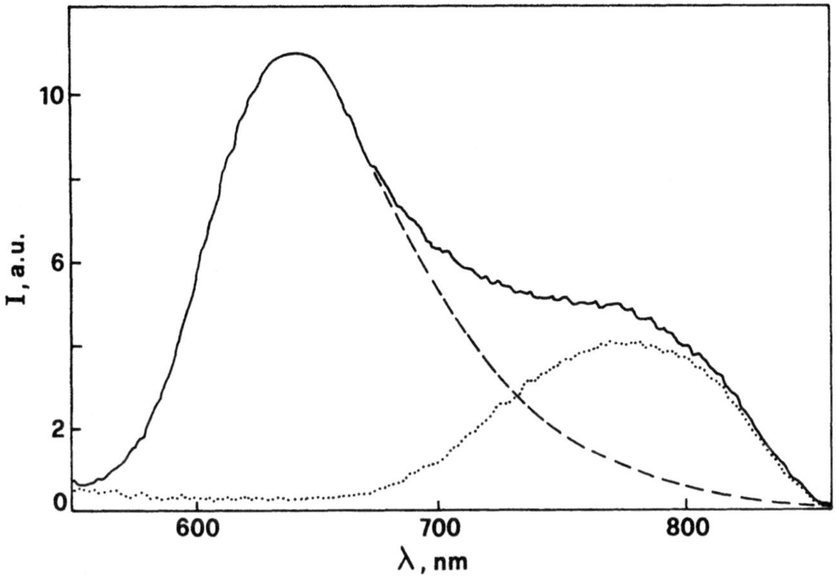

Fig. 5 - Luminescence spectra of isoabsorptive acetonitrile solutions (298 K) of: **1.Ru_2Os** (—), **1.Os_3** (···), and **1.Ru_3** (---). The last spectrum has been normalized to the maximum of the **1.Ru_2Os** spectrum. Excitation wavelength 461 nm.

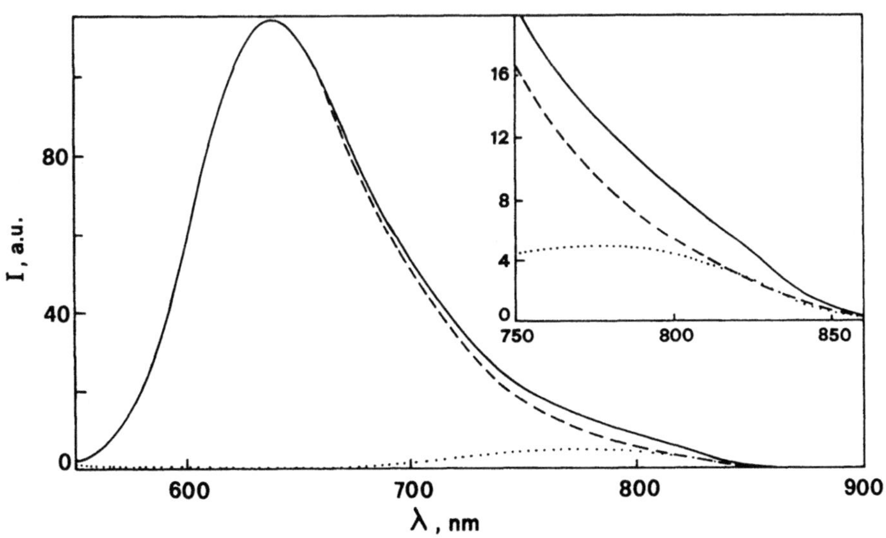

Fig. 6 - Luminescence spectra of isoabsorptive acetonitrile solutions (298 K) of: **2.Ru_2Os** (—), **2.Os_3** (···), and **2.Ru_3** (---). The last spectrum has been normalized to the maximum of the **1.Ru_2Os** spectrum. Excitation wavelength 461 nm.

3. Results

All the compounds examined were stable in the solvents used both in the dark and under laboratory light. In order to check the stability of the complexes towards metal exchange under the preparative conditions, equimolar amounts ($5x10^{-4}$ M) of **1.Ru$_3$** and **1.Os$_3$** were heated at 120 °C in ethylene glycol for 20 hours. Formation of mixed-metal species would have caused a decrease in the luminescence intensity at 640 nm (*vide infra*). No change, however, was detected.

The absorption spectra of the ligands **1** and **2** exhibit the characteristic bpy bands in the UV region. In all the complexes studied the ligand centered (LC) bands in the UV region are accompanied by the characteristic [12,13] metal-to-ligand charge transfer (MLCT) bands in the visible (Table I). For illustration purposes, the visible spectra of **1.Ru$_3$**, **1.Os$_3$**, and **1.Ru$_2$Os** are displayed in Fig. 3.

Table I. Absorption maxima, nm (ε, M^{-1} cm^{-1})

1.Ru				450(12400)		290(94000)
1.Ru$_2$				451(24700)		290(149100)
1.Ru$_3$				450(38100)		290(208900)
1.Os$_3$	642(9500)	588(9700)	478(33900)	450(34700)	370(30000)	290(247700)
1.Ru$_2$Os				450(36300)		289(218900)
2.Ru$_3$				450(37300)		287(230800)
2.Os$_3$	642(8800)	589(9700)	478(33500)	449(34100)	370(29600)	290(273300)
2.Ru$_2$Os				450(35330)		288(232900)

The electrochemical potentials of the investigated complexes are collected in Table II.

Table II. Electrochemical data[a]

	Redox potential, V (relative current intensity)		
	oxidation		reduction
	Ru	Os	
1.Ru	+1.30(1)	---	-1.21
1.Ru$_2$	+1.30(2)	---	-1.19
1.Ru$_3$	+1.31(3)	---	-1.18
1.Os$_3$	---	+0.870(3)	-1.14
1.Ru$_2$Os	+1.30(2)	+0.865(1)	-1.16
2.Ru$_3$	+1.31(3)	---	-1.17
2.Os$_3$	---	+0.865(3)	-1.15
2.Ru$_2$Os	+1.31(2)	+0.880(1)	-1.14

[a]acetonitrile solution; potential values vs SCE; room temperature; the second and third reduction processes are irreversible except for **1.Ru**

For illustration purposes, the cyclic voltammogram of **1.Ru$_2$Os** is shown in Fig. 7.

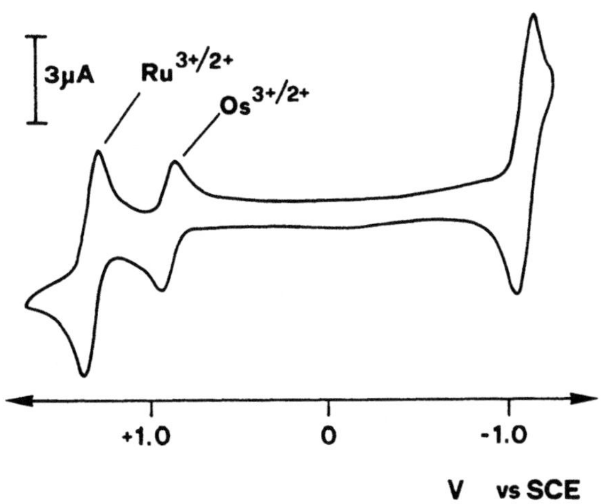

Fig. 7 - Cyclic voltammogram of **1.Ru$_2$Os** in acetonitrile solution at room temperature.

Table III. Luminescence properties

	298 K[a]						77 K[b]			
	Ru			Os			Ru		Os	
	λ_{max} nm	τ ns	I_{rel}	λ_{max} nm	τ ns	I_{rel}	λ_{max} nm	τ ms	λ_{max} nm	τ ms
1·**Ru**$_3$	640	200	100	--	--	--	595	4.2	--	--
1·**Os**$_3$	--	--	--	780	25	100	--	--	720	0.68
2/3(**1**·**Ru**$_3$)+1/3(**1**·**Os**$_3$)	640	190	64	c	c	c	595	3.5	720	
1·**Ru**$_2$**Os**	642	185	2.5	780	24	98	595	d	720	0.70
2·**Ru**$_3$	640	210	100	--	--	--	595	4.4	--	--
2·**Os**$_3$	--	--	--	780	.25	100	--	--	720	0.67
2/3(**2**·**Ru**$_3$)+1/3(**2**·**Os**$_3$)	640	200	63	c	c	c	595	4.0	720	
2·**Ru**$_2$**Os**	640	200	31	c	c	80	595	4.0	720	0.60

[a]Aerated acetonitrile solution; [b]propionitrile-butyronitrile solution; [c]not measurable because it is covered by the much more intense Ru-based emission; [d]not measurable because of strong overlap with the more intense Os-based emission.

All the complexes examined are luminescent both in rigid matrix at 77 K and in fluid solution at room temperature. Some luminescence spectra are displayed in Figs. 4-6. The luminescence properties of **1.Ru$_3$**, **1.Os$_3$**, **1.Ru$_2$Os**, **2.Ru$_3$**, **2.Os$_3$**, **2.Ru$_2$Os**, and of the 2:1 stoichiometric mixtures of **1.Ru$_3$** and **1.Os$_3$**, and **2.Ru$_3$** and **2.Os$_3$** are shown in Table III. The luminescent properties of **1.Ru** and **1.Ru$_2$** are exactly identical to those of **1.Ru$_3$** and therefore are not reported in Table III. The temperature dependence of the luminescence of **2Ru$_3$** and **2.Ru$_2$Os** is shown in Table IV.

Table IV. Temperature dependence of the luminescence[a]

	2.Ru$_3$		**2.Ru$_2$Os**		
T	λ_{max}	τ	λ_{max}	τ	$I_{(2.Ru2Os)}/I_{(2.Ru3)}$
K	nm	μs	nm	μs	
96	607	2.38	607	2.59	0.50
155	642	0.81	638	0.80	0.43
200	647	0.68	647	0.70	0.49
250	644	-	644	-	0.43
298	640	0.61	640	0.62	0.42

[a]Deaerated propionitrile-butyronitrile mixture. [b]Emission intensity ratio measured at 640 nm; excitation at 461 nm.

4. Discussion

4.1 *Intercomponent interactions*

Extensive investigations on mono- and oligonuclear Ru(II) and Os(II) complexes [2,4-7,9-13] have shown that: (i) oxidation is metal centered; (ii) Os(II) is easier to oxidize than Ru(II); (iii) reduction is ligand centered; (iv) the absorption bands in the visible region are due to spin-allowed metal-to-ligand charge transfer (MLCT) transitions; (v) luminescence takes place from the lowest energy excited state which is a formally triplet MLCT level. In oligonuclear complexes electronic interaction between the various components may range from very strong (with profound changes of the various properties on passing from mononuclear to oligonuclear species) to very weak (with almost equal properties for separated and bridged units).

An important thing to be noticed is that the first oxidation potential, the first reduction potential, the absorption maxima, and all the luminescence properties are identical (within the experimental errors) for **1.Ru**, **1.Ru_2**, and **1.Ru_3**. This suggests little or no electronic interaction between the identical metal-containing components that are present in **1.Ru_2** and **1.Ru_3**. Further confirmations of a very weak (if any) interaction come from: (a) the identical absorption, luminescence, and redox properties exhibited by **1.Ru_3** and **2.Ru_3** and, respectively, by **1.Os_3** and **2.Os_3**; (b) the identical first oxidation and first reduction potentials of **1.Os_3**, **1.Ru_2Os**, **2.Os_3** and **2.Ru_2Os**; (c) the identical absorption spectrum exhibited by the mixed-metal **1.Ru_2Os** and **2.Ru_2Os** complexes and by the 2:1 mixtures of their homometallic "parent" trinuclear complexes.

We can thus draw a general conclusion: the interaction of a $M(bpy)_2T^{2+}$ component (where T is the bpy-type coordinating site of the tripod ligands **1** and **2**) with any other (homo- or hetero-metallic) component which is present in the dinuclear and trinuclear supramolecular species is, at most, weak. As we will see later, the results obtained from electronic energy transfer experiments show that some electronic interaction does occur (at least between excited state and ground state components) and decreases with the increasing size of the spacer in the tripod ligand.

4.2 *Properties of the components*

Each metal-containing component can be viewed as a mixed-ligand complex since the coordination sites of the tripod ligands **1** and **2** have properties slightly different from those of bpy because of the presence of an electron withdrawing amido group.

This is the reason why the properties of $Ru(bpy)_2T^{2+}$ and $Os(bpy)_2T^{2+}$ are not identical to those of $Ru(bpy)_3^{2+}$[24] and, $Os(bpy)_3^{2+}$[13] respectively. These differences consist of: (a) a less negative first reduction potential, since the coordination sites of **1** or **2** are slightly easier to reduce than bpy; (b) a higher oxidation potential, since the metal ion in $M(bpy)_2T^{2+}$ is slightly more positive than in $M(bpy)_3^{2+}$ owing to the presence of the electron withdrawing substituent in T; (c) some slight differences in the shape of the absorption bands in the visible, caused by the presence of two MLCT transitions in the $M(bpy)_2T^{2+}$ components (M --->T at lower energy, and M --->bpy at higher energy); (d) a slight red shift of the luminescence in going from $M(bpy)_3^{2+}$ to $M(bpy)_2T^{2+}$, since in the latter species the lowest MLCT level involves the tripod ligand; (e) a slightly shorter lifetime at 77 K, because of the smaller energy gap between ground state and luminescent level; (f) a slightly longer lifetime in fluid solution at room temperature in the case of the Ru compounds, because of the slightly higher energy gap between luminescent level and the upper-lying, short-lived triplet ligand-field level; (g) a greater sensitivity to the environment as shown by the red shift of the luminescence band with increasing solvent polarity (for **1.Ru$_3$**, λ_{max} is 636, 640, and 651 nm in CH_2Cl_2, CH_3CN, and methylformamide, respectively).

4.3 *Luminescence quenching and sensitization*

In polypyridine complexes, Os(II) is easier to oxidize than Ru(II) (Table II). Therefore, for complexes of the same ligands the MLCT levels lie at higher energy in the Ru(II) than in the Os(II) complexes, as is also shown by the results obtained in this work (Table III). As a consequence, in Ru(II)-Os(II) mixed-metal supramolecular systems energy transfer from the Ru-based components to the Os-based components is thermodynamically allowed. For our compounds, the excited-state energy of the Ru-based components is 2.08 eV and that of the Os-based components is 1.72 eV [25]. Therefore, the free energy change (neglecting entropy changes) for energy transfer from an excited Ru-based component to a ground state Os-based component is approximately - 0.36 eV.

In principle, fast excited state quenching can also occur by electron transfer. From the excited state energy of the Ru-based components (2.08 eV) and the redox potentials shown in Table II, it can be estimated that the reductive quenching process of the excited Ru-based component by the ground state Os-based component is nearly isoergonic, and the oxidative quenching process is strongly endoergonic. Therefore, it seems unlikely that quenching by electron transfer, which implies a noticeable reorganizational energy in polar solvents, can be fast enough in our systems to compete with quenching by energy transfer.

As one can see from Fig. 4, in acetonitrile solution the luminescence intensity of the Ru-based components in **1.Ru$_2$Os** and **2.Ru$_2$Os** is quenched to 4% and 50%, respectively (compare the I_{rel} value of the two compounds with those of the corresponding 2:1 mixture of the parent trinuclear homometallic species, Table III). Following the procedure indicated in the experimental section and based on a comparison of the (corrected) luminescence intensities at 780 nm of **1.Ru$_2$Os** and **2.Ru$_2$Os** with those of **1.Os$_3$** and **2.Os$_3$** (Figs. 5 and 6), we have found (Table III) that *the quenching of the luminescence of the Ru-based components is accompanied by a parallel sensitization of the luminescence of the Os-based units* (for the complex of **2**, the experimental uncertainty is large because of the strong interference of the residual Ru-based emission, Fig. 6). This shows unequivocally that the quenching occurs by an energy transfer mechanism.

4.4-Energy transfer mechanism.

In principle, the observed electronic energy transfer can occur from Ru-based to Os-based units which belong to the same **1.Ru$_2$Os**, or **2.Ru$_2$Os**, species (*intramolecular intercomponent energy transfer, Fig. 8a*) or between units which belong to distinct **1.Ru$_2$Os**, or **2.Ru$_2$Os**, species (*intermolecular energy transfer, Fig. 8b*).

Intermolecular energy transfer can be excluded because it does not take place in the 2 : 1 **1.Ru$_3$-1.Os$_3$** and **2.Ru$_3$-2.Os$_3$** mixtures. In fact, the luminescence intensity of such mixtures is 2/3 that of isoabsorptive **1.Ru$_3$** and

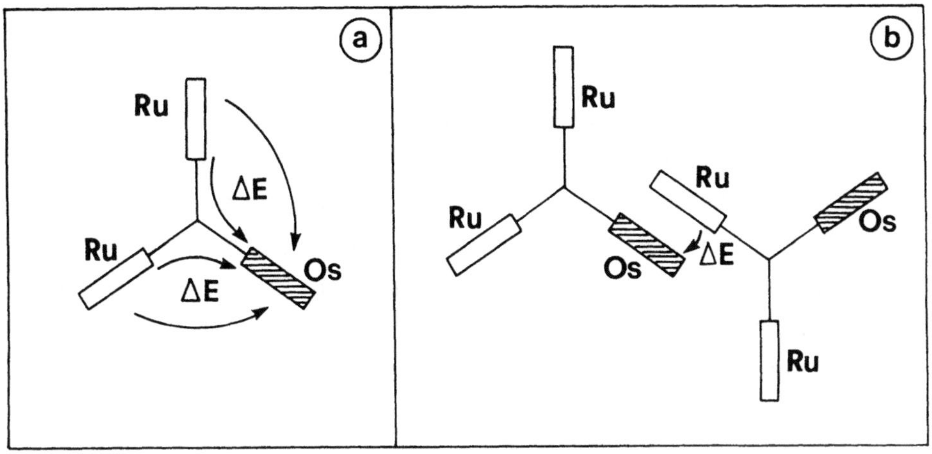

Fig. 8 - Schematic representation of intercomponent (a) and intermolecular (b) energy transfer.

2.Ru$_3$ solutions (Table III), as expected because the light absorbed (at the 461 nm isosbestic point of **1.Ru$_3$**, **1.Os$_3$**, **2.Ru$_3$**, and **2.Os$_3$**) by the Ru-based chromophoric units in the mixtures is 2/3 that absorbed by the reference **1.Ru$_3$** and **2.Ru$_3$** solutions. Furthermore, (i) the lifetime of the Ru-based luminescence in the mixtures is identical (within the experimental error) to that of **1.Ru$_3$** and **2.Ru$_3$** and (ii) the intensity quenching for **2.Ru$_2$Os** also takes place in rigid matrix at 96 K (see Table IV). The predominance of intercomponent energy transfer over intermolecular energy transfer is indeed an expected result because CPK molecular models show that the structure of our compound is very flexible, so that for both **1.Ru$_2$Os** and **2.Ru$_2$Os** the Ru-based and Os-based components of the same species can approach each other as much closely as can the components of two different species in an encounter.

In the case of **1.Ru$_2$Os**, the residual Ru-based luminescence intensity is 4% that of the 2:1 **1.Ru$_3$-1.Os$_3$** mixture. The luminescence lifetime, however, is almost the same in the two cases. In the case of **2.Ru$_2$Os**, the Ru-based luminescence intensity reduces to 50%, and the luminescence lifetime is again unchanged. The behaviour of **1.Ru$_2$Os** could be accounted by assuming that energy transfer is very fast and 100% efficient, and that the residual luminescence is due to the presence of small amounts of **1.Ru**, **1.Ru$_2$**, or **1.Ru$_3$** impurities. This, however, is unlikely in view of the synthetic and purification procedures used. In any case, the behavior of **2.Ru$_2$Os**, where the residual luminescence intensity is 50%, cannot be accounted for by the presence of impurities. Partial quenching of the luminescence intensity without quenching of the luminescence lifetime cannot be explained by a simple mechanism based on the occurrence of energy transfer in direct competition with the decay of the luminescent level. Mechanisms involving either two excited states or two conformers must be invoked. Energy transfer from a singlet MLCT level of the Ru-based components to a singlet MLCT level of the Os-based components in competition with ^{1}MLCT······>^{3}MLCT intersystem crossing in the Ru-based components could account for the observed results, but it does not seem likely because for $Ru(bpy)_3^{2+}$, deactivation of upper excited states to the lowest ^{3}MLCT level occurs in the picosecond time scale [26,27], and moreover there is no reason why energy transfer should not take place also from the ^{3}MLCT luminescent level.

In view of the lack of rigidity of the supramolecular array, we believe that the observed results are more likely due to the presence of conformers. A possibility could be a through-bond [28] energy transfer that can occur only in specific molecular conformations (note that the three metal-containing units are linked to the spacer by single bonds). It seems unlikely, however, that such conformers do not interchange in the time scale of the excited state lifetime (200 ns). More probable appears to be the involvement of conformers where the luminescent Ru-based units are either very close to (and therefore rapidly quenched through-space [28] by) or very far from (and therefore unquenched by) the Os-based unit. A statistical distribution among three conformers like those shown in Fig. 9a would leave 67% of residual luminescence intensity (as compared to the luminescence of the isoabsorptive 2:1 mixture of the parent homotrimetallic species). For **2.Ru$_2$Os**, however, the residual luminescence is about 50%. These data require the presence of other confor-

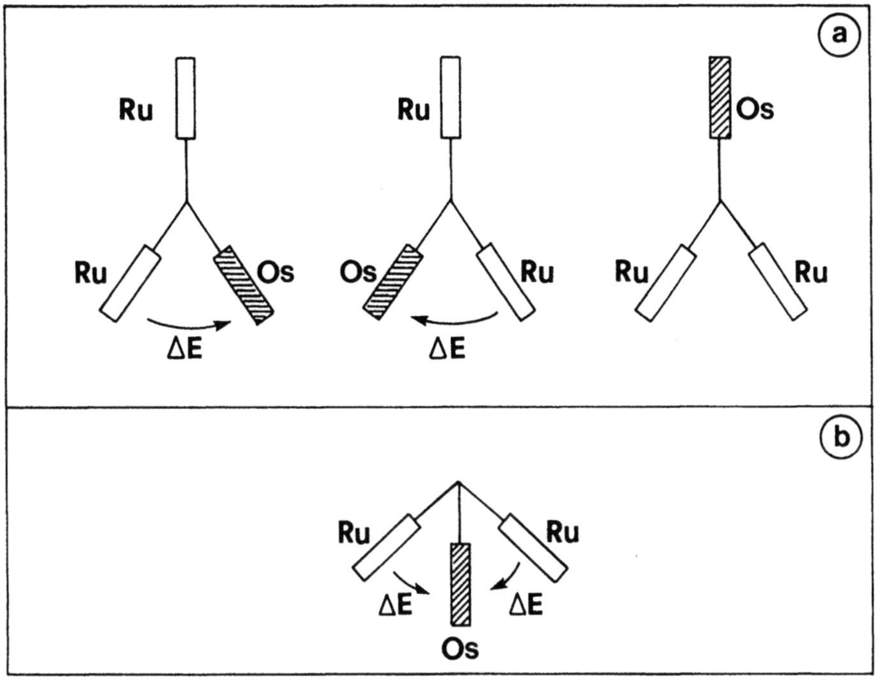

Fig. 9 - Possible conformers for the supramolecular species **1.Ru_2Os** and **2.Ru_2Os**. For more details, see text.

mers, like those schematized in Fig. 9b, where the three components are very close to one another, so that both the Ru-based luminescent components can be efficiently quenched by the Os-based component. The almost complete quenching of the luminescence intensity for **1.Ru_2Os** can be accounted for by considering that, because of the smaller spacer, the distance between the two Ru-based components and the Os-based component is so short to assure a fast and efficient quenching in most conformations.

For such a through-space energy transfer process, a Dexter-type mechanism[29] seems more likely than a Förster-type mechanism[30] because of the triplet character of the excited states involved and the close approach of the two partners.

An evaluation of the rate constant of the energy transfer process in the species which exhibit a conformation suitable for quenching can be obtained from eq. 1, where $I°$ and $\tau°$ are the luminescence intensity and lifetime of the

$$k_{et} = 1/\tau^{\circ} (I^{\circ}/I - 1) \tag{1}$$

species that can be quenched, and I is their residual luminescence intensity after quenching. The last quantity can be evaluated by comparing the total residual luminescence intensity (measured under stationary conditions by a fluorimeter) and the residual luminescence intensity of the unquenched species (measured from laser experiment at t > 100 ns). For **2.Ru$_2$Os** these two experimental values are equal (50 ± 2)% within the experimental error. Therefore, the residual luminescence intensity of the quenched species is at most 2%. Using $I^{\circ} = 50\%$, $I \leq 2\%$, and $\tau = 200$ ns, from eq. 1 one obtains $k_{et} > 1 \times 10^8 \ s^{-1}$.

Acknowledgments. We would like to thank Mr. G. Gubellini for technical assistance. This work was supported by MURST and CNR (Italy), Swiss National Science Foundation (Switzerland), and Bundesministerium für Forschung und Technologie, Project no. 0329120 A (Germany).

References and Notes

[1] Hader, D.P.; Tevini, M. *General Photobiology*; Pergamon Press: Oxford, U.K., 1987.

[2] Balzani, V.; Scandola, F. *Supramolecular Photochemistry*; Horwood: Chichester, U.K., 1991.

[3] Vögtle, F. *Supramolecular Chemistry*; Wiley: Chichester, U.K, **1991**.

[4] For reviews, see: (a) Ref. 2, chapter 6; (b) Scandola, F.; Indelli, M.T.; Chiorboli, C.; Bignozzi, C.A. *Topics Curr. Chem.* **1990**, *158*, 73. For some recent papers, see refs. 4-11.

[5] Ryu, C.K.; Schmehl, R.H. *J. Phys. Chem.* **1989**, *93*, 7961.

[6] Furue, M.; Yoshidzumi, T.; Kinoshita, S.; Kushida, T.; Nozakura, S.; Kamachi, M. *Bull. Chem. Soc. Jpn.* **1991**, *64*,1632.

[7] De Cola L.; Barigelletti, F.; Balzani, V.; Hage, R.; Haasnoot, J.G.; Reedijk, J.; Vos, J.G. *Chem. Phys. Letters* **1991**, *178*, 491.

[8] Sigman, M.E.; Closs, G.L. *J. Phys. Chem.* **1991**, *95*, 5012.

[9] Strouse, G..F.; Worl, L.A.; Younathan, J.N.; Meyer, T.J. *J. Am. Chem. Soc.* **1989**, *111*, 9101.

[10] Denti, G.; Serroni, S.; Campagna, S.; Ricevuto, V.; Balzani, V. *Inorg. Chim. Acta* **1991**, *182*, 127. Denti, G.; Campagna, S.; Serroni, S.; Ciano, M.; Balzani, V. *J. Am. Chem. Soc.*, in press, and references therein.

[11] Bignozzi, C.A.; Bortolini, O.; Chiorboli, C.; Indelli, M.T.; Rampi, M.A.; Scandola, F., *Inorg. Chem.* in press.

[12] Juris, A.; Balzani, V.; Barigelletti, F.; Campagna, S.; Belser, P.; von Zelewsky, A. *Coord. Chem. Rev.* **1988**, *84*, 85.

[13] Meyer, T.J. *Pure Appl. Chem.* **1986**, *58*, 1193, and refs. therein.

[14] De Cola, L.; Barigelletti, F.; Balzani, V.; Belser, P.; von Zelewsky, A.; Seel, C.; Frank, M.; Vögtle, F. *Coord. Chem. Rev.* **1991**, *111*, 255.

[15] Sullivan, B.P.; Salmon, D.J.; Meyer, T.J. *Inorg. Chem.* **1978**, *17*, 3334.

[16] Lay, P.A.; Sargeson, A.M.; Taube, H.; Chou, M.H.; Creutz, C. *Inorg. Synth.* **1986**, *24*, 294.

[17] Kröhnke, F.; Zecher, W. *Angew. Chem.* **1962**, *74*, 811. Kröhnke, F.; Gross, K.F. *Chem. Ber.* **1959**, *42*, 22.

[18] Kröhnke, F.; *Angew. Chem.* **1963**, *74*, 181. Huang, T.L.J.; Brewer, D.G. *Can. J. Chem.* **1981**, *59*, 1689.

[19] Black, G.; Depp, E.; Corson, B.B. *J. Org. Chem.* **1949**, *14*, 14.

[20] Grammenudi, S.; Frank, M.; Vögtle, F.; Steckhan, E. *J. Inclusion Phenom.* **1987**, *5*, 695.

[21] Ebmeyer, F.; Vögtle, F. *Angew. Chem.* **1989**, *101*, 95. Sendhoff, N.; Kissener, W.; Vögtle, F.; Franken, S.; Puff, H. *Chem. Ber.* **1988**, *121*, 2179.

[22] Juris, A.; Balzani, V.; Belser, P.; von Zelewsky, A. *Helv. Chim. Acta* **1981**, *64*, 2175.

[23] Sutin, N.; Creutz, C. *Adv. Chem. Ser.* **1978**, *168* , 1. Lin, C.T.; Boettcher, W.J.; Chou, M.; Creutz, C.; Sutin, N. *J. Am. Chem. Soc.* **1976**, *98*, 6536.

[24] De Cola, L.; Belser, P.; Ebmeyer, F.; Barigelletti, F.; Vögtle, F.; von Zelewsky, A.; Balzani, V. *Inorg. Chem.* **1990**, *29*, 495.

[25] Excited state energies have been estimated as the energy of the emission maximum at 77 K. Other methods [5,9] give practically equivalent results.

[26] (a) Bradley, P.C.; Kress, N.; Hornberger, B.A.; Dallinger, R.F.; Woodruff, W.H. *J. Am. Chem. Soc.* **1989**, *103*, 7441. (b) Carrol, P.J.; Brus, L.E.; *J. Am. Chem. Soc.* **1987**, *109*, 7613.

[27] (a)Yabe, T.; Anderson, D.R.; Orman, L.K.; Chang, Y.J.; Hopkins, J.B. *J. Phys. Chem.* **1989**, *93*, 2302. (b) Cooley, L.F.; Bergquist, P.; Kelley, D.F.; *J. Am. Chem. Soc.* **1990**, *112*, 2612.

[28] Ref. 2, chapter 5.

[29] Dexter, D.L. *J. Chem. Phys.*, **1953**, *21*, 836.

[30] Förster, Th.H., *Discuss. Faraday Soc.* **1959**, *27*, 7.

KINETIC AND SPECTROSCOPIC INVESTIGATION OF THE INFLUENCE OF CONFORMATION AND ORBITAL-SYMMETRY ON LONG-RANGE INTRAMOLECULAR DONOR-ACCEPTOR INTERACTION

J.W. VERHOEVEN*, J. KROON
Laboratory of Organic Chemistry, University of Amsterdam, Nieuwe Achtergracht 129, 1018 WS Amsterdam, The Netherlands

M.N. PADDON-ROW
Department of Chemistry, University of New South Wales, P.O. Box 1, Kensington, NSW 2033, Australia

J.M. WARMAN
Interfaculty Reactor Institute, Delft University of Technology, Mekelweg 15, 2629 JB Delft, The Netherlands

ABSTRACT. In conformationally well defined systems where an electron-donor (D) and acceptor (A) are interconnected by one ore more arrays of saturated carbon-carbon bonds the D/A interaction is influenced by changes in the length as well as in the configuration of these arrays. This not only affects the kinetics of (photoinduced) electron transfer between D and A, but also the electronic absorption and emission spectra are modified as a result of the presence of a new transition of the charge-transfer type.
It is shown that especially charge-transfer emission can yield detailed information about the conformational and orbital-symmetry dependence of D/A interaction, while the effect of these parameters on electron transfer kinetics tends to be masked by concomitant changes in Franck-Condon factors.

1. Introduction.

Exploring the degree and the mechanism of electronic interaction between functional groups connected by saturated hydrocarbon moieties has been a subject of many experimental and theoretical investigations. While it is obvious that significant interaction may result from direct orbital overlap in situations where the hydrocarbon "bridges" allow or even enforce close contact of the functional groups, the mechanism and the extent of interaction in the absence of direct orbital overlap has remained a topic of debate[1,2]. This may be due in part to the lack of a uniform conceptual basis for the classification of long-range interactions. A very important contribution to such a conceptual basis was made

V. Balzani and L. De Cola (eds.), Supramolecular Chemistry, 181–200.

over two decades ago by R. Hoffmann et al.[3,4] when they proposed description of the overall interaction between two functional groups in terms of "through-bond" and "through-space" interaction of orbitals localized on these groups. In particular the concept of through-bond interaction (TBI) has revolutionized the description of many long-range interaction phenomena although its adoption in a number of fields remains slow and incomplete especially where its advantage over more classical descriptions in terms of e.g. inductive- and field-effects is less obvious or even absent[1].

Most of the early experimental evidence for TBI appears to be derived from photo electron spectroscopy (PES) data obtained for systems in which interaction between two or more identical functional groups, held at a distance excluding through space interaction, leads to removal of the orbital energy degeneracy, a phenomenon not easily explainable via any alternative interaction mechanism[5]. While PES data provide quantitative information on the overall interaction energy in symmetrical systems (i.e. systems containing identical functional groups) its application to non-symmetrical systems is severely thwarted by the problems arising in the choice of appropriate non-interacting reference systems. Furthermore it has become evident that PES, as well as the related electron transmission spectroscopy (ETS)[6], cannot detect level splittings smaller than ~ 0.1 eV (800 cm^{-1}), which severely limits applicability for the detection of weak interactions.

2. Application of electronic absorption and emission spectroscopy in the detection of through-bond interaction.

Interestingly, it had been recognized at a very early stage that electronic absorption spectroscopy provides a powerful tool for the detection of interaction between two different functional groups especially if these have pronounced electron donor (D) and electron acceptor properties (A) respectively. Interaction between D and A groups is then expected to lead to one or more new electronic transitions of the charge transfer (CT) type, ($D....A + h\nu_{ct} \rightarrow D^{+}....A^{-}$) and the presence or absence of such CT transitions thus in principle provides a "zero background" detection method for interaction between D and A.

While CT absorption resulting from direct D/A contact (i.e. from through-space interaction), either inter- or intramolecularly, is a widely documented phenomenon[7], it may be less well known that some of the earliest evidence for TBI was also provided by the observation of CT absorption bands[8-11], which were, however, initially referred to as "sigma coupled transitions"[9]. Such transitions were especially detected in several conformationally restrained β-aminoketones and their CT nature as well as the conformational requirements for their occurrence were already noted before the introduction of the concept of TBI.

The number and scope of compounds displaying intramolecular CT absorption resulting from TBI were very much amplified in later studies by one of us[12-14] when it turned out that substitution of the amino group by other donating groups and especially substitution of the carbonyl group by a more efficient acceptor group could provide a plethora of systems displaying discrete long wavelength CT absorption bands resulting from TBI. Some striking examples from these early observations are shown in Fig. 1. Compounds **1** and **2** resemble the β-aminoketone scheme in that the D and A groups are separated by three sigma-bonds, and the data compiled in Fig. 1 evidence the remarkably high intensity of the through-bond mediated CT transition that can be achieved between such classically "isolated" D and A groups. Systems **3a** and **3b** extend the coupling distance to five sigma bonds[14].

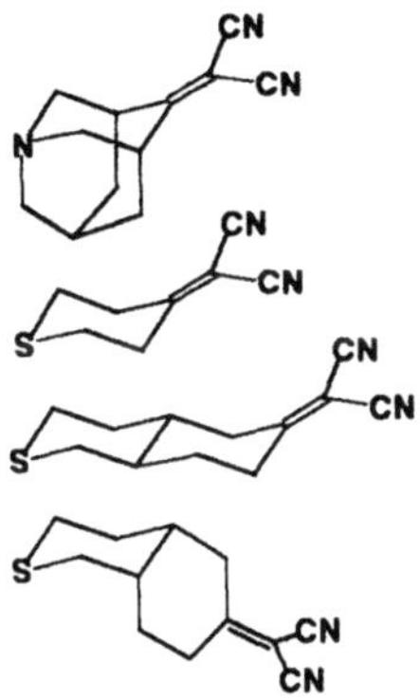

CT-absorption (solvent n-hexane)

	λ_{max} (nm)	ε_{max} (l/mol.cm)
1	314	4500
2	290	2860
3a	270	2600
3b	-	-

Figure 1. Examples of bifunctional systems that display discrete intramolecular CT-absorption as a result of through-bond D/A interaction across three bonds (**1** and **2**, see refs. 12 and 13) and across five bonds (**3**, see ref. 14).

The *trans* -fused system **3a** not only constitutes one of the earliest convincing demonstrations of TBI across such a distance, but the absence of detectable CT absorption in the *cis* -fused isomer **3b** also provided an early example of the conformational/configurational dependence of such interaction, an aspect that we discuss in more detail below.

When in parallel studies[15-19] we substituted one or both of the small D and A groups that do not absorb in the near UV ≥ 250 nm by more extended chromophores, it soon became evident that severe spectral overlap between the local transitions of such chromophores and the eventual CT absorption in general prohibits unequivocal detection of the latter. Some interesting exceptions to this rule are collected in Fig. 2.

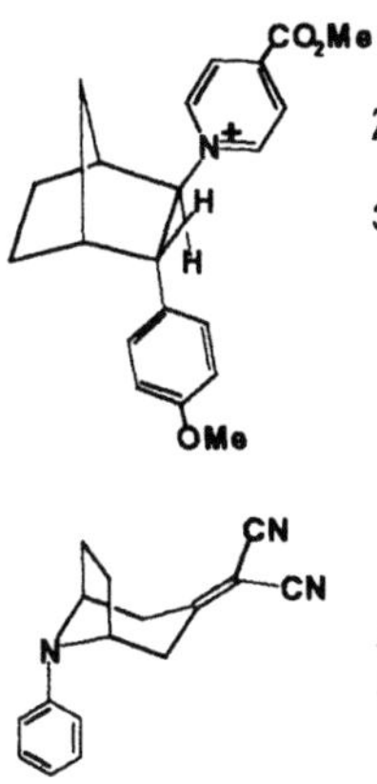

CT-absorption

	λ_{max} (nm)	ε_{max} (l/mol.cm)
4	327	745 (ethanol)
5	348	6950 (n-hexane)
	360	6100 (dichloromethane)

Figure 2. Systems that contain an aromatic electron donor and/or acceptor moiety and display discrete through-bond mediated CT-absorption (see ref. 15 for **4** and ref. 16 for **5**).

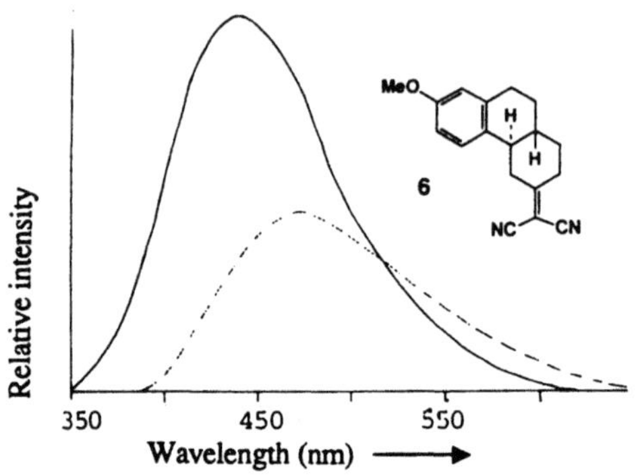

Figure 3. CT-fluorescence[17] of **6** in diethyl ether (——) and in chloroform (- - -).

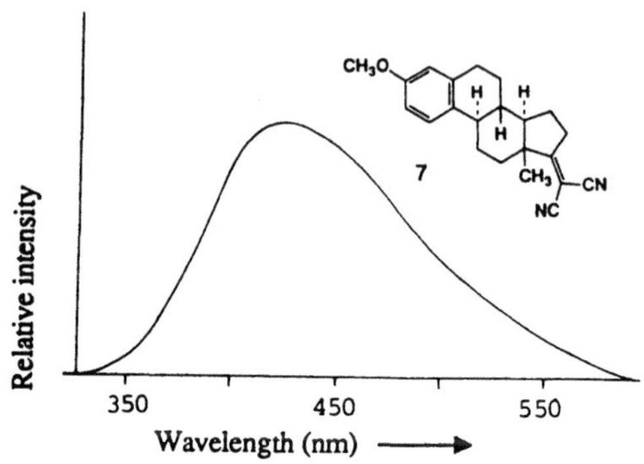

Figure 4. CT-fluorescence[18] of **7** in diethyl ether.

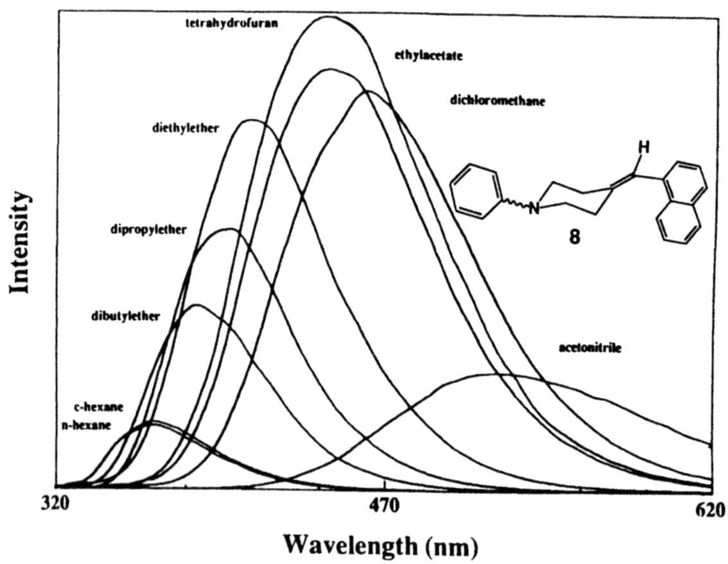

Figure 5. Typical solvatochromic CT-fluorescence of **8** in a series of solvents[19].

Nevertheless, for several rigid D-bridge-A systems displaying virtually no detectable CT absorption the presence of such a transition was unequivocally demonstrated from the observation[17-19] of a typical broad CT fluorescence that undergoes a large and strongly solvent dependent Stokes shift, which moves it to a region not overlapped by eventual emission bands of the individual D and A chromophores. Figures 3-5 show examples of compounds containing an aromatic donor moiety, for which no discrete CT absorption can be detected, but that nevertheless display an easily recognized CT-type emission resulting from radiative recombination of a charge separated state populated with high efficiency during or directly following photoexcitation (D^{+}-bridge-A^{-}→ D-bridge-A + $h\upsilon_{ct}$).

It's important to stress that, unlike the situation encountered in the well known intramolecular exciplex formation of flexibly bridged bichromophores, the structural rigidity of the compounds compiled in Figs. 3-5 prevents major conformational changes thus implying that the D/A interaction allowing charge separation and charge recombination to occur is contained within the ground-state conformation of these molecules.

3. Detection of long-range D/A interaction from electron-transfer kinetics.

While, as discussed above, photo-electron spectroscopy and electronic spectroscopy have been applied to detect TBI in many earlier studies, the focus of attention in studies aimed at detection and quantification of long-range D/A interactions appears to have largely shifted over the past decade to measurement of electron-transfer kinetics. Main reasons for this change are undoubtedly the larger dynamic range and especially the much lower limit of detectability available in such kinetic measurements. As already pointed out in the introduction the minimum level splitting detectable with PES is about $\Delta IP \approx 0.1$ eV (800 cm^{-1}). For weakly interacting symmetrical (D = A) systems $\Delta IP = 2H_{DA}$, where H_{DA} represents the electronic interaction integral. Thus for $H_{DA} \leq 400$ cm^{-1} detection of interaction by PES is problematic.

For detection of the CT absorption in D-bridge-A systems the situation is, in general, only marginally better. Applying the well-known Hush approximation[20] we may write:

$$H_{DA}(\text{in cm}^{-1}) = \frac{2.06\text{x}10^{-2}\sqrt{\varepsilon_{max}\upsilon_{max}\Delta\upsilon_{1/2}}}{R_c} \qquad (1)$$

In eqn (1) ε_{max} and υ_{max} represent the molar extinction coefficient (in $l.mol^{-1}.cm^{-1}$) and position (in cm^{-1}) of the CT absorption, $\Delta\upsilon_{1/2}$ its full width at half height and R_c the D/A center to center distance (in Å). For typical values of $\Delta\upsilon_{1/2} = 4000$ cm^{-1}, R_c = 10 Å and $\upsilon_{max} = 25000$ cm^{-1} an H_{DA} of 100 cm^{-1} is thus expected to produce a CT absorption with $\varepsilon_{max} \approx 200$. This will in general be close to the limit of detectability due to spectral overlap problems and also because the need to avoid intermolecular interaction requires that absorption measurements are made at low concentration. Nevertheless, in fortuitous cases when spectral overlap is negligible, H_{DA} values as low as ~25 cm^{-1} have been detected[25] via absorption spectroscopy, implying a sensitivity gain of about one order of magnitude as compared to PES. Considerably weaker D/A interaction is, however, in principle sufficient to produce a kinetically significant rate of electron transfer. A general expression for the rate of electron transfer (k_{et}) between weakly coupled D and A sites is provided[22] by the golden rule expression, eqn (2), where FC_{DA} is the so called Franck-Condon weighted density of states.

$$k_{et} = \frac{4\pi^2}{h} FC_{DA} H_{DA}^2 \tag{2}$$

As shown by Marcus[23], the Franck-Condon factor can be approximated for weak coupling in a high temperature limit by eqn (3), where λ represents the overall reorganization energy and ΔG^o the Gibbs free energy change ("driving force").

$$FC_{DA} = \frac{1}{\sqrt{4\pi\lambda k_B T}} \exp\left[\frac{-(\Delta G^0 + \lambda)^2}{4\lambda k_B T}\right] \tag{3}$$

The maximum rate attainable under barrierless conditions ($\Delta G^o = -\lambda$) is then:

$$k_{et}^{max} = \frac{2\pi^{3/2}}{h\sqrt{\lambda k_B T}} H_{DA}^2 \tag{4}$$

For a realistic value of $\lambda = 0.5$ eV and at room temperature this gives $k_{et} = 3.6 \times 10^8 (H_{DA})^2$ (k_{et} in s^{-1}, H_{DA} in cm^{-1}). Practical lower limits to k_{et} for detection of intramolecular electron-transfer are either set by the lifetime of the electronically excited state from which it occurs (i.e. for photoinduced electron-transfer) or by competition of intermolecular electron- transfer, which is often occurring at a diffusion controlled rate. The former typically limits the time window for observation of k_{et} to ≤ 10 ns, while at high dilution (~ 10^{-5} mol/l) the latter does not produce problems if the intramolecular electron transfer proceeds significantly within ≤ 10 µs. These constraints thus require k_{et} values $\geq 10^8$ s^{-1} and $\geq 10^5$ s^{-1} corresponding to $H_{DA} \geq 0.5$ cm^{-1} and $H_{DA} \geq 0.017$ cm^{-1} for photoinduced and thermal electron-transfer respectively, while the latter can be expanded to still much lower values if diffusional encounter is avoided; e.g. in solid matrices or for D/A pairs encapsulated in large protein envelopes.

4. Distance dependence of k_{et} and H_{DA}.

As discussed above, electron-transfer kinetics can yield information about extremely weak long-range interactions and the recognition of this possibility has triggered an avalanche of research which gained considerable momentum over the past decade. As we[24] and others[25] have occasionally pointed out, however, the gain in sensitivity resulting from studying long-range interactions via electron-transfer kinetics instead of via other techniques, such as PES or electronic spectroscopy, comes at the expense of direct and quantitative information about the actual value of the electronic interaction integral H_{DA}. This is because, as evident from eqn (2), k_{et} not only depends on H_{DA} but also on the Franck-Condon factor FC_{DA}, which cannot be determined independently and for which calculated values depend strongly on the method of approximation and its parametrization[23,25].

In many cases absolute values of H_{DA} are not required, the main point of interest being how H_{DA} changes as a function of e.g. D/A distance, the medium between D and A and their relative orientation etc. In such cases it may be assumed that the observed changes in k_{et} reflect variations in H_{DA} rather then in FC_{DA}, but it appears important to verify this assumption where possible. A crucial test was in fact realized by us earlier[24] with regard to distance dependence in the series of rigidly bridged D/A systems displayed in Fig. 6.

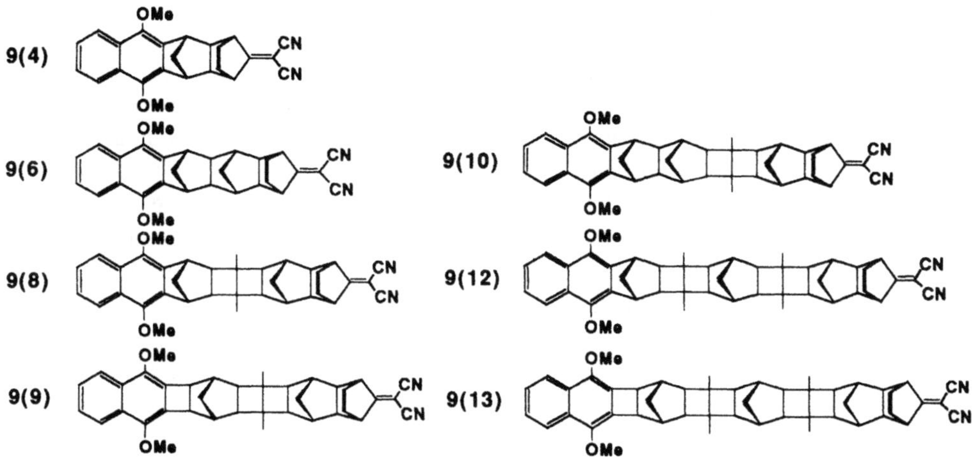

Figure 6. Series of rigid D-bridge-A systems.

For all members of that series the kinetics of charge-separation (k_{cs})[26,27] as induced by excitation of D and of charge-recombination (k_{cr})[28] have been studied. While, depending on the solvent, charge-separation and -recombination have widely different values, both show a virtually identical and exponential distance dependence $k \sim \exp(-\alpha n)$ with $\alpha \approx 0.9\text{-}1.0$ when distance is expressed as the number of bonds, n, separating D and A. Neglecting the distance dependence of FC_{DA} this implies that the distance dependence of the electronic coupling of the locally in D excited state with the charge separated state (H_{D*A,D^+A^-}, which we abbreviate to H_{D*A}) and that of the ground state with the charge separated state (H_{DA,D^+A^-}, which we abbreviate to H_{DA}) are very similar.

For $n \leq 10$ we were able to detect CT fluorescence accompanying the charge recombination[24,26]. From that fluorescence H_{DA} was determined via a method that is discussed in the appendix. The results showed that H_{DA} displays a closely exponential distance dependence $H_{DA} \sim \exp(-0.44\ n)$. Thus, within the limits of experimental uncertainty, the distance dependence of k_{cr} and H_{DA} indeed obey the quadratic interdependence predicted by eqn (2) for a fixed value of FC_{DA}. This appears a highly important result, but it should be stressed that, due to the limited sensitivity of the CT-fluorescence based determination of H_{DA}, we were unable to test the interdependence for $n > 10$, and furthermore that even so the absence of CT fluorescence in polar solvents made it impossible to perform the test in such solvents where major changes of the reorganization energy as a function of D/A distance are likely to occur.

Keeping these limitations in mind we will in the following sections investigate the influence of other factors than the distance, in particular that of the bridge configuration and of orbital symmetry, on both the electron-transfer kinetics and on H_{DA}, as determined from CT-fluorescence, for a number of rigid D-bridge-A systems with a limited D/A distance ($n \leq 8$).

5. Influence of the bridge configuration; the "all-*trans* rule".

Whereas the D/A distance is evidently a crucial parameter in determining both inter-, and intramolecular D/A interaction, the structure and organization of the intervening medium may also be expected to play an important role. Bridged D/A systems provide a valuable tool not only for the study of distance dependence but also for the study of medium effects[29]. At one hand rigidly bridged D/A systems allow one to change the surrounding medium (solvent) without changing the distance and relative orientation, on the other hand variation of the bridge structure itself allows one to vary in a very well defined way that part of the intervening medium.

In the present section we confine our attention to the latter aspect and especially to the influence of the bridge configuration. The data presented in Fig. 1 for the isomers **3a** and **3b** demonstrate the importance of bridge configuration and in fact constitute one of the earliest verifications of the "all-*trans* rule" (see ref. 30 for a similar demonstration via PES). This "rule" is based on the computational results of Hoffmann et al.[3,4], which indicate that optimal TBI requires an all-*trans* (anti periplanar) arrangement of the functional-group orbitals and the sigma-bond orbitals in the coupling path. Such an arrangement is found in **3a** but not in **3b**. The fact[14] that **3a** but not **3b** displays a detectable intramolecular CT absorption band thus underscores the validity of the all-*trans* rule and also demonstrates the sensitivity of the CT electronic transition probability for changes in bridge configuration.

Interestingly, also the rigidly bridged systems displayed in Fig. 6 contain all-*trans* coupling paths, which is likely to contribute to the very fast rates of photoinduced charge separation observed for these systems. We have therefore recently studied[31,32] charge separation kinetics for three systems (**10(6)**, **10(8)**, **11(8)**) which lack an all-*trans* coupling path. The structure of these systems is shown in Fig. 7 together with that of the related all-*trans* bridged systems **9(6)** and **9(8)** having a coupling path of the same overall length. While the absence of an all-*trans* path in **10(6)** and **10(8)** is immediately evident from their structures, it is useful to note that in **11(8)** only the near planarity of the five-membered ring[32], carrying the acceptor chromophore, destroys the all-*trans* nature of the overall coupling path.

Tables 1 and 2 compile rates of photoinduced charge separation (k_{cs}) and of charge recombination (k_{cr}) determined for the compounds of Fig. 7. Virtually without exception both k_{cs} and k_{cr} are diminished in the absence of an all-*trans* coupling path. In conjunction with eqn(2), this appears another convincing demonstration of the validity of the all-*trans* rule which implies that H_{DA} should be lowered when the all-*trans* nature of the coupling path is lost. Upon closer inspection, however, it is clear that also other factors

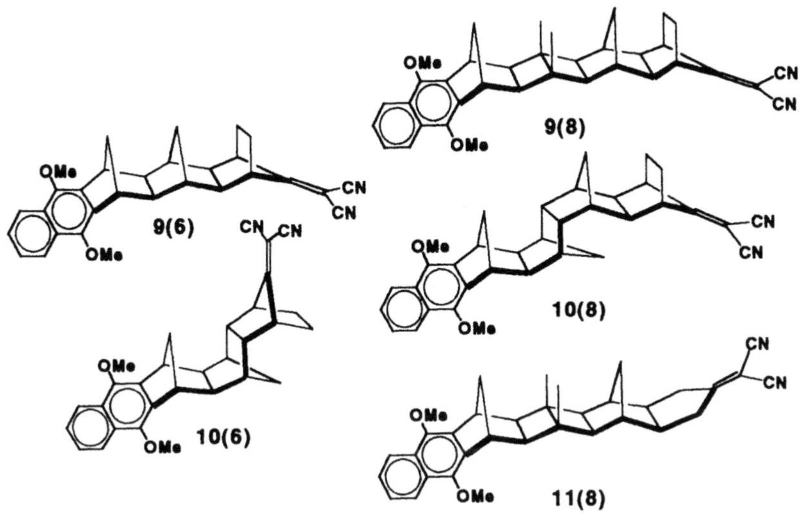

Figure 7. All-*trans* bridged systems **9(6)** and **9(8)** and systems with an equal overall bridge length lacking the all-*trans* configuration.

TABLE 1. Rates ($k_{CS} \times 10^{-8}$ s^{-1}) and rate ratios of photoinduced charge separation (see refs. 31 and 32 for experimental details)

solvent	9(6)	10(6)	9(8)	10(8)	11(8)	9(6)/10(6)	9(8)/10(8)	9(8)/11(8)
n-hexane	-	-	150	19	-	-	7.9	-
cyclohexane	-	-	210	29	-	-	7.2	-
benzene	-	-	520	180	261	-	2.9	2
di-n-butylether	-	-	470	62	114	-	7.6	4.1
diethylether	-	1250	470	72	132	-	6.5	3.6
ethylacetate	≥ 3300	770	450	89	148	≥ 4.3	5.1	3
tetrahydrofuran	≥ 3300	-	670	165	207	-	4.1	3.2
acetonitrile	~2500	360	300	22	59	~7	13.6	5.1

TABLE 2. Rates ($k_{cr} \times 10^{-6}$ s^{-1}) and rate ratios of charge recombination as measured from the lifetime of CT-fluorescence [or from time resolved microwave conductivity[28]].

solvent	9(6)	10(6)	9(8)	10(8)	11(8)	9(6)/10(6)	9(8)/10(8)	9(8)/11(8)
cyclohexane	23[26]	15[19]	31[36]	17[20]	-	1.5[1.4]	1.8[1.8]	-
benzene	167[167]	167[200]	26[36]	15[15]	4.5	1.0[0.8]	1.7[2.4]	5.8
di-n-butylether	-	-	23	10	5.5	-	2.3	4.2

must be involved. Thus, a significant solvent effect applies to the ratio between k_{cs} in an all-*trans* system and that in a comparable system lacking the all-*trans* structure (see Table 1). As we have pointed out earlier[31], when discussing the k_{cs} ratio of **1(8)** and **2(8)**, this can be explained by assuming that part of the overall interaction occurs through-solvent, making this part insensitive to modification of the bridge configuration. The solvent sensitivity of k_{cs} should then increase with decreasing through-bond interaction. Qualitative support for such a correlation is provided by the, admittedly limited, set of data now available for k_{cs} in **9(8)**, **10(8)** and **11(8)**. Clearly the solvent sensitivity of k_{cs} increases in the order **9(8)** < **11(8)** < **10(8)** and the rate in each individual solvent decreases in the same order indicating a smaller contribution of TBI.

While, as discussed in the previous section, the changes in the distance were found to have a very similar effect on k_{cs} and k_{cr}, the effects of changes in the bridge configuration appear to be different. Comparison of the data in Table 1 and in Table 2 suggests that k_{cr} is somewhat less sensitive to changes of the bridge configuration than k_{cs}. It should be realized, however, that this does not automatically imply that H_{DA} is less sensitive to the bridge configuration than H_{D*A}. In fact the interpretation of the response of k_{cr} to the bridge configuration is even more complex than that of k_{cs}. As for charge separation both H_{DA} and FC_{DA} may vary but furthermore two competing pathways of recombination are in principle available. Not only can recombination populate the electronic ground-state directly but, as evidenced from transient absorption data[33], also formation of a locally in the donor excited triplet state as an intermediate can occur to some extent (D^{+}-bridge-A^{-} $\rightarrow$ $^{3}D^{*}$-bridge-A).

In view of these complications we now decided to investigate the effect of bridge configuration on H_{DA} by quantitative determination of the radiative rate constant, k_F, for CT fluorescence accompanying direct recombination to the ground-state. Whereas, as mentioned in §4, this approach was applied[24] quite succesfully to determine the distance dependence in series **9(n)**, it's application to the systems **10(6)**, **10(8)** and **11(8)** is severely thwarted by the weakness of their CT fluorescence. In a limited number of solvents the CT-emission can be detected as a long-wavelength tail on the residual local donor fluorescence. Also it's lifetime can be measured, and thereby the lifetime of the charge-separ-

TABLE 3. CT-fluorescence data and H_{DA} values calculated therefrom via eqn (10).

	solvent	R_c [Å]	υ_{ct} [10^3 cm^{-1}]	τ_{cr} [ns]	Φ_{ct}	k_F [10^4 s^{-1}]	H_{DA} [cm^{-1}]	$(H_{DA})^2$ ratios
9(6)	cyclohexane	9	22.4	43	0.03	69.8	137	2.86
10(6)	cyclohexane	9.8	21.5	65	0.018	27.7	81	
9(8)	benzene	11.8	19.2	40	0.0031	7.75	36	
	di-n-butylether		19.9	44	0.0044	10	44	
10(8)	benzene	12.6	-	68	-	-	-	2.25
	di-n-butylether		-	100	-	-	-	2.47
11(8)	benzene	12.1	21.1	220	0.01	4.5	24	
	di-n-butylether		21.4	180	0.0094	5.2	28	

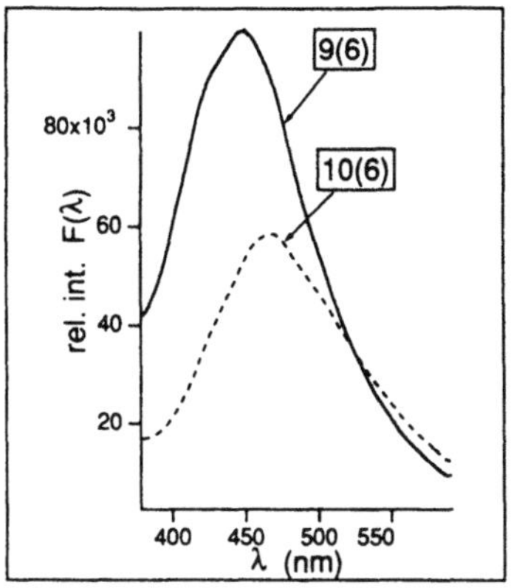

Figure 8. CT-fluorescence (deconvoluted from local donor emission) observed for **9(6)** and **10(6)** in cyclohexane.

ated state, but for **10(8)** the CT fluorescence was too weak to carry out a reasonably accurate determination of the quantum yield, and therefore the radiative rate constant, k_F, for this system could not be determined.

In Table 3 position υ_{ct}, lifetime τ_{cr}, and quantumyield Φ_{ct} data are compiled while in Fig. 8 CT-fluorescence spectra of **9(6)** and **10(6)**, obtained after correction for the contribution of residual donor fluorescence, are shown. In Table 3 also the H_{DA} values, calculated as described in the appendix, are tabulated. While only few data were obtained, the results consistently show that H_{DA} across an all-*trans* bridge is higher than across a bridge lacking such a configuration, thus confirming the validity of the all-*trans* rule. The quantitative correspondence between the squared ratio of H_{DA} for the pairs **9(6)/10(6)** and **9(8)/11(8)** (see last columns of Table 3) and the ratio of k_{cr} in these pairs (see Table 2) is rather poor, however. Within the context of eqn(2) this implies that the change of the bridge configuration not only affects H_{DA} but also brings about (minor) changes in FC_{DA} thereby (slightly) enhancing or diminishing the effect on the electron transfer kinetics. This demonstrates that one has to be very careful with the interpretation of the kinetic results of subtle structural changes in terms of H_{DA} only. This will become even more evident in the next section, where we investigate the effect of orbital symmetry on charge recombination kinetics.

6. Orbital symmetry effects on intramolecular D/A interaction.

In earlier research by one of us pronounced orbital symmetry effect on intramolecular charge-transfer absorption[34,35] and emission[36] spectra were reported from studies on cyclophanes containing aromatic D/A pairs in a close-contact sandwich type arrangement. The pairs **12(5,5)/12(4,8)** and **13(6,6)/13(3,8)** epitomize the plethora of D/A cyclophanes investigated in these studies (see Fig. 9).

The cyclophanes **12(5,5)** and **12(4,8)**, that contain the powerful 1,4-dimethoxyben-

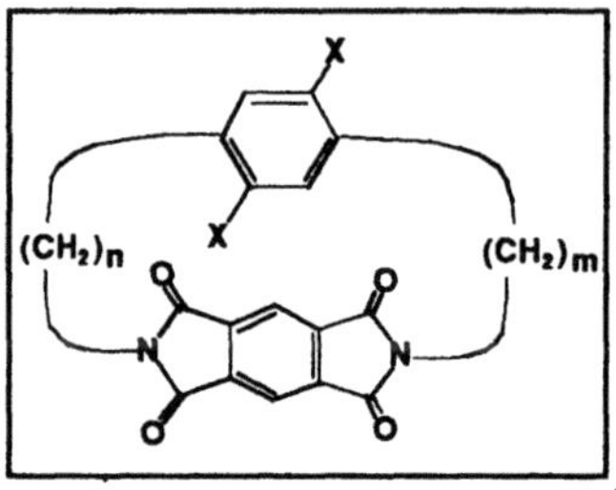

	X	n	m	CT-absorption (in CH_2Cl_2) λ_{max} [nm]	ε_{max} [l/mol.cm]			
12(5,5)	OMe	5	5	410sh	37			
12(4,8)	OMe	4	8	450	200			
				CT-fluorescence (in cyclohexane) λ_{max} [nm]		Φ	τ_{cr} [ns]	k_F [s^{-1}]
13(6,6)	Me	6	6	497		0.01	24	$0.4x10^6$
13(3,8)	Me	3	8	505		0.027	23.5	$1.15x10^6$

Figure 9. Structure and electronic absorption and emission data for D/A cyclophanes.

zene donor moiety, display a discrete intramolecular CT absorption which is much stronger in the asymmetrically bridged isomer. While these cyclophanes display no CT-fluorescence, **13(6,6)** and **13(3,8)**, which contain a weaker donor, do. In these, however, the corresponding absorption is overlapped[37] by other stronger bands. Clearly also the radiative rate constant is significantly higher in the asymmetric **13(3,8)** than in the symmetric **13(6,6)** indicating a stronger D/A coupling in the former (note, however, that the charge recombination rates, $1/\tau_{cr}$, are virtually identical!).

Although the flexibility of the bridges allows for considerable fluctuation of conformation and distance it is evident that for equal bridge lengths (i.e. in **12(5,5)** and **13(6,6)**) the centers of D and A will be effectively superimposed leading to an overall C_2 symmetry with the twofold axis perpendicular to the D and A π-planes. Since the frontier orbitals ($HOMO_D$ and $LUMO_A$) have different symmetries (see Fig. 10) with respect to

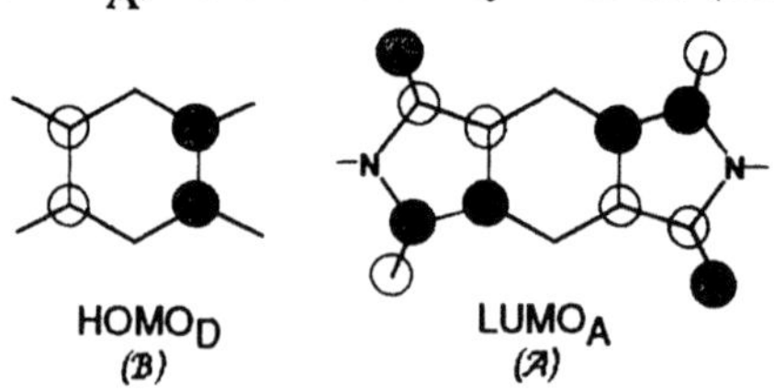

Figure 10. Relative orbital coefficients in the frontier orbitals of the D and A moieties incorporated in the cyclophanes of Fig. 9. The symmetry labels 𝒜 and ℬ refer to C_2 symmetry (see text).

twofold rotation, the D/A interaction in such a conformation is "symmetry forbidden" thereby rationalizing the remarkable increase in both CT-absorption and CT-emission probability when the C_2 symmetry is broken as the D and A groups are shifted relative to each other in **12(4,8)** and **13(3,8)**.

In view of the results presented above it appeared of considerable interest to investigate the effect of orbital symmetry in rigidly linked D-bridge-A systems. For this purpose we have now compared[37] the 8-bond bridged molecules **9(8)**, **14(8)**, **15(8)** and **16(8)** compiled in Fig. 11. Importantly all of these belong to the C_s point group with the mirror plane bisecting D and A. In Fig. 12 the relevant frontier orbitals of D and A are drawn qualitatively. Evidently $HOMO_D$ and $LUMO_A$ are of different symmetry in **9(8)** but of equal symmetry in the other systems. Under C_s symmetry, the highest occupied π-MO of the donor ($HOMO_D$) transforms as $\mathcal{A}''$ and the lowest unoccupied π^*-orbital of the acceptor ($LUMO_A$) as $\mathcal{A}'$ in **9(8)** and as $\mathcal{A}''$ in **14(8)-16(8)**, thereby leading to $\mathcal{A}''$ and $\mathcal{A}'$ symmetry of the charge separated state respectively.

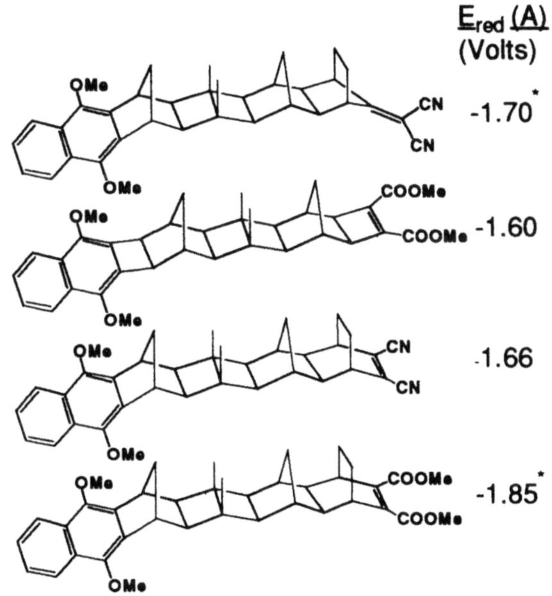

9(8)

14(8)

15(8)

16(8)

Figure 11. Structure of rigidly bridged D/A systems used to investigate orbital symmetry effects. The electrochemical reduction potentials of the acceptor groups incorporated are given inVolts vs. SCE in acetonitrile; an asterisk indicates that the electrochemical reduction is irreversible.

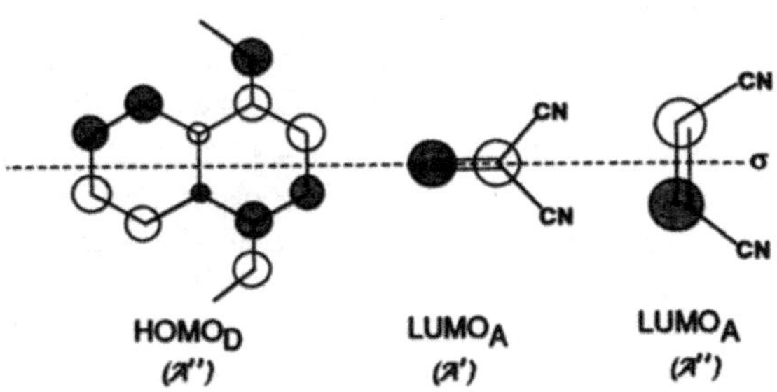

Figure 12. Relative orbital coefficients in the frontier orbitals of the D and A moieties incorporated in the D-bridge-A systems of Fig. 11. The symmetry labels $\mathcal{A}'$ and $\mathcal{A}''$ refer to C_s symmetry (see text).

TABLE 4. CT-fluorescence data in benzene (and in di-n-butylether) and H_{DA} values calculated therefrom via eqn (10).

	ν_{ct} [10^3 cm^{-1}]	τ_{cr} [ns]	Φ_{ct}	k_F [10^4 s^{-1}]	H_{DA} [cm^{-1}]	$\left[\frac{H_{DA}}{H_{DA}(\mathbf{9(8)})}\right]^2$
9(8)	19.23	40	0.0031	7.75	36	1
	(19.88)	(44)	(0.0044)	(10)	(44)	1
14(8)	20.00	33	0.0143	43	83	5.3
	(20.62)	(26)	(0.016)	(60)	(107)	(5.9)
15(8)	21.14	47	0.052	111	129	12.8
	(21.41)	(48)	(0.046)	(96)	(133)	(9.1)
16(8)	21.05	19	0.023[a]	95	120	11.1
	(21.51)	(29)	(0.018)[a]	(62)	(106)	(5.8)

a) Corrected for incomplete charge separation, (60% in benzene, 46% in di-n-butylether)

Because vibrational coupling between closely spaced excited states is likely to contaminate the symmetry of the first donor excited state[38,39], we defer discussion of the possible effect of changes in symmetry on k_{cs} and H_{DA}. Clearly, however, charge recombination to the $\mathcal{A}'$ ground state is symmetry 'forbidden' in **9(8)** and 'allowed' in **14(8)-16(8)**.

Interestingly, the charge recombination rate, $1/\tau_{cr}$ (see Table 4), responds weakly or not at all to the symmetry change. Thus the maximum rate enhancement from 'forbidden' to 'allowed' occurs between **9(8)** and **16(8)** with no more than a twofold decrease in the lifetime of the charge separated state! As pointed out above, this may be due to compensating changes in the Franck-Condon factors, FC_{DA}, which is even more likely here than for the systems described in §4 and §5 where the structure of D and A was kept constant. Furthermore it cannot *a priori* be excluded, that in competition with recombination to the electronic ground state formation of a locally excited triplet of D and/or A occurs[33] with different efficiencies in the series.

Fortuitously, for all compounds studied here charge recombination is accompanied by weak, long-wavelength charge-transfer (CT) fluorescence, albeit only in a limited number of rather nonpolar solvents. Quantitative data on this emission, whose decay time τ_{cr} provides a convenient measure for the lifetime of the charge separated state, are compiled in Table 4, while representative spectra are shown in Fig. 13.

Table 4 also compiles H_{DA} values calculated via eqn(10) (see appendix) from the CT-fluorescence data employing a uniform value R_c = 11.4 Å. Inspection of the results shows that the H_{DA} values for the symmetry "allowed" charge recombination in **14(8)-16(8)** are consistently higher than that for the symmetry "forbidden" recombination in **9(8)**, irrespective of whether the acceptor is stronger (**14(8)**), weaker (**16(8)**) or about equally strong (**15(8)**), compared to that in **9(8)**. The squared ratio of the H_{DA} values

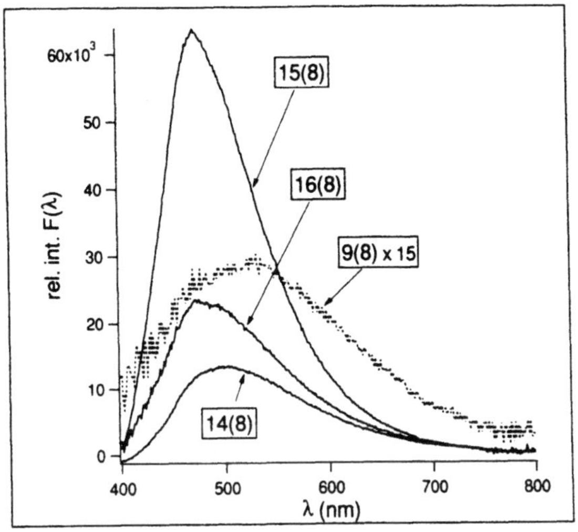

Figure 13. CT-fluorescence spectra of the compounds **9(8)** and **14(8)-16(8)** in benzene solution. The spectra have been corrected for the contribution of overlapping local donor emission in the short wavelength region.

ranges from 5.3, for **14(8)** versus **9(8)** in benzene, to 12.8 for **15(8)** versus **9(8)** in the same solvent. It should be realized that the incomplete charge separation in **16(8)** causes an extra source of uncertainty in the determination of H_{DA} for this compound.

Estimation of symmetry effects through comparison of the calculated H_{DA} value for **14(8)** with those of the other molecules in the series is complicated by the fact that the bridge in **14(8)** is quite different from that in the other molecules. Thus, the chromophores are fused to cyclobutyl rings in **14(8)**, whereas they are fused to norbornyl rings in the other systems. Thus, the most meaningful comparison yet, available for gauging symmetry effects, is that between the isomeric systems **9(8)** and **15(8)**, which leads to a best estimate of the pure symmetry effect on H_{DA} of 3.3±0.3. Since for nonadiabatic electron transfer the rate is proportional to the square of the electronic coupling (see eqn (2)) this implies that such symmetry factors could in principle change k_{cr} by about one order of magnitude. However, this rate effect is largely masked in the present series of molecules because of concomitant changes in Franck-Condon factors and/or in the contribution of competing recombination pathways.

Significant differences in Franck-Condon factors are indeed indicated by the response of υ_{ct} to changes in the acceptor reduction potential. While *e.g.* $E_{red}(A)$ is more negative by ~0.2 V in **16(8)** than in **15(8)** (see Fig. 11) this hardly changes υ_{ct}, thus implying non-negligible differences in reorganization energies and/or solvation even between these structurally closely related systems.

In conclusion, the presence of detectable CT fluorescence for **9(8)**, **14(8)**, **15(8)** and **16(8)** allowed us to evaluate directly the effect of structural modifications leading to a change in orbital symmetry, on the electronic coupling relevant for charge recombination. The results indicate that in the present type of rigidly bridged systems orbital symmetry effects could affect the rate of electron transfer by about one order of magnitude. This is considerably less than the two or three orders of magnitude recently deduced[40] from an analysis of charge separation rates in two systems related to **9(8)** and **14(8)**. In that study, however, the experimentally observed rate ratio was only 5±2, and a huge "correction" for differences in Franck-Condon factors was thus applied, which in view of the present results is likely to have led to a quite exaggerated estimate of the actual symmetry effect on H_{DA}.

The symmetry effect on $(H_{DA})^2$ now determined from CT-fluorescence very significantly exceeds that on k_{cr} (see Table 4). The discrepancy appears larger than in the cases where H_{DA} is modified by changing the bridge configuration (see §5) or the bridge length (see §4). It should be realized that in these latter cases the structures of D and A were kept virtually constant whereas that of A had now to be modified to change its orbital symmetry. While, as stipulated above, these changes in structure lead to significant changes in the Franck-Condon factor that can either enhance or decrease the changes in k_{cr} brought about by the symmetry related change of H_{DA}, it's a bit surprising that in all cases the latter appears to apply. We therefore wish to point out that in studying symmetry effects an extra complication might in fact arise. When relating CT-absorption or CT-emission data with electron-transfer kinetics it is implicitly assumed that H_{DA} does not depend critically upon the differences in the point of the configurational space where these processes take place. Especially for the rigid systems studied here this assumption appears quite acceptable since their rigidity evidently strongly limits the configurational space available. Where symmetry effects are involved, however, relatively minor configurational changes, such as those occurring between the configuration of maximum CT emissivity and that allowing for a maximum rate of charge recombination, could effectively destroy the symmetry. This is expected to increase especially the rate of a "symmetry forbidden" process, thereby lowering effectively the ratio of the rates of "allowed" and "forbidden" electron transfer.

7. Appendix; determination of H_{DA} from CT-fluorescence data.

The radiative rate constant k_F for an electronic transition from an upper state u to a lower state l is equal to the Einstein coefficients $\mathcal{A}$ for spontaneous emission summed over the complete fluorescence spectrum as expressed in eqn (5) where the summation is over the vibronic levels m of l.

$$k_F = \mathcal{A}_{u0\to l} = \sum_m \mathcal{A}_{u0\to lm} \qquad (5)$$

As discussed by Birks[41] this can be expanded to eqn (6):

$$k_F = \frac{64\pi^4 n^3}{3hc^3} \left|\bar{M}_{ul}\right|^2 \left\langle \nu_f^{-3} \right\rangle_{Av}^{-1} \qquad (6)$$

In (6) $\bar{M}_{ul}$ designates the mean value of the electronic transition moment, n the solvent refractive index and $(<\upsilon_f^{-3}>_{Av})^{-1}$ is the reciprocal of the mean value of υ^{-3} over the fluorescence spectrum given by eqn (7):

$$\langle\upsilon_f^{-3}\rangle_{Av}^{-1} = \frac{\int F(\upsilon)d\upsilon}{\int \upsilon^{-3}F(\upsilon)d\upsilon} = \frac{\int \lambda^{-4}F(\lambda)d\lambda}{\int \lambda^{-1}F(\lambda)d\lambda} \tag{7}$$

In (7) the integrals are taken over the frequency (wavelength) range of the fluorescence spectrum with the intensity measured in terms of relative number of quanta per frequency interval ($F(\upsilon)$) or per wavelength interval ($F(\lambda)$).

For a charge transfer transition between weakly interacting D and A systems at a center to center distance R_c, the electronic transition moment can furthermore be approximated employing a two state model as[42]:

$$\bar{M}_{ul} = \frac{eR_cH_{ul}}{\Delta E_{ul}} \tag{8}$$

In (8) H_{ul} and ΔE_{ul} are the electronic coupling and the energy gap between the DA and D^+A^- states.

Equations (6-8) relate the electronic coupling for charge recombination ($H_{ul}=H_{DA}$) and the CT emission properties of weakly interacting D-bridge-A systems with a well defined center to center separation[43]. Thus we may now write from (6) and (8) and upon substitution of $\Delta E_{ul} = \Delta E_{DA} = h\upsilon_{ct}$, where υ_{ct} corresponds to the frequency at the CT emission maximum:

$$H_{DA} = \sqrt{\frac{3h^3c^3\upsilon_{ct}^2k_F}{64\pi^4n^3e^2R_c^2\langle\upsilon_f^{-3}\rangle_{Av}^{-1}}} = \sqrt{\frac{1.4 \times 10^5\upsilon_{ct}^2k_F}{n^3R_c^2\langle\upsilon_f^{-3}\rangle_{Av}^{-1}}} \tag{9}$$

In eqn (9) H_{DA} and υ are in cm^{-1}, R_c is in Å and k_F in s^{-1}. If $(<\upsilon_f^{-3}>_{Av})^{-1} \approx \upsilon_{ct}^3$, which turns out to be a good approximation in the present case (see below), eqn (9) can be simplified to eqn (10), which was employed to calculate H_{DA} from the CT fluorescence data.

$$H_{DA} = \sqrt{\frac{1.4 \times 10^5\, k_F}{n^3\, R_c^2\, \upsilon_{ct}}} \tag{10}$$

8. Acknowledgements

We wish to acknowledge the contribution of many colleagues and coworkers who contributed to the research described above. We are particularly indebted to A.J. de Gee, R.L.J. Zsom, L.G. Schroff†, J.H. Borkent, P. Pasman, L.B. Krijnen, R.M. Hermant, H. Oevering, D. Bebelaar, A.M. Oliver, E. Cotsaris, D.C. Craig, N.S. Hush, K.J. Smit and M.P. de Haas.

Our thanks are furthermore due to R.P.H. Rettschnick (UvA) for helpful discussions about the relation between electronic coupling and CT-emission.

The present investigations were supported in part by the Netherlands Foundation for Chemical Research (SON), with financial aid from the Netherlands Organization for the Advancement of Research (NWO), by the Australian Research Council, by the Australian Federal Government, and by the Netherlands Ministry of Economic Affairs.

9. References and notes.

1. a) R.D. Topsom, *Accounts Chem. Res.* **16** (1983) 292;
   b) R.D. Topsom, *Progr. Phys. Org. Chem.* **12** (1976) 1.
2. a) W. Adcock, V. Sankar Iyer, *J. Org. Chem.* **50** (1985) 1538;
   b) W. Adcock, A.N. Abeywickiema, G.B. Kok, *J. Org. Chem.* **49** (1984) 1387.
3. R. Hoffmann, A. Imamura, W.J. Hehre, *J. Am. Chem. Soc* . **90** (1968) 1499.
4. R. Hoffmann, *Accounts Chem. Res.* **4** (1971) 1.
5. a) R. Gleiter, *Pure & Appl. Chem* . **59** (1987) 1585;
   b) M.N. Paddon-Row, K.D. Jordan, in: "Molecular structure and energetics", Vol.6, J.F. Liebman and A. Greenberg (Eds.), VCG Publishers, New York (1988).
6. V. Balaji, L. Ng, K.D. Jordan, M.N. Paddon-Row, H.K. Patney, *J. Am. Chem. Soc.* **109** (1987) 6957.
7. R. Foster, "Organic Charge-Transfer Complexes", Academic Press, London (1969).
8. K. Wiesner, H.W. Brewer, D.L. Simmons, D.R. Babin, F. Bickelhaupt, J. Kallor, T. Borgi, *Tetrahedron Lett.* (1960) 17.
9. R.C. Cookson, J. Henstock, J. Hudec, *J. Am. Chem. Soc.* **88** (1966) 1061.
10. J. Kuthan, J. Palecek, L. Musil, *Z. Chem.* **8** (1968) 229; *Coll. Czech. Chem. Commun.* **39** (1974) 750.
11. J. Hudec, *Chem. Commun* . (1970) 829.
12. A.W.J.D. Dekkers, J.W. Verhoeven, W.N. Speckamp, *Tetrahedron* **29** (1973) 1691.
13. P. Pasman, F. Rob, J.W. Verhoeven, *J. Am. Chem. Soc.* **104** (1982) 5127.
14. P. Pasman, J.W. Verhoeven, Th.J. de Boer, *Tetrahedron Lett.* (1977) 207.
15. A.J. de Gee, W.J. Sep, J.W. Verhoeven, Th.J. de Boer, *J. Chem. Soc.* , *Perkin Trans. II* (1975) 670.

16. a) L.B. Krijnen, H.B. Beverloo, J.W. Verhoeven, *Recl. Trav. Chim. Pays-Bas* **106** (1987) 135;
b) B. Krijnen, H.B. Beverloo, J.W. Verhoeven, C.A. Reiss, K. Goubitz, D. Heijdenrijk, *J. Am. Chem. Soc.* **111** (1989) 4433.
17. P. Pasman, J.W. Verhoeven, Th.J. de Boer, *Chem. Phys. Lett.* **59** (1978) 381.
18. P. Pasman, N.W. Koper, J.W. Verhoeven, *Recl. Trav. Chim. Pays-Bas* **101** (1982) 363.
19. R.M. Hermant, N.A.C. Bakker, T. Scherer, B. Krijnen, J.W. Verhoeven, *J. Am. Chem. Soc.***112** (1990) 1214.
20. N.S. Hush, *Coord. Chem. Rev.* **64** (1985) 135.
21. C.A. Stein, N.A. Lewis, G. Seitz, *J. Am. Chem. Soc.* **104** (1982) 2596.
22. R.R. Dogonadze, in: "Reactions of molecules at electrodes", N.S. Hush (Ed.), Wiley-Intersciences, New York, (1971).
23. R.A. Marcus, N. Sutin, *Biochim. Biophys. Acta* **811** (1985) 265.
24. H. Oevering, J.W. Verhoeven, M.N. Paddon-Row, J.M. Warman, *Tetrahedron* **45** (1989) 4751.
25. G.L. Closs, L.T. Calcaterra, N.J. Green, K.W. Penfield, J.R. Miller, *J. Phys. Chem.* **90** (1986) 3673.
26. H. Oevering, M.N. Paddon-Row, M. Heppener, A.M. Oliver, E. Cotsaris, J.W. Verhoeven, N.S. Hush, *J. Am. Chem. Soc.* **109** (1987) 3258.
27. J.W. Verhoeven, *Pure & Appl. Chem.* **62** (1990) 1585.
28. J.M. Warman, K.J. Smit, M.P. de Haas, S.A. Jonker, M.N. Paddon-Row, A.M. Oliver, J. Kroon, H. Oevering, J.W. Verhoeven. *J. Phys. Chem.* **95** (1991) 1979.
29. J. Kroon, J.W. Verhoeven, M.N. Paddon-Row, A.M. Oliver, *Angew. Chem. Int. Ed. Engl.* **30** (1991) 1358.
30. a) F.S. Jorgensen, M.N. Paddon-Row, H.K. Patney, *J. Chem. Soc. Chem. Commun.*(1983) 573;
b) M.N. Paddon-Row, L.M. Englehardt, B.W. Skelton, A.H. White, F.S. Jorgensen, H.K. Patney, *J. Chem. Soc. Perkin Trans. II* (1987) 1835.
31. A.M. Oliver, D.C. Craig, M.N. Paddon-Row, J. Kroon, J.W. Verhoeven, *Chem. Phys. Lett.***150** (1988) 366.
32. J.M. Lawson, D.C. Craig, M.N. Paddon-Row, J. Kroon, J.W. Verhoeven, *Chem. Phys. Lett.* **164** (1989) 120.
33. H. Oevering, Dissertation, University of Amsterdam (1988).
34. L.G. Schroff, A.J.A. van der Weerdt, D.J.H. Staalman, J.W. Verhoeven, Th.J. de Boer, *Tetrahedron Lett.* (1973) 1649.
35. R.L.J. Zsom, L.G. Schroff, C.J. Bakker, J.W. Verhoeven, Th.J. de Boer, J.D. Wright, H. Kuroda, *Tetrahedron* **34** (1978) 3225.
36. J.H. Borkent, J.W. Verhoeven, Th.J. de Boer, *Chem. Phys. Lett.* **42** (1976) 50.
37. A.M. Oliver, M.N. Paddon-Row, J. Kroon, J.W. Verhoeven, submitted.
38. J.R. Reimers, N.S. Hush, *Chem. Phys.* **146** (1990) 105.
39. J.R. Reimers, N.S. Hush, D.M. Sammeth, P.R. Callis, *Chem. Phys. Lett.* **169** (1990) 622.

40. Y. Zeng, M.B. Zimmt, *J. Am. Chem. Soc* . **113** (1991) 5107.
41. J.B. Birks, "Photophysics of Aromatic Molecules", chap. 4, Wiley-Interscience, London (1970).
42. D.N. Beratan, J.J. Hopfield, *J. Am. Chem. Soc.* **106** (1984) 1584.
43. We note that eqn (10) is the equivalent of the well-known Hush expression[20] (see eqn (1)), for relating CT absorption ("intervalence absorption") and H_{DA}.

SUPRAMOLECULAR ARRAYS FOR THE EFFICIENT CONVERSION OF LIGHT INTO CHEMICAL ENERGY IN THE SOLID STATE

Michael R. Wasielewski, George L. Gaines, III, Michael P. O'Neil,
Mark P. Niemczyk, and Walter A. Svec
Chemistry Division
Argonne National Laboratory
Argonne, IL 60439 USA

ABSTRACT. We recently solved the problem of achieving high quantum yield charge separation in the solid state using porphyrin-based donor-acceptor molecules. This has lead us to prepare supramolecular systems that are capable of high quantum yield, long-lived charge separation in the solid state. We report here on a pair of photosynthetic reaction center models consisting of a zinc porphyrin primary electron donor, ZP, positioned between a naphthoquinone electron acceptor, NQ, and an N,N,N,N-tetraalkyl-p-phenylenediamine secondary electron donor, TAPD. The resulting rigid structures place the TAPD donors at fixed 23 Å and 18 Å center-to-center distances from the NQ acceptors. Moreover, the orientation of the π systems of TAPD and NQ differ substantially between the two molecules. Excitation of ZP within each molecule in 2-methyltetrahydrofuran at 5 K results in two-step sequential electron transfer: TAPD-^{1*}ZP-NQ -> TAPD-ZP^{+}-NQ^{-} -> TAPD^{+}-ZP-NQ^{-}. The final TAPD^{+}-ZP-NQ^{-} radical pairs each live for several milliseconds and exhibit spin-polarized EPR spectra characteristic of spin-correlated radical pairs. The EPR spectra of these long-lived, spin-polarized radical ion pairs closely mimic the bacteriochlorophyll dimer cation - quinone anion radical pair produced in photosynthetic reaction centers. One of these molecules stores 1.85 eV of energy for 4 ms with a 67% quantum yield in the solid state.

1. Introduction

Recent interest in the role of the solvent in electron transfer reactions has focused on ultrafast photoinduced electron transfers and theoretical modeling of solvation.[1-

V. Balzani and L. De Cola (eds.), Supramolecular Chemistry, 201–218.

12] The vast majority of work in this area involves donor-acceptor systems in which the donor and acceptor either freely diffuse in solution to form a complex following excitation, or are already complexed in their ground state prior to excitation. Many difficulties in the analysis of electron transfer rates as a function of reaction free energy arise from translational degrees of freedom within such complexes.

We have employed an alternative approach to developing an understanding of the effects of solvent on electron transfer processes. This approach involves using a structurally related series of donor-acceptor molecules in which the donor and acceptor possess fixed distances and restricted orientations relative to one another. We have employed this approach successfully in a number of previous studies of photoinduced charge separation and dark charge recombination reactions.[13-15] Our most recent work focuses on the influence of solvent polarity both in liquid and solid solution on the free energy dependence of these reactions.[16,17]

1.1 Charge Separation in Solid Solution

In contrast to photosynthesis, which exhibits a quantum yield of primary charge separation near unity at cryogenic temperatures, most photosynthetic model systems based on chlorophyll or porphyrin electron donors exhibit significantly reduced efficiencies of light-initiated, singlet state electron transfer whenever they are dissolved in rigid glass media. This reduction occurs because solvent dipoles reorient around an ion-pair in a polar liquid, decreasing the energy of the ion-pair, while solvent dipoles cannot reorient around an ion-pair produced within a frozen solvent, and thus, provide little stabilization of the ion-pair. As a result, the energy level of the ion-pair is much higher in the rigid glass than in the liquid. In fact, the ion-pair state energy may be so high that it lies above the energy of the excited state, in which case photoinduced electron transfer cannot occur.

To obtain a quantitative picture of the dependence of charge separation rate on free energy of reaction in the rigid glass, we recently measured electron transfer rate data for 14 porphyrin-triptycene-acceptor molecules, Figure **1**, which possess sufficiently large, negative free energies for charge separation to allow electron transfer to compete with excited singlet state decay in glassy 2-methyltetrahydrofuran (MTHF) at 77 K.[17] Electron transfer rate constants, k_{cs}, were determined using picosecond transient absorption and emission techniques along with fluorescence quenching measurements. Since ion-pair recombination in porphyrin-acceptor molecules is non-radiative, the free energy of charge separation, ΔG_{cs}, in the rigid glass is difficult to obtain. On the other hand, in polar liquids the free energies of charge separation in these molecules, $\Delta G'_{cs}$, can be estimated using the one-electron oxidation and reduction potentials of the donor and acceptor, respectively,

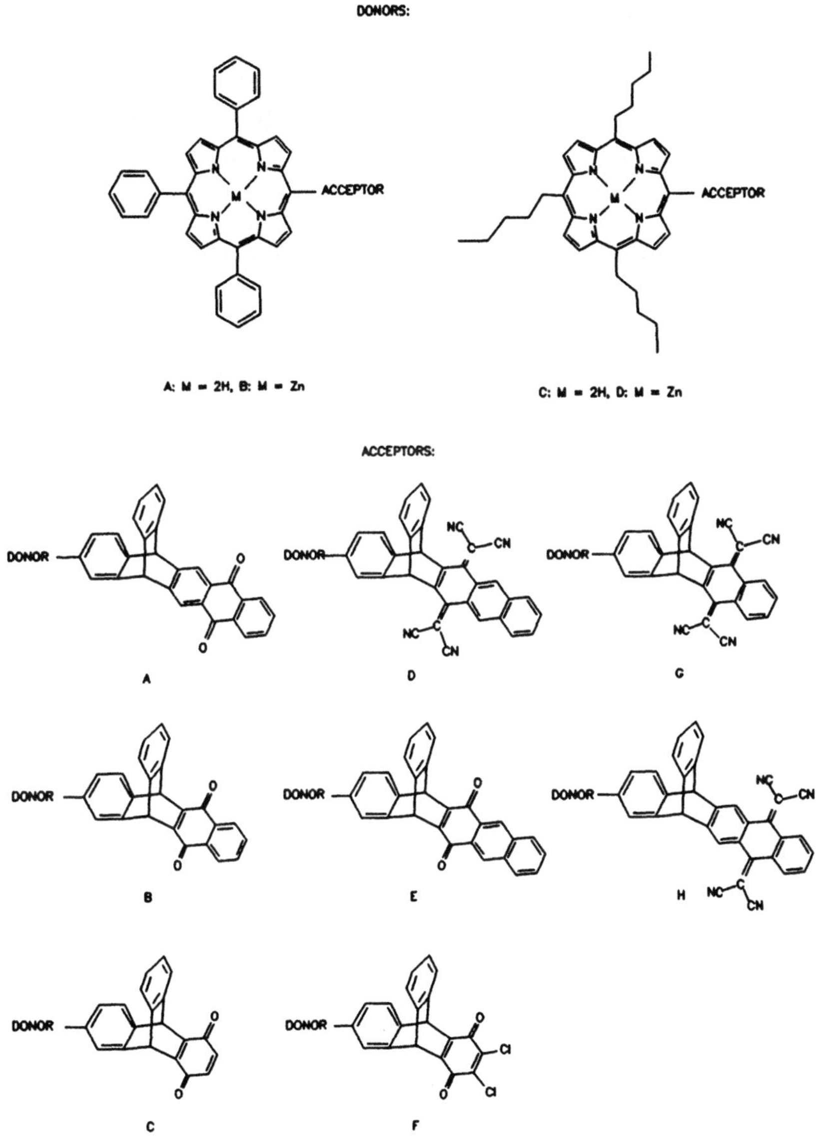

Figure 1. Structures of porphyrin-triptycene-acceptor molecules. Each molecule consists of a donor (**A-D**) covalently attached to an acceptor (**A-H**).

and the coulomb stabilization of the ion-pair.[18] We determined $\Delta G'_{cs}$ for the 14 compounds. Thus, the plot of ln k_{cs} *vs* $-\Delta G'_{cs}$ in Figure **2** *is based solely on experimentally determined quantities with no corrections applied.* Although the plot of ln k_{cs} *vs* $-\Delta G'_{cs}$ can be used as a predictive tool, $\Delta G'_{cs}$ obtained from polar solution data differs significantly from ΔG_{cs} in the rigid glass. Fortunately, we can determine the precise relationship between these two free energy scales. When $\Delta G_{cs} \approx 0$, thermal repopulation of the lowest excited singlet state from the ion-pair state in **BC** and **DB** yields biphasic fluorescence decays. These data can be analyzed to yield ΔG_{cs} directly. Assuming a linear free energy relationship the $\Delta G'_{cs}$ energy scale can be calibrated to yield ΔG_{cs} in the rigid glass using data from **BC** and **DB**, Figure **2**. Thus, the energies of the ion-pair states in these molecules are destabilized by $\Delta G_d = 0.80 \pm 0.05$ eV relative to their energies in polar liquids. The measured value of this destabilization energy agrees remarkably well with predictions based on the dielectric continuum model of the solvent.[18]

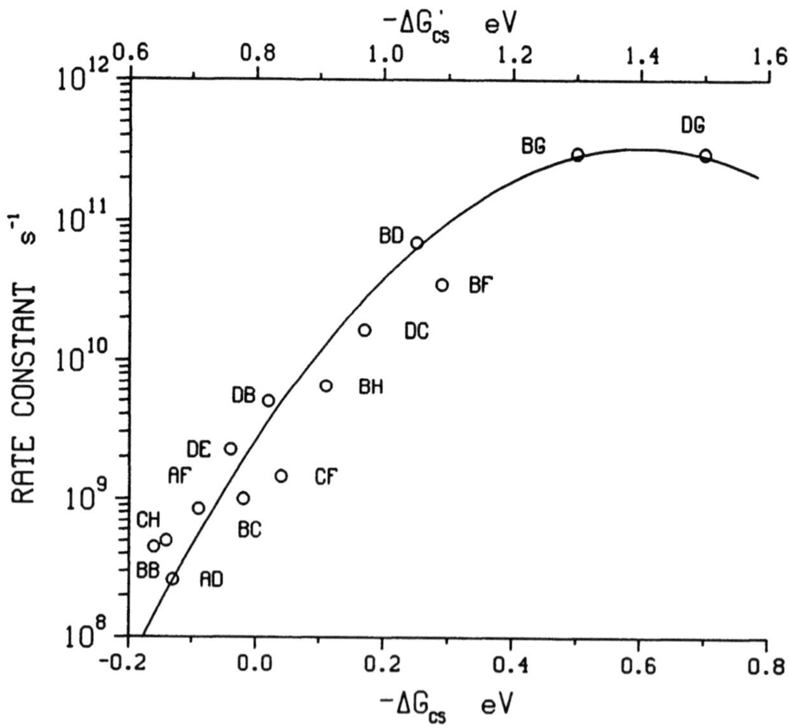

Figure **2**. Rate *vs* free energy relationship for selected molecules in MTHF at 77 K. The labels correspond to the compounds depicted in Figure **1**.

1.2 The Relationship between Long-Lived Charge Separation and Spin-Polarization

Now that we understand the energetic requirements to achieve high quantum yield charge separation in the solid state, we can use this information to design supramolecular systems to efficiently convert light into chemical potential in a solid state system. To achieve this goal we need to increase the lifetime of the charge separation by reducing the electronic coupling between the radical ion pairs that are produced following excitation of the donor-acceptor pair. This can be attained by careful design of the spacer groups linking the donor and acceptor and by using more than a single electron transfer step to increase the distance between the separated charges as is done in natural photosynthesis.

Photoinduced, multi-step charge separation in bacterial photosynthetic reaction centers proceeds from the lowest excited singlet state of the dimeric bacteriochlorophyll electron donor in two steps to yield a weakly interacting dimer cation - quinone anion radical pair, P^+-Q^-, separated by 28 Å.[19] The chromophores within the reaction center are positioned at precise distances and orientations to insure that the electronic coupling between P^+ and Q^- is sufficiently weak to allow P^+-Q^- to live for about 100 ms.[20] Semi-classical electron transfer theory predicts that the rate constant for charge recombination depends both on the electronic coupling matrix element, V, between the radicals within the ion-pair and the Franck-Condon weighted density of states, FCWD.[21] The FCWD term depends on the free energy of the recombination reaction, as originally given by Marcus [22]:

$$k_{cr} = (2\pi/\hbar)\, V^2 \cdot \mathrm{FCWD} \qquad (1)$$

For an optimized free energy of reaction, equation 1 predicts that a radical ion pair which lives for milliseconds should possess $V < 0.001$ cm^{-1}. Under these conditions the electron-electron exchange interaction between the radicals, $2J$, which is on the same order of magnitude as V, is sufficiently weak that differences in local magnetic fields surrounding each radical result in S-T_0 mixing of the radical pair spin sublevels.[23] This mixing produces a non-Boltzmann population of the spin sublevels of the radical pair and may result in the appearance of spin-polarized EPR spectra. Such spectra have been reported extensively for both bacterial and green plant reaction centers,[24-26] but have not been observed previously in rigid model systems in the solid state. Thus, a benchmark for achieving precise control over the electronic coupling parameters that lead to long-lived, efficient charge separation in the solid state, analogous to that observed in photosynthesis, would be the observation of long-lived, spin-polarized radical ion pairs in a supramolecular, biomimetic model system.

To observe the important structure-dependent anisotropic spin-spin interactions, such as the dipolar interaction, *D*, within radical pairs and to prevent spin lattice relaxation from destroying the spin polarization, it is necessary to examine the radical pairs in the solid state at low temperatures. Thus, our work on achieving high quantum yield charge separation in low temperature solids gives us the necessary tools to produce the appropriate supermolecule. We recently reported preliminary results on 2-TAPD-ZP-2-NQ in Figure 3, the first molecule to exhibit spin-polarized radical ion pair formation in the solid state.[27]

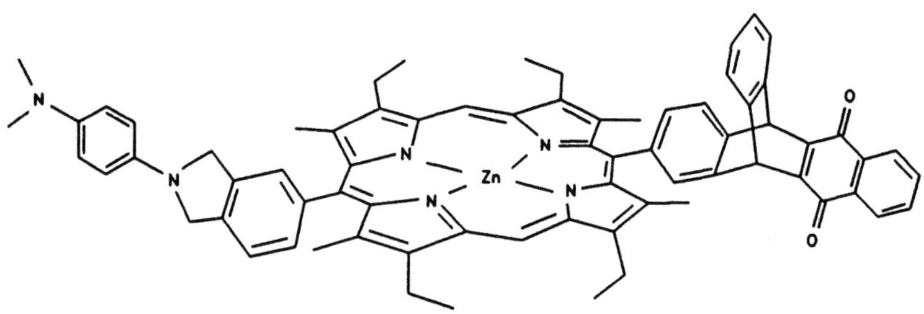

2–TAPD–ZP–2–NQ

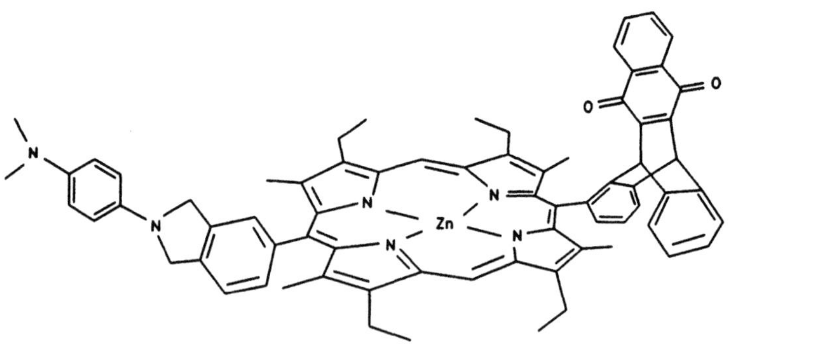

2–TAPD–ZP–1–NQ

Figure 3. Structures of 2-TAPD-ZP-2-NQ, **1**, and 2-TAPD-ZP-1-NQ, **2**.

2. Experimental

The compounds in Figure **3** were synthesized by methods described earlier[27] and the porphyrin syntheses generally follow methodology developed earlier[28]. In addition, reference compounds **3** and **4**, which lack the quinone and the secondary donor, respectively were also prepared. The details of their preparation will be presented in a future publication.

2-Methyltetrahydrofuran (Aldrich) was freshly distilled from lithium aluminum hydride before each experiment.

UV-visible absorption spectra were taken on a Shimadzu UV-160. Fluorescence spectra were obtained using a Perkin-Elmer MPF-2A fluorimeter interfaced to an IBM personal computer. All samples for fluorescence were purified by preparative TLC on Merck silica gel plates. Samples for fluorescence measurements were 10^{-7} M in 1 cm cuvettes. The emission was measured 90° to the excitation beam. Fluorescence quantum yields were determined by integrating the digitized emission spectra from 600 to 800 nm and referencing the integral to that for either free base or zinc porphyrin in toluene.[29]

Redox potentials for each donor-acceptor molecule were determined in MTHF containing 0.1M tetra-n-butylammonium tetrafluoroborate using a Pt disc electrode at 21°. These potentials were measured relative to a saturated calomel electrode using ac voltammetry. Both the one electron oxidations and reductions of these molecules exhibited good reversibility.

Ultrafast transient absorption spectra and kinetics were obtained with a pump-probe apparatus with 1 ps resolution previously described.[30] Solutions with an absorbance of about .3 at 610 nm (2 mm pathlength cells) were used. Kinetic parameters were obtained by iterative reconvolution of the data with least squares fitting using the Levenberg-Marquardt algorithm.

EPR spectra were obtained on a Varian E-9 system equipped with an Air Products liquid helium low temperature dewar and controller. Samples concentrations were 5×10^{-4} M in MTHF. Samples were prepared in sealed tubes following 3 freeze-pump-thaw cycles. The samples were irradiated with light > 540 nm from a filtered xenon-arc lamp modulated at 500 Hz. The time-resolved EPR signals were detected with a lock-in amplifier. The microwave frequencies given in the figure captions were determined with an HP frequency counter. *g*-Factors were determined by calibration with weak samples of DPPH.

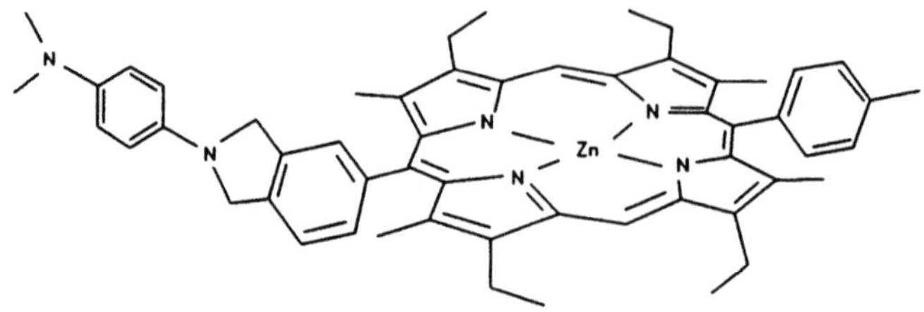

3

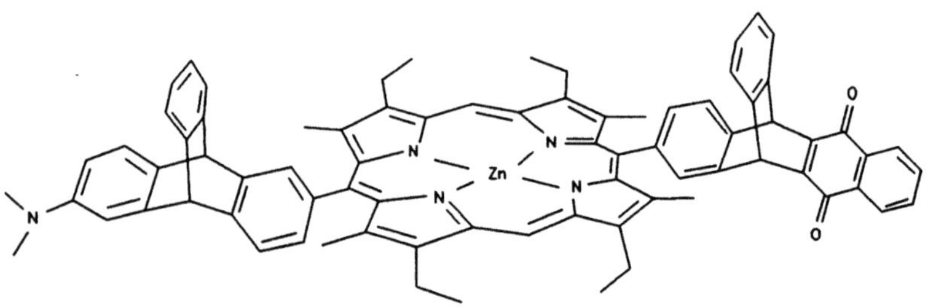

4

Figure 4. Structures of reference compounds 3 and 4.

3. Results

3.1 Rates and Energetics

In polar liquids the free energies of photoinduced charge separation in a donor-acceptor molecule, $\Delta G'_{cs}$, can be estimated with reasonable accuracy using the one-electron oxidation, E_{ox}, and reduction, E_{red}, potentials of the donor and acceptor, respectively, and the coulomb stabilization of the ion-pair:

$$\Delta G'_{cs} = E_{ox} - E_{red} - e_0^2/\epsilon_s r_{12} - E_s \qquad (2)$$

where e_0 is the charge of the electron, ϵ_s is the static dielectric constant of the high polarity medium, r_{12} is the center-to-center distance between the ions, and E_s is the energy of the lowest excited singlet state of the porphyrin donor determined from the frequency of the (0,0) band of its fluorescence spectrum. The free energy for the subsequent dark secondary charge separation is also determined from equation 2 by setting $E_s = 0$. We determined $\Delta G'_{cs}$ for compounds **1** and **2** using data obtained in MTHF at 295 K containing 0.1M tetra-n-butylammonium tetrafluoroborate. Since $r_{12} \geq 11$ Å for the ion pairs within these compounds, the coulombic term is ≤ 0.065 eV and is neglected. The energy levels of the various excited and ionic states of interest in compounds **1** and **2** are given in Figure **5a**.

Since our previous results on determining ΔG_{cs} for ion pair formation in solid solution showed that the simple dielectric continuum model of solvation is applicable, we used Weller's expression [18] for the destabilization of the ion-pairs as a function of solvent dielectric constant, equation 3.

$$\Delta G_d = e_o^2 [(1/2r_1)+(1/2r_2)-(1/r_{12})]/\epsilon_s - e_o^2[(1/2r_1)+(1/2r_2)]/\epsilon_s' \qquad (3)$$

where ΔG_d is the energy by which the ion-pair is destabilized in a solvent with an arbitrary value of ϵ_s and high frequency dielectric constant ϵ_0, if the redox potentials of the donor and acceptor are measured in a medium with a high dielectric constant, ϵ_s', r_1 and r_2 are the radii of the two ions, and the remaining parameters are defined above. Equation 3 was used to calculate the energy levels of the ion pairs for compounds **1** and **2** in low temperature MTHF, where $\epsilon_s = 2.6$ [31]. This data is presented in Figure **5b**.

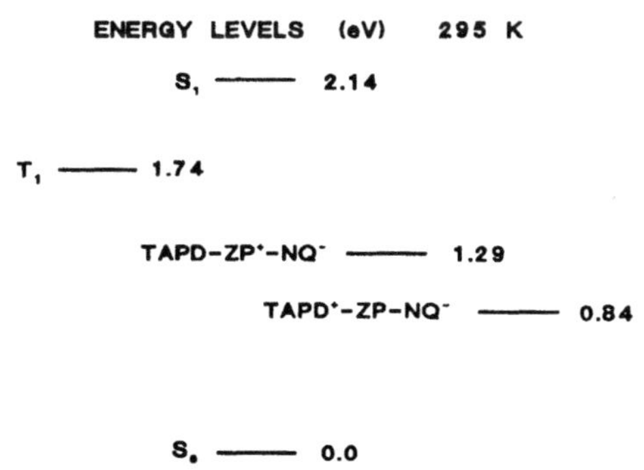

Figure **5a**. Energy levels at 294 K for compounds **1** and **2**.

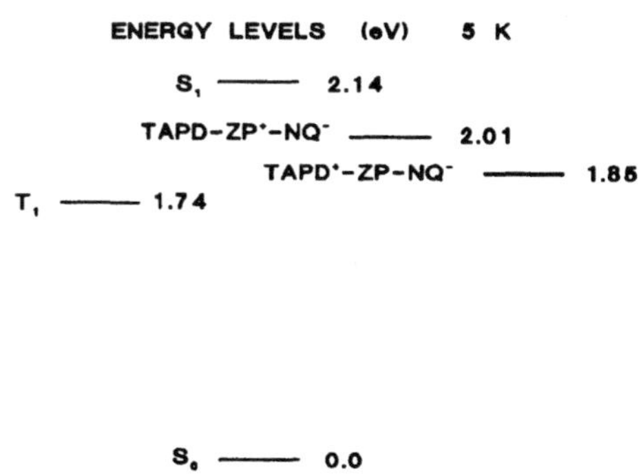

Figure **5b**. Energy levels at 5 K for compounds **1** and **2**.

Picosecond transient absorption and emission measurements with no applied magnetic field show that the primary reaction, 2-TAPD-^{1*}ZP-2-NQ -> 2-TAPD-ZP$^+$-2-NQ$^-$, occurs in τ = 5 ps, while the secondary reaction, 2-TAPD-ZP$^+$-2-NQ$^-$ -> 2-TAPD$^+$-ZP-2-NQ$^-$, occurs with τ = 530 ps. Based on these kinetics the quantum yield of 2-TAPD$^+$-ZP-2-NQ$^-$ formation is 67%. The absorption of TAPD$^+$ is easily monitored at 600 nm [32]. Charge recombination within the long-lived ion-pair, 2-TAPD$^+$-ZP-2-NQ$^-$, occurs with a 4 ms time constant, Figure 6. The corresponding kinetics for the two charge separation steps in 2-TAPD-ZP-1-NQ are 2.5 ps and 600 ps. The lifetime of 2-TAPD$^+$-ZP-1-NQ$^-$ is 3 ms and the quantum yield of its formation is 15 %.

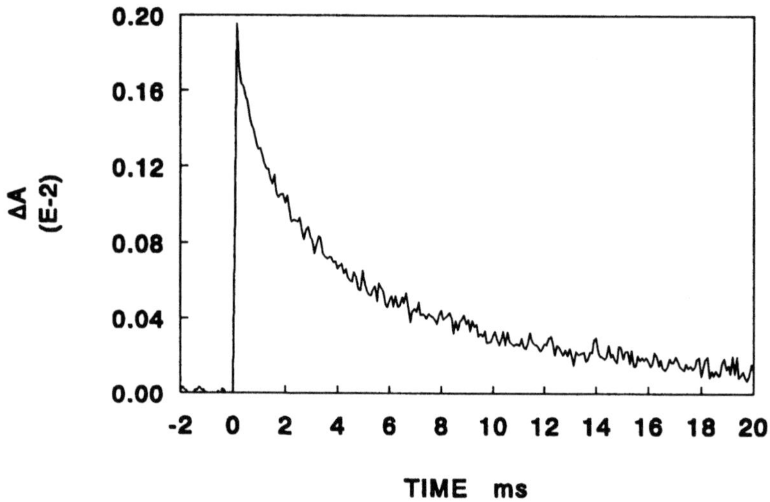

Figure 6. Decay of 2-TAPD$^+$-ZP-2-NQ$^-$ monitored at 600 nm.

3.2 Spin-Polarized EPR Spectra.

Time resolved EPR spectra of the photochemically generated radical ion pairs were obtained using a modulated xenon arc lamp and phase sensitive detection of the signal.[33] The EPR spectrum of 2-TAPD$^+$-ZP-2-NQ$^-$ is compared with those of reference compounds **3** and **4** obtained under identical conditions in Figure **7**. These spectra are displayed in the first derivative mode. It is readily seen that irradiation of **1** produces an intense EPR signal which consists of an emissive, **E**, low field line and an absorptive, **A**, high field line. Extensive control experiments were performed

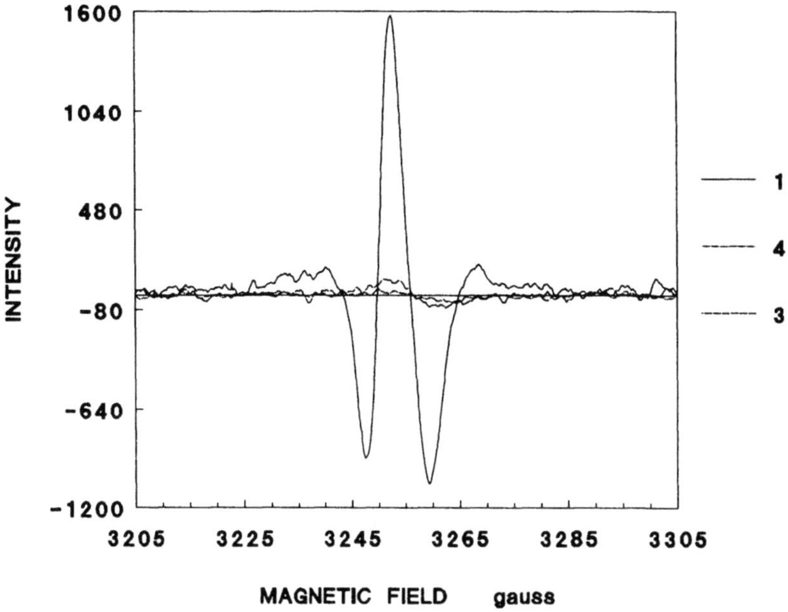

Figure 7. EPR signals from compounds **1**, **3**, and **4** in MTHF at 5 K generated using 540 nm, 500 Hz modulated light with phase-sensitive detection. Microwave frequency = 9.13 GHz, power = 2 mW.

to assign the origin of this signal. At 5 K the $TAPD^{+}$ and NQ^{-} free radicals display gaussian lines: $TAPD^{+}$, g_{iso} = 2.0030, FWHM = 35 G; NQ^{-}, g_{iso} = 2.0047, FWHM = 5 G. Removing the Zn atom from the porphyrin in **1** makes the initial electron transfer reaction endergonic in the solid state. Thus, in the absence of subsequent electron transfer excitation of the free base porphyrin results in formation of the porphyrin lowest excited triplet state, which shows a strong EPR signal with zero field splitting parameters $|D|$ = 0.0451 cm^{-1} and $|E|$ = 0.0041 cm^{-1}. Replacement of the NQ in **1** by a p-tolyl group, **3**, results in no observed EPR signals at 5 K. Replacing the TAPD moiety by N,N-dimethylaniline, **4**, also results in no photoinduced EPR signals. Since the N,N-dimethylaniline moiety oxidizes at 0.78 V vs SCE,[34] the secondary electron transfer in **4** is endergonic and does not occur. Thus, both sequential electron transfer reactions are required to observe a spin-polarized radical pair.

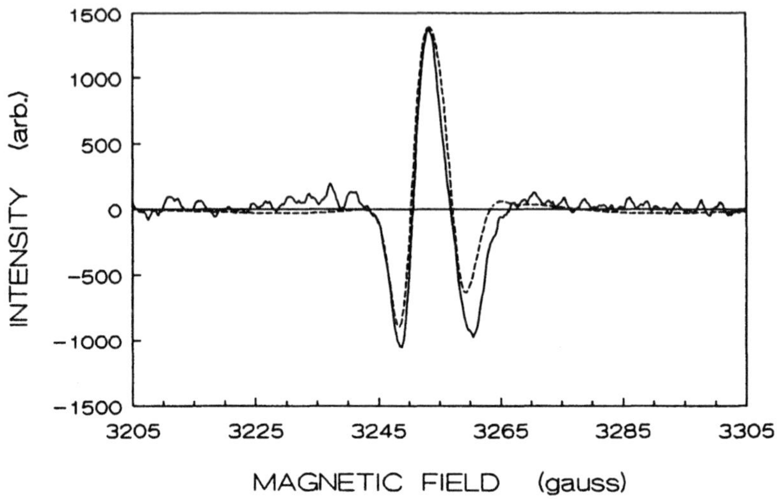

Figure 8. Solid line: spin-polarized EPR spectrum of 2-TAPD$^+$-ZP-2-NQ$^-$ in MTHF at 5 K, other conditions as in Figure 7. Dotted line: simulation using the parameters given in the text.

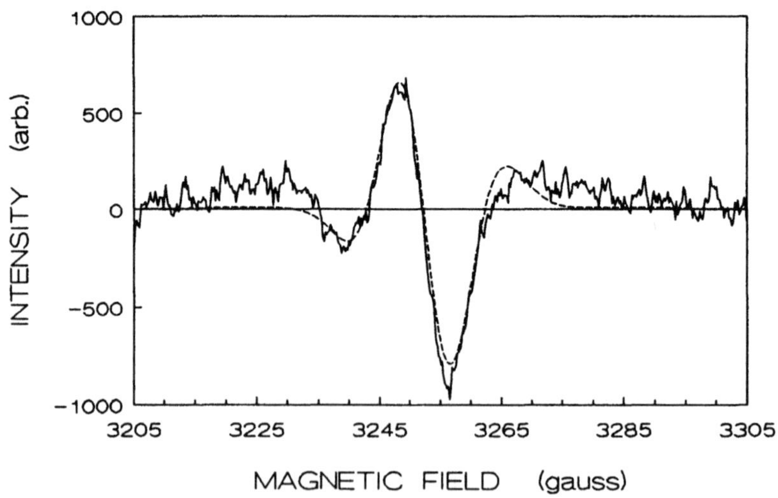

Figure 9. Solid line: spin-polarized EPR spectrum of 2-TAPD$^+$-ZP-1-NQ$^-$ in MTHF at 5 K, other conditions as in Figure 7. Dotted line: simulation using the parameters given in the text.

Figures **8** and **9** show the spin-polarized EPR spectra obtained from 2-TAPD$^+$-ZP-2-NQ$^-$ and 2-TAPD$^+$-ZP-1-NQ$^-$, respectively, along with simulations of these spectra. The differences in both radical pair distance and orientation between these two molecules is reflected in changes in the EPR spectra of their radical pairs. Thus, the radical pair interactions are a sensitive probe of structure.

4. Discussion

The EPR spectra of both 2-TAPD$^+$-ZP-2-NQ$^-$ and 2-TAPD$^+$-ZP-1-NQ$^-$ can be attributed to 2 radicals, TAPD$^+$ with a broad linewidth at lower *g*-factor and NQ$^-$ with a narrow linewidth at higher *g*-factor. Spin-polarization is observed on a millisecond time scale because the spin-lattice relaxation times of the radicals are long at 5 K. Spin polarization in these radical pairs can result from two mechanisms. The first mechanism is the usual radical pair mechanism, RPM, of CIDEP.[35] S-T_0 mixing in TAPD-ZP$^+$-NQ$^-$ is followed by polarization transfer to a non-interacting radical pair TAPD$^+$-ZP-NQ$^-$, i.e. $2J = 0$. The second mechanism assumes that TAPD$^+$-ZP-NQ$^-$ is itself an interacting spin correlated radical pair, i.e. $|2J| \sim 0.1 - 10$ G and $|D| \sim 1 - 5$ G.[36-38] In this case S-T_0 mixing in TAPD$^+$-ZP-NQ$^-$ can also produce polarized spectra.

If spin polarization in TAPD$^+$-ZP-NQ$^-$ results from the RPM of CIDEP, all polarization produced from differences in hyperfine interactions due to ZP$^+$ within the precursor radical TAPD-ZP$^+$-NQ$^-$ i.e. "multiplet effect" is lost, and only "net" polarization due to Δg differences between ZP$^+$ and NQ$^-$ remains. Thus, if $2J < 0$, $\Delta g > 0$, and the initial state is a singlet, the RPM of CIDEP predicts that the NQ$^-$ signal should be absorptive, while the TAPD$^+$ signal should be emissive. However, the spectra in Figures **8** and **9** clearly show that the high *g*-factor radical assigned to NQ$^-$ is emissive, while TAPD$^+$ is primarily absorptive. Thus, the spectra in Figures **8** and **9** cannot be described by the RPM of CIDEP. The RPM of CIDEP also requires that the initial radical pair lives sufficiently long for polarization to develop. The 0.5 ns lifetime of TAPD-ZP$^+$-NQ$^-$ may be too short for this to occur.[35]

On the other hand, if the correlated radical pair mechanism is operative, a pair of partially overlapping anti-phase doublets is expected.[36-38] The polarization pattern observed, **E A E**, is similar to that observed for P700$^+$ - A_1^- in Photosystem I of green plants,[24] and P865$^+$-Q$^-$ in bacterial reaction centers.[25] Recently, Stehlik et al. [39] have developed a simple theoretical model that can be used to simulate these spectra. This model focuses on the influence of *J*, *D*, and *g*-anisotropy on the EPR spectra of radical pairs. The latter two quantities are particularly useful in determining the distance between the radicals and their mutual orientation. The dipolar coupling, *D*, between the two radicals is strictly analogous to a classical dipole-dipole interaction, and thus, depends on the distance and orientation of the radicals relative to one another. In addition, this theory explicitly includes the anisotropies of the *g* tensors of each radical. In both 2-$TAPD^+$-ZP-2-NQ^- and 2-$TAPD^+$-ZP-1-NQ^- the *g* tensor of NQ^- is quite anisotropic because about 80% of the spin density in NQ^- is centered on the oxygen atoms. The anisotropy of the *g* tensor for $TAPD^+$ is unknown. This anisotropy should be somewhat less than that of NQ^-, but may remain non-negligible because of significant spin density on nitrogen. In fact, the broad wings of the spectra in Figures **8** and **9** are due to the large linewidth of $TAPD^+$, which is in turn due to the high spin density on nitrogen.

We employed Stehlik's theoretical description of the spin-spin interactions within radical pairs to produce the simulations shown in Figures **8** and **9**. The key parameters that were included in the simulation are given in Table **1**.

Table 1. Parameters used in Simulations of Radical Pair Spectra

	2-$TAPD^+$-ZP-2-NQ^-	2-$TAPD^+$-ZP-1-NQ^-
J (G)	-3	-3
D (G)	-2	-8
ϕ	30	0
θ	30	90
$TAPD^+$:		
ΔH (G)	35	35
g_{xx}	2.003	2.003
g_{yy}	2.003	2.003
g_{zz}	2.003	2.003
NQ^-:		
ΔH (G)	4	4
g_{xx}	2.0058	2.0058
g_{yy}	2.0051	2.0051
g_{zz}	2.0023	2.0023

J and *D* are the exchange and dipolar spin-spin interactions, respectively; ϕ and θ are the polar angles that describe the orientation of the anisotropic *g* tensor of NQ^- relative to a vector that connects the center of the π systems of $TAPD^+$ and NQ^-.

The simulations fit the observed spectra reasonably well. The overall breadth of the spectra and the signs of the polarization for the radicals are all adequately simulated. There are some problems with simulating the wings of the spectra. However, since we do not include the g anisotropy of $TAPD^+$, a poor fit in the wings of the spectra is expected. It is important to note that the distances and orientations between the NQ^- π system relative to that of $TAPD^+$ used in the simulations were determined independently from the structures of **1** and **2**. The structures of these molecules are fixed by synthesis. The triptycene spacer groups that hold the TAPD and NQ components cannot rotate about the single bond attaching them to the 5 and 15 positions of the porphyrin. This fact has been confirmed by dynamic NMR measurements.[14] Thus, the appropriate distances and angles between the TAPD π system and that of NQ were determined directly from space-filling molecular models.

The dipolar couplings, D = 2 and 8 G, respectively, for 2-$TAPD^+$-ZP-2-NQ^- and 2-$TAPD^+$-ZP-1-NQ^- were calculated using the classical expression for the magnetic dipole-dipole interaction[40], the radical-radical distances and orientations, and their spin distributions. The angles θ and ϕ were obtained from the orientation of the oxygen-oxygen axis of NQ relative to a vector connecting the center of the π systems of TAPD and NQ. Using this information the simulations matched the observed spin-polarized spectra reasonably well. Although the theory upon which the simulations are based is not complete, e.g. it does not consider the kinetics of radical pair formation,[41] it does predict an excellent correlation between the spectra and the structure of the radical pair. This agreement gives us confidence that such theoretical descriptions of the anisotropic spin-spin interactions can be used to determine the structure of radical pairs for which independent detailed structural information is lacking.

This is precisely the situation that occurs for radical pairs produced in photosynthetic organisms. Since the spin-polarized EPR signal from 2-$TAPD^+$-ZP-1-NQ^-, Figure **9**, closely resembles those from photosynthetic organisms, [24,25] and since we know the structure of this molecule, the magnitudes of the spin-spin interactions and the relative orientations of the donors and acceptors in the proteins should be similar to those in 2-$TAPD^+$-ZP-1-NQ^-. Further comparisons of the EPR signals from our supramolecular arrays with those from the donor-acceptor arrays within photosynthetic reaction center proteins for which no x-ray structures are known will yield information concerning the distances and orientations of the donors and acceptors within these proteins.

5. Conclusions

Our results show that porphyrin-triptycene-acceptor molecules possess ion-pair states that are destabilized by 0.8 eV in rigid glasses relative to their energies determined from electrochemical measurements in polar liquids. This destabilization energy agrees remarkably well with the predictions of the dielectric continuum model of the solvent.

This information has been used successfully to design supermolecules to separate charge efficiently in the solid state. In one such molecule two electron transfer steps, one photodriven, one dark, are sufficient to give a 67% yield of a radical pair that lives for 4 ms, about 10x longer than observed in any previous supramolecular system. This molecule exhibits a spin-polarized EPR spectrum which suggests that the electronic coupling between the radicals within the ion-pair is only about 0.001 cm^{-1}. This spin-polarized EPR spectrum is very similar to those observed previously for P^+ - Q^- radical pairs produced in both green plant and bacterial photosynthetic reaction centers.

Our observation of spin-polarized EPR spectra depends on achieving a delicate balance of structure dependent electronic interactions between the donors and acceptors within the supramolecular array. This strongly suggests that we now know how to precisely mimic the interactions found to date only in natural photosynthesis. With this knowledge we will be able to design efficient biomimetic supermolecules for the efficient conversion of photon energy into chemical energy in the solid state.

6. Acknowledgement

This work was supported by the Division of Chemical Sciences, Office of Basic Energy Sciences, United States Department of Energy under contract No. W-31-109-Eng-38. The authors wish to thank Dr. C. Bock and Prof. D. Stehlik for copies of their simulation program, and Drs. M. C. Thurnauer, J. R. Norris, and H. Levanon for helpful discussions on radical pair phenomena.

7. References

1. Brunschwig, B.; Ehrenson, S.; Sutin, N. *J. Phys. Chem.*, **1987**, *91*, 4714.
2. Castner, Jr., E. W.; Bagchi, B.; Maroncelli, M.; Webb, S. P.; Ruggiero, A. J.; Fleming, G. R. *Ber. Bunsenges. Phys. Chem.* **1988**, *92* 363.
3. Maroncelli, M.; Fleming, G. R. *J. Chem. Phys.* **1988**, *89*, 875.
4. Hynes, J. T. *J. Phys. Chem.* **1986**, *90*, 3701.
5. Rips, I.; Klafter, J.; Jortner, J. *J. Chem. Phys.*, **1988**, *89*, 4288.

6. Wolynes, P. *J. Chem. Phys.* **1987**, *86*, 5133.
7. Sparpaglione, M.; Mukamel, S. *J. Phys. Chem.* **1987**, *91*, 3938.
8. Kahlow, M. A.; Jarzeba, W.; Kang, T. J.; Barbara, P. F. *J. Chem. Phys.* **1988**, *90*, 151.
9. Bashkin, J. S.; McLendon, G.; Mukamel, S.; Marohn, J. *J. Phys. Chem.*, **1990**, *94*, 4757.
10. Heitele, H.; Pollinger, F.; Weeren, S.; Michel-Beyerle, M. E. *Chem. Phys.* **1990**, *143*, 325.
11. Mataga, N. in "The Exciplex", M. Gordon and W. R. Ware, Eds. Academic, New York, 1975, p. 113.
12. Rehm, D.; Weller, A. *Israel J. Chem.*, **1970**, *8*, 259.
13. Wasielewski, M. R.; Niemczyk, M. P. *J. Am. Chem. Soc.* **1984**, *106*, 5043.
14. Wasielewski, M. R.; Niemczyk, M. P.; Svec, W. A.; Pewitt, E. B. *J. Am. Chem. Soc.* **1985**, *107*, 5583.
15. Wasielewski, M. R.; Niemczyk, M. P. in "Porphyrins, Excited States and Dynamics", Gouterman, M.; Rentzepis, P.,Straub, K. D., Eds. ACS Symposium Series, 1986, p. 154.
16. Gaines, III, G. L.; O'Neil, M. P.; Svec, W. A.; Niemczyk, M. P.; Wasielewski, M. R. *J. Am. Chem. Soc.* **1991**, *113*, 719.
17. Wasielewski, M. R.; Gaines, III, G. L.; O'Neil, M. P.; Svec, W. A.; Niemczyk, M. P.; Prodi, L.; Gosztola, D. in "Dynamics and Mechanisms of Photoinduced Electron Transfer and Related Phenomena", N. Mataga, T. Okada, and H. Masuhara, Eds., Elsevier, Amsterdam, 1991, pp. 133-148.
18. Weller, A. *Z. Phys. Chem. N.F.* **1982**, *133*, 93.
19. Deisenhofer, J.; Epp, O.; Miki, K.; Huber, R.; Michel, H. *J. Mol. Biol.* **1984**, *180*, 385.
20. Kirmaier, C.; Holten, D. *Photosyn. Res.* **1987**, *13*, 225.
21. Jortner, J. *J. Chem. Phys.* **1976**, *64*, 4860.
22. Marcus, R. A. *J. Chem. Phys.* **1956**, *24*, 966.
23. Thurnauer, M. C.; Katz, J. J.; Norris, J. R. *Proc. Natl. Acad. Sci. USA* **1975**, *72*, 3270.
24. Thurnauer, M. C.; Gast, P. *Photobiochem. Photobiophys.* **1985**, *9*, 29.
25. Feezel, L. L.; Gast, P.; Smith, U. H.; Thurnauer, M. C. *Biochim. Biophys. Acta* **1989**, *974*, 149.
26. Hoff, A. J. *Photochem. Photobiol.* **1986**, *43*, 727.
27. Wasielewski, M. R.; Gaines, III, G. L.; O'Neil, M. P.; Svec, W. A.; Niemczyk, M. P. *J. Am. Chem. Soc.* **1990**, *112*, 4559.
28. Lindsey, J. S.; Wagner, R. W. *J. Org. Chem.*, **1989**, *54*, 828.
29. Seybold, P. G.; Gouterman, M. *J. Molec. Spectroscopy* **1969**, *31*, 1.
30. Wasielewski, M. R.; Johnson, D. G.; Niemczyk, M. P.; Gaines, III, G. L.; O' Neil, M. P.; Svec, W. A. *J. Am. Chem. Soc.* **1990**, *112*, 6842.

31. Furutsuka, T.; Imura, T.; Kojima, T.; Kawabe, K. *Technol. Rep. Osaka Univ.* **1974**, *24*, 367.
32. Jeanmaire, D. L.; Van Duyne, R. P. *J. Electroanal. Chem.* **1975**, *66*, 235.
33. Levanon, H.; Vega, S. *J. Chem. Phys.* **1974**, *61*, 2265.
34. Wasielewski, M. R.; Niemczyk, M. P.; Svec, W. A.; Pewitt, E. B. *J. Am. Chem. Soc.* **1985**, *107*, 5583.
35. CIDEP = Chemically Induced Dynamic Electron Polarization. For a review see: F. J. Adrian, in "Chemically Induced Magnetic Polarization", edited by L. T. Muus, P. W. Atkins, K. A. McLauchlan, and J. B. Pedersen (Reidel, Dordrecht, 1977), Chap. 5, pp. 77-105.
36. Thurnauer, M. C.; Norris, J. R. *Chem. Phys. Lett.* **1980**, *76*, 557.
37. Buckley, C. D.; Hunter, D. A.; Hore, P. J.; McLauchlan, K. A. *Chem. Phys. Lett.* **1987**, *135*, 307.
38. Closs, G. L.; Forbes, M. D. E.; Norris, J. R. *J. Phys. Chem.* **1987**, *91*, 3592.
39. Stehlik, D.; Bock, C. H.; Petersen, J. *J. Phys. Chem.* **1989**, *93*, 1612.
40. Wertz, J. E.; Bolton, J. R. "Electron Spin Resonance", Mc Graw-Hill, New York, 1972, pp. 223-230.
41. Norris, J. R.; Morris, A. L.; Thurnauer, M. C.; Tang, J. *J. Chem. Phys.* **1990**, *92*, 4239.

INTO THE THIRD DIMENSION OF COORDINATION CHEMISTRY: TOWARDS STARBURST ARRAYS

E.C. CONSTABLE *, A.M.W. CARGILL THOMPSON
Cambridge Centre for Molecular Recognition,
University Chemical Laboratory,
Lensfield Road,
Cambridge CB2 1EW
United Kingdom

D.A. TOCHER
Department of Chemistry,
University College London,
20 Gordon Street,
London WC1H 0AJ,
United Kingdom

ABSTRACT. The coordination properties of the tridentate ligand 2,2':6',2"-terpyridine make it an ideal structural unit to incorporate into monomers to be used for the assembly of coordination polymers. The electronic properties of metal complexes may be subtly controlled by the incorporation of electron withdrawing or electron releasing substituents upon the 2,2':6',2"-terpyridine. The synthesis of a range of such ligands is described, and a detailed study of the electronic and solution properties of their homoleptic and heteroleptic ruthenium(II) complexes is reported. *Linear* coordination oligomers and polymers are readily prepared by the use of *bis*(2,2':6',2"-terpyridine) ligands as linkers, and the synthesis of materials derived from two such linkers is reported. The electronic properties of these compounds may be controlled by the introduction of suitably substituted terminator ligands. The synthesis of a novel *tris*(2,2':6',2"-terpyridine) ligand is reported which forms an extended three-dimensional coordination polymer. The preparation and characterisation of a single triruthenium unit from such a polymer is reported.

1. Introduction

The coordination chemistry of the oligopyridines 2,2'-bipyridine (bpy) and 2,2':6',2"-terpyridine (tpy) has been intensively investigated during the past century [1]. This class of ligands forms complexes with most of the elements in the periodic table, but those with transition metal ions have been most widely studied to date. In general, bpy acts as a chelating bidentate N_2 and tpy as a chelating tridentate N_3 donor, although isolated exceptions to these behaviours are known. Substituted derivatives of 2,2'-bipyridines have been widely investigated as ligands, but there has been less interest in substituted 2,2':6',2"-terpyridines until recently. The introduction of substituents into these oligopyridine ligands allows control over the redox or photophysical properties of the metal centre and also permits the further structural development of the system by specific covalent or non-covalent peripheral interactions of the substituents. There has been considerable recent interest in the

V. Balzani and L. De Cola (eds.), Supramolecular Chemistry, 219–233.

use of substituted 2,2'-bipyridines for the assembly of polynuclear coordination oligomers and polymers possessing specific electron transfer and photochemical properties. In this article we wish to present evidence that 2,2':6',2"-terpyridines possess significant advantages as metal-binding structural units for such assembly processes, and that the control of the coordination properties by the use of suitable substituted 2,2':6',2"-terpyridines is facile.

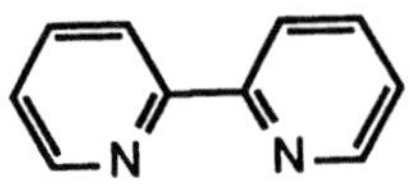

bpy

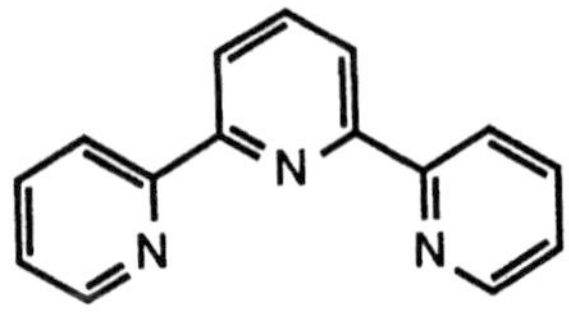

tpy

2. Coordination Properties of Substituted Bpy and Tpy Ligands

An octahedral metal centre may coordinate to *three* bpy ligands or to *two* tpy ligands. In the complex cations $[M(bpy)_3]^{n+}$ the three bpy ligands are mutually orthogonal, and approximately define the *xy*, *xz* and *yz* planes, whilst in $[M(tpy)_2]^{n+}$ cations the two ligands are orthogonal but only occupy (say) the *xy* and *yz* planes. Furthermore, within the D_{2d} $[M(tpy)_2]^{n+}$ cation the two ligands are simply related by an inversion through the metal centre followed by a 90° rotation. Regardless of the site of substitution, monosubstituted Xbpy ligands give rise to both *mer* and *fac* isomers upon formation of $[M(Xbpy)_3]^{2+}$, whereas substituted Xtpy ligands give *single* geometrical isomers of $[M(Xtpy)_2]^{n+}$ (Figure 1). Although single isomers are obtained with *symmetrically* disubstituted X_2bpy ligands, such systems are not suitable for subsequent controlled structural development.

If substituted Xbpy units are used for the formation of coordination polymers using octahedral metal centres, each metal is a branch point with a connectivity of at least three (and six if the polymer is developed at each position of a disubstituted X_2bpy). This results in an uncontrolled 'chain-type' polymerisation, with complex geometrical relationships between adjacent metal centres. In contrast, monosubstituted Xtpy ligands give rise to highly ordered and geometrically simple coordination polymers in which each metal centre shows a connectivity of two. In the specific case of a 4'-substituted Xtpy, a linear connectivity about the metal is found, with the planes defined by the two orthogonal ligands.

In order to assemble *non-linear* polymers, it is necessary to have centres which have a connectivity of three or greater. The most obvious way to achieve this is to use substituted bpy ligands, although we have previously noted that this gives rise to systems which contain varying amounts of possible isomers and possess complex geometrical relationships between the various metal centres. However, it is not necessary for the branch point with a connectivity of greater than two to be the metal centre. A 1,3,5-trisubstituted benzene provides the core for a *ligand* branch point [2]. A 1,3,5-(terpyridinyl)substituted benzene allows the assembly of a three-dimensional coordination which maintains the simplicity of linear connectivity about the metal [2].

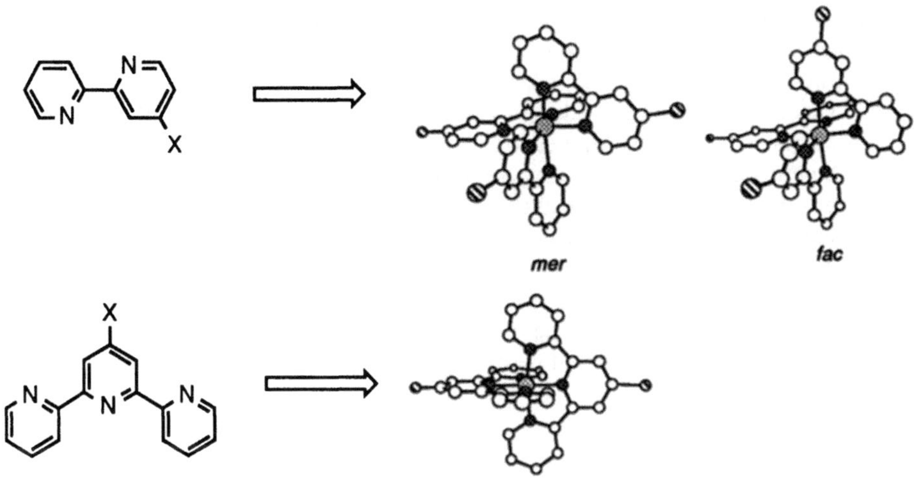

Figure 1. The formation of *pseudo*octahedral $[M(Xbpy)_3]^{2+}$ and $[M(Xtpy)_2]^{n+}$ complexes.

3. Synthesis of 4'-Substituted 2,2':6',2"-Terpyridines

We have embarked upon a systematic investigation of 4'-substituted 2,2':6',2"-terpyridines as a prelude to using them to control the properties of coordination polymers, and have specifically been interested in the six compounds tpy, Cltpy, Phtpy, EtOtpy, HOtpy and Me_2Ntpy, which have been prepared as shown in Figure 2.

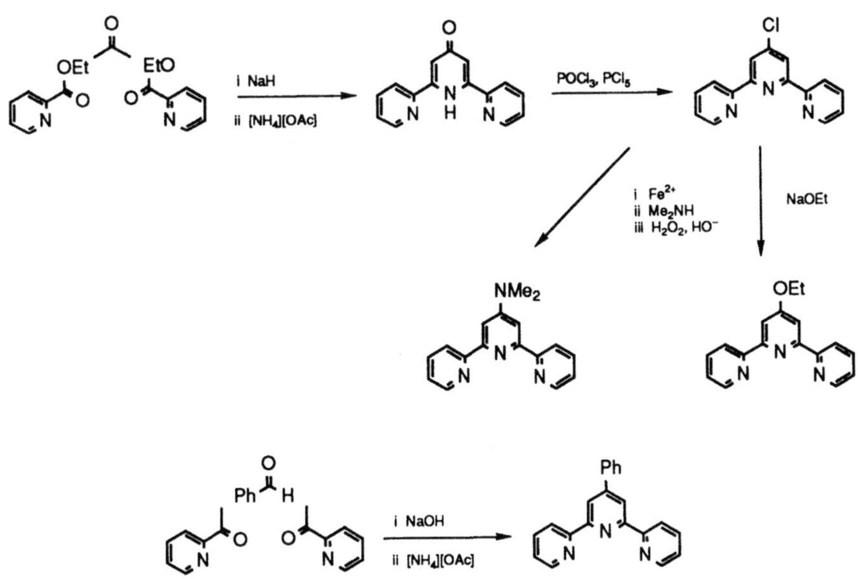

Figure 2. Synthesis of 4'-substituted Xtpy ligands

These specific ligands were chosen as they represent a range of donor properties varying from the electron-withdrawing substituent in Cltpy to the strongly electron-releasing one in Me_2Ntpy. The ligands are simple to prepare, and were all obtained in yields between 60 and 100% [3]. The synthetic methods adopted are routine, with the exception of the conversion of Cltpy to Me_2Ntpy. This interconversion could not be achieved by the direct reaction of Cltpy with dimethylamine under a variety of reaction conditions, but activation of the Cltpy by coordination to an iron(II) centre allowed a smooth displacement of halide by Me_2NH [4,5]. Demetallation was readily achieved by treatment with alkaline hydrogen peroxide [5,6]. The compound HOtpy is of interest as the free ligand exists as the 2,2':6',2"-terpyridin-4'(1'*H*)-one tautomer, but coordination to a metal forces it to adopt the tautomeric 4'-hydroxy-2,2':6',2"-terpyridine form. These complexes are acidic, and deprotonated forms are readily accessed by treatment with mild bases..

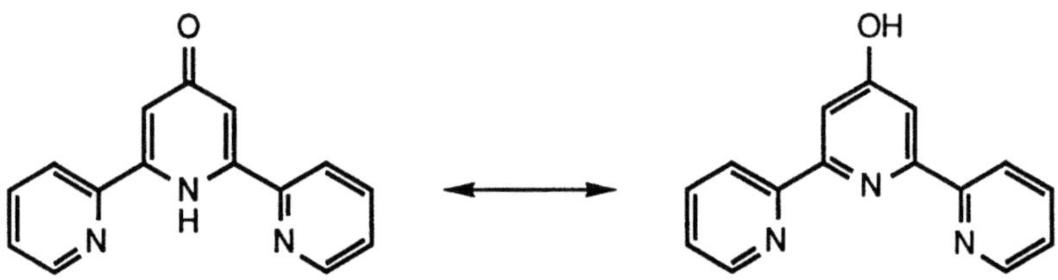

The crystal structure of Me_2Ntpy confirms the molecular structure, and reveals that the dimethylamino substituent is approximately coplanar with the central pyridine ring to which it is bonded. In the solid state the molecules are coplanar and exhibit graphitic stacking interactions. These interactions appear to be charge-transfer in nature, and are associated with the overlap of the central rings of molecules in adjacent layers. The central rings are mutually twisted by 180° within the planes, maximising their dipolar interactions (Figure 3).

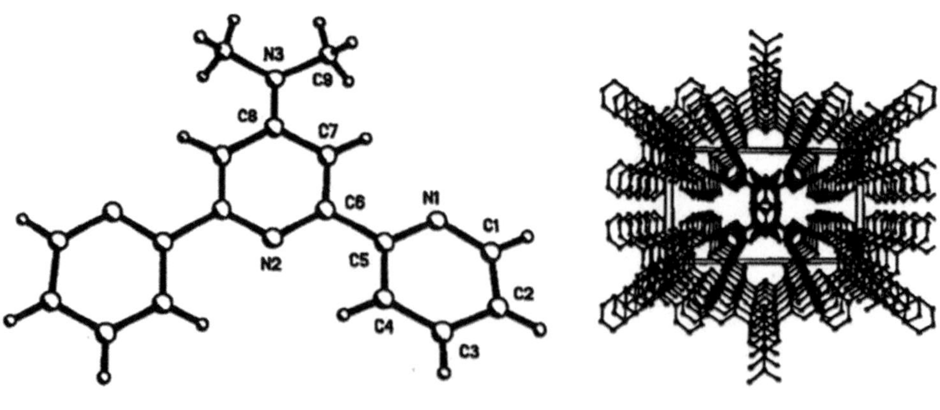

Figure 3. Molecular structure of Me_2Ntpy and the stacking shown in the solid state

4. Synthesis of Homoleptic and Heteroleptic $[Ru(Xtpy)(Ytpy)]^{2+}$ Complexes

The six Xtpy ligands could give rise to a total of twenty one heteroleptic and homoleptic $[M(Xtpy)(Ytpy)]^{n+}$ complexes with a given metal ion. We have prepared all twenty one of the possible $[Ru(Xtpy)(Ytpy)]^{2+}$ complexes. These are readily prepared by sequential syntheses in which the complex $[Ru(Xtpy)Cl_3]$ is prepared as an intermediate.

$$RuCl_3.3H_2O + Xtpy \longrightarrow [Ru(Xtpy)Cl_3]$$

$$[Ru(Xtpy)Cl_3] + Ytpy \longrightarrow [Ru(Xtpy)(Ytpy)]^{2+}$$

In the second step, the reactions are performed in the presence of *N*-ethylmorpholine to achieve the reduction of ruthenium(III) to ruthenium(II). In general the choice of Xtpy and Ytpy is not important, and the preparation of $[Ru(Xtpy)(Ytpy)]^{2+}$ may be achieved by both possible methods, although the yields may vary considerably. Side-products of the reactions include the 'scrambled' $[Ru(Xtpy)_2]^{2+}$ and $[Ru(Ytpy)_2]^{2+}$ ions; in the case of the electron releasing Me_2Ntpy ligands significant amounts of these products may be formed. The complexes were isolated in 60-80% yield as their hexafluorophosphate salts. All of the products are various shades of red or orange, with those complexes containing Me_2Ntpy being the darkest.

The ^1H n.m.r. spectra of solutions of all twenty one complexes are sharp and well-resolved, and there is no evidence for interligand interactions within the heteroleptic species. Specifically, the ^1H n.m.r. spectrum of $[Ru(Xtpy)(Ytpy)]^{2+}$ is essentially a superimposition of those of $[Ru(Xtpy)_2]^{2+}$ and $[Ru(Ytpy)_2]^{2+}$ (Figure 4).

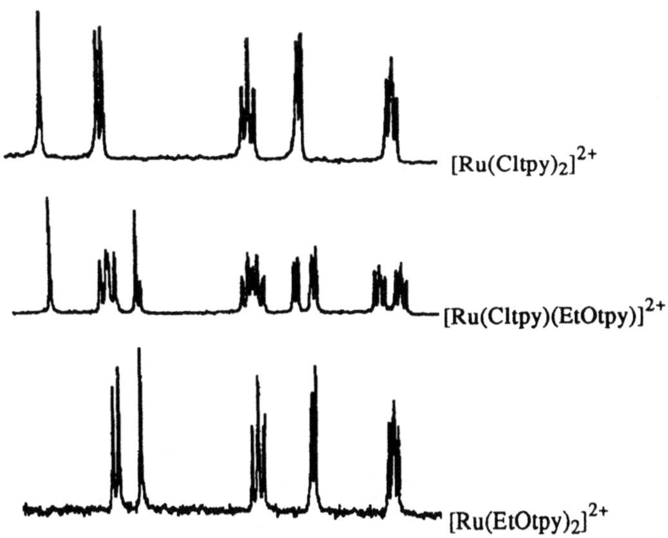

Figure 4. ^1H n.m.r. spectra of CD_3COCD_3 solutions of $[Ru(Cltpy)_2]^{2+}$, $[Ru(Cltpy)(EtOtpy)]^{2+}$ and $[Ru(Cltpy)_2]^{2+}$

The complexes containing HOtpy are readily deprotonated, and this is associated with marked changes in the ^1H n.m.r. spectra. The changes are essentially limited to the resonances of the HOtpy ligand in the heteroleptic complexes (Figure 5).

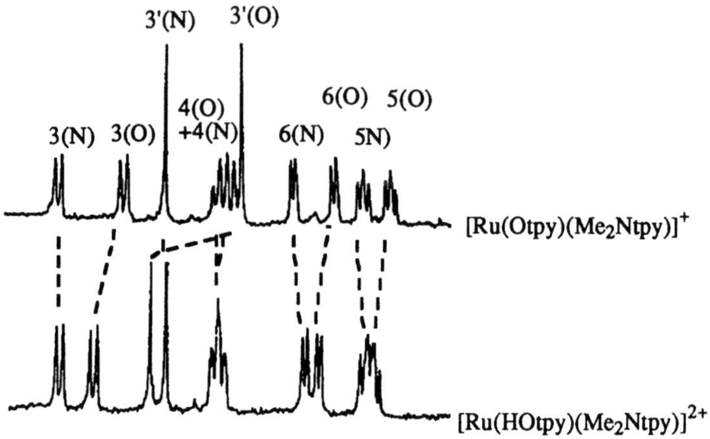

Figure 5. ^1H n.m.r. spectra of CD_3OD solutions of $[Ru(HOtpy)(Me_2Ntpy)]^{2+}$ before and after deprotonation

Although the *deprotonation* of the coordinated HOtpy ligand is facile, we have been unable to *protonate* a coordinated Me_2Ntpy ligand in any of the complexes. There are no changes in the ^1H n.m.r. spectra of any of the Me_2Ntpy complexes upon the addition of CF_3CO_2D.

5. Electrochemical Properties of Homoleptic and Heteroleptic Complexes

All of the ruthenium(II) complexes described above are electrochemically active, and exhibit one reversible Ru(II)/Ru(III) process, and a varying number of reversible or *pseudo*reversible ligand-based reductions. The Ru(II)/Ru(III) oxidative processes span a considerable range of potentials from ~0.4 to ~1.0V *versus* Fc/Fc^+. In general, the less positive potentials are associated with the complexes containing the more electron-releasing substituents, with the lowest oxidation potentials being found with the Me_2Ntpy ligands. Once again, we have observed additive behaviour; the oxidation potentials of the heteroleptic complexes are at the weighted average of the two homoleptic complexes.

In an attempt to quantify the variation in redox potentials, we have developed a Hammett-type correlation. The Ru(II)/Ru(III) oxidation potential data are concerned with a process in which positive charge is developing in the transition state. Accordingly it is most appropriate to use Hammett σ^+ parameters in any correlation. In practice, we obtain similar correlations with the use of either σ or σ^+ parameters. We next defined a mean σ^+ value for any specific complex in terms of the σ^+ parameters associated with the substituents on the Xtpy and Ytpy ligands:

$$\text{Mean } \sigma^+ = 0.5(\sigma^+(X) + \sigma^+(Y))$$

The mean σ^+ values are plotted against the observed Ru(II)/Ru(III) oxidation potentials for all twenty one complexes in Figure 6.

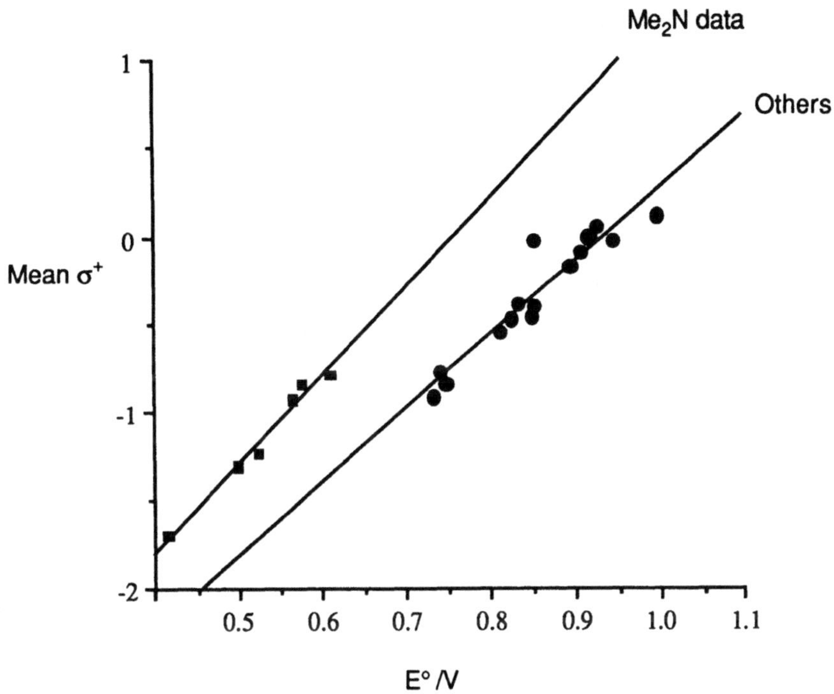

Figure 6. Correlation of mean σ^+ parameters with redox data for [Ru(Xtpy)(Ytpy)] and [Ru(Xtpy)$_2$] complexes

It is evident that the data fall into two distinct sets - those for complexes containing Me$_2$Ntpy ligands constituting one set, and the remainder the other. The observation of these Hammett correlations allows us to state that the properties of the ruthenium centre are controlled by interaction with *both* of the ligands, although we observe no interligand effects in the ^{1}H n.m.r. spectra of the complexes. Why do the Me$_2$Ntpy ligands give a different correlation? We believe that it is associated with changes in the conformation of the Me$_2$Ntpy group with respect to the central pyridine ring. We would anticipate that the ruthenium(III) complex would require greater electron-donation from the ligand, and that this could be achieved in a π-bonding manner by greater overlap of the Me$_2$Ntpy lone pair with the pyridine ring, resulting in more coplanarity. However, these naive predictions will be complicated by changes in the π-back-bonding from the metal to the ligand. Regardless of the precise changes in donor properties, we can say that conformational changes associated with the change in oxidation state of the metal render a single σ^+ value for the Me$_2$N substituent suspect. It is significant that we can find no single σ^+ value for Me$_2$Ntpy which fits the data to a single correlation. This is consistent with a varying value of mean σ^+ depending upon the relative interplay of σ- and π-bonding effects of the two ligands at the metal centre.

6. Molecular Structure of $[Ru(Me_2Ntpy)_2][PF_6]_2$

In an attempt to further probe the conformational changes associated with coordination of the Me_2Ntpy ligand, we have determined the crystal and molecular structure of $[Ru(Me_2Ntpy)_2][PF_6]_2$, and the molecular structure of the cation is presented in Figure 7 [5].

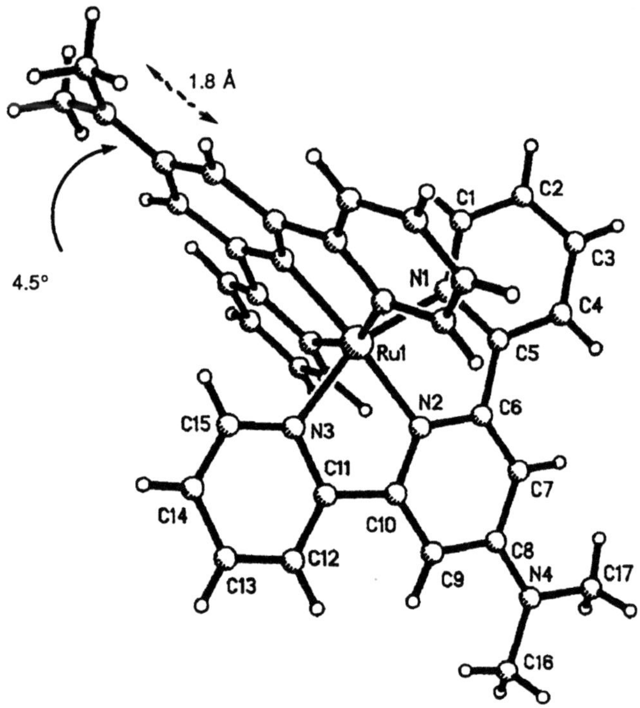

Figure 7. Molecular structure of the $[Ru(Me_2Ntpy)_2]^{2+}$ cation in $[Ru(Me_2Ntpy)_2][PF_6]_2$

The dimethylamino substituent is approximately coplanar with the central pyridine ring of the Me_2Ntpy ligand, and makes an interplanar angle of 4.5° with it. This is comparable to the interplanar angle observed in the free ligand.

7. *Bis*(2,2':6',2"-terpyridines) for the Synthesis of Linear Coordination Polymers.

We have designed two *bis*(2,2':6',2"-terpyridine) ligands which can give rise to coordination oligomers or coordination polymers (Figure 8). The first of these ligands consists of two tpy ligands directly linked through the 4'-position. Over the past ten years we have investigated about twenty routes for the synthesis of this ligand, but finally achieved success with a non-catalytic nickel(0) coupling of Cltpy. Nickel(0) couplings of halopyridines are not successful if only catalytic amounts of nickel(0) are used as the oligopyridine products effectively sequester the transition metal, giving rise to catalytic turnovers of one! In contrast, the use of stoichiometric amounts of nickel(0) allows a convenient isolation of the products as their nickel(II) complexes, from which the free ligands may be easily obtained by cyanide demetallation. In the second ligand the two tpy moieties are linked through the 4-positions by a 1,4-phenylene spacer. This compound was obtained by the Kröhnke method [7]. Both ligands are obtained in respectable yields.

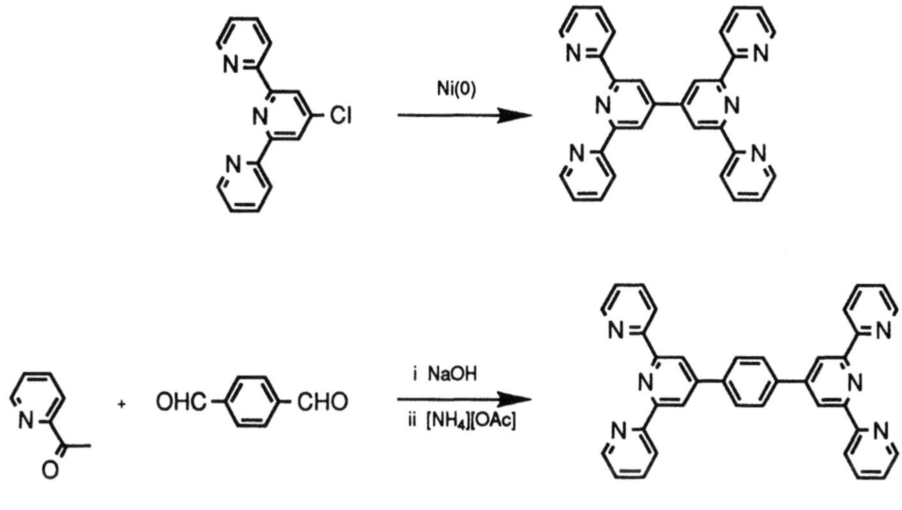

Figure 8. Syntheses of *bis*(2,2':6',2"-terpyridines) for linear coordination polymers

These *bis*(2,2':6',2"-terpyridines) may be used for the asembly of linear coordination polymers or discrete linear oligomers (Figure 9). Insoluble deep blue coordination polymers are obtained cleanly upon simply mixing solutions of either of the *bis*(2,2':6',2"-terpyridines) with solutions of $[Fe(H_2O)_6][BF_4]_2$. In contrast, stepwise reaction with suitable $[Ru(Xtpy)Cl_3]$ building blocks allows the controlled assembly of discrete oligomeric units with terminator Xtpy ligands which will confer specific electronic properties to the compounds.

We have prepared a wide variety of discrete dimeric ruthenium(II) complexes of the type $[(Xtpy)Ru(bistpy)Ru(Xtpy)]^{4+}$ and can make a few general observations regarding their properties. Firstly, all of the binuclear complexes, exhibit a single ruthenium(II)/ruthenium(III) redox process. The redox properties of the complexes incorporating the 1,4-phenylene-linked ligand are identical to those of mononuclear analogues containing Phtpy. We take this as evidence for essentially non-interacting metal centres in these compounds. Our attempts to control metal-metal interactions through the conjugated aromatic linkers appear to have been unsuccessful. We attribute this to conformations about the interannular C-C bonds linking the pyridine rings to the phenylene rings such that the π-conjugation is minimised. Support for this comes from the molecular structures of Phtpy and its nickel(II) complex, which demonstrate that the phenyl and pyridine rings are not coplanar [3]. We have yet to complete studies on the heteroleptic [(Xtpy)Ru(*bis*tpy)Ru(Ytpy)]$^{4+}$ complexes.

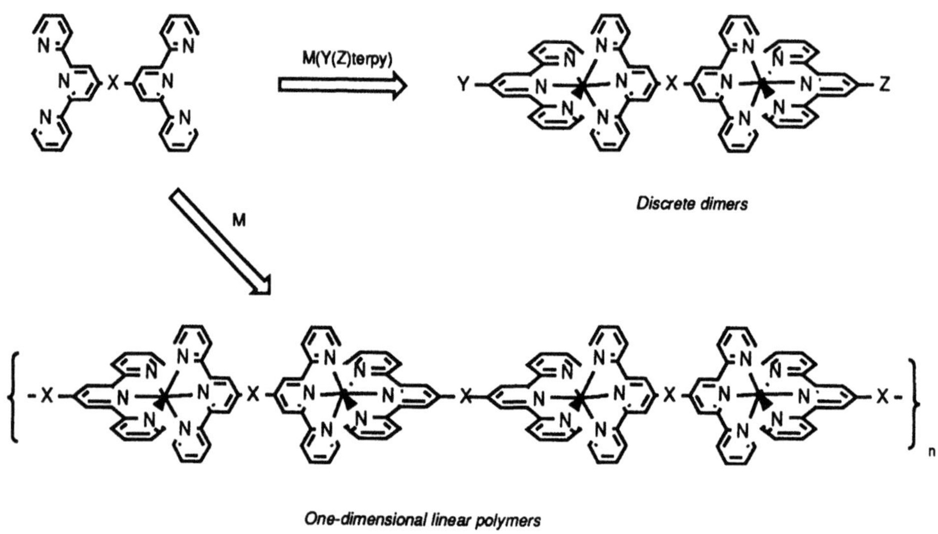

Figure 9. Preparation of coordination oligomers and polymers with *bis*(2,2':6',2"-terpyridine) ligands

We are currently investigating the electronic propertes of these new materials, and also extending the synthetic efforts to the preparation of biphenylene linked and chiral ligands. We are also investigating solid-state dynamic processes involving the linker groups.

8. A New Ligand for Starburst Polymer Assembly

We have introduced the concept of using a ligand rather than a metal as the branch point for polymer assembly. Thus far, we have concentrated upon the assembly of linear arrays by utilising *bis*(2,2':6',2"-terpyridines) which give rise to both metal and ligand centres with connectivities of two. Poly(2,2'-bipyridine) ligands have been used for the assembly of coordination polymers and oligomers in which the metal centres exhibit connectivities of three. We have now developed a *tris*(2,2':6',2"-terpyridine) which assembles coordination oligomers and polymers in which the metal centres exhibit connectivities of two and the ligand centres exhibit connectivities of three.

The key structural feature of this ligand is a 1,3,5-substituted benzene, which acts as a planar trigonal structure former for the assembly. The specific ligand we have prepared is 1,3,5-*tris*(6-(2,2':6',2"-terpyridinyl))benzene (Figure 10), which we have prepared by a Kröhnke-type synthesis. The key intermediate is 1,3,5-benzenetrialdehyde, which is readily prepared by the selective reduction of 1,3,5-benzenetricarbonyl chloride with

Li[AlH(^tBuO)$_3$]. Treatment of this with an excess of 2-acetylpyridine in aqueous ethanolic base yields a *tris*(chalcone) in acceptable yield. Finally, reaction of the *tris*(chalcone) with a 2-acetylpyridine enolate equivalent in the presence of ammonium acetate yields the desired ligand as an off-white solid (Figure 11).

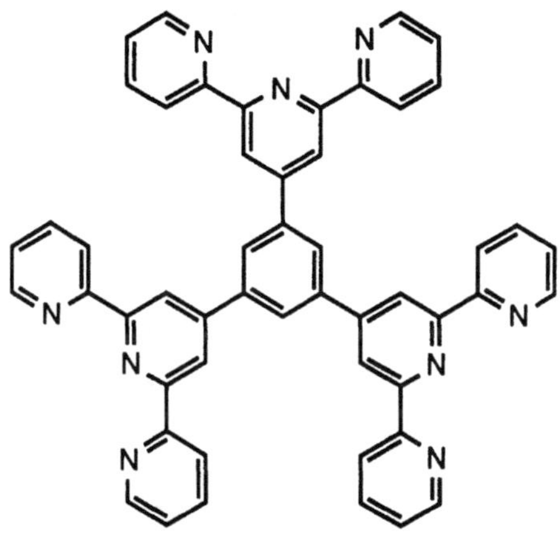

Figure 10. The new starburst *tris*(2,2':6',2"-terpyridine) ligand

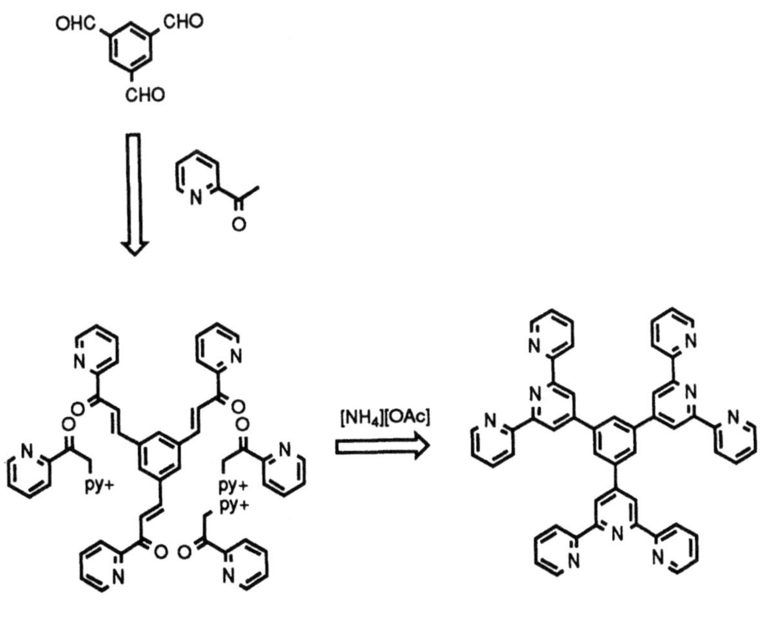

Figure 11. Synthesis of the *tris*(2,2':6',2"-terpyridine) ligand

The ligand exhibits a parent ion in its mass spectrum at *m/z* 771 and is surprisingly soluble in organic solvents. Our previous experience of oligopyridine ligands has been that they are very insoluble species, and we have attributed this to the efficient stacking of the coplanar molecules in the solid state. We believe that the increased solubility of the 1,3,5-*tris*(6-(2,2':6',2"-terpyridinyl))benzene reflects the inability of the compound to adopt a completely planar conformation. The compound is sufficiently soluble to allow characterisation by ^{1}H n.m.r. spectroscopy, and the ^{1}H n.m.r. spectrum of a $CDCl_3$ solution is presented in Figure 12. The compound exhibits the expected high molecular symmetry on the n.m.r. time-scale, with all three 2,2':6',2"-terpyridinyl groups equivalent, and only six proton environments are observed.

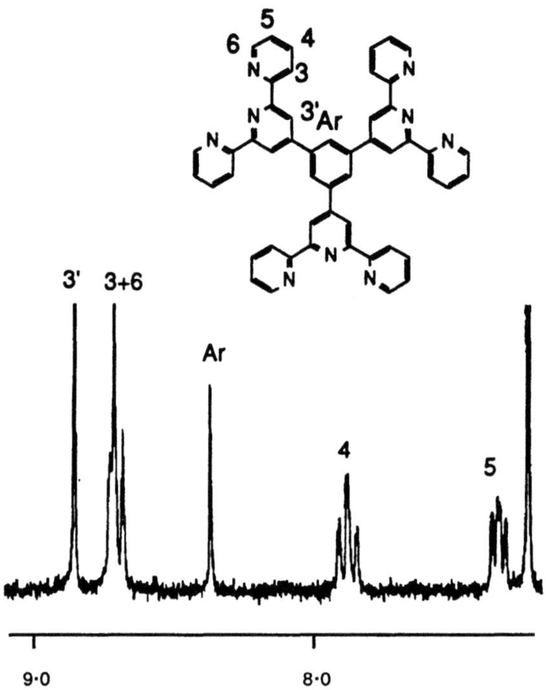

Figure 12. ^{1}H n.m.r. spectrum of a $CDCl_3$ solution of the *tris*(2,2':6',2"-terpyridine) ligand

The reaction of this new ligand with [Ru(tpy)Cl_3] in the presence of *N*-ethylmorpholine proceeds smoothly to yield the orange-red trinuclear complex [{(tpy)Ru}$_3$(*tris*tpy)]$^{6+}$ in high yield. Strangely, this is the only isolable product from the reaction when one, two or three equivalents of [Ru(tpy)Cl_3] are used. We have no explanation for this observation at present. We had anticipated difficulties in isolating complexes with two ruthenium(II) centres, let alone three! The new hexacationic trinuclear ruthenium complex is isolated as its [PF_6] salt which is readily soluble in MeCN or Me_2CO. The F.A.B. mass spectrum of the complex exhibited intense ion clusters at *m/z* 2355 and 2210 ([{(tpy)Ru}$_3$(*tris*tpy)][PF_6]$_4$ and [{(tpy)Ru}$_3$(*tris*tpy)][PF_6]$_3$ respectively). The complex is electrochemically active, and shows a single fully reversible ruthenium(II)/ruthenium(III) oxidation process at +0.900V, a reversible reduction at −1.685V and an irreversible reduction at −1.829V all *versus* Fc/Fc^+ (Figure 13). Once again, the conformational properties of the central ligand seem to result in the electronic isolation of the three metal centres.

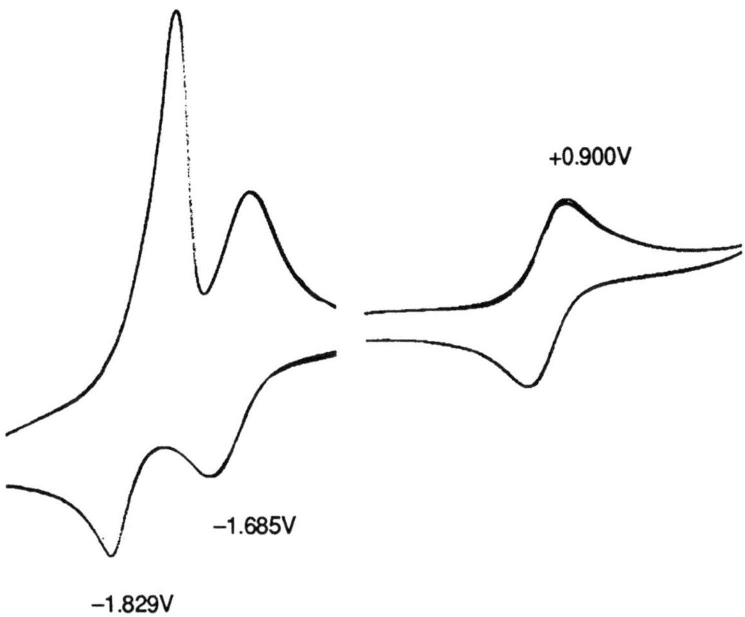

Figure 13. Cyclic voltammogram of $[\{(tpy)Ru\}_3(\textit{tris}tpy)][PF_6]_6$ (MeCN, $[^nBu_4N][PF_6]$ supporting electrolyte, potentials *versus* Fc/Fc^+)

The ^1H n.m.r. spectrum of a solution of $[\{(tpy)Ru\}_3(\textit{tris}tpy)][PF_6]_6$ in CD_3CN is sharp and well-resolved, and clearly illustrates the high symmetry of the complex (Figure 14). Each one of the tpy ligands is equivalent, as are the three 2,2':6',2"-terpyridinyl moieties of the *tris*tpy ligand. The resonances assigned to the tpy ligand are almost unshifted with respect to the corresponding resonances for $[Ru(tpy)_2]^{2+}$. This is a clear indication that the ruthenium environment in this discrete trinuclear complex, and by implication the metal environment in the coordination polymers formed from the *tris*tpy ligand, is essentially unperturbed from that in $[M(tpy)_2]^{n+}$

We are currently attempting to define conditions for the sequential synthesis of mononuclear and binuclear complexes of the *tris*tpy ligand, together with the preparation of complexes with Xtpy terminator ligands.

Simply mixing solutions of *tris*tpy and $[Fe(H_2O)_6][BF_4]_2$ results in the precipitation of a deep blue coordination polymer insoluble in all solvents investigated. Modelling studies indicate that the polymers consist of interlinked cyclic supercages. This topology is a direct consequence of the linear two-connectivity and 90° phase-changes at each metal centre and the trigonal configuration and three-connectivity at each ligand

Future studies will include solid state n.m.r. and other structural studies of these coordination polymers, linked with attempts to intercalate other molecular species within the supercages.

9. Conclusions

2,2':6',2"-Terpyridines provide a versatile metal-binding substructure for the development of new monomers for the preparation of coordination polymers and oligomers. The ligands are easily prepared, and the metal complexes are readily characterised. Structurally developed ligands allow the control of metal-centred properties, and allow a variety of connectivities to be defined by the ligand whilst the metal is constrained to a connectivity of two. This provides a useful complement to related 2,2'-bipyridine ligands.

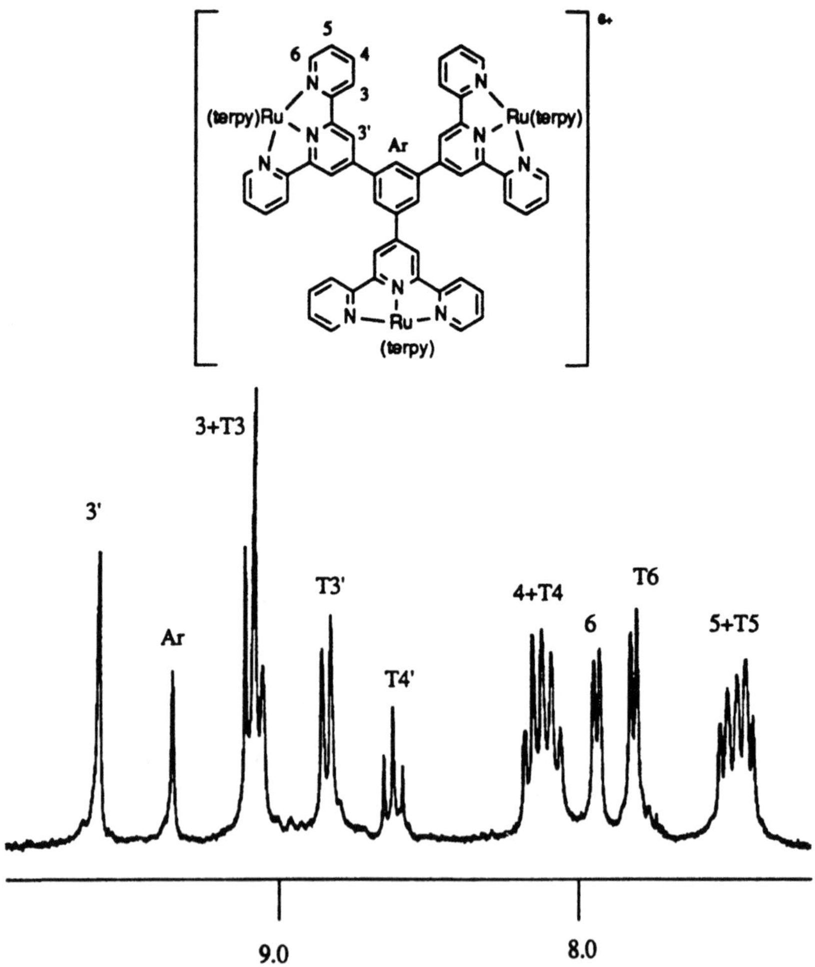

Figure 14 ^{1}H n.m.r. spectrum of a CD_3COCD_3 solution of $[\{(tpy)Ru\}_3(\textit{tris}tpy)][PF_6]_6$

Acknowledegements

We should like to thank the S.E.R.C., The Royal Society of Great Britain and The Isaac Newton Trust for support.

References

1. Constable, E.C. (1987) 'The coordination chemistry of 2,2':6',2"-terpyridine and higher oligopyridines', *Adv. Inorg. Chem. Radiochem.*, **30**, 69-121; Constable, E.C. (1989) 'Homoleptic complexes of 2,2'-bipyridine', *Adv. Inorg. Chem.*, **34**, 1-63
2. Constable, E.C., and Cargill Thompson, A.M.W. (1992) 'A new starburst ligand for the assembly of coordination polymers; 1,3,5-*tris*(6-(2,2':6',2"-terpyridinyl))benzene and a trinuclear ruthenium(II) complex', *J. Chem. Soc., Chem. Commun.*, manuscript submitted; Vögtle, F., Lüer, I., Balzani, V., and Armaroli, N. (1991) 'Endoreceptors with convergent phenanthroline units: a molecular cavity for guest molecules', **30**, 1333-1336
3. Constable, E.C., Lewis, J., Liptrot, M.C.and Raithby, P.R. (1990) 'The coordination chemistry of 4'-phenyl-2,2':6',2"-terpyridine; the Synthesis, crystal and molecular structures of 4'-phenyl-2,2':6',2"-terpyridine and bis(4'-phenyl-2,2':6',2"-terpyridine) nickel(II) chloride decahydrate', *Inorg. Chim. Acta*, **178**, 47-54; Constable, E.C., and Ward, M.D. (1990) 'Synthesis and co-ordination behaviour of 6',6"-bis(2-pyridyl)-2,2':4,4":2",2"'-quaterpyridine; 'back-to-back' 2,2':6',2"-terpyridine', *J. Chem. Soc., Dalton Trans.*, 1405-1409
4. Constable, E.C. (1984) 'Nucleophilic attack upon coordinated heterocycles; definitive evidence for the enhanced electrophilicity of coordinated pyridines', *Inorg. Chim. Acta*, **82**, 53-57; Constable, E.C. and Leese, T.A. (1988) 'Ligand reactivity in polypyridine complexes; $[Ru(bipy)_3]^{n+}$ analogs incorporating pendant polyamine substituents', *Inorg. Chim. Acta*, **146**, 55-58
5. Constable, E.C., Cargill Thompson, A.M.W., and Tocher, D.A. 'Synthesis, characterisation and spectroscopic properties of ruthenium(II)-2,2':6',2"-terpyridine coordination triades. X-ray structures of 4'-(N,N-dimethylamino)-2,2':6',2"-terpyridine and bis(4'-(N,N-dimethylamino)-2,2':6',2"-terpyridine)ruthenium(II) hexafluorophosphate', *New. J. Chem.*, Manuscript submitted
6. Constable, E.C., Ward, M.D. and Corr, S. (1988) 'A convenient, high-yield synthesis of 2,2':6',2"-terpyridine and its iron(II) complex' *Inorg. Chim. Acta*, **141**, 201-203
7. Kröhnke, F. (1976) 'The specific synthesis of pyridines and oligopyridines', *Synthesis*, 1-24

ANTENNA EFFECTS AND PHOTOINDUCED ELECTRON TRANSFER IN POLYNUCLEAR METAL COMPLEXES.

F. SCANDOLA, R. ARGAZZI, C. A. BIGNOZZI, C. CHIORBOLI,
M. T. INDELLI, M. A. RAMPI
Dipartimento di Chimica dell'Università, Centro di Fotochimica C.N.R., 44100 Ferrara, Italy

ABSTRACT. Interesting light-induced functions, such as antenna effect and charge separation, can be performed by suitably designed supramolecular systems. Metal polypyridine complexes are good building blocks for the design and synthesis of such supramolecular systems. Photoinduced electron transfer and charge shift processes have been observed in Ru(II)-Rh(III) and Rh(III)-DQ dyads (DQ = N,N'-cycloalkyl bpy derivative). The possibility to obtain related triads for photoinduced charge separation is discussed. Trinuclear cyano-bridged Ru(II) complexes feature very efficient antenna effects and can be profitably used, following appropriate functionalization, in spectral sensitization of semiconductors. An extension towards larger one-dimensional antenna systems is discussed.

1. Introduction.

1.1 SUPRAMOLECULAR SYSTEMS AND LIGHT-INDUCED FUNCTIONS

As is often the case for new and rapidly developing areas, it is difficult to give an objective and truly comprehensive definition of *supramolecular chemistry* [1]. Generally speaking, the field spans an "intermediate" range of chemical complexity, greater than that of simple molecular species, but smaller than that of extended systems such as biological structures or solid state materials. In broad terms, supramolecular systems can be defined in terms of their constituents, namely, *molecular components* held together by intercomponent interactions. It is mainly the type of intercomponent linkage (which may range from weak intermolecular forces to true covalent bonding) which provides differentiation within the field. Research on supramolecular chemistry is substantially powered by the idea that supramolecular systems are better suited than simple molecular species to achieve interesting chemical or physical *functions* (molecular recognition, self-assembling, pre-organization of reactants, multi-electron catalysis, signal transduction, switching processes, etc.).

Within the general subject of supramolecular chemistry, a particular but important field is that of *supramolecular photochemistry* [2], i.e., the study of the interaction of light with supramolecular systems. In this context, light is important from at least three viewpoints: (i) it is an useful tool (both in absorption and emission) to probe supramolecular systems (particularly with regard to intercomponent interactions and related effects); (ii) it can be used as an energy input (enabling the supramolecular system to perform an energy-demanding function); (iii) it can be used to address specific components of a supramolecular system (so as to, e.g., trigger a

V. Balzani and L. De Cola (eds.), Supramolecular Chemistry, 235–248.

specific function). In cases (ii) and (iii) the supramolecular system performs a *light-induced function* and can be regarded as a "photochemical molecular device" [1b,2].

This article deals with two main types of light-induced functions: (i) *photoinduced charge separation* ; (ii) *antenna effect*. The antenna effect (Fig. 1a) relies on light absorption by several photosensitizer molecular components and intercomponent energy transfer to a common acceptor component, which can use the energy under the form of emission (the component is in this case a "luminophore") or for a useful chemical reaction (the component is in this case a "reaction center"). Photoinduced charge separation (Fig 1b, showing one out of several possible schemes) involves light absorption by a chromophoric molecular component (photosensitizer), electron transfer from the excited photosensitizer to a primary acceptor, and a further electron transfer step (in competition with charge recombination between reduced primary acceptor and oxidized photosensitizer) to a secondary acceptor component.

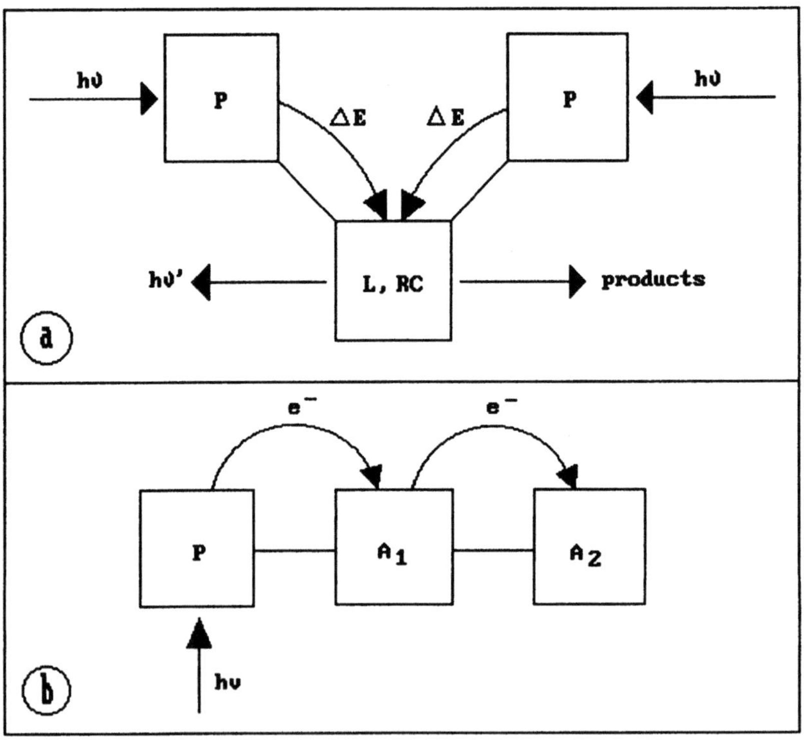

Figure 1. Schematic representation of three-component supramolecular systems featuring (a) antenna effect and (b) photoinduced charge separation.

Recent work on *polynuclear metal complexes* [3], i.e., supramolecular systems containing covalently linked transition metal complexes as molecular components, is described in this article. Such systems have been designed in order to study intercomponent processes relevant to photoinduced charge separation and antenna functions. The possibility to use supramolecular antenna systems in the spectral sensitization of wide-bandgap semiconductors is also discussed in some detail.

1.2 BUILDING BLOCKS

The transition metal complexes used as building blocks for the supramolecular systems described in this article are of the metal-polypyridine type (Fig 2a, polypyridine = 2,2'-bipyridine, ancillary ligands not shown). Metal polypyridine complexes have been intensively studied and their photochemistry and photophysics is now well characterized [4]. The metal centers used in this work are Ru(II), Re(I), and Rh(III), and the polypyridine ligands are 2,2'-bipyridine (bpy) or 1,10-phenanthroline (phen) and their 4,4'- or 4,7-substituted derivatives. In this section, the basic properties of the molecular components used are briefly recalled.

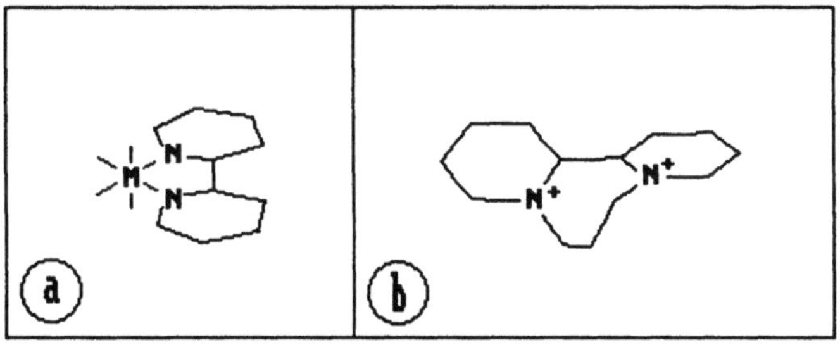

Figure 2. Two types of building blocks used in this work: (a) metal polypyridine (e.g., bipyridine) complexes; (b) quaternarized bipyridine derivatives (e.g., with an N,N'-bridged $-(CH_2)_3-$ chain).

In Ru(II) and Re(I) polypyridine complexes [4], the lowest excited state is a metal-to-ligand (e.g., M ⟶ bpy) charge transfer (MLCT) state of formal triplet multiplicity (Ru, and especially Re, have a high degree of spin-orbit coupling). The energy of these states depends on the redox properties of the metal and of the polypyridine ligand, which can be easily determined by electrochemical oxidation and reduction, respectively, of the complex. For a given metal, the oxidation potential depends strongly on the nature (electron donating and accepting power) of the ancillary ligands. The reduction potential of the polypyridine ligand depends not so much on the structure (e.g., bpy or phen) but rather on the nature of the substituents on the rings. All these effects can be used to obtain a fair degree of synthetic control on ground-state redox properties, MLCT energies, and, therefore, also excited-state redox potentials [5]. Generally speaking, Ru(II) and Re(I) polypyridine complexes are poor electron acceptors and donors in the ground state, but are very good excited-state reductants and mild (Ru) to good (Re) excited-state oxidants. The lifetime of the MLCT states depends in a complex way on their energy: the absolute energy is relevant (because of the energy-gap-law for radiationless deactivation), but the energy gap to the ligand field states of the metal is also important (because of the possibility of thermally activated decay through this states)[6]. Lifetimes in the range 50-1000 ns are quite typical for Ru(II) and Re(I) polypyridine complexes in fluid solution at room temperature. The MLCT excited states can be easily monitored by their phosphorescent emission.

In their ground state, Rh(III) polypyridine complexes behave as moderately good electron acceptors. Reduction seems to involve the metal center rather than the polypyridine ligand, although the energy difference between the two redox sites is probably small. The lowest-energy excited state is a ligand-centered (LC) state of triplet multiplicity. Its energy can be slightly tuned by substitutions on the polypyridine ligand, but is quite independent of the nature of the other ancillary ligands. Typical lifetimes in fluid solutions are in the 10-1000 ns range, depending on the energy of close-lying metal-centered states (which may intervene in thermally activated decay paths) [7]. In room temperature fluid solution, the LC state is very slightly emissive but can be easily monitored by transient absorption measurements. Excited Rh(III) polypyridine complexes exhibit strong oxidizing power, but no appreciable reducing character [8].

Besides these metal-containing building blocks, a type of fully organic component is also used in this work. Quaternarized bipyridine derivatives (Fig 2b, with a cyclic N,N'-three-methylene chain) are appropriate as electron acceptors. The length of the N,N'-alkylating chain affects the degree of planarity of the fragment, and thus the reduction potential. If intercomponent electron transfer processes take place, these units are very convenient for monitoring purposes as their one-electron reduced forms display prominent visible absorption bands [9].

1.3 SYNTHETIC STRATEGIES

Using metal polypyridine units, there are two main possibilities to create intercomponent connections: (i) have a conventional bridging ligand in one of the ancillary positions to bind a second metal center; (ii) make covalent organic linkages between the polypyridine ligands of two metal centers. The two strategies used in the present work are schematized in Fig 3a (type (i) with cyanide as bridging ligand) and 3b (type (ii) with a two-methylene linkage between the 4-positions of two bipyridine ligands). Each of these two modes has peculiar advantages and problems.

Several ambidentate ligands and appropriate synthetic routes are available to implement mode (i). The choice, however, is somewhat limited by the fact that the bridging ligand, as any other ancillary ligand, may strongly affect the properties of the metal-polypyridine fragment. For example, for the above-mentioned reasons, bridging ligands of weak ligand field must be avoided to have long excited-state lifetimes (this is one of the reasons for choosing cyanide as a convenient bridging ligand). Furthermore, practical models of the molecular components may not be available in some cases (e.g., a mononuclear complex modelling the N-bonded metal center of Fig 3a is difficult to find). Furthermore, the intercomponent electronic coupling provided by some bridging ligands (e.g., cyanide) may be sufficiently strong that the properties of a molecular component are substantially perturbed by attachment of the second metal. Generally speaking, the ideal picture of molecular components with well-defined intrinsic properties, linked by connectors with purely structural role, is not quite accurately met in this coupling mode.

This ideal picture is better approached using the second strategy, where the intercomponent interaction is likely to be much smaller. Here, mononuclear models of the metal polypyridine fragments are readily available, and their properties are likely to be strictly maintained in the polynuclear system. Possible problems, on the other hand, are related to the synthetic sequence needed (first link together two free polypyridine ligands, and then bind stepwise the two metals). Furthermore, the non-rigid nature of the binuclear complex obtained

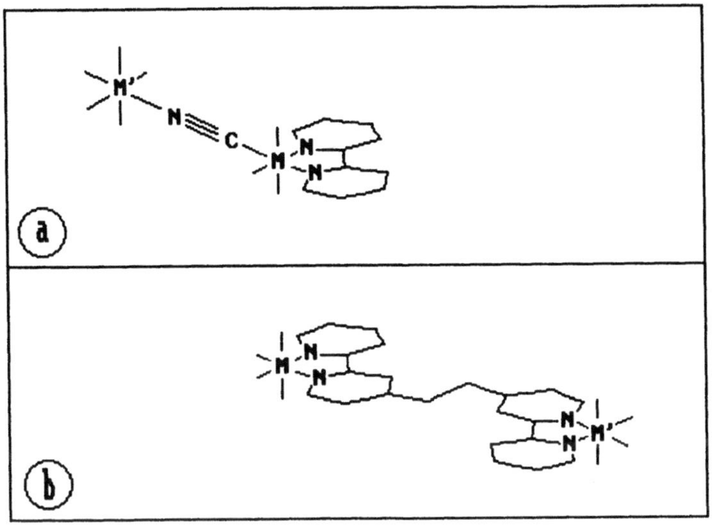

Figure 3. Two basic strategies followed in this work to assemble metal-containing polypyridine molecular components: (a) via bridging cyanides; (b) via covalently linked bipyridines.

(with the polymethylene linkage of Fig 3b, and with most other practical alternatives) may complicate the interpretation of the results obtained.

2. Towards photoinduced charge separation

2. 1 CHARGE SHIFT.

This section deals with a dyad designed for the study of a charge shift process (Fig. 4a). The practical problems which must be solved in the study of such a system are related to *initiation* and *observation* of the charge shift process. Limitations on the possibility to observe the intramolecular electron transfer step are essentially of technical nature, and can in principle be overcome by the use of an appropriate experimental apparatus (in terms of sensitivity and time resolution). The problems related to the initiation step, on the other hand, are of non-trivial nature. First of all, the initiation step must be faster than the intercomponent electron transfer step. Since the initiation step is reduction by an external source, it may not be easy to find experimental conditions that would satisfy this requirement. Moreover, in the initiation the initial reduction of the dyad must take place at the thermodynamically *less* favoured site. This is hard to achieve by chemical reduction, unless extremely powerful external reductants, acting in a nearly statistical fashion towards the two redox sites, are used (e.g., solvated electrons and other reducing radicals used in pulse radiolysis) [10]. The "trick" used here to overcome this problem has been to *photoinitiate* the process. By selective excitation of the A_1 component, this component can be selectively reduced by an external electron donor (unable to reduce both A_1 and A_2 in their ground states).

The dyad designed to this purpose is shown in Fig. 4b [11]. This dyad is thereafter designated as $Rh(NN)_3^{3+}$–DQ^{2+}, where the $Rh(NN)_3^{3+}$ component is (4,4'-dimethyl-2,2'-bipyridine)-bis(1,10-phenanthroline)rhodium(III) and the DQ^{2+} component is a three-methylene

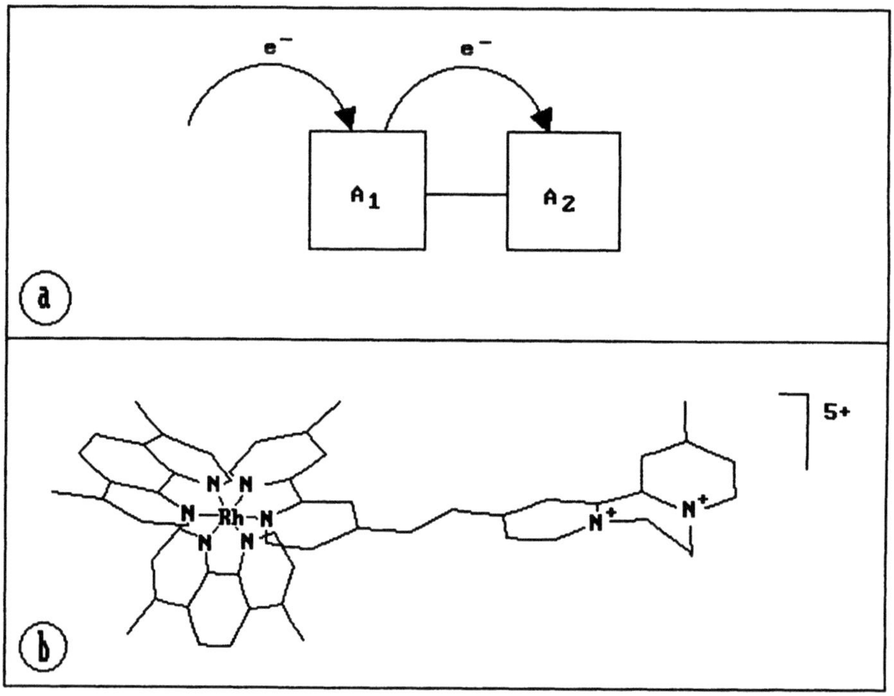

Figure 4. Block diagram (a) and experimental example (b) of charge shift in a two-component (dyad) system.

N,N'-bridged diquaternarized 4,4'-dimethyl-2,2'-bipyridine). In this dyad, $Rh(NN)_3^{3+}$ is designed to play the role of A_1, and DQ^{2+} that of A_2 (Fig 4a). The external donor D is in this case 1,2,3-trimethoxybenzene (thereafter indicated as TMB). Upon pulsed laser excitation of the $Rh(NN)_3^{3+}$ component (eq 1), the following sequence of events takes place: (i) bimolecular reductive quenching of the $Rh(NN)_3^{3+}$ excited state by TMB (eq 2); (ii) intercomponent charge shift (eq 3); (iii) bimolecular back electron transfer regenerating the starting situation (eq 4). The occurrence of the charge shift step is unequivocally demonstrated by the appearance of the typical absorption spectrum of the reduced DQ^+ component. By using high TMB concentrations, the primary reduction can be made sufficiently fast that the rate of the charge shift step (~ 3×10^7 s^{-1} in acetonitrile) can be directly monitored. The relatively slow rate of the charge shift

$$Rh(NN)_3^{3+}\text{–}DQ^{2+} + h\nu \longrightarrow {}^*Rh(NN)_3^{3+}\text{–}DQ^{2+} \quad (1)$$

$$^*Rh(NN)_3^{3+}\text{–}DQ^{2+} + TMB \longrightarrow Rh(NN)_3^{2+}\text{–}DQ^{2+} + TMB^+ \quad (2)$$

$$Rh(NN)_3^{2+}\text{–}DQ^{2+} \longrightarrow Rh(NN)_3^{3+}\text{–}DQ^+ \quad (3)$$

$$Rh(NN)_3^{3+}\text{–}DQ^+ + TMB^+ \longrightarrow Rh(NN)_3^{3+}\text{–}DQ^{2+} + TMB \quad (4)$$

process is most probably related to the small driving force of the process (estimated as ~ 0.2 eV) [11].

2.2 PHOTOINDUCED ELECTRON TRANSFER

In the most common scheme (Fig 5a), photoinduced electron transfer in a dyad involves (i) excitation of a photosensitizer molecular component , (ii) electron transfer from the excited photosensitizer to an electron acceptor component (often called "charge separation"), (iii) back

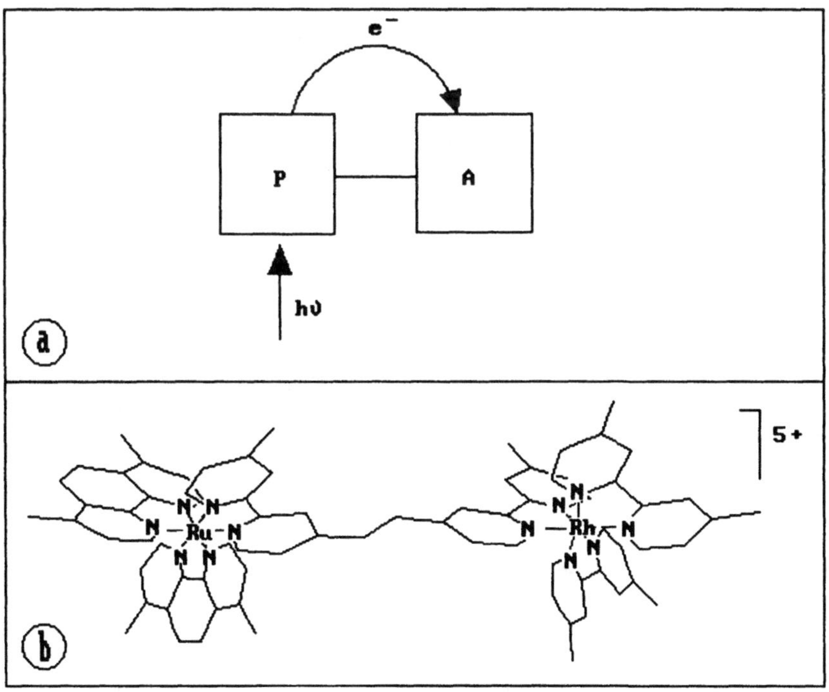

Figure 5. Block diagram (a) and experimental example (b) of photoinduced electron transfer in a dyad system.

electron transfer from the reduced acceptor to the oxidized photosensitizer (often designated as "charge recombination", not shown in the figure) [12,13]. The practical consequences of this sequence of processes may vary from system to system. Quenching of the excited photosensitizer is always observed (usually from emission intensity and lifetime measurements). The formation and disappearance of the charge separated state can in principle be monitored by fast spectroscopic techniques. The possibility of observation depends on both instrumental factors (sensitivity and time resolution) and on kinetic considerations (charge recombination must be slower than charge separation).

The dyad designed to this purpose is shown in Fig 5b [14]. It is thereafter indicated as $Ru(NN)_3^{2+}$–$Rh(NN)_3^{3+}$, where $Ru(NN)_3^{2+}$ represents (4,4'-dimethyl-2,2'-bipyridine)-bis(4,7-dimethyl-1,10-phenanthroline)ruthenium(II) and $Rh(NN)_3^{3+}$ represents tris(4,4'-dimethyl-2,2'-bipyridine)rhodium(III). The $Ru(NN)_3^{2+}$ unit is designed to play the role of P and the $Rh(NN)_3^{3+}$ unit that of A. Upon excitation of the $Ru(NN)_3^{2+}$ unit in the dyad (eq 5), the typical MLCT emission of this unit is strongly quenched with respect to that of a free $Ru(NN)_3^{2+}$ model, indicating the occurrence of efficient excited-state electron transfer (eq 6). The rate constant of

$$Ru(NN)_3^{2+}\text{–}Rh(NN)_3^{3+} + h\nu \longrightarrow {}^*Ru(NN)_3^{2+}\text{–}Rh(NN)_3^{3+} \qquad (5)$$

$$ {}^*Ru(NN)_3^{2+}\text{–}Rh(NN)_3^{3+} \longrightarrow Ru(NN)_3^{3+}\text{–}Rh(NN)_3^{2+} \qquad (6)$$

$$Ru(NN)_3^{3+}\text{–}Rh(NN)_3^{2+} \longrightarrow Ru(NN)_3^{2+}\text{–}Rh(NN)_3^{3+} \qquad (7)$$

this process (~ $2x10^8$ s^{-1} in acetonitrile) can be measured from the emission lifetime. For the rate constant of the back electron transfer process (eq 7), a lower limiting value ($\geq 5x10^7$ s^{-1}) can only be given at the present stage. The forward rate constant is consistent with a slightly exergonic process (estimated $\Delta G \approx 0.2$ eV). The back electron transfer process is much more exergonic (estimated $\Delta G \approx 1.9$ eV) and should presumably lie in the "Marcus inverted region" [15], although it is not presently possible to check this prediction on experimental kinetic data [14].

2.3 FROM DYADS TO TRIADS

Since the acceptor component of the dyad used for photoinduced electron transfer (Fig 5b)is very similar to the primary acceptor component of the dyad involved in charge shift (Fig 4b), an obvious next step is to design a triad for photoinduced charge separation (Fig 1b) by fusing together two dyads of this type. In practice, two synthetic routes are available, both of which would lead to tetrads rather than triads ($[Ru(NN)_3^{2+}]_2$–$Rh(NN)_3^{3+}$–DQ^{2+} or $Ru(NN)_3^{2+}$–$Rh(NN)_3^{3+}$–$[DQ^{2+}]_2$). Considering for simplicity the basic $Ru(NN)_3^{2+}$–$Rh(NN)_3^{3+}$–DQ^{2+} triad structure, the key to obtain efficient charge separation would be an efficient competition of the charge shift process (eq 11) over primary back electron transfer (eq 10). Given the uncertainty in the actual value of the rate constant of the last process, it is not possible to predict with certainty

$$Ru(NN)_3^{2+}\text{–}Rh(NN)_3^{3+}\text{–}DQ^{2+} + h\nu \longrightarrow {}^*Ru(NN)_3^{2+}\text{–}Rh(NN)_3^{3+}\text{–}DQ^{2+} \qquad (8)$$

$$ {}^*Ru(NN)_3^{2+}\text{–}Rh(NN)_3^{3+}\text{–}DQ^{2+} \longrightarrow Ru(NN)_2^{3+}\text{–}Rh(NN)_3^{2+}\text{–}DQ^{2+} \qquad (9)$$

$$Ru(NN)_3^{3+}\text{–}Rh(NN)_3^{2+}\text{–}DQ^{2+} \longrightarrow Ru(NN)_3^{2+}\text{–}Rh(NN)_3^{3+}\text{–}DQ^{2+} \qquad (10)$$

$$Ru(NN)_3^{3+}\text{–}Rh(NN)_3^{2+}\text{–}DQ^{2+} \longrightarrow Ru(NN)_3^{2+}\text{–}Rh(NN)_3^{3+}\text{–}DQ^{+} \qquad (11)$$

the performance of a triad (or tetrad) made up of the above-mentioned molecular components. An obvious way to improve the chance of success is to increase the rate of the charge shift step by changing the DQ^{2+} component so as to make it a better electron acceptor (e.g., by having a two- instead of a three-methylene chain bridging the N,N'-positions). These possibilities are currently being explored.

3. Antenna effect

3.1 ANTENNA EFFECT IN CYANO-BRIDGED Ru(II) COMPLEXES

The antenna effect is schematically depicted in Fig 1a. The essential requirements for a supramolecular system to perform this function are as follows: (i) the excited state of the energy collecting component must be lower in energy than those of the other light absorbing units (possibly only slightly lower, in order to preserve most of the absorbed energy) ; (ii) the intercomponent energy transfer processes must be efficient, i.e., must be faster than excited-state decay. In the design of such a system, the energetic requirement can be usually met by a careful

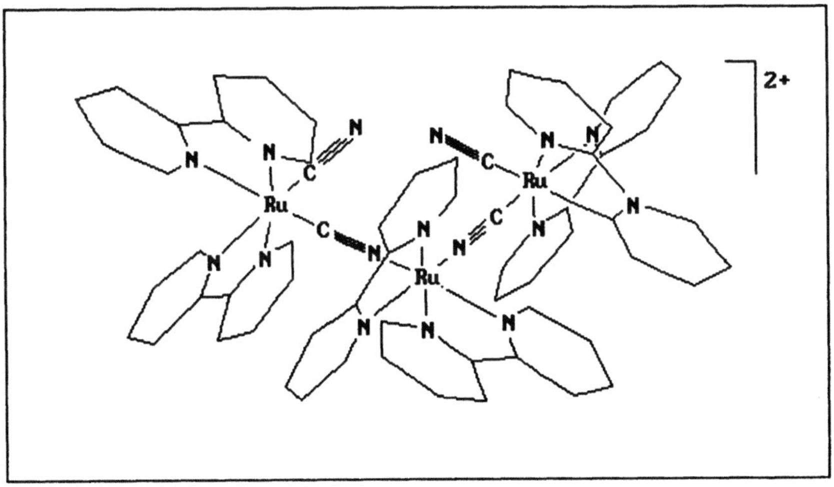

Figure 6. Cyano-bridged polynuclear complex with three $-Ru(bpy)_2^{2+}-$ units.

selection of molecular components. The kinetic problem is more difficult to tackle, as this would require a general knowledge of how the energy transfer processes depend on driving force, distance, and nature of connecting groups. As energy transfer processes usually involve small reorganizational energies, small driving force values seem to be required to have fast processes. With excited states connected to the ground state by transitions of high oscillator strength (e.g. singlet states of organic molecules), resonance energy transfer could be efficient over large distances and with highly "insulating" connectors. With forbidden excited states (e.g., triplet states of organic molecules and, as in the present case, lowest excited states of coordination compounds), efficient exchange energy transfer is likely to require shorter distances and more "conducting" connectors.

The systems described in this section make use of -Ru(bpy)$_2^{2+}$- units as molecular components, and of cyanide as both terminal and bridging ligand. A representative example of this type of supramolecular system is the trinuclear complex NC-Ru(bpy)$_2$-CN-Ru(bpy)$_2$-NC-Ru(bpy)$_2$-CN^{2+}, schematically represented in Fig 6 [16]. In complexes of this type, cyanide, besides being a convenient connector, provides a simple way to introduce excited-state energy differences between the (otherwise identical) -Ru(bpy)$_2^{2+}$ components. In NC-Ru(bpy)$_2$-CN-Ru(bpy)$_2$-NC-Ru(bpy)$_2$-CN^{2+}, the bridging mode (N-bonded to the central metal, C-bonded to the terminal ones) is imposed by the synthetic route used. Because of the already noticed influence of the ancillary ligands on the energy of the MLCT states of Ru(II) polypyridine complexes, the excited state of the central unit is lower by ca. 1600 cm^{-1} than those of the terminal ones.

In NC-Ru(bpy)$_2$-CN-Ru(bpy)$_2$-NC-Ru(bpy)$_2$-CN^{2+}, a single emission, attributable to the lowest energy chromophore, is observed (λ_{max} 714 nm) [16]. Although no selective population of the various MLCT excited states is possible due to partial overlapping of absorption bands, the exact correspondence between excitation and absorption spectra points towards a very efficient energy-transfer process from the higher-energy chromophores to the lowest energy emitting one (eqs 16,17). Thus, in this trinuclear complex the terminal and the central units play the role of P

$$\text{NC-Ru(bpy)}_2\text{-CN-Ru(bpy)}_2\text{-NC-Ru(bpy)}_2\text{-CN}^{2+} + h\nu \longrightarrow \text{NC-*Ru(bpy)}_2\text{-CN-Ru(bpy)}_2\text{-NC-Ru(bpy)}_2\text{-CN}^{2+} \quad (12)$$

$$\text{NC-Ru(bpy)}_2\text{-CN-Ru(bpy)}_2\text{-NC-Ru(bpy)}_2\text{-CN}^{2+} + h\nu \longrightarrow \text{NC-Ru(bpy)}_2\text{-CN-Ru(bpy)}_2\text{-NC-*Ru(bpy)}_2\text{-CN}^{2+} \quad (13)$$

$$\text{NC-Ru(bpy)}_2\text{-CN-Ru(bpy)}_2\text{-NC-Ru(bpy)}_2\text{-CN}^{2+} + h\nu' \longrightarrow \text{NC-Ru(bpy)}_2\text{-CN-*Ru(bpy)}_2\text{-NC-Ru(bpy)}_2\text{-CN}^{2+} \quad (14)$$

$$\text{NC-*Ru(bpy)}_2\text{-CN-Ru(bpy)}_2\text{-NC-Ru(bpy)}_2\text{-CN}^{2+} \longrightarrow \text{NC-Ru(bpy)}_2\text{-CN-*Ru(bpy)}_2\text{-NC-Ru(bpy)}_2\text{-CN}^{2+} \quad (15)$$

$$\text{NC-Ru(bpy)}_2\text{-CN-Ru(bpy)}_2\text{-NC-*Ru(bpy)}_2\text{-CN}^{2+} \longrightarrow \text{NC-Ru(bpy)}_2\text{-CN-*Ru(bpy)}_2\text{-NC-Ru(bpy)}_2\text{-CN}^{2+} \quad (16)$$

$$\text{NC-Ru(bpy)}_2\text{-CN-*Ru(bpy)}_2\text{-NC-Ru(bpy)}_2\text{-CN}^{2+} \longrightarrow \text{NC-Ru(bpy)}_2\text{-CN-Ru(bpy)}_2\text{-NC-Ru(bpy)}_2\text{-CN}^{2+} + h\nu'' \quad (17)$$

and L in Fig 1a, respectively. The lack of appreciable risetime in the emission places a lower limit on the rate of intercomponent energy transfer ($k \geq 1 \times 10^9\ s^{-1}$). As far as the mechanism of energy transfer is concerned, a singlet-singlet process is unlikely in view of the fast and efficient intersystem crossing that characterizes Ru(II)-polypyridine complexes. For the more plausible triplet-triplet pathway, the mechanism is expected to be of an exchange type. The high rate of energy transfer is understandable as cyanide bridges are known to provide a relatively strong electronic coupling [16].

3.2 ANTENNA EFFECTS IN SPECTRAL SENSITIZATION OF SEMICONDUCTORS

Dye sensitization, i.e., charge injection from an electronically excited adsorbed dye, is a well established technique [17,18] that permits to drive photoelectrochemical and photocatalytic processes on wide-bandgap semiconductors using sub-bandgap excitation. This feature is of obvious relevance to the use of semiconductors in solar energy conversion [19]. The main

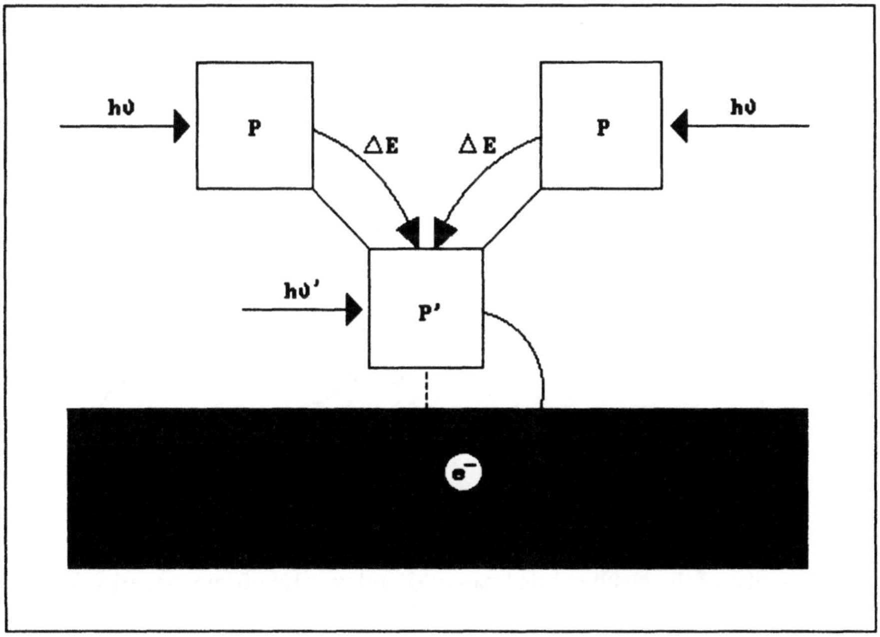

Figure 7. Schematic representation of an antenna-sensitizer molecular device for the spectral sensitization of semiconductors.

drawback of this technique is that, at a monolayer coverage, light absorption by the dye is often inefficient. On the other hand, multilayer adsorption does not help, as the inner layers tend to act as insulators with respect to the outer ones [20]. Thus, the only successful strategy to obtain good light harvesting efficiency of sensitized semiconductors has been so far that of increasing the surface area by using colloidal suspensions or films [21-23].

As an alternative (or an addition) to the obvious strategy of increasing the surface area (e.g., by using colloidal suspensions or colloidal films), a possible way to increase the light absorption efficiency of a sensitized semiconductor is to replace the sensitizer molecule at the semiconductor-solution interphase with a *sensitizer-antenna* molecular device (Fig 7)[24]. This device is similar to the antenna system of Fig 1a, in which the energy collecting unit is now a good photosensitizer for charge injection. In this way, the light energy absorbed by all the components can be used to effect charge injection, with an increase in the overall cross-section for light absorption compared to the case of a simple molecular sensitizer. How this increase is actually distributed over the action spectrum depends on the spectral characteristics of the antenna and sensitizer chromophores (subject to the obvious condition that hv' < hv).

In an attempt to demonstrate this effect, the trinuclear complex described in the previous section has been modified so as to have carboxylate groups in the 4,4' positions of the bipyridine ligands of the central unit [24]. The NC-Ru(bpy)$_2$-CN-Ru[(COO)$_2$-bpy]$_2$-NC-Ru(bpy)$_2$-CN^{2-} complex is schematically depicted in Fig 8. The photophysical behavior of this complex is almost

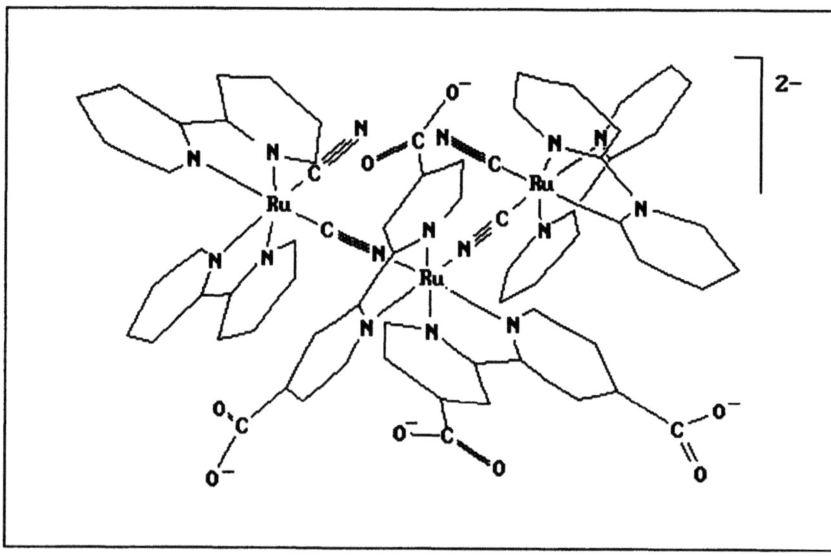

Figure 8. Antenna-sensitizer trinuclear complex.

identical to that of the unsubstituted parent complex, with efficient energy funnelling to the central molecular component (which again has the lowest excited state of the system) and emission from this unit. The presence of the carboxylate groups is essential to electrostatically anchor the complex, via its central component, to a TiO_2 surface (the groups are negatively charged and the surface is positively charged in aqueous solutions at $3.5 \leq pH \geq 6$). In experiments carried out using TiO_2-coated electrodes (aqueous solution, pH 3.5, NaI as electron donor) photocurrents were obtained on irradiation with visible light [24]. The photocurrent spectrum reproduces closely the absorption spectrum of the complex, indicating that the efficiency of conversion of absorbed light to electrons is constant throughout the spectrum, regardless of whether the incident light is absorbed by the central unit or by the terminal ones. This indicates that the light absorbed by the peripheral units is efficiently transferred to the central one, where it is used for electron injection. Thus, the $[Ru(bpy)_2(CN)_2]_2Ru(bpyCOO)_2)_2^{2-}$ trinuclear complex appears to perform indeed as an *antenna-sensitizer molecular device* on the surface of TiO_2 as schematized in Fig. 7a. Remarkably high efficiencies of conversion of incident light to electricity have been recently obtained by Graetzel [25], using this trinuclear complex on colloidal TiO_2 films.

The idea of antenna systems for sensitization of semiconductors can be extended towards larger systems. Highly branched Ru(II) polypyridine complexes of high nuclearity (presently up to 10 units) are already known [26]. On the other hand, one-dimensional systems that could stack perpendicularly to the surface (with the various antenna units working in series rather than in parallel) would be interesting in order to increase the ratio of chromophoric components to occupied surface area. Of course, the maximum size (in terms of number of components) of a one-dimensional antenna system depends very critically on the driving force demand of the single energy transfer steps involved. To tackle this problem, new cyano-bridged supramolecular species are being developed, as exemplified by the pentanuclear Re(I)-Ru(III) complex [27]

$$Re(CO)_3(phen)\text{-}CN\text{-}Ru(bpy)_2\text{-}CN\text{-}Ru(bpy)_2\text{-}CN\text{-}Ru(bpy)_2\text{-}CN\text{-}Ru(bpy)_2\text{-}CN^{4+}$$

Due to the *cis-* configuration at Ru(II), the metal-metal chain in this complex is presumably coiled in a more or less extended conformation. From independent information, the lowest MLCT states of the various chromophores are expected to occur in the energy ordering $Re(CO)_3(phen)$-CN-$^+$ > -CN-$Ru(bpy)_2$-CN- > -CN-$Ru(bpy)_2$-CN. The results available suggest quantitative energy migration from every absorbing chromophore to the terminal -CN-$Ru(bpy)_2$-CN unit, followed by emission. This indicates that in cyano-bridged supramolecular systems of this type exchange energy transfer is efficient even between isoenergetic chromophores, providing good perspectives for the synthesis of extended one-dimensional antenna systems [27].

4. Conclusions

The examples discussed show that coordination compounds, and in particular metal polypyridine complexes, are versatile building blocks for supramolecular systems of photochemical interest. In such supramolecular systems, thermodynamic and kinetic control of intercomponent energy and electron transfer can be achieved by careful choice of molecular components and connectors. By appropriate supramolecular design, simple but interesting light-induced functions can be obtained.

Acknowledgments. The work described in this article has been supported by the Ministero della Università e Ricerca Scientifica and by the Consiglio Nazionale delle Ricerche.

References.

[1] (a) Lehn, J.-M. *Science* **1985**, *227*,0 84.
(b) Balzani, V.; Moggi, L.; Scandola, F. In: *Supramolecular Photochemistry*; Balzani, V., Ed.; Reidel, 1987, p. 1.
(c) Lehn, J.-M. *Angew. Chem. Int. Ed. Engl.* **1988**, *27*, 89
(d) Ringsdorf, H.; Schlarb, B.; and Venzmer, J. *Angew. Chem. Int. Ed. Engl.* **1988**, *27*, 113.
(e) Vogltle, F. *Supramolekulare Chemie*; Teubner: Stuttgart, 1989.

[2] Balzani, V.; Scandola, F. *Supramolecular Photochemistry*; Horwood, 1991.

[3] Scandola, F.; Indelli, M. T.; Chiorboli, C.; Bignozzi, C. A. *Top. Curr. Chem.* **1990**, *158*, 73.

[4] Kalyanasundaram, K. *Photochemistry of Polypyridine and Porphyrin Complexes*; Academic, 1992.

[5] Juris, A.; Barigelletti, F.; Campagna, S.; Belser, P.; von Zelewsky, A. *Coord. Chem. Rev.* **1988**, *84*, 85.

[6] Meyer, T. J. *Pure Appl. Chem.* **1986**, *58*, 1193.

[7] Indelli, M. T.; Scandola, F. *Inorg. Chem.* **1990**, *29*, 3056.

[8] Indelli, M. T.; Ballardini, R.; Scandola, F. *J. Phys. Chem.* **1984**, *88*, 2547.

[9] Cooley, L. F.; Headford, C. E. L.; Elliott, C. M.; Kelley, D. F. *J. Am. Chem. Soc.* **1988**,*110*, 6673.

[10] Closs, G. L; Miller, J. R. *Science* **1988**, *240*, 440.

[11] Indelli, M. T.; Polo, E.; Bignozzi, C. A.; Scandola, F. *J. Phys. Chem.* **1991**, *95*, 3889.

[12] Connolly, J. S.; Bolton, J. R. In: *Photoinduced electron transfer. Part D*; Fox, M. A. and Chanon, M., Eds.; Elsevier, 1988; p.303.

[13] Ref. 2, Chapter 5.

[14] Indelli, M. T.; Bignozzi, C. A.; Scandola, F., manuscript in preparation.

[15] Marcus, R. A.; Sutin, N. *Biochim. Biophys. Acta* **1985**, *811*, 265.

[16] Bignozzi, C. A.; Roffia, S.; Chiorboli, C.; Davila, J.; Indelli, M. T.; Scandola, F. *Inorg. Chem.* **1989**, *28*, 4350.

[17] Gerischer, H.; Willig, F. *Topics Curr. Chem.* **1976**, *61*, 31

[18] Memming, R. *Prog. Surface Sci.* **1984**, *17*, 7

[19] Graetzel, M., Ed., *Energy resources through photochemistry and catalysis.*; Academic, 1983

[20] Gerischer, H. *Ber. Bunsenges. Phys. Chem.* **1973**, *77*, 771

[21] Desilvestro, J.; Graetzel, M.; Kavan, L.; Moser, J. *J. Am. Chem. Soc.* **1985**, *107*, 2988.

[22] Furlong, D.N.; Welles, D.; Sasse, W.H.F. *J. Phys. Chem.* **1986**, *90*, 1107.

[23] Graetzel, M. In: *Photochemical energy conversion.* Norris, J.R., Jr. and Meisel, D., Eds; Elsevier, 1989.

[24] Amadelli, R.; Argazzi, R.; Bignozzi, C.A.; and Scandola, F. *J. Am. Chem. Soc.* **1990**, *112*, 7099.

[25] O'Regan, B.; Graetzel, M. *Nature* **1991**, *353*, 737.

[26] Serroni, S.; Denti, G.; Campagna, S.; Ciano, M.; Balzani, V. *J. Chem. Soc., Chem Commun.* **1991**, 944.

[27] Argazzi, R.; Bignozzi, C. A., work in progress.

LONG-RANGE ELECTRON AND ENERGY TRANSFER IN MOLECULAR ASSEMBLIES

Wayne E. Jones, Jr., Steven M. Baxter, Sandra L. Mecklenburg, Bruce W. Erickson, Brian M. Peek and Thomas. J. Meyer*
Department of Chemistry
University of North Carolina at Chapel Hill
Campus Box 3290, Venable Hall
Chapel Hill, North Carolina 27599-3290

Abstract

Long-range photoinduced electron and energy transfer are being explored in molecular assemblies of soluble polymers and peptides.

Introduction

We are exploring the preparation of molecular assemblies where photoinduced electron or energy transfer will occur over tens of nanometers. Past the fundamental chemistry involved, our interest in these molecules is in their possible incorporation in molecular devices or chemical arrays where it may be possible, for example, to achieve artificial photosynthesis. In these studies we have utilized both pre-formed polymers derivatized with light absorbers and electron transfer sites and polypetides prepared from Boc-protected amino acids by Merrifield solid-phase peptide synthesis.[1]

The photochemistry is based on polypyridyl complexes of Ru^{II}, Os^{II} and Re^{I}. They have well-defined excited states for which photoinduced electron and energy transfer are well documented processes.[2,3] These processes continue to occur when the complexes are incorporated into soluble polymers or derivatized peptides. This fact and the molecular architecture offered by polymers and peptides has provided the basis for our studies on long-range electron and energy transfer.

In order to summarize the current state of progress in this area, it is convenient to review the principles behind photoinduced electron and energy transfer and then to turn to our investigations of these processes in soluble polymers and peptides.

Photoinduced Electron and Energy Transfer

The photochemical and photophysical properties of polypyridyl complexes of Ru^{II}, Os^{II} and Re^{I} and their derivatives in solution are well understood. They absorb light in the visible due to metal-to-ligand charge transfer (MLCT) transitions.

$$\underset{(d\pi)^6}{[Ru^{II}(bpy)_3]^{2+}} \xrightarrow{h\upsilon} \underset{(d\pi)^5(\pi^*)^1}{[Ru^{III}(\bar{b}py)(bpy)_2]^{2+*}} \qquad (1)$$

Because of the electron-hole pair produced by the excitation, the resulting excited states are capable of both donating and accepting electrons. The excited states are reasonably long

V. Balzani and L. De Cola (eds.), Supramolecular Chemistry, 249–258.

lived (nanoseconds to microseconds) and kinetic studies in solution have shown that electron or energy transfer processes that involve them can occur in the presence of appropriate quenchers.[1,2]

$$
\begin{aligned}
\mathrm{Ru(bpy)_3^{2+*}} + \mathrm{PTZ} &\longrightarrow \mathrm{Ru(bpy)_3^{+}} + \mathrm{PTZ^{\cdot+}} \\
+ \mathrm{PQ^{2+}} &\longrightarrow \mathrm{Ru(bpy)_3^{3+}} + \mathrm{PQ^{\cdot+}} \\
+ \mathrm{An} &\longrightarrow \mathrm{Ru(bpy)_3^{2+}} + {}^3\mathrm{An}^*
\end{aligned}
\qquad (2)
$$

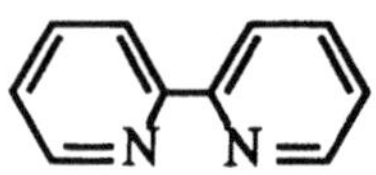

(bpy)

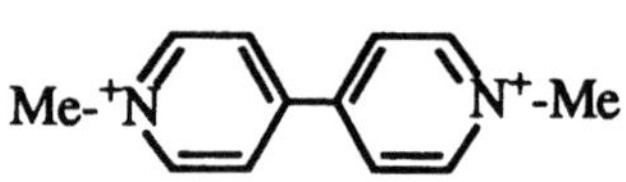

(PQ^{2+})

(1)

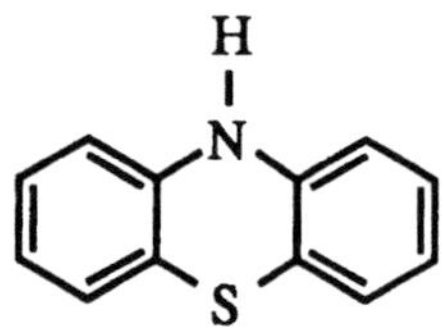

(PTZ)

(2)

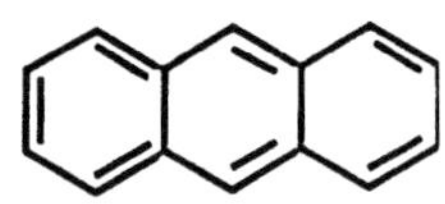

(An)

(3)

"Chromophore-quencher" complexes have been prepared in which the same processes occur but intramolecularly, e.g., eq. 3.[4] The sequence of events that

$$[(\overset{\bar{\cdot}}{\mathrm{bpy}})\mathrm{Re^{II}(CO)_3(py\text{-}PTZ)}]^{+*} \longrightarrow [(\overset{\bar{\cdot}}{\mathrm{bpy}})\mathrm{Re^{I}(CO)_3(py\text{-}PTZ^{+\cdot})}]^{+} \qquad (3a)$$

$$[\mathrm{Ru^{III}}(\overset{\bar{\cdot}}{\mathrm{bpy}}\mathrm{\text{-}CH_2OCH_2An)(bpy\text{-}CH_2OCH_2An)_2}]^{2+*} \longrightarrow \qquad (3b)$$

$$[\mathrm{Ru^{II}(bpy\text{-}CH_2OCH_2\text{-}{}^3An^*)(bpy\text{-}CH_2OCH_2An)_2}]^{2+}$$

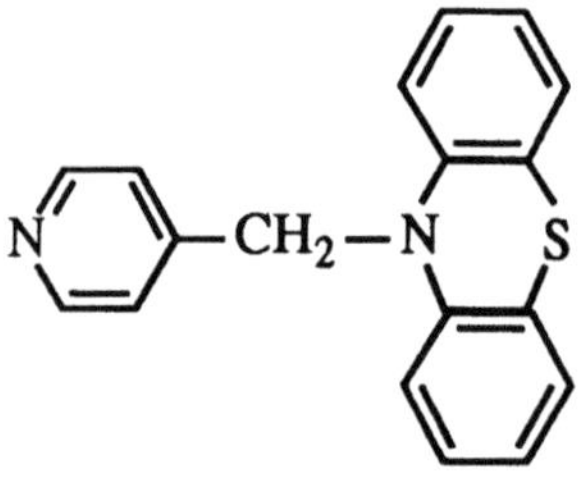

(py-PTZ)

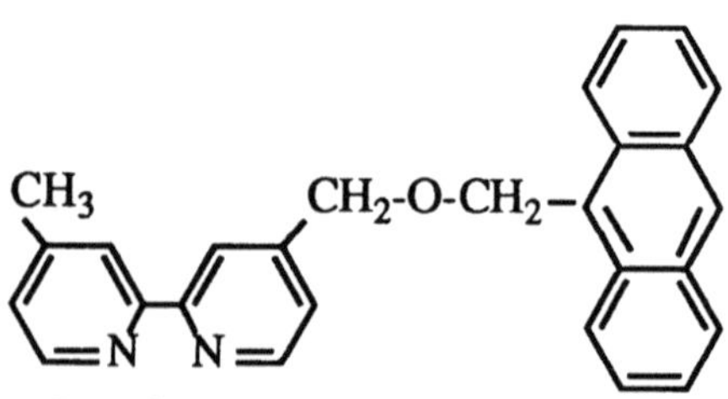

$(bpy\text{-}CH_2OCH_2An)$

occurs following excitation of these complexes has been monitored by fast transient laser techniques. The changes in electronic distribution that accompany electron or energy transfer typically result in significant changes in absorptivity and these changes can be wavelength and time-resolved to obtain spectra and rate constants.

The photochemical events shown in equations 2 and 3 provide a starting point for achieving long-range electron or energy transfer. In more complex assemblies photoproduced redox or excited state equivalents can be transferred to adjacent donors or acceptors and then transported over long distances by electron or energy transfer hopping. The efficiencies of these processes are determined by a series of competitions. They are outlined for electron transfer in Scheme 1 where: 1) A_1 and A_2 are electron transfer acceptors, 2) A_2 is a better acceptor than A_1 and, 3) Ru^{II} is a derivatized polypyridyl complex of Ru^{II}.

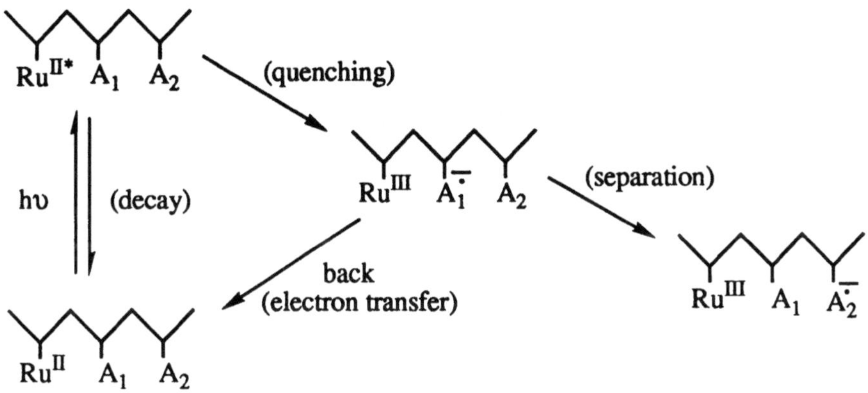

Scheme 1

In order to achieve high efficiencies for photoinduced electron transfer, it is necessary that the rate constant for quenching exceed that for excited state decay and that the rate constant for separation be greater than that for back electron transfer. The separation of the photoproduced redox equivalents relies on the existence of a free energy cascade, $Ru^{II*} \rightarrow A_1 \rightarrow A_2$, with $\Delta G^o < 0$ for each of the electron transfer steps. Following forward electron transfer, A_1 acts as a spacer which inhibits back electron transfer from $A_2^{\bar{\cdot}}$ to Ru^{III} by decreasing electronic coupling between sites. Also, back electron transfer by reversal of the electron transfer chain, $A_2^{\bar{\cdot}} \rightarrow A_1 \rightarrow Ru^{III}$, is disfavored since $\Delta G^o > 0$ for the first step, $A_2^{\bar{\cdot}} \rightarrow A_1$.

In a more complex assembly, long-range electron transfer can be achieved by electron transfer hopping amongst additional sites A_2 until the excited electron reaches a final acceptor site, A_3 in eq. 4.

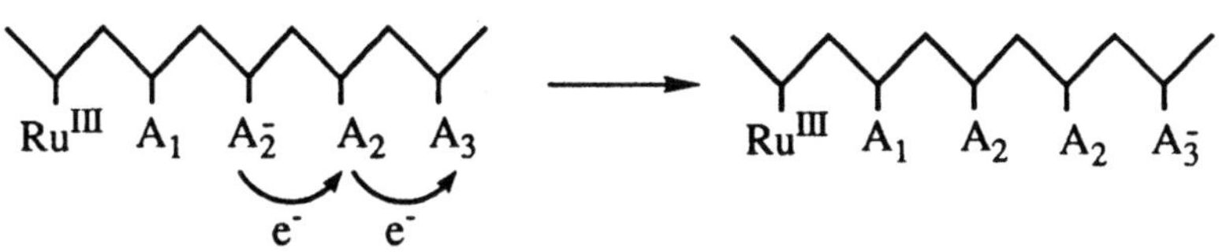

(4)

A closely related scheme can be written based on reductive electron transfer quenching by electron transfer donors, $Ru^{II*} \leftarrow D_1 \leftarrow D_2$, which would result in longe-range hole transport.

All of these electron transfer processes are subject to the fundamental rules of electron transfer.[5] The rate constants depend on ΔG^o, the extent of electronic coupling between sites, and the structural and solvent reorganizational energies that are associated with an electron transferring from one redox site to another.

For long-range energy transfer the existence of a free energy cascade is also required. In this case energy transfer by site to site hopping must be more rapid than the rate constant for excited state decay at the carrier site, A_1 in Scheme 2.

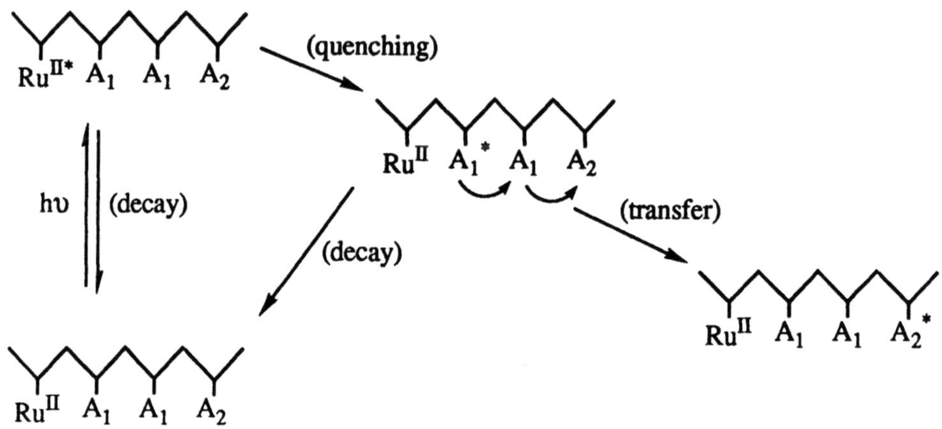

Scheme 2

The rules for energy transfer are different in detail but depend on the same factors: ΔG^o, the reorganizational energies, and, for triplet energy transfer, the extent of electronic coupling.[6]

Long-Range Electron and Energy Transfer in Derivatized Polystyrene

The extended, multisite character of polymers make them attractive candidates for the study of long-range electron or energy transfer.[7] Long-range energy transfer has been observed in polymers derivatized by attaching organic chromophores.[8] Synthetic strategies also have been developed for attaching polypyridyl complexes of Ru^{II} to polymers with the resulting materials used in studies on photoinduced electron transfer.[7e-h]

We have developed a general synthetic strategy for the attachment of a variety of chromophores and electron or energy transfer carriers to a 1:1 co-polymer of styrene/*p*-chloromethylstyrene. The polymer was prepared by free radical (AIBN) polymerization, which gave a polymer of polydispersity ($M_w/M_n \sim 2.7$) and average strands that consisted of ~30 repeating units.[9] In the derivatization chemistry, nucleophilic displacement at the C-Cl bond of the *p*-chloromethylstyrene units by alkoxides was exploited to give ethers, carboxylates to give esters, or amides to give amines, e.g., eq. 6.[9-11]

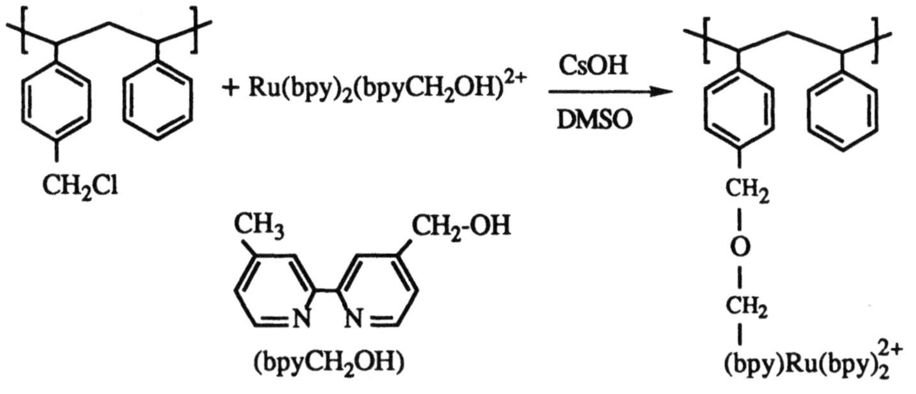

(5)

Since these reactions are essentially quantitative, the degree of loading on an average strand can be controlled by controlling the stoichiometries of the reagents. Following a purification step, additional components can be added to the strands by cycling back through the substitution chemistry.

We have used this approach to prepare derivatized polymers that have varying compositions in polypyridyl complexes of Ru^{II} or Os^{II} and electron or energy transfer donors or acceptors based on derivatives of PQ^{2+} (1), PTZ (2), or An (3).[9-11] In abbreviating these polymers, it is convenient to indicate the extent of derivatization as the number of derivatized sites per 30 available, e.g., $[PS\text{-}Ru^{II}_3](PF_6)_6$ where the number of Ru^{II}, Os^{II}, etc. sites on an average strand are indicated by the subscripts.

The photophysical properties of the polypyridyl complexes attached to the polymers have been investigated by transient emission measurements.[12] Excited state decay of the Ru-based MLCT excited states in $[PS\text{-}Ru^{II}_3](PF_6)_6$ in polar organic solvents is very similar to decay of $[Ru(bpy)_2(bpyCH_2OH)]^{2+*}$ under the same conditions. For the polymer where a complex is attached to each available site, $[PS\text{-}Ru^{II}_{30}](PF_6)_{60}$, there is clear evidence in the photophysical properties for multiphoton effects. Excited state decay is nonexponential and the degree of nonexponentiality depends on the light intensity and the solvent. A contribution to the nonexponentiality comes from polarization of the surrounding solvent and local polymeric backbone by the change in electronic distribution associated with forming the excited state. If a second excited state is formed within an interaction distance of the first, it influences the local environment and thus modifies excited state decay. The decay properties of MLCT excited states are known to be medium dependent.[2]

In the fully loaded polymer, the diameter of the individual complexes (~14 Å) is significantly greater than the unit repeat distance along the polymeric backbone (5-6 Å). Because of the relatively large excluded volume of the complexes, molecular modeling studies (which treat the complexes as di-cationic spheres of diameter 14 Å) show that the polymers adopt extended structures. In energy minimized structures the internuclear separation is ~21 Å and the average separation distance between the peripheries of adjacent complexes is ~7 Å .

We have explored intrastrand electron and energy transfer in polymers that contain combinations of Ru^{II} and Os^{II}. In the mixed Ru^{II}/Os^{II} polymer, $[PS\text{-}Ru^{II}_{25}Os^{II}_5)](PF_6)_{60}$, the energy of the Ru^{II*}-based MLCT excited state is 2.1 eV, the energy of the Os^{II*}-based excited state is 1.7 eV, and $Ru^{II*} \rightarrow Os^{II}$ energy transfer is spontaneous by 0.4 eV. Following laser flash photolysis, emission from both Ru^{II*} and Os^{II*} can be observed. By time resolving the transient decays, it was shown that energy transfer from Ru^{II*} to Os^{II}

was rapid, k(25°, CH_3CN) > 2 x 10^8 s^{-1} but that $Ru^{II*} \rightarrow Ru^{II}$ energy self-exchange was slow k < 10^6 s^{-1}.[13] These conclusions are summarized in Scheme 3.

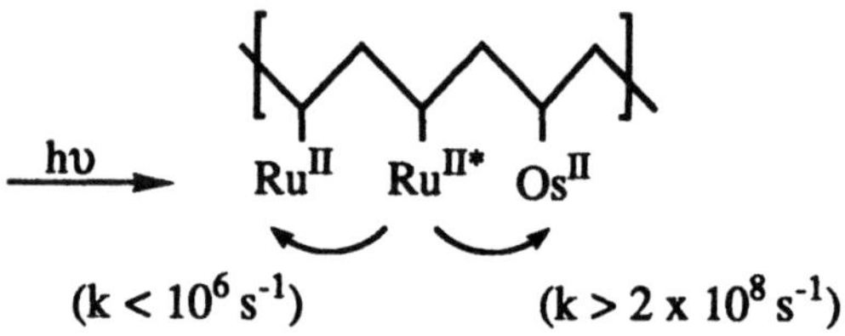

Scheme 3

Intrastrand electron transfer has been investigated in the mixed Ru^{II}/Os^{II} polymer. The approach taken was to conduct a series of laser flash photolysis experiments on solutions containing the polymer and an excess of the irreversible oxidative quencher [p-$MeOC_6H_4N_2$](BF_4). In these solutions, excitation ($Ru^{II} \rightarrow Ru^{II*}$) and oxidative quenching ($Ru^{II*} \xrightarrow{-e^-} Ru^{III}$) are followed by oxidation of Os^{II} by Ru^{III} which is favored by 0.4 eV. For the majority of redox events that occur, the oxidative equivalent at Ru^{III} must migrate to Os^{II} by electron transfer hopping first from Ru^{II} to Ru^{III}. This sequence of electron transfer events that occur is shown in Scheme 4.[13] Based on these experiments, k(25°, CH_3CN) > 5 x 10^6 s^{-1} for $Ru^{II} \rightarrow Ru^{III}$ site to site electron transfer hopping on the individual polymeric strands.

• CH_3CN, 298°, [PS-$Os_5{}^{II}Ru_{25}{}^{II}$]$(PF_6)_{60}$ + 35 m<u>M</u> (p-$MeOC_6H_4N_2$)(BF_4)

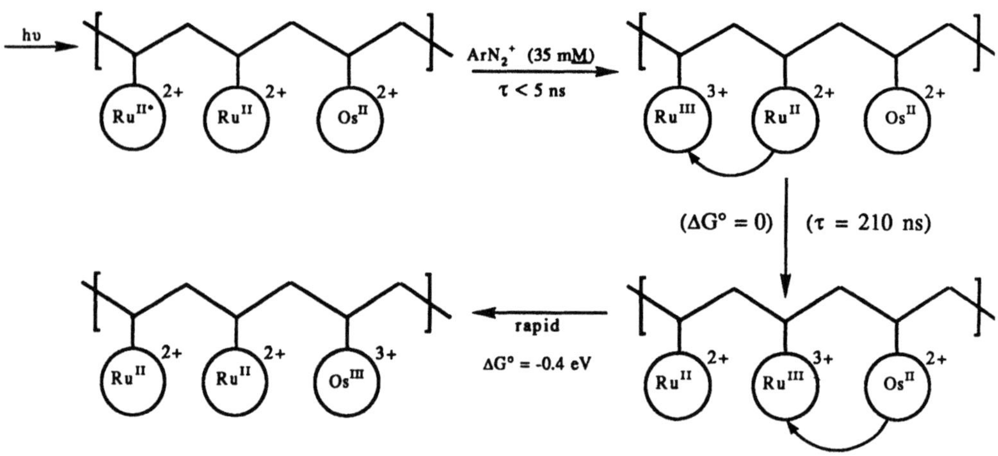

Scheme 4

Intrastrand energy transfer has also been investigated in polymers containing combinations of Ru^{II}, Os^{II} and An.[14, 15] In these polymers the energy of the triplet excited state for the attached anthracene is 1.8 eV which lies between the energies of the Ru^{II*} (2.1

eV) and Os^{II*} (1.7 eV) MLCT excited states. Experiments on mixed Ru^{II}-An polymers have shown that $Ru^{II*} \rightarrow$ An energy transfer occurs following excitation at Ru^{II}. Energy transfer can occur by at least two mechanisms. In the lightly loaded sample [PS-$Ru^{II}_3An_6$]$(PF_6)_6$ the anthracene triplet appears following laser flash excitation at Ru^{II} with k(25°, $ClCH_2CH_2Cl$) = 7.6 x 10^6 s^{-1}. From the dependence of k on solution viscosity, energy transfer appears to occur by local segmental rotations and intrastrand coiling which brings the excited state and quencher sites into close proximity.[14] In the polymer [PS-$Ru^{II}_3An_{21}$]$(PF_6)_6$ each Ru^{II} finds an anthracene or anthracenes as nearest neighbors. In this case $Ru^{II*}\rightarrow$An energy transfer is rapid (k > 2 x 10^8 s^{-1}) based on the quenching of the MLCT emission and the prompt appearance of the anthracene triplet as shown by transient absorption measurements.

In the polymer [PS-$Ru^{II}_3Os^{II}_3$]$(PF_6)_6$ there is no evidence for $Ru^{II*}\rightarrow Os^{II}$ energy transfer following excitation at Ru^{II}. Presumably the sites are sufficiently spread along the individual strands and intrastrand coiling sufficiently inhibited that energy transfer is slow compared to the excited state lifetimes. However, long range energy transfer between Ru^{II*} and Os^{II} was achieved in [PS-$Ru^{II}_3An_{18}Os^{II}_3$]$(PF_6)_6$ where a high percentage of the intervening sites were functionalized with anthracenes. In this case the intervening anthracene groups were used as an energy transfer "bridge" between Ru^{II*} and Os^{II}. The net effect was to create a "molecular light pipe" for long range energy transfer, Scheme 5.[14,15]

• 1,2-DCE, 298°, [PS-$Ru_3^{II}Os_3^{II}An_{17}$]$(PF_6)_{12}$

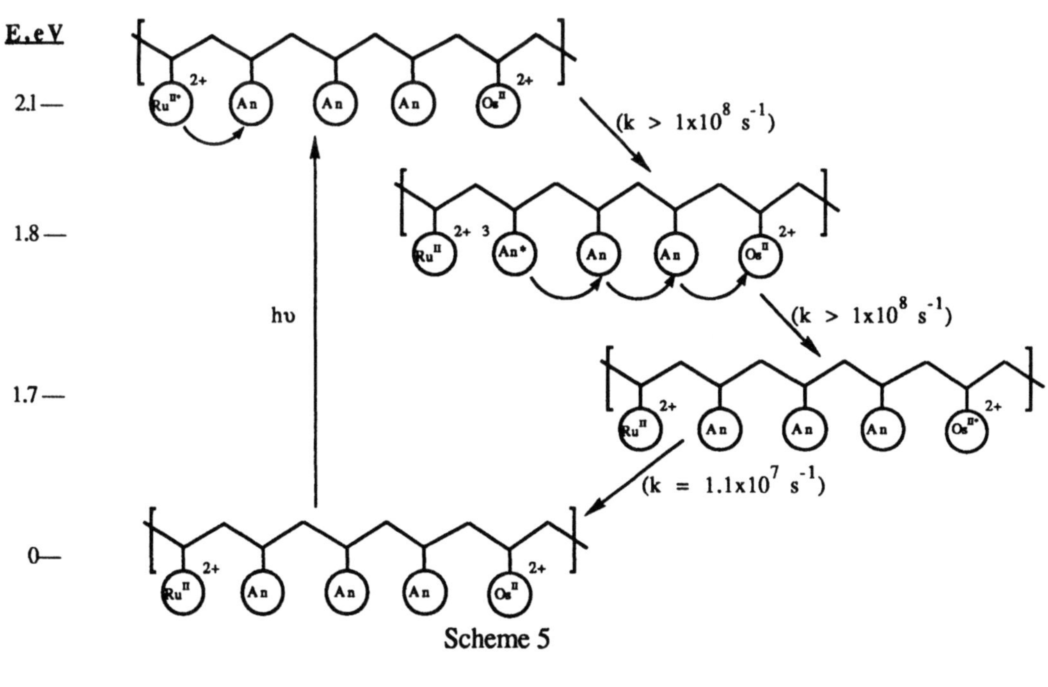

Scheme 5

Photoinduced Electron and Energy Transfer in Modified Amino Acids and Peptide Assemblies

Although soluble polymers offer the advantage of multiple sites for bonding combinations of chromophores and quenchers, there is no control, in a spatial sense, over the relative position of sites. The attachment chemistry leads to statistically random distributions. Peptide based assemblies prepared by solid phase synthesis also offer a multisite capability but with structural control by taking advantage of linear syntheses and stepwise additions. The linkage chemistry in this case involves the formation of amide bonds through the coupling of caboxylic acid groups with amines.

We have prepared modified amino acids for studying photoinduced electron transfer.[16] For the lysine derivative in eq. 6, initial excitation at Ru^{II} is followed by a

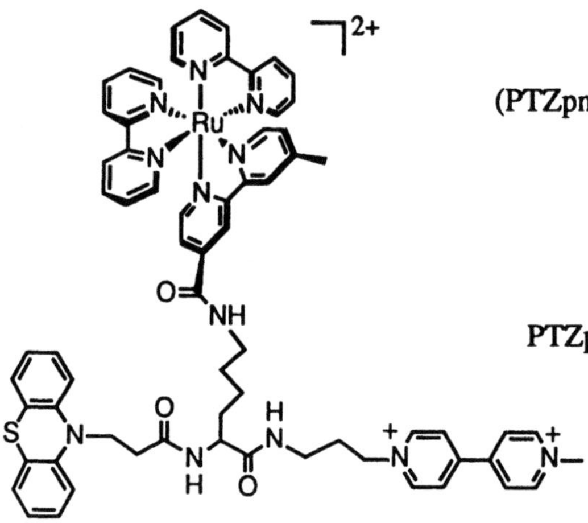

$$\mathrm{PTZpn\text{-}Lys(Ru^{II}b_2m)^{2+}\text{-}NH\text{-}prPQ^{2+}} \xrightarrow{h\nu} \mathrm{(PTZpn^{\cdot +})\text{-}Lys(Ru^{II}b_2m)^{2+}\text{-}NH\text{-}pr(PQ^{\cdot +})} \qquad (6)$$

series of electron transfers leading to the redox-separated state shown in eq. 6. The initial electron transfer chemistry is dominated by electron transfer from phenothiazine to the excited state. This is followed by electron transfer to the pyridinium to give the final redox-separated state.[17]

In anticipation of building more complex peptide-based structures, we have prepared a variety of "redox modules" based on polypyridyl complexes of Ru^{II} or Os^{II} and electron transfer donors or acceptors based on PTZ, pyridiniums, quinones, and ferrocenes. Synthetically, the emphasis has been on assembling these modules into complex structures in order to study photoinduced electron and energy transfer. The important issues initially are to explore those factors that maximize the efficiency of energy transfer or redox separation and those which are important in achieving long range energy or electron transfer in extended arrays.

In one sequence of experiments we have prepared the "triad" molecule Anq-$\mathrm{Lys(Ru^{II}b_2m)^{2+}}$-NH-prPTZ, where the electron acceptor is based on anthraquinone (Anq) rather than PQ^{2+}.[18] The abbreviation for this compound follows from the abbreviation for the structure in eq. 6. In this case, laser flash excitation leads to light-induced electron transfer and formation of the redox-separated state, $(\mathrm{An}\overline{\dot{\mathrm{q}}})$-$\mathrm{Lys(Ru^{II}b_2m)^{2+}}$-NH-(prPTZ$^{\cdot +}$) which decays to the ground state with k(25°, CH_3CN) = 5×10^6 s^{-1}.

We are currently attempting to prepare more elaborate examples where the number of complexes that intervene between the terminal anthraquinone and PTZ groups can be varied systematically as in Anq-$\mathrm{[Lys(Rub_2m)^{2+}\text{-}(Ala)_z]_n}$-$\mathrm{Lys(Rub_2m)}$-NH-prPTZ (n=1,2,3,...) with alanine spacers between the complexes. The complexes are sufficiently large compared to the repeat spacing on the peptide backbone that these materials will have extended structures, thus restricting intrastrand contacts. We hope to prepare "molecular rods" where light-induced electron transfer leads to stored redox equivalents as $PTZ^{\cdot+}$ and $An\bar{\dot{q}}$ which are separated by a series of metal complex spacers.

We have also begun to synthesize more complex molecules such as tetrads and pentads where free energy gradients will be exploited to achieve long-range redox separation by sequential electron transfer. An example is Bzq-Lys(Anq)-Lys($Ru^{II}b_2m)^{2+}$-Lys(PTZpr)-Lys(Fcc)-OMe where the teminal sites are a benzoquinone derivative (Bzq) and a ferrocene derivative (Fcc). Following visible excitation and intrastrand electron transfer, 1.1 eV will be stored in a redox-separated state in which there are three spacers between the final oxidative and reductive equivalents.

$$\begin{array}{c} Bz\bar{\dot{q}}\text{-Lys(Anq)-Lys(}Ru^{II}b_2m)^{2+}\text{-Lys(PTZpn)-Lys(}Fcc^{\cdot+}\text{)-OMe} \\ \Big\uparrow h\upsilon \\ \text{Bzq-Lys(Anq)-Lys(}Ru^{II}b_2m)^{2+}\text{-Lys(PTZpn)-Lys(Fcc)-OMe} \end{array} \qquad (7)$$

Peptide assemblies are especially appealing as vehicles for preparing extended molecular arrays. Based on this chemistry it should be possible to achieve long range electron and energy transfer and to develop such features as antennae arrays and redox modules for energy conversion. They are also appealing because of the possiblity of exploiting tertiary structures and helical arrays to control the relative dispostions of the various redox components in three dimensions. This an area where some interesting materials could appear and one that we have begun to map out for possible targets.

Acknowledgement

Funding from the Department of Energy under grant no. DE-FG05-86ER13633, the National Science Foundation under grant no. CHE-9022493 and the Public Healsh Service for grant GM- 42031 for support of this research is gratefully acknowledged.

References

1. Merrifield systhesis

2. a) Meyer, T. J. (1986) Pure Appl. Chem. 58, 1193. b) Krausz, E., Ferguson, J., (1989) Prog. Inorg. Chem. 37, 293. c) Watts, R.J. (1983) J. Chem Educ. 60, 834. d) Crosby, G. A., Highland, K. A., and Truesdell, K. A. (1985) Coord. Chem. Rev. 64, 41. e) DeArmond, M. K., Hanck, K. W., and Wertz, D.W. (1985) Coord. Chem. Rev. 64, 65. f) Juris, A., Balzani, V., Barigelletti, F., Campagna, S., Belser, P., and Von Zelewsky, A. (1988) Coord. Chem. Rev. 84, 85. g) Kalyanasundaran, K. (1982) Coord. Chem. Rev. 46, 159. h)Sutin, N. and Creutz, C. (1980) Pure Appl. Chem. 52, 2127. h) Sutin, N., (1979) J. Photochem. 10, 19.

3. a) Balzani, V., Scandola, F., (1991) "Supramolecular Photochemistry." Ellis Harwood, New York. b)Meyer, T. J. (1989) Acc. of Chem. Res., 22, 163.

4. a) Chen, P., Duesing, R., Graff, D. K., and Meyer, T. J. (1991) J. Phys. Chem., 95, 5850. b) Boyde, S., Strouse, G. F., Jones, W. E., Jr. and Meyer, T. J. (1989) J. Am. Chem. Soc. 111, 7448.

5. a) Marcus, R. A., Sutin, N. (1985). Biochim. Biophys. Acta, 811, 265-322. b) Sutin, N. (1983). Prog. Inorg. Chem., 30, 441. c) Meyer. T. J., Taube, H. (1987). "Comprehensive Coordination Chemistry, The Synthesis, Reactions, Properties, and Applications of Coordination Compounds", Wilkinson, Sir G., Gillard, R. D., McCleverty, J. A., eds., (1987). Pergammon Press, Oxford

6. (a) Ulstrup, J., Jortner, J. (1975). J. Chem. Phys., 63, 4358. (b) Orlandi, G., Monte, S., Barigelletti, F., Balzani, V. (1980). Chem. Phys., 52, 313.

7. a) Rabani, J. in "Photoinduced Electron Transfer", Part B: Fox, M. A.; Chanon, M.; Eds.; (1988) Elsevier: New York. b) Guillet, J., "Polymer Photophysics and Photochemistry"; (1985) Cambridge University Press:Cambridge. c) Phillips, D. ed. "Polymer Photophysics", Chapman and Hall (1985) New York. d) Winnik, M. A., ed. "Photophysical and Photochemical Tools in Polymer Science: NATO-ASI Series;" (1986) Reidel: Dordrecht.

8. a) Webber, S. E. (1990) Chem. Rev., 90, 1469. b) Chatterjee, P. K.; Kamioka, K.; Batteas, J. D.; Webber, S. E. (1991) J. Phys. Chem., 95, 960. c) Morsihima, Y.; Kobayashi, T.; Furui, T.; Nozakuros, S. I. (1987) Macromolecules, 20, 1707. d) Sienicki, K. (1990) Macromolecules, 23, 5025. e) Förster, R. J.; Vos, J. G. (1990) Macromolecules, 23, 4372. f) Kaneko, M.; Yamada, A.; Tsuchida, E.; Kurimura, Y. (1984) J. Phys. Chem., 88, 1061. g) Furue, M.; Sumi, K.; Nozakura, S. I. (1981) Chem. Lett., Jpn., 1349. h) Ennis, P. M.; Kelly, J. M. (1989) J. Phys. Chem., 93, 5735.

9. a) Arshady, R.; Reddy, B.S. R., George, M.H. (1984) Polymer,25, 716. b) Younathan, J. N.; McClanahan, S. F.; Meyer, T. J. (1989) Macromolecules, 22, 1048.

10. Olmsted, J. L., McClanahan, S. F., Danielson, D., Younathan, J. M. and Meyer, T. J. (1987) J. Am. Chem. Soc., 109, 3297.

11. Worl, L. A., Strouse, G. F., Younathan, J. N., Baxter, S. M. and Meyer, T. J. (1990) J. Am. Chem. Soc., 112, 7571.

12. Danielson, E., Worl, L. A., Murtaza, Z., Meyer, T. J., manuscript in preparation.

13. Jones, W. E., Jr., Baxter, S., Meyer, T. J., manuscript in preparation.

14. Jones, W. E., Jr., Baxter, S., Strouse, G. F., Younathan, J. Meyer, T. J., manuscript in preparation.

15. Strouse, G. F., Worl, L. A., Younathan, J. N. and Meyer, T. J. (1989) J. Am. Chem. Soc., 111, 9101.

16. Peek, B. M., Ross, G. T., Edwards, S. W., Meyer, G. J., Meyer, T. J. and Erickson, B.W. (1991) Int. J. Peptide Protein Res., 38, 114.

17. Mecklenburg, S. L., Peek, B. M., Erickson, B. W. and Meyer, T. J. (1991) J. Am.Chem. Soc., 113, 8540.

18. Peek, B. M., Edwards, S. W., Erickson, B.W., Mecklenburg, S. L. and Meyer, T. J. (1992) In "Peptides: Chemistry, Structure & Biology", Smith, J. A, & Rivier, J. E., ed., ESCOM: Leiden.

SYNTHETIC MOLECULAR KNOTS

C.O. DIETRICH-BUCHECKER and J.-P. SAUVAGE
Laboratoire de Chimie Organo-Minérale, UA 422 au CNRS
Faculté de Chimie
Université Louis Pasteur
67000 Strasbourg
France

ABSTRACT. Knots and interlaced designs have been part of human artistry and culture since the earliest times. In chemistry, knots have been the focus of theoretical investigations for several decades. In parallel, a few experimental approaches have been attempted by synthetic chemists. Until recent years, the only preparative routes pursued used the tools of classical organic chemistry. Despite their intellectual elegance, they have not succeeded. By taking advantage of the three-dimensional template effect of a transition metal (copper I), it has recently been possible to interlace two molecular threads prior to cyclisation and formation of a dimetallic trefoil knot. The demetallated knotted molecule and its di-copper(I) precursor have been fully characterized and studied. The X-ray structure of the dimetallic trefoil knot has been solved. It confirms the topology of the system.

1. Introduction

1.1. KNOTS AND INTERLACED THREADS IN ART AND LIFE

Knots and interlaced designs are pagan or religious symbols of continuity and eternity and have been used abundantly in ancient asiatic and eastern mediterranean art. First utilised in the decoration of stone, wood or metal, interlacing knotwork, arising probably from the imitation of the three dimensional arts of plaiting, weaving and basketry, developed rapidly and reached its zenith in the art of the Celtic culture. The magnificent sacred embellishments found in the famous Celtic manuscript "The Book of Kells" [1] give evidence about the fascination that braids, wreaths and knots exerted on the 8th century Irish monks.

This fascination, related to beauty and perfection, has lasted over the centuries, spreading from pure art, through folkart, into everyday life. For instance, the superb interlacing knotwork designs created by Leonardo da Vinci and by Albrecht Dürer during the 16th century were engraved and printed for the use of painters, goldsmiths, weavers, damaskeeners and needleworkers.[2]

Fine examples of rich ornamental metalwork or jewelry are found in numerous museums, but an attentive stroller may also discover many interesting, even if less

V. Balzani and L. De Cola (eds.), Supramolecular Chemistry, 259–277.

sophisticated, interlacing knotwork designs in handicrafts.[3]

Finally, this short survey on knots in art and life cannot be complete without recalling that modern art has also devoted special attention to knotted threads. In particular, one should mention the Dutch artist Cornelius Escher [4] to whom we owe some austere but highly beautiful representations of the trefoil knot (the simplest knot consisting of three lobes and three crossing points or nodes).

But if, as discussed below, interlacing knotwork designs find their origins in early civilizations, the recent advances achieved in biology show that the most ancient knots are indisputably those created by Nature. In given conditions, circular single or double stranded DNA forms knotted systems ranging from the simple trefoil knot to the most intricate knotworks. These knotted forms of DNA, generated by a whole class of specific enzymes (topoisomerases) are today commonly recognized as essential in the replication, site-specific recombination and transcription processes of circular DNA.[5]

In mathematics, knots were in vogue during the 19th century, but fell into disfavour and it is only recently that there has been a revival of interest [6], thanks to the rapid development of graph theory and topology. This important field is obviously connected to other fields of scientific investigation, chemistry and molecular biology being among the most active ones.

1.2. KNOTS IN CHEMISTRY - CHEMICAL TOPOLOGY

The synthetic chemist is usually highly interested in the aesthetics of the molecule he wants to assemble. In fact, the search for aesthetically attractive molecules has been a goal since the very origin of chemistry. What are the criteria which will lead the observer to find a molecule beautiful or not? Very often, symmetrical shapes analogous to the traditional geometrical objects (tetrahedron, dodecahedron, etc...) have been considered as the ultimate goal by prestigious laboratories. Dodecahedrane [7] is an example whose importance stems both from the amount of synthetic work involved and from the cultural significance of the geometrical shape reached. The aesthetic aspect of dodecahedrane is closely connected to Euclidian geometry : the object is conveniently represented by points and lines, the metric properties (length of a segment, angles, etc...) being of utmost importance. In this case, the object cannot be put out of shape. However, another interesting facet of beauty rests in the topological properties of the object. The latter can be deformed as much as desired, provided edges are not cleaved. Among the most fascinating objects displaying non-trivial topological properties, knots occupy a special position and their synthesis has been a challenging problem to chemists for several decades. But before describing the various routes towards a molecular knot attempted by several groups, the notion of "**topologically non-trivial**" deserves further detailed explanation.

Graph theory when applied to chemistry has been termed Chemical Topology. However, it must be stressed that it is only very recently that a rigorous mathematical treatment has been applied to molecules. Walba's use of graph theory for describing molecular systems now provides a unified view.[8]

Before being put into rigorous theory, Chemical Topology received much interest from many scientists. The book written by Schill in 1971 [9] is considered as the real bible of the topologist. In addition to interesting theoretical considerations regarding interlocked rings (catenanes) and knots, it contains very rich information dealing with the numerous experimental approaches used by Lüttringhaus and, later, Schill and their coworkers in order to prepare such topologically novel systems.

The first theoretical written discussion to appear on "Chemical Topology" (this was the title of the paper) was a publication by Frisch and Wasserman.[10] This general article seems to be the actual foundation stone of the field and it contains, expressed in a very chemical and accessible language, most of the notions which constitute the background of Chemical Topology. For instance, the idea of topological isomers was introduced. Later, the expression **topological stereoisomers** was proposed.[8] It is best exemplified using a single closed curve : a normal (topologically trivial) or knotted cycle (the most simple knot being the trefoil knot). The three objects of Figure 1 are topological stereoisomers : although they may consist of exactly the same atoms (vertices) and chemical bonds connecting these atoms (edges), they cannot be interconverted by any type of continuous deformation in three-dimensional space. In addition, the compounds of Figures 1b and 1c are **topological enantiomers** since the mirror image of any presentation of 1b is identical to a given presentation of 1c.

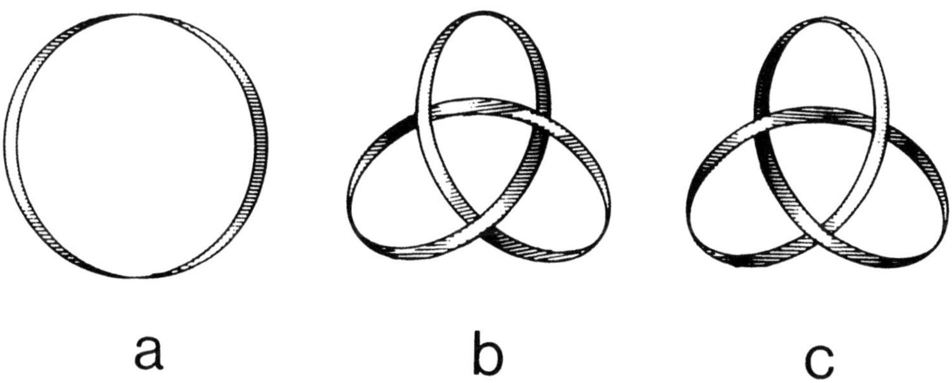

Figure 1. a, b and c are topological stereoisomers ; the two knots b and c are topological enantiomers.

In fact, "topologically non-trivial" can now be put in a more rigorous form. The molecular graph of a system (vertices and edges) has to be non-planar for the chemical species to look non-trivial. Another way of stating the same idea is that any topologically interesting and novel molecular system will be impossible to draw on a sheet of paper without crossing points, notwithstanding the deformations that may undergo bond lengths and angles. Early examples of such systems are, of course, the catenanes [9,11]. The preparation of the first molecular Möbius strip in 1982 [12] also represents a particularly significant accomplishment.

Clearly, the building of molecules displaying novel topological properties and the isolation and study of their topological stereoisomers are challenges to synthetic chemists. This is especially true for interlocked rings and knots.

2. The Various Routes Towards A Molecular Knot

2.1. APPROACHES INVOLVING RANDOMNESS

In principle, the most obvious way to acceed to a molecular knot relies on the direct random knotting occuring in a difunctionalized single molecular strand before its cyclisation (Figure 2). Despite the simplicity of the concept, this approach has, to our knowledge, never been attempted due to the low probability a chain has to tie a knot before its ends A and B find each other and connect. Frisch and Wassermann [10] could indeed estimate that for a cyclising $-(CH_2)_{60}-$ saturated hydrocarbon chain, the probability that the formed ring would exist as a trefoil knot is about 10^{-3} - 10^{-2}. This probability would of course increase with chain length.

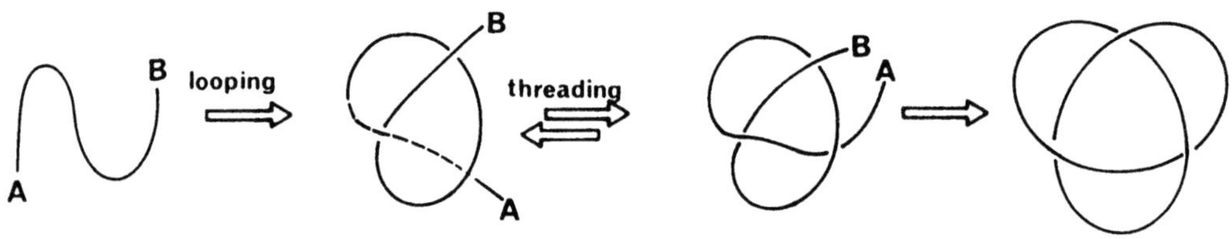

Figure 2. Tying a knot in the course of the cyclisation reaction of a linear molecule A-B after statistical threading.

On the contrary, the "Möbius strip" approach, depicted in Figure 3, also suggested by Frisch and Wasserman in 1961 [10], appeared much more realistic.

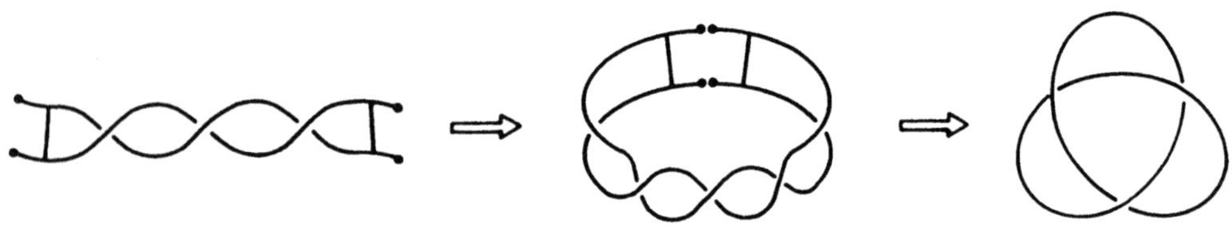

1 2 3

Figure 3. Principle of the Möbius strip approach to a trefoil knot.

In this approach, it is envisaged that the target molecule **3** can be obtained, after cleavage of the vertical rungs of compound **2**, which in turn is prepared from the ladder-shaped molecule **1**, the ends of which are able to twist prior to bimacrocyclisation. In order to achieve a synthesis of a trefoil knot by this strategy, Walba and his co-workers [13] prepared the 4-rung diol-ditosylate molecular ladder **4** (derived from tetrahydroxymethylethylene = THYME) and submitted it to cyclisation (Figure 4). Among the different compounds formed in this synthesis, only the untwisted molecular cylinder **5** could be fully characterized. Twisted isomers (**6**, **7**) and specially the three half-twist isomer **8** could not be identified with certainty.

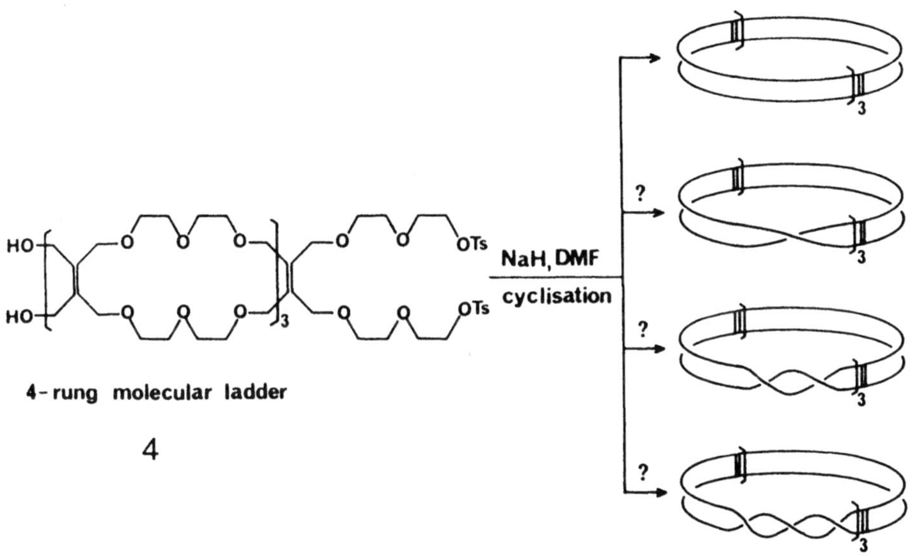

Figure 4. Towards a trefoil knot starting from a 4-rung molecular ladder as conceived by Walba's group.[13]

Randomness, strongly related to the threefold twisting required for precursors like **8**, may be reduced or even eliminated if the starting ladder-shaped molecule contains three half-twists prior to cyclisation. In the latter case, the Möbius strip approach may become a true directed synthesis.

2.2. DIRECTED SYNTHESES

A directed knot synthesis relying upon the Möbius strip principle was conceived by Schill and his co-workers [14] who synthesized the doubly bridged tetraamino-p-benzoquinone **9**. Connection of three such molecules by long chains should give the molecular ladder **10**, precursor of the three half-twist containing Möbius strip **11** (Figure 5). Nevertheless, despite the elegance of its principle, this synthesis did not yield a trefoil knot, because of the large number of steps required. The same authors also attempted the directed synthesis of a trefoil which relied on the use of a benzo-acetal central core.[15] These syntheses are closely related to the ones which this group had already used successfully to prepare various [2]- and [3]-catenanes.[16-19] Starting from compound **12** (Figure 6) the diaminoacetal **13** was synthesized in a nine-step reaction sequence. Compound **13** appears as a key intermediate : Euclidian geometry (bond angles, bond lengths, tetrahedral structure of the ketal carbon atom, lengths of the alkyl chloride chains) will dictate that intramolecular macrocyclisation occurs in a predetermined way, so that only four monomeric fourfold bridged 5,6-diamino-1,3-

benzodioxole derivatives are to be expected. Among them, only the cyclisation product **14** may lead to a trefoil knot **15** after selective cleavage of the aryl-nitrogen bonds and hydrolysis of the ketal. Unfortunately, after cyclisation of diamino compound **13**, three isomeric products were obtained in a total yield of 1.7% and the pre-knot **14** could not be identified with certainty.

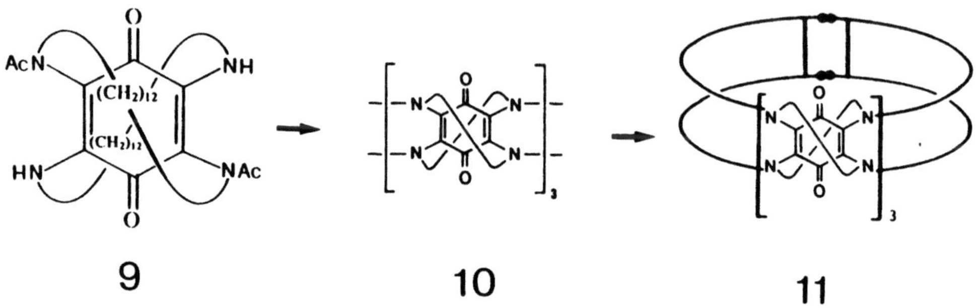

Figure 5. A way of introducing three half-twists in a molecular ladder as envisaged by Schill and his group.[14]

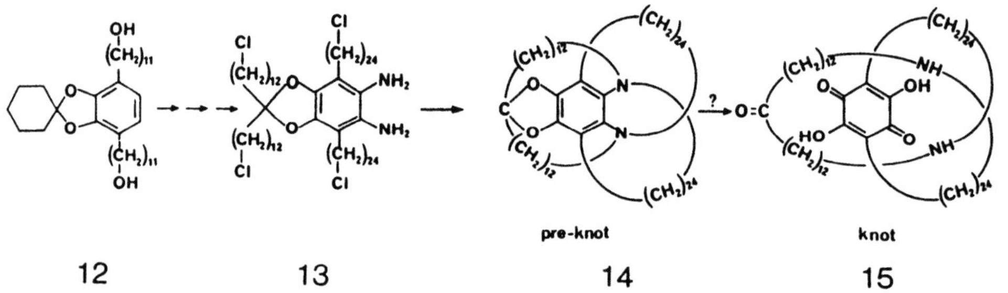

Figure 6. The 6-centre attempt to a trefoil knot .[15]

As seen above, randomness in the Möbius strip approach and numerous difficult steps in Schill's directed approach, appear as highly limiting factors in a trefoil knot synthesis. Both these major obstacles may be circumvented by the use of an unambiguous templated synthesis procedure.

2.3. TEMPLATED SYNTHESIS

2.3.1. *Based On One Metal.* If three half-twists are very difficult to introduce in a Möbius ladder for statistical reasons, one half-twist, by contrast, occurs much more frequently as evidenced clearly by Walba's high yield first synthesis of a molecular Möbius strip.[12] On the basis of this observation, the present authors conceived an ingenious strategy for the synthesis of knotted rings which requires only one half-twist of the ladder, additional torsion being present already in the starting compound. The principle of such an approach, called the "hook and ladder" approach, is given in Figure 7.[20]

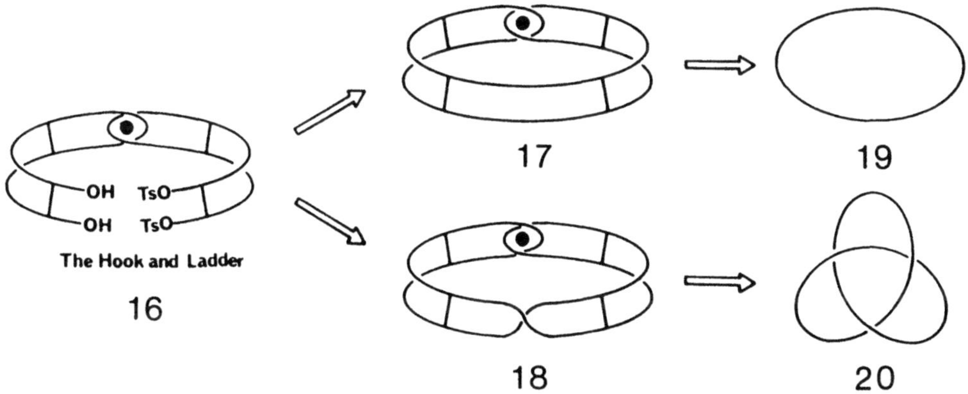

Figure 7. The hook and ladder approach to a trefoil knot as proposed by Walba.[20]

A molecular hook can be obtained by the entwining of two 1,10-phenanthroline derivatives around a copper(I) ion. Connection of the latter complex to THYME polyether ladders should lead to the desired "hook and ladder" precursor **16**. Intramolecular cyclisation of the twisted ends of **16**, followed by the cleavage of the vertical rungs, and demetalation of the "hook" part would give the target trefoil **20**. Studies and experimental work concerning this approach are presently underway. Another very interesting approach towards a trefoil knot can be found in a review by the Russian chemist Sokolov.[21] The principle of the synthesis imagined by this author as early as 1973 is given in Figure 8. Three bidentate chelates disposed in a suitable fashion around an octahedral transition metal used as a matrix may, after connection of their ends, lead to a molecular knot. Obviously, the probability that the six ends will connect (two by two) in the required fashion, is quite low. Nevertheless, strict geometrical control of the involved coordinated fragments or slight changes in the above proposed scheme may one day give access to a knot using this strategy.

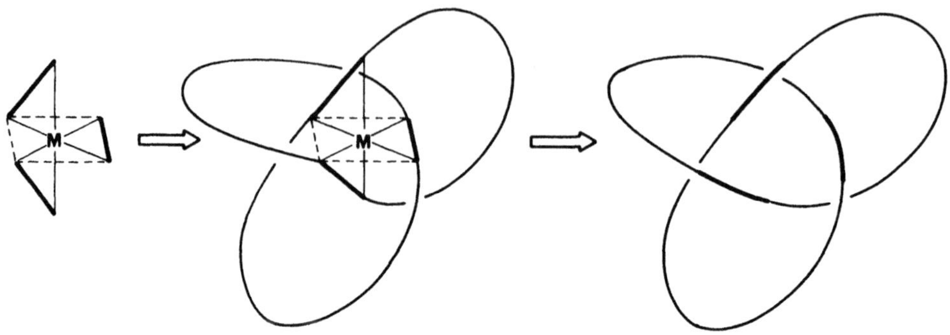

Figure 8. Sokolov's strategy for constructing a trefoil knot on an octahedral tris-chelate complex.[21]

2.3.2. *Based On Two Metals.*

Principle.

For our part, towards the goal of preparing a trefoil knot, we extended the synthetic concept which had already proved successful for the synthesis of catenanes [22] (Figure 9, top). In the strategy towards a trefoil knot, we exploited the three-dimensional template effect induced by two transition metals. As shown in Figure 9 (bottom), two bis-chelating molecular threads $\mathbf{A_2}$ can be interlaced on two transition metal centers, leading to a double helix $\mathbf{B_2}$. After cyclisation to $\mathbf{C_2}$ and demetalation, a knotted system $\mathbf{D_2}$ should be obtained. An important prerequisite for the success of this approach is the formation of a helicoidal dinuclear complex. Of utmost importance is the stability of the double helix $\mathbf{B_2}$, which has to resist the cyclisation conditions used to obtain $\mathbf{C_2}$. The choice of the chemical link between the two chelating subunits of each $\mathbf{A_2}$ fragment is thus crucial. Besides chemical stability, steric parameters and flexibility are factors that also have to be considered. In particular, during the reaction of $\mathbf{A_2}$ with the transition metal ion, the formation of $\mathbf{B_2}$ must be favoured over the formation of polymeric complexes or the formation of a mononuclear complex produced from folding of $\mathbf{A_2}$ onto one metal center. It is interesting to note that only a few dinuclear helicoidal complexes were reported in the literature up to the 1980's ; since then, this domain has developed so rapidly that numerous beautiful dinuclear double helices are now known.[23-29] Although the preparation of double helices from various transition metals and bis-chelate ligands is very likely to have occured long ago, it is only relatively recently that the first such system was recognized and characterized.[23] Moreover, the scientific interest of these arrangements was not at all obvious. One of the earliest dinuclear helical complex was identified by Fuhrhop and his co-workers in 1976.[23]

In the last decade, several laboratories have prepared and investigated double-stranded helical complexes, the systems containing either pyrrolic ligands [24,25] and derivatives [26,27] (with Zn^{2+}, Ag^{+}, Cu^{+}) or oligomers of 2,2'-bipyridine.[28,29] "Helicates" [28] may consist of up to five copper(I) centers and these self-assembling systems are reminiscent of the DNA double helix.

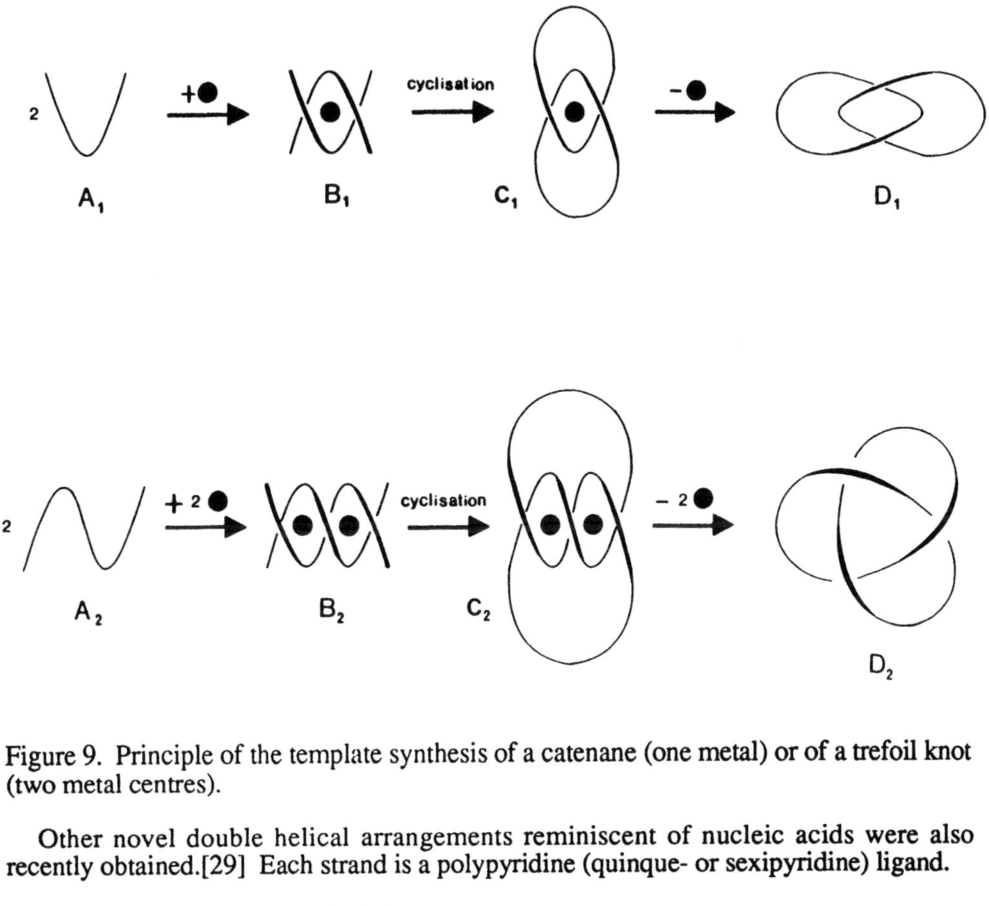

Figure 9. Principle of the template synthesis of a catenane (one metal) or of a trefoil knot (two metal centres).

Other novel double helical arrangements reminiscent of nucleic acids were also recently obtained.[29] Each strand is a polypyridine (quinque- or sexipyridine) ligand.

A Synthetic Molecular Trefoil Knot.

Following the strategy illustrated in Figure 9, we will now describe the synthesis of a molecular trefoil knot.[30] The project was started in 1984 and it took several years before the first positive results were obtained.

After many attempts with various linkers, we found that 1,10-phenanthroline nuclei connected *via* their 2-positions by a $-(CH_2)_4-$ linking unit will indeed form a double helix when complexed to two copper(I) centers. Recent work using $-(CH_2)_2-$ and $-(CH_2)_6-$ linkers has also been carried out. The hexamethylene fragment seems presently to be among the most promissing connectors. In addition, by introducing appropriate functions at the 9-positions, the strategy of Figure 9 could be followed to achieve the synthesis of a molecular knot of the D_2 type. The precursors used and the reactions carried out are represented in Figure 10.

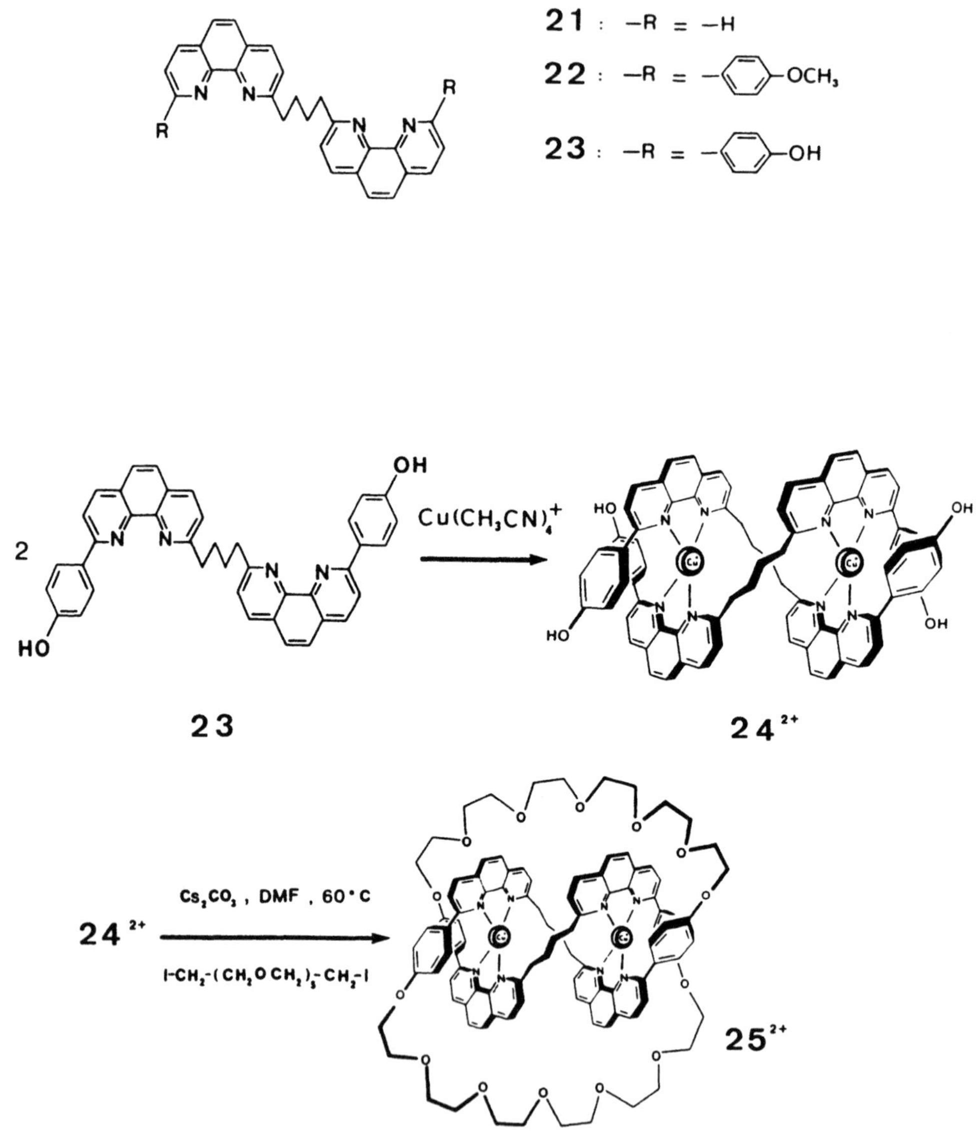

Figure 10. Precursors and reaction scheme leading to the di-copper(I) trefoil knot 25^{2+}.

Compound **21** was prepared in 86% yield from Li-$(CH_2)_4$-Li [31] and 1,10-phenanthroline. Addition of p-lithio anisole to **21**, followed by water hydrolysis and MnO_2 treatment, gave **22** in 83% yield. Demethylation of **22** to **23** was carried out quantitatively under drastic conditions : HCl-pyridine at 210°C.[32] The double helix **24**$^{2+}$ was prepared by addition of a stoichiometric amount of $Cu(CH_3CN)_4.BF_4$ to **23**.

^{1}H NMR studies showed that the double helicoïdal complex is formed together with an important proportion of other copper(I) complexes. The complex mixture containing **24**$^{2+}$ was reacted under high dilution conditions with two equivalents of the di-iodo derivative of hexaethyleneglycol $ICH_2(CH_2OCH_2)_5CH_2I$ in the presence of Cs_2CO_3 in DMF at 60°C. A small amount (3% yield) of the knotted system **25**$^{2+}$ containing two copper centres could be isolated and purified.

This bis-copper complex **25**$^{2+}$.$(BF_4^-)_2$ is a dark red crystalline solid (needles). The high resolution mass spectrum (FAB) of **25**$^{2+}$.$(BF_4^-)_2$ shows a molecular peak at 1816,74 (calculated molecular weight for **25**$^{2+}$ = 1817.10). The ^{1}H NMR spectrum of **25**$^{2+}$ is in complete agreement with its structure. Since **25**$^{2+}$ contains a double helix on two copper(I) atoms, it is chiral. This was demonstrated in the presence of Pirkle's reagent.[33] In order to prove the knotted topology of **25**$^{2+}$, we had to consider the various possible compounds obtained in the cyclisation reaction. From a tetrafunctional double helicoïdal precursor, several 2+2 connections are possible, with the most probable ones being indicated in Figure 11.

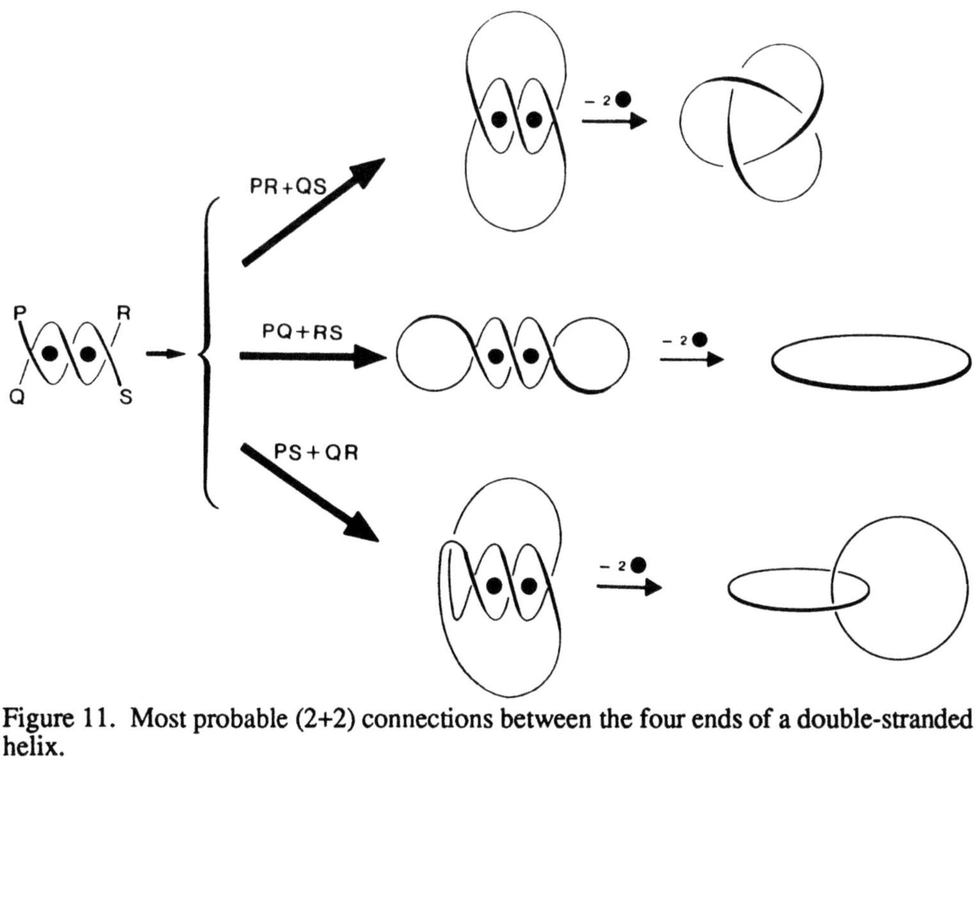

Figure 11. Most probable (2+2) connections between the four ends of a double-stranded helix.

If the three di-metal complexes of Figure 11 display classical chirality because of the presence of a double helix as core of the system, this is no longer true for the products obtained after demetallation. The trefoil knot is the only chiral demetallation product, its rotation being determined by the right- or left-handed nature of its double helix precursor, as shown in Figure 12.

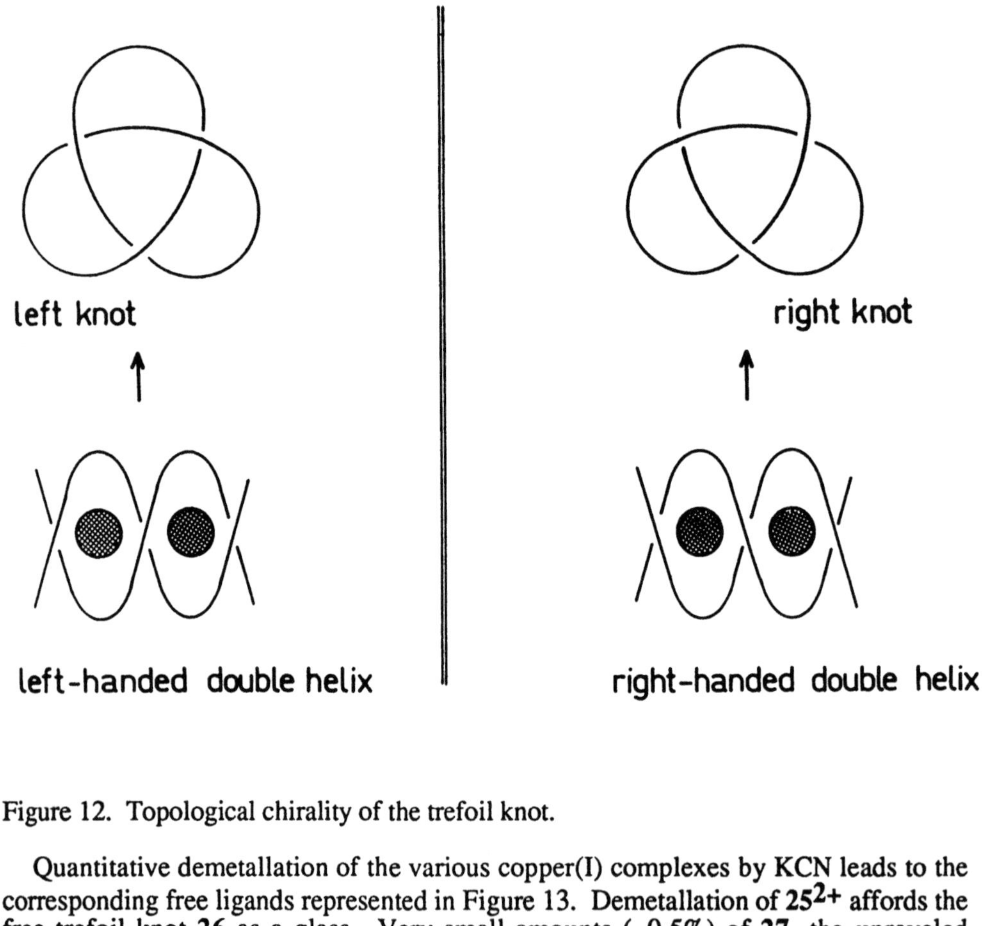

Figure 12. Topological chirality of the trefoil knot.

Quantitative demetallation of the various copper(I) complexes by KCN leads to the corresponding free ligands represented in Figure 13. Demetallation of **25**$^{2+}$ affords the free trefoil knot **26** as a glass. Very small amounts (~0.5%) of **27**, the unraveled topological isomer of **26** could also be isolated.

A comparison between the two isomers **26** and **27** brought strong evidence for the knotted topology of **26**. Both compounds have identical mass spectra (FAB) : molecular peaks appear at the same value. The general pattern and the various fragmentations of both compounds are strikingly similar. By contrast, the ^{1}H NMR of **26** and **27** are dramatically different. The broad spectrum of **26** reflects the raveled nature of its organic skeleton undergoing slow conformational changes.

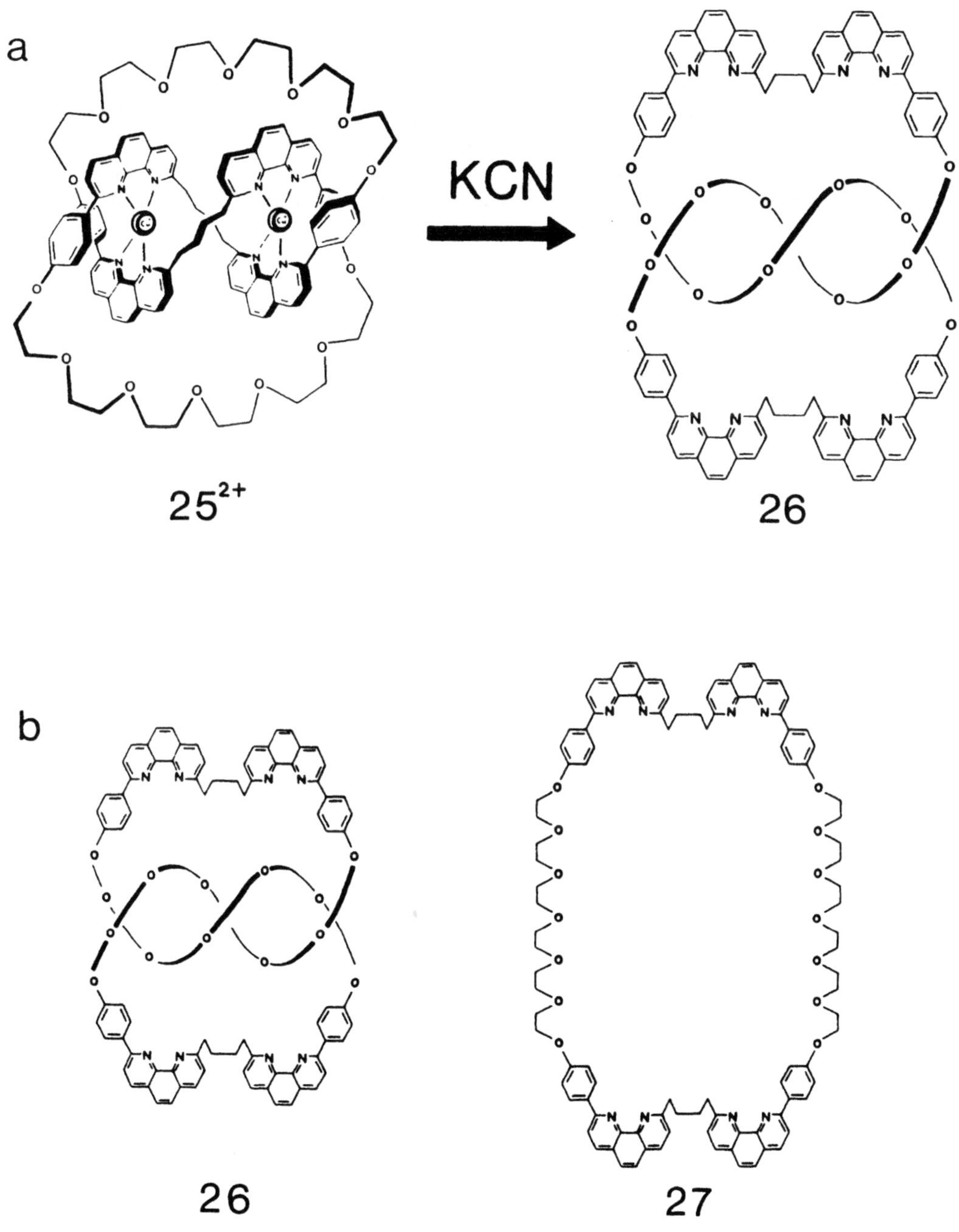

Figure 13. a) Obtention of the free trefoil knot after demetallation. b) **26** and **27** are true topological stereoisomers.

Addition of $K^+PF_6^-$ to **26** led to a drastic change in its ^1H NMR spectrum, evidence for an interaction between K^+ and the polyether oxygen atoms of **26**. In the sharp well resolved ^1H-NMR spectrum of **26** resulting from K^+ complexation, clear unambiguous and large splittings of peaks were observed upon addition of Pirkle's chiral reagent. This was a convincing demonstration that the molecular system **26** is chiral, in agreement with the topological chirality of a trefoil knot. A recent X-ray diffraction study fully confirmed those results.[34] The molecular structure of $\mathbf{25^{2+}}$ is represented in Figure 14.

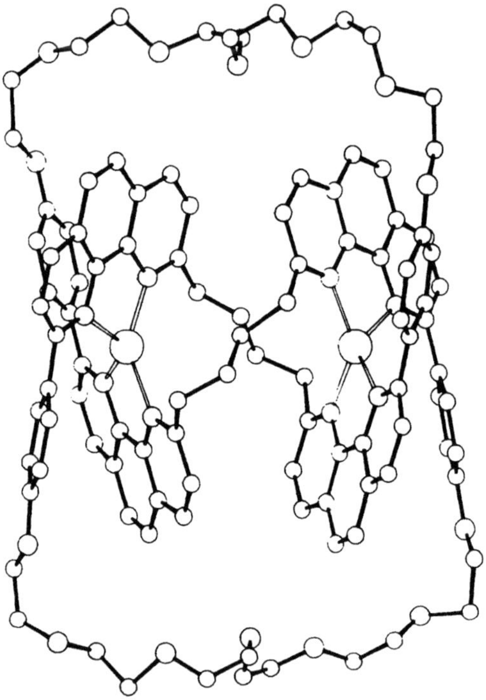

Figure 14. X-ray structure of the dicopper trefoil knot $\mathbf{25^{2+}}$.[34]

Although several decades have ellapsed between the first mention of chemical knots in the literature [10] and their actual realisation [30], much remains to be done. Presently, the synthetic procedure is far from being preparative : a skilful synthetic chemist would require several months to prepare 100 mg. of $\mathbf{25^{2+}}$ or **26**. Clearly, if the physical properties of knotted molecules are to be studied and compared to those of their topological stereoisomers, the efficiency of the synthetic method will have to be greatly improved. The key step to the success of such a preparative route is the formation of the double helix. This reaction is very dependent on the nature of the chelate used and on the geometrical properties of the linker connecting the chelating fragments. As mentioned above, the $-(CH_2)_2-$ and $-(CH_2)_6-$ linkers have already been used, the latter leading to significantly improved yields of trefoil knot.[35]

3. Conclusion and Prospect

At first, trying to tie molecular threads into knots may seem to be a bizare scientific activity. However, such a pursuit is not without fundamental importance in its ability to contribute to many questions either in chemistry or in other areas of human endeavour. It may even also find various long term applications.

(1) There is a continuous challenge to construct artificial molecular systems reminiscent of those that Nature has built in such a magisterial fashion, though she has achieved this within an exceedingly long period of time. Knotted DNA [5] has important and precise biological functions that no synthetic molecular system will be able to replace. However, when considered as a topological object, the first synthetic knot discussed in the present paper may display some analogies with knotted DNA.

(2) In parallel, the challenge continues for the chemist to prepare compounds whose properties are similar to objects belonging to other fields of human science or culture. Platonic solids have been a target in synthetic chemistry for decades. Among them, the dodecahedron [7] is certainly one of the most beautiful examples. Molecular-level construction of a tetrahedron [36] or a cube [37] also represent remarkable accomplishements. Buckminsterfullerene C_{60} and all its derived structures represent an explosive field of research in which aesthetics and applications are intricately linked.[38]

Topological novelty has also attracted synthetic efforts from various organic chemists. Arborols or Starburst dendrimers [39-41] are especially appealing objects. They are new multibranched chemical species displaying fascinating properties at a basic level, but also having important potentialities from a practical point of view.[41]

Interlocked rings (catenanes) continue to occupy a privileged position in chemical topology. They still exert a real fascination on synthetic chemists as witnessed by their recent development [22,42] despite the long standing interest of numerous laboratories in these compounds.[9,10]

Knotted molecules are certainly among the most fascinating topologically novel entities of chemistry. They belong also to the realm of mathematics where highly elaborated modern theories have recently been developed for their description.[43]

(3) The concept of using polynuclear double helices as precursors to more complex interlocked and knotted systems can be generalized.[44] The principle of the strategy (today still an interesting intellectual dream) is explained schematically in Figure 15. It is easy to see that depending on the number of metal centres used, two families of compounds are formed. If the number of metal centres is **even**, the number of crossing points in the molecular graph after cyclisation is **odd**, and a single closed knotted curve is formed (trefoil, pentafoil, heptafoil, etc...). If the double helix is built around an **odd** number of templating metals, there will be an **even** number of crossing points which will result in multiply interlocked [2]-catenanes.

(4) Coordinating knots can be regarded as novel ligands, leading to new and interesting multimetallic systems. As already mentioned, a trefoil, pentafoil, heptafoil... knot provides an ideal coordinating system for a set of two, four, six... metals respectively. The metal centres, could be either aligned or disposed in another type of arrangement, depending on the linkers between the 4,8,12,... chelates forming the coordinating knot.

Another particularity of knotted molecules is that the double-helical multimetallic complex that has been used as a templating core will be more or less frozen. In other words, the stability of the double helix will be insured by tying the molecular string used to make the knot in such a way as to make ligand exchange, helicity inversion, etc...

practically impossible.[5]

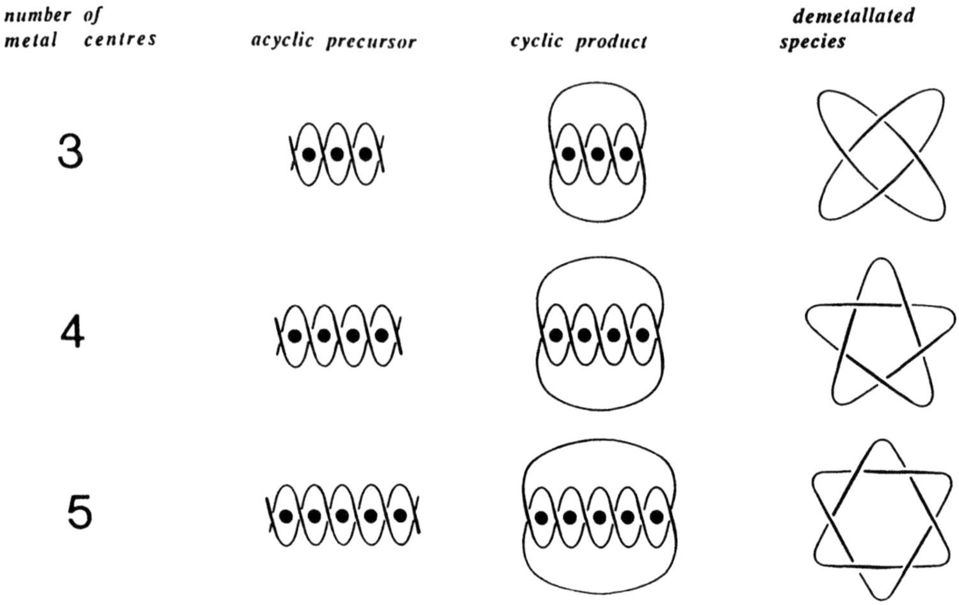

Figure 15. A possible strategy for synthesizing multiply interlocked rings or molecular knots from polymetallic double helices.

(5) The topological properties of the knots and, in particular, the non-invertible nature of most of them are very important in relation with chemical chirality. It is clear that the simplest knots containing an odd number of crossing points are not equivalent to their mirror images, regardless of the distorsion that the system may undergo. A trefoil knot is either left-handed or right-handed and this property will be maintained throughout the lifetime of the knot. This is also true for a pentafoil or a heptafoil knot, contrary to a figure eight knot (4 crossing points) which is topologically achiral.

Topological chirality [8,45] implies that there is no other way for the molecule to racemize than breaking a chemical bond. Knots represent beautiful examples of such systems. In addition to their fascinating topological properties related to chirality, they may also serve as interesting chiral systems for **enantioselective** interactions and processes (electron transfer ; interaction with other molecules, in particular, DNA ; incorporation of knots into the structure of molecular catalysts ; etc...). Consequently, and as a first step, it is particularly challenging to separate the left- and right-handed forms of our synthetic trefoil knot.

(6) Molecular dynamics and conformational properties of knotted rings should be totally different from those of their non-knotted topological stereoisomers. It will thus be of special interest to examine the dynamic properties of these systems. From now on, it is clear that the conformational changes occuring in knotted cycles will be much slower than for classical macrocyclic systems. In our 86-membered knotted ring **26**, fast conformational equilibrium is only reached at high temperature (~150°C). Ring size is expected to be a factor of utmost importance.

Conformational studies on knots may be relevant to polymer chemistry (interpenetrating network polymers[43]) as well as to biology (motions in proteins or DNA).

(7) It would be frustrating to confine our (or other's) contribution to the first member of the family, the trefoil knot. As mentioned above, it is easy (on paper!) to extend the synthetic concept to more complex knots, either using octahedra [21] (see Figure 8) or tetrahedra (see Figure 9). The "hook and ladder" approach proposed by Walba [20] is also of special interest since it could allow to prepare knots containing an **even number** of crossing points (like the figure eight knot), contrary to our strategy based on multimetallic double-stranded helices. However, by connecting in an appropriate fashion two entwined transition metal complexes it should also be possible to obtain a figure eight knot. Both routes indicated in Figure 16 are topologically-speaking very similar to each other, although in practice they may be very different in their realisation.

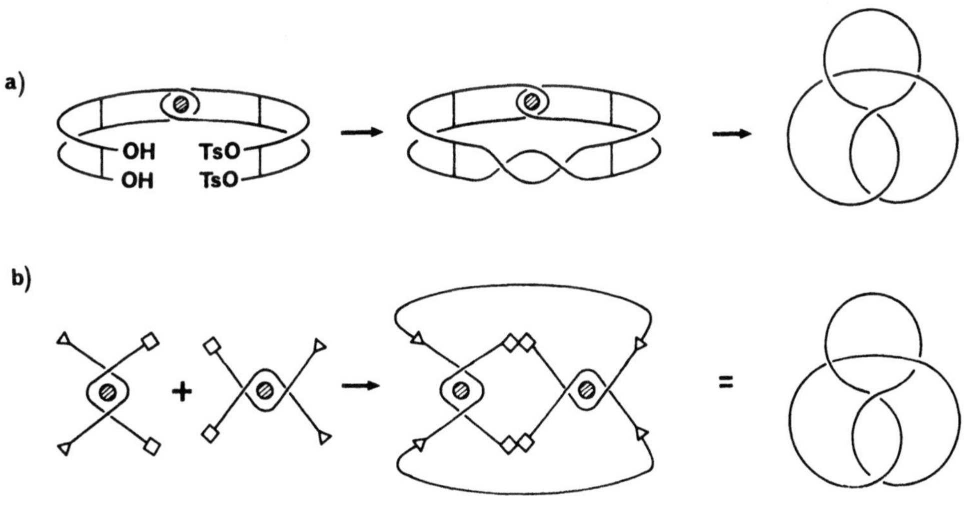

Figure 16. a) The "hook and ladder" approach to a figure eight knot, as proposed by Walba [20]. b) The same knot is obtained by connecting two bis-chelate complexes in an orthogonal fashion. The squares represent chemical functions to be interconnected ; the triangles must also be linked in between themselves.

4. Acknowledgement

We thank the CNRS for its constant financial support.

5. References

[1] Bain, G. (1973) Celtic Art. The Methods of Construction, Dover Publications Inc., New York.

[2] In Dürer, A. (1971) Master Printmaker, Department of Prints and Drawings, Museum of Fine Arts, Boston, Massachusetts.

[3] Klein, G. 'Notre Pain Quotidien', Exposition au Musée Alsacien, Strasbourg, 14.12.1975 to 29.02.1976.

[4] Locher, J.C. (1971) The world of M.C. Escher, H.N. Abram Inc. Publishers, New York.

[5] Wasserman, S.A. and Cozzarelli, N.R.(1986) Science 232, 951.

[6] Michel, F. and Weber, C. (1988) 'Noeuds et Entrelacs' Publication of the Institut de Mathématiques, University of Geneva, Switzerland.

[7] Woodward, R.B., Fukunaga, T., and Kelly, R.C. (1964) J. Am. Chem. Soc. 86, 3162. Eaton, P.E. (1979) Tetrahedron 35, 2189. First synthesis of Dodecahedrane : Ternansky, R.J., Balogh, D.W., and Paquette, L.A. (1982) J. Am. Chem. Soc. 104, 4503. Crystallographic study : Galluci, J.C., Doecke, C.W., and Paquette, L.A. (1986) J. Am. Chem. Soc. 108, 1343

[8] Walba, D.M. (1985) Tetrahedron 41, 3161.

[9] Schill, G. (1971) Catenanes, Rotaxanes and Knots, Academic Press, New York and London.

[10] Frisch, H.L., and Wasserman, E. (1961) J. Am. Chem. Soc. 83, 3789.

[11] Wasserman, E. (1960) J. Am. Chem. Soc. 82, 4433.

[12] Walba, D.M., Richard, R.M., and Haltiwanger, R.C. (1982) J. Am. Chem. Soc. 104, 3219.,

[13] Walba, D.M., Armstrong, J.D., Perry A.E., Richards, R.M., Homan, T.C., and Haltiwanger, R.C. (1986) Tetrahedron 42, 1883.

[14] Schill, G. and Tafelmair, L. (1971) Synthesis 10, 546. Schill, G., Keller, U., and Fritz, H.(1983) Chem. Ber. 116, 3675.

[15] Schill, G., Doerjer, G., Logemann, E., and Fritz, H. (1979) Chem. Ber. 112, 3603. Schill, G. and Boeckmann, J. (1974) Tetrahedron 30, 1945.

[16] Schill, G. and Lüttringhaus, A. (1964) Angew. Chem. 76, 567. Schill, G. (1967) Chem. Ber. 100, 2021.

[17] Schill, G. and Zollenkopf, H. (1969) Liebigs Ann. Chem. 721, 53.

[18] Schill, G. (1971) Catenanes, Rotaxanes and Knots, Academic Press, New York and London, pp. 115. Schill, G. and Zürcher, C. (1971) Naturwissenschaften 58, 40.

[19] Schill, G. and Zürcher, C. (1969) Angew. Chem. 81, 996. Schill, G. and Zürcher, C. (1977) Chem. Ber. 110, 2046.

[20] Walba, D.M. (1987) in R.B. King and D.H. Rouvray (eds.), Graph Theory and Topology in Chemistry, Elsevier.

[21] Sokolov, V.I.(1973) Russian Chemical Reviews 42, 452.

[22] Sauvage, J.-P. and Dietrich-Buchecker, C.O. (1987) Chem. Rev. 87, 795. Sauvage, J.-P. and Dietrich-Buchecker, C.O. (1990) Tetrahedron 46, 503.

[23] Fuhrhop, J.H., Struckmeier, G., and Thewalt, U. (1976) J. Am. Chem. Soc. 98, 278.
[24] Sheldrick, W.S. and Engel, J. (1980) J. Chem. Soc., Chem. Commun. 5.
[25] Bonfiglio, J.V., Bonnett, R., Buckley, D.G., Hamzetash, D., Hursthouse, M.B., Malick, K.M.A., McDonald, A.F., and Trotter, J. (1983) Tetrahedron 39, 1865.
[26] Van Stein, G.C., Van der Poel, H., and Van Koten, G. (1980) J. Chem. Soc., Chem.Commun. 1016. Van Stein, G.C., Van Koten, G., Vrieze, K., Brevard, C., and Spek, A.L. (1984) J. Am. Chem. Soc. 106, 4486.
[27] Van Stein, G.C., Van Koten, G., Passenier, H., Steinebach, O., and Vrieze, K. (1984) Inorg. Chim. Acta 89, 79.
[28] Lehn, J.-M., Rigault, A., Siegel, J., Harrowfield, J., Chevrier, B., and Moras, D. (1987) Proc. Natl. Acad. Sci. U.S.A. 84, 2565. Lehn, J.-M. and Rigault, A. (1988) Angew. Chem. Int. Ed. Engl. 27, 1095.
[29] Constable, E.C., Drew, M.G.B., and Ward, M.D. (1987) J. Chem. Soc., Chem. Commun. 1600. Barley, M., Constable, E.C., Corr, S.A., McQueen, R.C.S., Nutkins, J.C., Ward, M.D., and Drew, M.G.B. (1988) J. Chem. Soc. Dalton Trans. 2655. Constable, E.C. and Ward, M.D. (1990) J. Am. Chem. Soc. 112, 1256.
[30] Dietrich-Buchecker, C.O. and Sauvage, J.-P. (1989) Angew. Chem., Int. Ed. Engl. 28, 189.
[31] West, R. and Rochow, E.G. (1953) J. Org. Chem. 18, 1739. McDermott, J.X., White, J.F., and Whitesides, G.M. (1976) J. Am. Chem. Soc. 98, 6529.
[32] Dietrich-Buchecker, C.O. and Sauvage, J.-P. (1983) Tetrahedron Lett. 24, 5091. Curphey, T.J., Hoffman, E.J., and McDonald, C. (1967) Chem. and Ind. (London) 1138.
[33] Pirkle, W. and Hoekstra, M.S. (1976) J. Am. Chem. Soc. 98, 1832.
[34] Dietrich-Buchecker, C.O., Guilhem, J., Pascard, C., and Sauvage, J.-P. (1990) Angew. Chem., Int. Ed. Engl. 29, 1154.
[35] Dietrich-Buchecker, C.O., Nierengarten, J.-F., and Sauvage, J.-P. to be published.
[36] Maier, G., Pfriem, S., Schäfer, U., and Matusch, R. (1978) Angew. Chem., Int. Ed. Engl. 17, 520.
[37] Eaton, P.E. and Cole, T.W. (1964) J. Am. Chem. Soc. 86, 3157.
[38] The number of publications appearing each month on C_{60} and its homologues is such that it is impossible to give any list. Some particularly significant results obtained by various groups have been highlighted : Stoddart, J.F. (1991) Angew. Chem. Int. Ed. Engl. 30, 70.
[39] Buhlein, E., Wehner, W., and Vögtle, F. (1978) Synthesis 155.
[40] Newkome, G.R., Yao, Z.-Q., Baker, G.R., and Gupta, V.K. (1985) J. Org. Chem. 50, 2003.
[41] Tomalia, D.A., Naylor, A.M., and Goddard III, W.A. (1990) Angew. Chem., Int. Ed. Engl. 29, 138 and references.
[42] Ashton, P.R., Goodnow, T.T., Kaiffer, A.E., Reddington, M.V., Slawin, A.M.Z., Spencer, N., Stoddart , J.F., Vicent , C., and Williams D.J. (1989) Angew. Chem., Int. Ed. Engl. 28, 1396.
[43] Fox, R.H. (1962) Topology of 3-manifolds and Related Topics, Fort Ed.
[44] Sauvage, J.-P. (1990) Acc. Chem. Res. 23, 319.
[45] Mitchell, D.K. and Sauvage, J.-P. (1988) Angew. Chem., Int. Ed. Engl. 27, 930.
[46] Mark, J.E. (1985) Acc. Chem. Res. 18, 202 and references.

PROTON TRANSFER PHOTOREACTIONS IN SUPRAMOLECULAR SYSTEMS

Michael G. KUZMIN
Department of Chemistry,
Moscow University,
Moscow, SU-117234 Russia

ABSTRACT. Kinetics of proton transfer photoreactions in simple model systems is analyzed from the point of view of reaction kinetics in microphases. Protolytic photodissociation of some hydroxyaromatic compounds ArOH (1- and 2-naphthol, chlorosubstituted naphthols) was studied in micellar solutions and phospholipid vesicles by fluorescence spectra and kinetics. Experimental results give evidence of at least two localization sites of naphthols in the microphase of these systems. In lipid bilayer membranes of vesicles there are two comparable fractions of ArOH molecules, one of which undergo photodissociation, but another do not dissociate. In micelles only minor fraction (few per cent) of ArOH molecules do not take part in excited-state proton transfer reaction. These phenomena reflect heterogeneous structure and dynamic properties of lipid bilayer membranes and micelles. A correlation between proton transfer rate constants and equilibrium constants in microphases similar to that in homogeneous solutions is observed. Microphase approach give a possibility to discuss reactions in dynamical organized molecular systems in terms of classical chemical kinetics.

1. Introduction

Proton transfer reactions are one of principal chemical processes and they are of primary importance in chemistry and biology. Protolytic reactions are the main type of acid-base (heterolytic) reactions. Mitchell's chemiosmotic hypothesis especially stimulated the investigation of proton transfer reactions in biochemistry.

Kinetics and mechanisms of protolytic reactions in homogeneous media are known in detail. Knowledge of general

V. Balzani and L. De Cola (eds.), Supramolecular Chemistry, 279–294.

behavior of proton transfer reactions in heterogeneous and organized molecular systems (in membranes, molecular films, molecular aggregates, at interfaces, etc.) is needed to understand the mechanisms of natural proton transport and to design model and artificial systems for molecular technologies, solar energy conversion, etc. [1]. Investigations of ground and excited state proton transfer reactions [1-4] supplies quite important and unique information about the local structure and dynamics of supramolecular systems.

Promising approach to charge transport phenomena in supramolecular systems is physicochemical and kinetical consideration of these reaction in terms of local properties of the media and reactants (local permittivity, microviscosity, local concentrations, pH, pK_a, ΔG, etc.). Relationships between the local properties and the structure of the systems can be analyzed separately.

Here we shall analyze the mechanisms and kinetic regularities of proton transfer photoreactions in the simple model systems - in micelles and vesicles from the point of view of classical chemical kinetics. The energetics and dynamics of chemical processes in microphases, and energetics and dynamics of diffusion steps in microphases and between them are considered.

2. Protolytic photodissociation in micelles and vesicles

We carried out systematic investigation of various protolytic photoreactions (both pseudounimolecular and bimolecular) in anionic, cationic and nonionic micelles and in lecithin vesicles.

To calculate rate constants of protolytic dissociation of hydroxyaromatic compounds in the singlet excited state in the micellar phase we studied the dependence of fluorescence quantum yields and fluorescence kinetics of undissociated (φ) and dissociated (φ') forms of hydroxyaromatic compounds in the micellar solutions on the concentration of hydroxonium ions in the solution. Similarly to aqueous solutions proton transfer photoreactions in micellar solutions can be described by simple kinetic scheme:

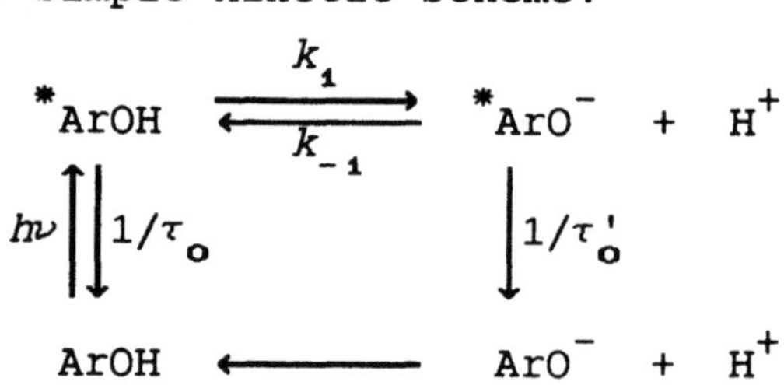

(1)

This scheme yields the equation for the ratio of relative

fluorescence quantum yields:

$$\varphi\varphi_o'/\varphi'\varphi_o = 1/k_1\tau_o + k_{-1}\tau_o'[H^+]/k_1\tau_o \qquad (2)$$

In contrast to the aqueous solutions in the case of micellar solutions this equation gives rather apparent rate constants of protolytic dissociation and equilibrium constant $K_a^* = k_1/k_{-1}$ in the micellar phase than true rate constants of protolytic reactions. Equilibrium constant K_a^* and protonation rate constant k_{-1}, calculated using pH of the solution (i.e. concentration of H^+ in aqueous phase), is some effective value including the distribution coefficient of H^+ between micellar and aqueous phases [4]. These apparent rate and equilibrium constants (some of them are given in the Table 1) can sometimes include also rate constants of the

TABLE 1. Rate constants of some protolytic photoreactions in aqueous (aq) and micellar (m) phases

Substance (Surfactant)	pK_a^* aq	pK_a^* m	k_1 $10^7\ s^{-1}$ aq	k_1 $10^7\ s^{-1}$ m	k_{-1} $10^9\ M^{-1}s^{-1}$ aq	k_{-1} $10^9\ M^{-1}s^{-1}$ m
CTAB						
2-Naphthol	2.81	1.01	7.3	10.5	47	1.1
2-Naphthol-8-sulfonate	1.42	0.71	300	29	79	1.5
2-Naphthol-6-sulfonate	1.66	1.19	77	7.1	35	1.1
1,7-Dihydroxynaphthalene	3.13	1.9	3.2	4.2	43	3.6
2,7-Dihydroxynaphthalene	2.39	1.33	12.5	8.1	31	2.0
2-Hydroxyanthracene	2.3	1.1	14	33	28	5.6
2-Naphthol-3,6-disulfonate	0.88	1.1	340	37	26	4.7
8-Hydroxypyrene--1,3,6-trisulfonate	0.32	0.92	890	48	19	4.0
1-Chloro-2-naphthol	0.90	1.07	76	21	6	3.0
Brij-56						
2-Hydroxyanthracene	2.3	2.45	14	0.75	28	2.0
1-Chloro-2-naphthol	0.9	2.3	76	(2.5)	6	8
1,4-Dichloro-2-naphthol	0.19	1.0	120	40	1.9	(4)
Decylsulfate						
1-Chloro-2-naphthol	0.9	2.6	76	2.2	6	8.4
1,4-Dichloro-2-naphthol	0.19	2.5	120	12	1.9	37
Dodecylsulfate						
1-Chloro-2-naphthol	0.9	2.5	76	1.0	6	3.5
1,4-Dichloro-2-naphthol	0.19	2.9	120	5.8	1.9	40

reaction product exit from the micellar phase to the volume one [4].

Deuterium isotope effect observed for rate constants k_1 (Table 2) has value similar to homogeneous aqueous solution [5, 6, 7]. This fact gives evidence that the photodissociation rate is limited by the proton transfer step (but not by the diffusion of excited molecules to micellar interface) and the reaction take place in micellar phase [8, 9]. Fluorescence lifetimes of 1-naphthol and 4-chloro-1-naphthol anions ArO^- are sensitive to the nature of the solvent [7, 10].

TABLE 2. Rate constants of protolytic photoreactions of the the hydroxyaromatic compounds in microgeterogeneous systems in H_2O (k_1 и k_{-1}) and in D_2O (k_1^D)

Substance	System	k_1 10^8 s^{-1}	$k_{-1}/10^8$ $M^{-1}s^{-1}$	k_1^D 10^8 s^{-1}	k_1/k_1^D
	Micelles				
2-Naphthol	DTAB	0.91	-	0.4	2.3
2-Naphthol	Brij 35	0.11	-	-	-
1-Naphthol	Brij 35	3.6	-	-	-
1-Naphthol	SDS	3.5	-	0.57	6.1
1-Naphthol	Triton X-305	4.0	-	0.46	8.8
4-Chloro-1-naphthol	SDS	4.4	-	0.76	5.8
1-Naphthol	Tween 60*	2.0	-	0.37	5.4
	Vesicles				
1-Naphthol	EL	5.5	-	1.8	3.1
1-Naphthol	DPPC(40°C)	4.0	-	-	-
4-Chloro-1-naphthol	EL	8.1	-	3.9	2.1
4-Chloro-1-naphthol	DPPC(40°C)	6.1	-	-	-
1-Chloro-2-naphthol	EL	2.1	16.7	-	-
1,4-Dichloro-2-naphthol	EL	3.5	4.4	-	-

*Microemulsion: Tween 60 (48.2%), *n*-pentanol (25.2%) and hexadecane (26.6%).

The essential difference in fluorescence decay times (τ'_0) of ArO^- in nonionic micelles and in water, observed for 1-naphthol and 4-chloro-1-naphthol (Table 3), gives another evidence for the formation of the photoreaction product inside the micelles and confirms that the proton transfer reaction proceeds in micellar phase. Shortwavelength shift of the emission band of $*ArO^-$ in Triton X-305 solution comparatively to aqueous solution also give evidence of the product formation in the micellar phase.

TABLE 3. Fluorescence of 1-naphthol and 4-chloro-1-naphthol in aqueous solutions, micelles and vesicles.

Medium	λ_{A^-}, nm	I_{A^-}/I_{AH}	τ'_o, ns
	1-Naphthol		
H_2O, pH≈6	490	»10	7.5
SDS, pH≈6	470	1.13	7.7
Triton X-305, pH≈6	465	4.4	15.4
Triton X-305, pH≈13	490		8.0
EL Vesicles, pH≈6	460	1.52	12.3
	4-Chloro-1-naphthol		
H_2O, pH≈6	500	»10	9.4
SDS, pH≈6	480	1.42	9.4
Triton X-305, pH≈6	470	4.6	14.0
EL Vesicles, pH≈6	475	2.0	11.8

No corrections on spectral sensitivity of the registration system are made in the values I_{A^-}/I_{AH}.

For protolytic photodissociation of hydroxyaromatic compounds we may expect that in anionic micelles the excited anions of hydroxyaromatic compounds having the same sign of charge will leave the micelles, and in the cationic micelles - hydrogen ions will leave micelles. Exit rate constants of excited anions of hydroxyaromatic compound from the micelles were determined using the non-solubilized fluorescence quenchers. Simultaneously, we have proved that it is in the micellar phase that protolytic dissociation does proceed and not as a result of the preliminary exit of excited molecules of hydroxyaromatic compounds from the micellar phase to an aqueous one.

Nitrate anion for anionic micelles and cesium cation for cationic micelles turned out to be convenient specific quenchers for this purpose. If we compare fluorescence quenching constants of the reaction product in the micellar and aqueous solutions, we can determine a fraction of the reaction product of the molecules coming out to the aqueous phase. And now we can easily find the rate constants k_{out} of their exit from the micelles:

$$\varkappa_Q^w/\varkappa_Q^m = 1 + 1/k_{out}\tau'_o \qquad (3)$$

where $\varkappa_Q^w$ and $\varkappa_Q^m$ are Stern-Volmer quenching constants in aqueous and micellar solutions respectively.

As can be seen from the Table 4, about 20 to 90 per cent of excited anions formed in the investigated anionic micelles come out from them. In the neutral micelles the fraction of anions which are coming out is less than 10-15 per cent, and in the cationic - less than 2-3 per cent.

TABLE 4. Stern-Volmer fluorescence quenching constants ($\varkappa$) by specific quencher NO_3^- and exit rate constants ($k_{out}/10^7$ s^{-1}) of $*ArO^-$

Surfactant	1-Chloro- -2-naphthol $\varkappa$	k_{out}	1,4-Dichloro- -2-naphthol $\varkappa$	k_{out}	2-Hydroxy- anthracene $\varkappa$	k_{out}
H_2O	10		5.4		0.7	
Decylsulfate	2.8	2.8	1.2	2.6		
Dodecylsulfate	7.0	7.5	2.4	6.3		
Cetylsulfate			5.0	16.5		
Brij-56	0.8		0.1		0.1	

3. Kinetics of Proton Transfer Photoreactions in Micelles

Fluorescence kinetics of ArOH and ArO^- in micellar solutions of surfactants with various charges was studied by conventional single photon counting method using numerical deconvolution of decay curves. Concentrations of anionic surfactant - SDS (0.1 M), cationic surfactant - DTAB (0.05 M) and nonionic surfactant - Triton X-305 and Brij 35 (0.01 M) were chosen by such a way, that more than 99% of hydroxycompounds were solubilized in micellar phase and protolytic photoreaction take place nearly completely in this phase [11-14].

In neutral aqueous solutions protonation reaction of $*ArO^-$ can be neglected, since its pseudounimolecular rate constant (calculated as a product of the diffusion rate constant 10^{10} $M^{-1}s^{-1}$ and the proton concentration 10^{-7}-10^{-6} M) is much less than decay rate constant of $*ArO^-$ ($1/\tau'_0 \approx 10^8$ s^{-1}). In such case fluorescence kinetics of ArOH (F) and ArO^- (F') at excitation by δ-pulse follows to the monoexponential and biexponential laws respectively:

$$F(t) = F_0 \exp(-t/\tau) \tag{4}$$

$$F'(t) = F'_0[\exp(-t/\tau') - \exp(-t/\tau)], \tag{5}$$

where $1/\tau = 1/\tau_0 + k_1$. The possibility to describe the fluorescence kinetics of ArOH and ArO^- in water by equations

(5) and (6) has been demonstrated in many works [6, 15, 16].
Fluorescence decay of ArOH in some micellar systems in contrast to aqueous solutions was found to be slightly biexponential (Table 5):

$$F(t) = A_1\exp(-t/\tau_1) + A_2\exp(-t/\tau_2) \qquad (6)$$

Fluorescence kinetics of ArO^- as usually is a difference of two exponents and rise time of ArO^- fluorescence (τ) in most cases coincides with the lifetime of the faster component (τ_1) of *ArOH decay (Table 5).

Second exponential term of *ArOH decay has intensity about few per cent of the main component and greater lifetime (which is different also from the observed lifetime τ'_0 of $*ArO^-$). This second term cannot be attributed to the presence of some impurities, since the fluorescence decay curves of the same compounds in aprotic solvents are precisely monoexponential. Registration wavelength was carefully selected by monochromator to exclude the fluorescence of the anion. Neither first nor second term cannot be attributed also to the dissociation of unsolubilized ArOH since the proton transfer rate constants in water are much greater [4-6, 15-17].

4. Kinetics of proton transfer photoreactions in vesicles.

Kinetics of protolytic photoreactions in the membranes of vesicles was studied for naphthols in the egg lecithin vesicles. In aqueous solution 1-naphthol and its chloroderivatives are well known to dissociate in the excited state with the rate constant greater than $10^{10}s^{-1}$. Only an anion emission band is observed in the fluorescence spectrum.

In vesicular solution at pH = 7.1 two emission bands are present in the fluorescence spectra: undissociated form and naphtholate anion. At the conditions used in our experiments almost all ArOH molecules are localized in the lipid membrane of vesicles. Described earlier microphase model of microheterogeneous systems can also be applied to vesicles. In such case the term "lipid membrane" means not only lipid bilayer itself but also head group region and adjacent water layers.

The localization of ArOH in lipid membrane of vesicles was proved by fluorescence quenching by hydrophylic ions. The ratio of observed Stern-Volmer quenching constant in vesicle solution to the same constant in aqueous solution gives evaluation of lower limit of the fraction of the investigated compound located in the aqueous phase [11]. This method can be applied most easy to the compounds which

TABLE 5. Fluorescence decay parameters of hydroxyaromatic compounds in microheterogeneous systems.

Substance	System	Undissociated form, ArOH				Dissociated form, ArO^-		
		τ_1, ns	τ_2, ns	$(1-\alpha)/\alpha$	χ^2	τ_1, ns	τ_2, ns	χ^2
	Micelles							
2-Naphthol	DTAB, H_2O	3.75	9.9	0.02	1.07	3.51	9.09	1.33
2-Naphthol	DTAB, D_2O	4.24	15.1	0.007	1.32	4.08	10.6	1.36
1-Naphthol	SDS, H_2O	1.57	6.0	0.01	1.20	1.30	7.7	1.61
1-Naphthol	SDS, D_2O	4.0	17.3	0.007	1.02	2.0	23.3	1.20
1-Naphthol	Triton X-305, H_2O	2.0	8.7	0.02	1.28	1.83	15.4	2.30
1-Naphthol	Triton X-305, D_2O	4.05	10.4	0.06	1.24	4.0	20.9	1.44
4-Chloro-1-naphthol	SDS, H_2O	1.24	7.2	0.02	1.32	1.07	9.43	1.16
4-Chloro-1-naphthol	SDS, D_2O	2.24	9.9	0.02	1.30	1.11	15.1	1.46
	Vesicles							
1-Naphthol	EL, H_2O	1.35	4.44	0.66	1.17	1.15	12.3	1.44
1-Naphthol	EL, D_2O	2.62	5.56	0.74	1.12	2.42	16.1	1.45
1-Naphthol	DPPC, H_2O(40°C)	1.56	4.22	0.65	1.18	0.56	13.1	1.63
4-Chloro-1-naphthol	EL, H_2O	0.85	3.29	0.75	1.07	0.7	11.8	1.17
4-Chloro-1-naphthol	EL, D_2O	1.31	3.79	0.61	1.44	1.11	13.2	1.44
4-Chloro-1-naphthol	DPPC, H_2O(40°C)	1.33	3.60	0.66	1.15	1.16	12.5	1.37
1,4-Dichloro-2-naphthol	EL, H_2O	1.45	4.12	0.54	1.12	1.21	9.15	1.75

Fluorescence decay curves for ArOH and ArO^- are calculated in the form $I(t) = I_0[\alpha\exp(-t/\tau_1)+(1-\alpha)\exp(-t/\tau_2)]$ by deconvolution with apparatus function; $\chi^2 = \Sigma w_i(I_i^{calc} - I_i^{exp})^2/(n-p)$; n is a number of points; p is a number of parameters; $w_i = 1/I_i^{exp}$.

just slightly dissociate in the excited state like 1-hydroxypyrene and 2-naphthol. Apparent fluorescence quenching constants of these compounds by NO_3^- and IO_3^- ions in EL vesicles were found to be smaller than 10^7 $M^{-1}s^{-1}$. These values are approximately two orders of magnitude smaller than in aqueous solution. This fact gives evidence that more than 99 % of these hydroxyaromatic compound are located at these conditions in lipid membrane. Similar distribution between vesicular membranes and aqueous phase can be expected for 1-naphthol and chlorosubstituted naphthols.

Proceeding of the protolytic photoreaction in lipid bilayer was proved also by the fluorescence spectra and the lifetimes of the hydroxyaromatic anions formed in the reaction, which are different from that in the aqueous solution. For example, fluorescence lifetimes of anion of 1-naphthol are 7.5 ns in water and 12.3 ns in the EL vesicles. These data indicate on the formation of $*ArO^-$ directly in the photoreaction inside the vesicular membrane. It is important to emphasize that fluorescence decay kinetics in the both cases is well monoexponential as at the beginning as after at least two orders of magnitude decay. In contrast to naphtholate anions formed in the photoreaction, $*ArO^-$ formed in the alkaline solution of vesicles (pH ≈ 13) are localized (both in the ground and excited singlet states) in aqueous phase and have the same fluorescence lifetime as in aqueous solution (Table 5). This means that migration of $*ArO^-$ formed in vesicles into aqueous phase is too slow comparatively to its lifetime.

Fluorescence decay of ArOH in vesicles can not be described by monoexponential dependence and follows to biexponential law with comparable intensities of both components: τ_1= 1.35 ns; τ_2= 4.44 ns; A_2/A_1= 0.66; χ^2= 1.17. The lifetime of *ArOH in the low polarity solvent (pentanol), where protolytic photodissociation do not take place, and the slow decay time (τ_2) in vesicles are close to each other, as well as the fast decay time (τ_1) and the rise time of $*ArO^-$ emission (Table 5). Fluorescence curve for ArO^- is described well by Eq.(5). Decay time of $*ArO^-$ emission (τ_2) is typical for fluorescence lifetimes (τ_0') of the respective anions. Treatment of the decay curve of *ArOH for 1-naphthol as a sum of three exponentials gives lower value of χ^2= 1.11 and the following values of the decay times: τ_1= 1.34 ns; τ_2= 4.31 ns; τ_3= 16.2 ns. But the amplitude of third exponential is very low (~1%). This result points on the validity of biexponential approximation of *ArOH decay curves.

Such behavior of hydroxyaromatic compounds in vesicles and

micelles can be explained by the existence of at least two types of localization sites in the lipid membrane or in the micelle, i.e. by kinetic unequality of ArOH located in this membrane. In principle one can expect some continuous distribution of ArOH molecules on protolytic reaction rate constants. But experimental data show that all hydroxyaromatic compound molecules localized in the membrane can be subdivided quite well on two separate fractions. The first fraction of these molecules can undergo excited state proton transfer reaction with rate constants considerably (some orders of magnitude) smaller than in aqueous solutions (Table 1). The second fraction do not dissociate, i.e. their dissociation rate constant smaller than 10^{6} s^{-1}. Observed photoprocesses can be described by the following kinetic scheme:

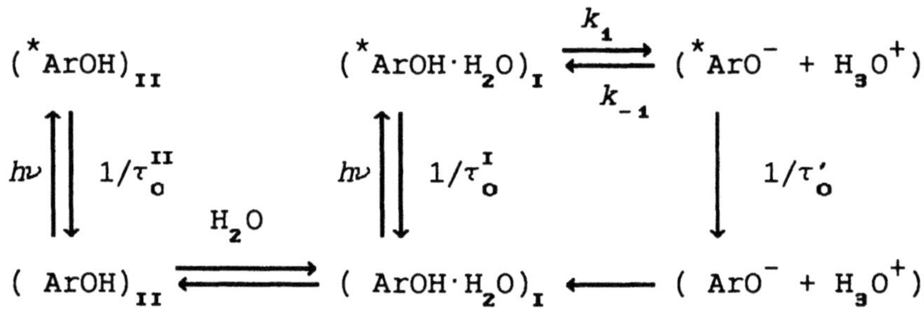

(7)

where τ_o^{I} and τ_o^{II} are lifetimes of the decay of two kinds of *ArOH molecules. According to scheme (7) the fluorescence kinetics of ArOH (F) and ArO^- (F') are described by the following equation:

$$F(t) = F_o[\alpha \exp(-t/\tau) + (1-\alpha) \exp(-t/\tau_o^{II})] \quad (8)$$

$$F'(t) = F_o'[\exp(-t/\tau_o') - \exp(-t/\tau)] \quad (9)$$

where $1/\tau = 1/\tau_o^{I} + k_1$; α is a fraction of hydroxyaromatic compound molecules capable to dissociate. The values of α determined from the ratio of pre-exponential factors in fluorescence decay curves $A_2/A_1 = (1-\alpha)/\alpha$, are in the range 0.57-0.65 for vesicular membranes and 0.94-0.99 for micelles. It was found that α weakly depends on the nature of the hydroxyaromatic compound studied and the fractions of naphthol molecules taking part in bimolecular proton transfer photoreaction with acetate anion and in protolytic photodissociation are the same [18].

A comparison of the rise time of naphtholate anion fluorescence in water and in vesicles shows that the rate constants of protolytic photodissociation in the membrane and in water have the same order of magnitude.

5. Bimolecular proton transfer photoreactions

Efficiency of proton transfer reactions in micelles and vesicles can be improved, if proton acceptors are used which are stronger bases than water. For example addition of acetate anion or hydroxylamine increases essentially the degree of photodissociation of 1-hydroxypyrene decreasing the quantum yield of its fluorescence and increasing the yield of anion fluorescence.

Rate constants were determined for a number of bimolecular photoreactions of hydroxy- and aminoaromatic compounds and heteroaromatic compounds with carboxylate anions and organic acids, respectively, in various micelles. Such bimolecular photoreactions were found to improve the efficiency of forward proton transfer reactions and to reduce reverse recombination of the primary products (up to 2 orders of magnitude) comparatively to aqueous solutions.

Kinetical unequality was observed in photoreaction of 2-naphthol with acetate anion in EL vesicles, similar to that in unimolecular photodissociation.

6. Dependences of the protolytic reactions rate constants on pK of the compounds and on the nature of the system

Now we shall analyze the regularities in rate constants of protolytic photodissociation of hydroxyaromatic compounds in the micellar phase and in the membranes of vesicles (Tables 1 and 5). For one and the same compounds in cationic micelles the acidity constants are usually greater than in the aqueous solution (pK_a^* decreases), and dissociation rate constants are the same or somewhat lower. In neutral micelles the acidity constants are lower and dissociation rate constants are lower 1.5-2.5 orders of magnitude. In anionic micelles the acidity constants are lower 1.5-2.5 orders, and dissociation constants are lower about 1.5 order of magnitude.

It should be mentioned also that solubilization in cationic micelles produces an equalization effect on the acidity constant and dissociation constants. For the compound studied the variation range of pK_a^* and dissociation constants in the micelles is markedly smaller than in the aqueous one.

Smaller values of protolytic dissociation equilibrium and rate constants in non-ionic micelles comparatively to aqueous solutions can be caused as by smaller polarity of the micellar phase as by lower activity of water in this phase. Effects of the charge of micelles on the equilibrium and rate constants are related to the influence of this charge on solubilization energies of the ions involved in the reaction.

In cationic micelles the factors of polarity and the electrostatic potential act in opposite directions and weaken the action of one another. In anionic micelles these factors act in the same direction and reduce the equilibrium constants and rates of photodissociation.

The dependence of rate constants of protolytic photodissociation and backward protonation reaction of excited anions on the pK_a^* value in aqueous solutions and micelles are presented on Figure 1. Protolytic dissociation for the compounds under discussion is an endergonic process and logarithms of their rate constants decrease linearly with the increase of pK_a^* both in aqueous, and in micellar solutions. Backward protonation reaction is limited by diffusion, and its rate constant does not depend on pK_a^*, but they are smaller about one order of magnitude in micellar phase relatively to aqueous phase. It is significant that for the same pK_a^* values (which refer to different compounds in aqueous and micellar phases) the values of the rate constants for both the direct and back reactions in micellar and aqueous phases differ to the same value - a little bit more than one order.

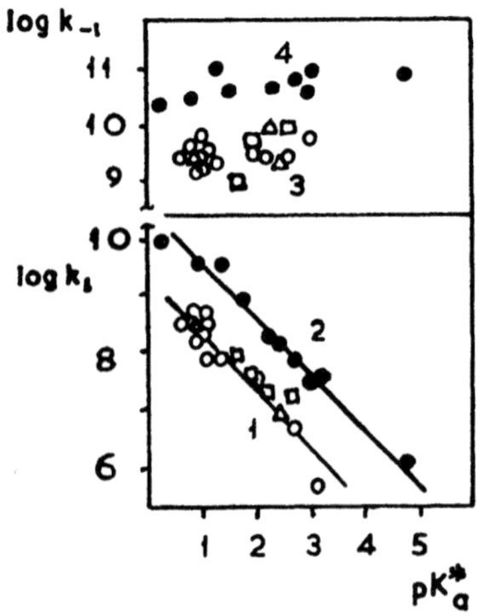

Figure 1. Dependence of protolytic reactions rate constants k_1 (1 and 3) and k_{-1} (2 and 4) from pK^* in aqueous solutions (3 and 4) and in micelles (1 and 2).

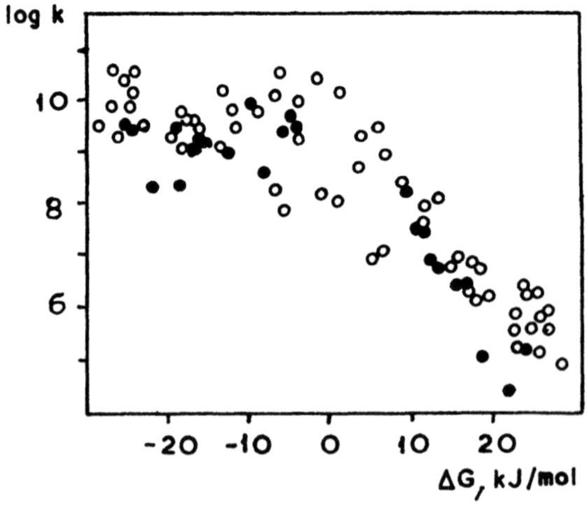

Figure 2. Dependence of protolytic reactions rate constants on Gibbs energy of the reaction ΔG in aqueous solutions (●) and in micelles and vesicles (o).

It is interesting to note that for different types of micelles - cationic, anionic and non-ionic we have obtained the same dependence of $\log k_1$ vs. pK_a^*. The existence of the common dependence gives evidence for the absence of the effect of electrostatic potential of micelles on the photodissociation process.

The dependencies of $\log k_1$ vs. the Gibbs energy ΔG of proton transfer reaction in micelles, vesicles and aqueous solutions coincide (Figure 2), if we take into account the electrostatic energy of separation of two charges in aqueous solution, and that in micelles it equals nil. This is typical dependence well known for homogeneous systems, which can be described by empirical Weller equation

$$\Delta G^{\ddagger} = (\Delta G/2) + [\ (\Delta G_0^{\ddagger})^2 + (\Delta G/2)^2\ ]^{1/2} \qquad (10)$$

or theoretical Marcus equation

$$\Delta G^{\ddagger} = \Delta G_0^{\ddagger}(\ 1 + \Delta G/4\Delta G_0^{\ddagger}\ )^2 \qquad (11)$$

where $\Delta G^{\ddagger}$ and $\Delta G_0^{\ddagger}$ are activation Gibbs energies at given ΔG and $\Delta G = 0$.

If the length of alkyl radical of the detergent increases in the decylsulfate-dodecylsulfate-cetylsulfate series, the rate constant of 1,4-dichloro-2-naphthol photodissociation decreases, and the exit rate constant of the anion formed from micelles, on the contrary, increases. This means that the exit rate from micelles depends on the solubilization free energy of the molecules which are coming out, and photodissociation rate constant is related to Gibbs energy of this reaction in the micellar phase. Solubilization of 1,4-dichloro-2-naphthol anion decreases with the extension of alkyl group of detergent.

6. Conclusion

Various organized molecular systems can be subdivided on two kinds - rigid systems and dynamical systems. Internal structure of the rigid systems is invariable along with the lifetime of the system. Such systems can have quite non-equilibrium internal structure and can be constructed by various methods, in particular by chemical synthesis.

Dynamical systems have quasi equilibrium internal structure. Very simple example of dynamical systems is a micellar solution. Micellar structure is spontaneously formed in the solutions of surfactants and reagents, which are present in the solution, are specifically distributed between volume phase and micellar phase. Reagents are incorporated into the micelles or any other organized system in definite order, determined by their properties (hydrophobity, electric charge, etc.) and by local properties of the system. Such systems are selforganized systems and moreover they are able to maintain their structure by some kind of "reparation", i.e. by restoration of equilibrium concentration distribution through the whole structure of the complex organized system which is inhomogeneous, but quasi equilibrium system. Most of natural systems are dynamical systems, which are in dynamical equilibrium (better to say in stationary state) with their surrounding.

In contrast to dynamical systems the rigid systems have more definite structure which can be constructed and can be determined and defined clearly. Construction of dynamic organized systems is considerably more complex from theoretical point of view and sometimes can be much more simple practically than synthetic construction of rigid organized molecular systems.

In dynamical systems we concerned with statistical structure, which can be characterized by some distribution of the substances and relaxation times. Certain relaxation time can be small comparatively to the lifetime of the system (of the experiment) and large comparatively to the resolution time of some experimental method.

Simple example of this situation is the existence of two kinds of localization of naphthols in the membranes of vesicles discussed earlier. We concerned with the equilibrium distribution of the aromatic compound molecules between various localization sites, but during lifetime of the excited molecules they can be treated as fixed in the definite local environment.

At equilibrium all substances are distributed through the microphases of the microheterogeneous system according to their local electrochemical potentials. But chemical reactions, especially if they are induced by photoexcitation, will generate non-equilibrium concentrations of the products and can produce flows of the substances in the system.

To consider the kinetics of chemical reactions in dynamical systems the idea of local properties of the reactants and products in local environment can be successfully used. At the interphase equilibrium the electrochemical potential of each substance (*i*) in each microphase (*m*) and volume phase (*v*) are the same:

$$\bar{\mu}_{im} = \mu^{o}_{im} + RT\ln a_{im} + z_i F\Phi_m = \bar{\mu}_{iv} = \mu^{o}_{iv} + RT\ln a_{iv} \qquad (12)$$

where $\bar{\mu}_{im}$ and $\bar{\mu}_{iv}$, μ^{o}_{im} and μ^{o}_{iv}, a_{im} and a_{iv} are the electrochemical potentials, standard chemical potentials and activities of a substance (or an ion) *i* in the microphase *m* and volume phase, respectively; *z* is electric charge of the ion and ϕ_m is the electrostatic potential of the microphase with respect of the bulk phase. Local concentrations of each substance i will depend on its overall concentration in the system and on the difference of μ^o_{im} and μ^o_{iv} (that is, to a marked degree by its hydrophobic balance), its electric charge, by the ratio of its activity coefficients in the microphase and bulk phase and by electrostatic potential of the microphase with respect of the bulk phase.

Gibbs energy and equilibrium constant of a chemical reaction depends on the local environment. For example, the equilibrium constant K^m_a for protolytic reaction

$$AH^{(n)} \rightleftharpoons A^{(n-1)} + H^+ \tag{13}$$

in microphase can be expressed as follows:

$$\ln K^m_a = \ln K^m_a + \{ \Delta\mu^o(H^+) + \Delta\mu^o(A^-) - \Delta\mu^o(AH) \}/RT \tag{14}$$

where K^m_a is the equilibrium constant in the volume phase;

$$\Delta\mu^o(i) = \mu^o_{iv} - \mu^o_{im}. \tag{15}$$

Equilibrium constants and Gibbs energies of the reactions in microphases can be expressed also in terms of local distribution coefficients of the reagents and products . For the reaction (13):

$$pK^m_a = -\log\{[H^+]_v[A^-]_m/[AH]_m\} = pK^v_a + \log(\rho/\rho') \tag{16}$$

where ρ and ρ' are the distribution coefficients of the acid AH and conjugated base A^-, respectively. Therefore so-called pK shift in microphases depends on the ratio of the local distribution coefficients of the acid and conjugated base.

Other kinds of chemical reactions (i.e. electron transfer, etc.) in the microphases can be treated in terms of local properties of the reagents and products by similar way.

Dynamical organized molecular systems can be imagined as a set of microreactors, combined in certain order. Chemical reactions in each of them obey classical laws of physical chemistry and chemical kinetics, but interconnections of these microreactors create specific regularities of overall processes such as nonlinear phenomena, etc. The advantage of such approach which discusses the kinetics in terms of local chemical potentials and local mobilities of the media in contrast to frequently used for interpretation "mechanical" and solely structural models is a possibility to give a quantitative description of the phenomena.

References

1. Fendler, J. H. "Membrane Mimetic Chemistry"; Wiley-Interscience: New York, 1982.
2. Fendler, J. H.; Fendler, E. J. "Catalysis in Micellar and Macromolecular Systems", Academic Press: New York, 1975.
3. Kalyanasundaram, K. "Photochemistry in Microheterogeneous System"; Academic Press: London, 1987.
4. Kuzmin, M. G.; Zaitsev, N. K. In: "The Interface Structure and Electrochemical Processes at the Boundary between Two Immiscible Liquids", Kazarinov, V.E., Ed.; Springer-Verlag: Berlin, Heidelberg, 1987, 207-244.
5. Lee, J.; Robinson, G. W.; Webb, S. P.; Philips, L. A.; Clark, J. H. J. Am. Chem. Soc. 1986, **108**, 6538-6542.
6. Shizuka, H.; Ogiwara, T.; Narita, A.; Sumitani, M.; Yoshihara, K. J. Phys. Chem. 1986, **90**, 6708-6714.
7. Kishi, T.; Tanaka, J.; Kouyama, T. Chem. Phys. Lett. 1976, **41**, 497-499.
8. Il'ichev Yu. V., Demyashkevich A. B., Kuzmin M. G. to be published.
9. Bayer, M.; Zaitsev, A. K.; Kuzmin, M. G. Chim. Physika (USSR) 1988, **7**, 172-174.
10. Zaitsev N.K., Demyashkevich A.B., Kuzmin M.G. Chimiya Visokich Energii (USSR) 1978, **12**, 430-435.
11. Zaitsev, A. K.; Il'ichev, Yu. V.; Gorelik, O. Ph.; Zaitsev, N. K.; Kuzmin, M. G. Chim. Physika (USSR) 1985, **4**, 1384-1389.
12. Abou-Al Einin, S.; Zaitsev, A. K.; Zaitsev, N. K.; Kuzmin, M. G. Chim. Physika (USSR) 1986, **5**, 219-225.
13. Abou-Al Einin, S.; Zaitsev, A. K.; Zaitsev, N. K.; Kuzmin, M. G. J. Photochem. Photobiol. 1988, **41**, 365-373.
14. Kuzmin, M. G.; Zaitsev, N. K.; Zaitsev, A. K.; Abou-Al Einin, S. Chim. Physika (USSR) 1988, **7**, 492-498.
15. Laws, W. R.; Brand, L. J. Phys. Chem. 1979, **83**, 795-802.
16. Webb, S. P.; Yeh, S. H.; Philips, L. A.; Tolbert, M. A.; Clark, J. H. J. Am. Chem. Soc. 1984, **106**, 7286-7288.
17. Weller, A. Prog. React. Kinet. 1961, **1**, 189-214.
18. Il'ichev Yu. V., Demyashkevich A. B., Kuzmin M. G. Chimiya Visokich Energii (USSR) 1989, **23**, 435–439.

THE OPTIMIZATION OF PHOTOCHEMICAL ENERGY CONVERSION: SYNTHETIC SUPRAMOLECULAR DEVICES VIS À VIS PHOTOSYNTHESIS

T. A. MOORE, D. GUST AND A. L. MOORE
Department of Chemistry and Biochemistry,
Center for the Study of Early Events in Photosynthesis
Arizona State University
Tempe, Arizona
USA 85287

ABSTRACT. A series of supramolecular compounds are described which are designed to undergo photochemical and subsequent dark intramolecular electron transfer reactions which result in electron transfer over a long distance and production of a long lived charge separated species. This requires a multistep electron transfer strategy in which intermediate species are formed. The decay of these intermediate species is partitioned between forward and recombination electron transfer steps. The rates of these steps are controlled by thermodynamic and electronic terms which must be optimized in order for the yield of the forward process to be high. An approach for making such an optimization is described and the results are compared to the natural photosynthetic electron transfer system.

Introduction

One elaborate example of a biological supramolecular structure is the reaction center from a photosynthetic bacterium. It is interesting to view this structure as a nano-scale machine which conducts vectorial photoinitiated electron transport and directed singlet - singlet and triplet - triplet energy transfer. In this manuscript, we will define some of the prerequisites for the design and synthesis of supramolecular structures which mimic the functions of the reaction center. We begin with a brief description of the energy conversion process that it carries out.

Photosynthetic Electron Transfer

Photosynthesis in photosynthetic bacteria involves light driven electron transfer across a bilayer lipid membrane which converts the light energy into chemical potential. After transmembrane charge separation, the chemical potential is in the form of reducing equivalents on the cytoplasmic side of the membrane, oxidizing equivalents on the periplasmic side and a membrane potential of perhaps 180 mV which is negative on the cytoplasmic side. From the crystal structure of the reaction center of *Rb. sphaeroides* (Yeates et al. (1988)) it is possible to construct an illustration of the supramolecular structure which accomplishes this process (Fig. 1).

Some parameters which characterize this energy conversion process are the quantum yield and the standard redox potentials of the reduced and oxidized species. Under average conditions the quantum yield of transmembrane electron transfer has been found to be essentially 100%, and the

V. Balzani and L. De Cola (eds.), Supramolecular Chemistry, 295–311.

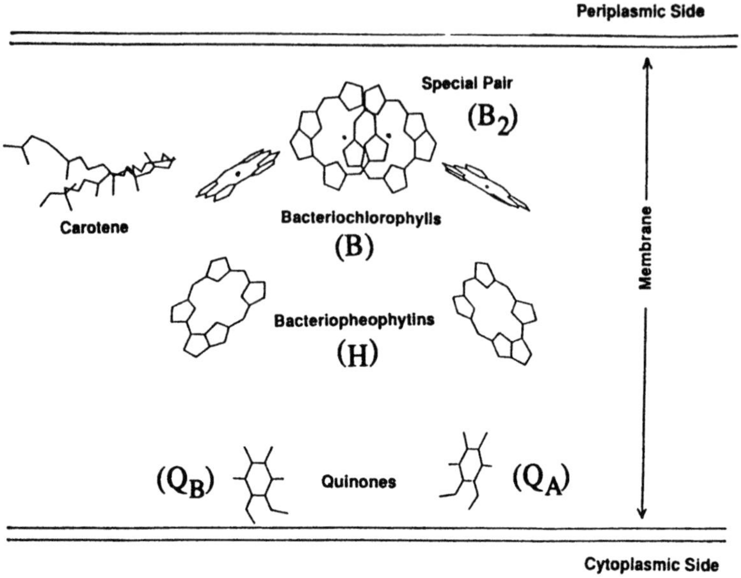

Figure 1. Location of the electron transfer cofactors in the photosynthetic reaction center of *Rb. sphaeroides*.

redox potential between the reduced ($Q_B^{-\cdot}$) and oxidized ($B_2^{+\cdot}$) species is approximately 500 mV (Feher et al. (1989), Parson (1991)). Considering that excitation at the red edge of the bacteriochlorophyll reaction center absorption spectrum requires a photon of 1.4 eV, only about one third of the excitation energy is conserved as redox potential at the state in which the charge separation can be considered to be transmembrane. Figure 2 illustrates the energies of the initial, intermediate and, for our purposes, final states formed upon excitation of the special pair (B_2) to the singlet excited state and subsequent charge separation (Feher et al. (1989).

Also indicated in Fig. 2 are the time constants associated with the various electron transfer processes. There are two views on the initial electron transfer process. In one view, initial charge separation takes place in about 3 ps to yield an intermediate having the electron on the accessory bacteriochlorophyll (B) on the A branch of the reaction center (Holzapfel et al. (1990)). Then, in an even more rapid step of less than one ps, the electron is transferred to the bacteriopheophytin (H) on the A branch. In the other limit, charge separation is viewed as moving an electron directly to bacteriopheophytin (H) in about 3.5 ps (Kirmaier and Holten (1987)). In any case, the intermediate $H^{-\cdot}$ is relatively well characterized in terms of its energy and lifetime since it can be trapped. With a time constant of 200 ps the electron is transferred to the quinone, Q_A, located near the membrane water interface on the cytoplasmic side. Finally, the electron is transferred laterally to Q_B in 200 μs (Feher et al. (1989), Parson (1991)).

As can be seen from Fig. 2, electron transfer is driven by the exergonic redox gradient of the cofactors. But, at each intermediate state there is a branch point at which the forward electron transfer process competes with the more exergonic charge recombination reaction which yields the ground state. Initially, the time scale for the electron transfer processes is set by the competition between charge transfer to yield $B_2^{+\cdot}B^{-\cdot}$ from $^1B_2$ and fluorescence, intersystem crossing and internal conversion, which together have a time constant of ca. 350 ps (Schenck et al. (1981)). To

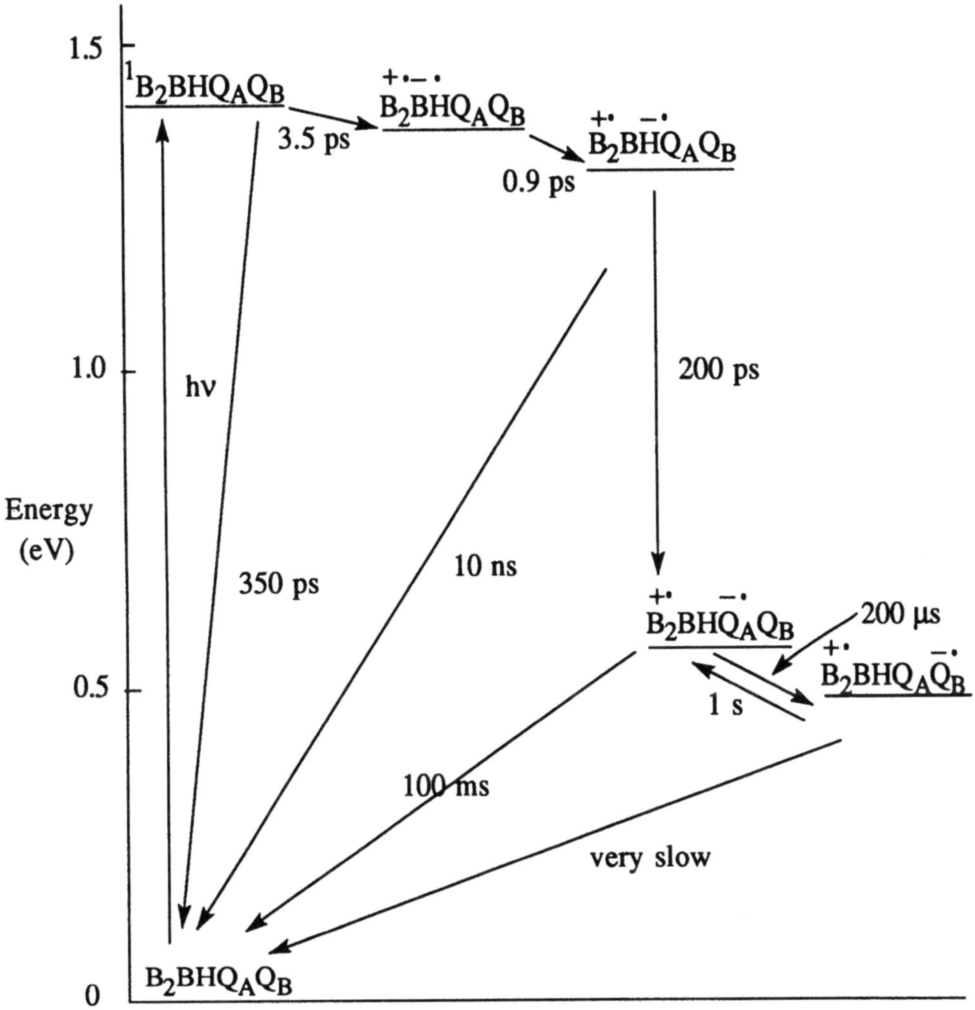

Figure 2. Energies and lifetimes of transient species in bacterial photosynthetic electron transfer.

date, it has not been possible to measure the recombination rate from the first intermediate state, $B_2^{+\cdot}B^{-\cdot}$. The time constant for recombination from the second intermediate state, $B_2^{+\cdot}BH^{-\cdot}$, is ca. 10 ns (Schenck et al. (1982)) and from the state in which Q_A bears an electron, ca. 100 ms (Feher et al. (1989)). Since at each point the forward step is much faster than the recombination step, it is easy to see that the quantum yield of formation of $B_2^{+\cdot}BHQ_AQ_B^{-\cdot}$ is ca. 1.

Returning to the initial photochemical event involving electron transfer from the excited singlet state of B_2 to B, the primary acceptor, let us consider whether this process could in principle occur across the membrane in one step, rather than the series of steps described above. It is generally agreed that in the simplest case, the distance dependence of the electron transfer rate constant is dominated by an exponential term which arises from the distance dependence of the overlap between the tails of the wavefunctions of the donor and acceptor species (Marcus and Sutin

(1985)). An equation which has been found to describe the distance dependence in many systems is

$$k = \alpha e^{-\beta (r - r_0)} \quad (1)$$

where α is the rate at van der Waals contact ($r=r_0$) and β is ca. 1 $Å^{-1}$. Within the framework of Marcus theory (Marcus and Sutin (1985)), α will have its maximum value when $\Delta G^o=\lambda$, the reorganization energy. Interestingly, both measurements and extrapolations to this condition for a wide variety of donor acceptor systems yields a value for α of ca. 10^{13} s^{-1} (Moser et al. (1991)). Using this value in equation 1 and an edge to edge separation of 10 Å, which would be required between two optimally oriented chlorophylls on opposite sides of a membrane for electron transfer across the membrane, the rate constant would be ca. 5×10^8 s^{-1}. Certainly, such a slow process could not compete effectively with the photophysical processes that depopulate the excited singlet state of the bacteriochlorophyll special pair in ca. 350 ps.

It is worth noting, however, that by involving orbitals of the material intervening between the donor and acceptor it is possible to have fast, long distance electron transfer (Bixon et al. (1989)). This through bond or superexchange mechanism can be very effective when the energies and spatial arrangements of the HOMO and LUMO orbitals of the intervening or linking material are appropriate for interaction with the donor and acceptor. For example, in specific systems where the excited state of the donor is more than 3 eV above the ground state, fast electron transfer over distances of nearly 15 Å has been demonstrated (Warman et al. (1991)). But, in a similar system long distance electron transfer from excited tetrapyrroles having singlet energies of around 2 eV to quinone acceptors is not efficient (Antolovich et al. (1991)).

There is another way to view the single step transmembrane electron transfer. In 1959, before the pigment and quinone composition of the reaction center was known, it was imagined that in photosynthesis a carotenoid polyene acted as a conductor of electrons across a membrane by making use of a charge transfer state of a carotene interposed across a membrane between a donor and acceptor (Platt (1959)). Along these same lines, a series of papers have recently appeared which report carotenoids linked to electron donors and acceptors with a focus on transmembrane communication and artificial photosynthesis (Kugimiya et al. (1991)). Although they are not single step devices, triads of the type carotene - porphyrin - quinone which do have long lived charge transfer states involving the radical cation of the carotenoid species have been shown to conduct charge across both planar bilayer lipid membranes and across liposome membranes (Seta et al. (1985), Moore et al. (1987)).

The above-mentioned results notwithstanding, the evidence that transmembrane electron transfer in photosynthesis involves a series of steps is overwhelming. The observation that the quantum yield for the series of forward steps is near the theoretical maximum while the standard state redox energy yield is quite low raises a basic question. Is the loss of approximately 1 eV of energy as heat necessary to ensure that the quantum yield is high? In order to answer this question it is necessary to explore the factors controlling the competition between the two electron transfer processes at a branch point. Toward this goal, we have synthesized a series of multicomponent compounds in which this competition can be studied (Gust and Moore (1989), Gust and Moore (1991a), Gust and Moore (1991b)).

Electron Transfer in Supramolecular Model Systems

a) DYADS

In the context of mimicking photosynthetic electron transfer, by 1979 several compounds featuring a porphyrin covalently linked to a quinone had been reported (Kong and Loach (1978), Tabushi et al. (1979)). A typical example of such a P-Q dyad, **1**, reported by us in 1987, (Gust et al. (1987)), will be used to illustrate the structure and photochemistry of these compounds. Dyad **1** consists of a tetraarylporphyrin joined via an amide linkage to a benzoquinone moiety. Excitation of **1** in solution with light absorbed by the porphyrin populates the porphyrin lowest excited singlet state, which can decay by the usual routes of fluorescence, intersystem crossing and internal conversion. By using the fluorescence decay to report the time dependence of the singlet state population, the sum of the rates of these processes can be determined by measuring the fluorescence lifetime (τ_0) of a porphyrin similar to the one incorporated in **1** but without the attached quinone. Alternatively, the attached quinone can be in the hydroquinone form, so that electron transfer cannot occur. Then, by measuring the fluorescence lifetime (τ) of **1**, in which there is the additional decay pathway of electron transfer to yield $P^{+\cdot}$-$Q^{-\cdot}$, the rate constant for electron transfer can be calculated from equation 2.

$$k_2 = 1/\tau - 1/\tau_0 \qquad (2)$$

From the fluorescence lifetime of 100 ps for **1** and 8 ns for τ_0, k_2 is found to be 9.9×10^9 s^{-1} in methylene chloride solution. The quantum yield of the charge separated state is calculated from the expression $\tau \cdot k_2$ and is 0.99. The energy level diagram shown in Fig. 3 illustrates these steps; note that the charge separated state preserves about 1.4 eV of the original 1.9 eV porphyrin singlet state. The energies of the transient species are estimated from spectroscopic or electrochemical measurements.

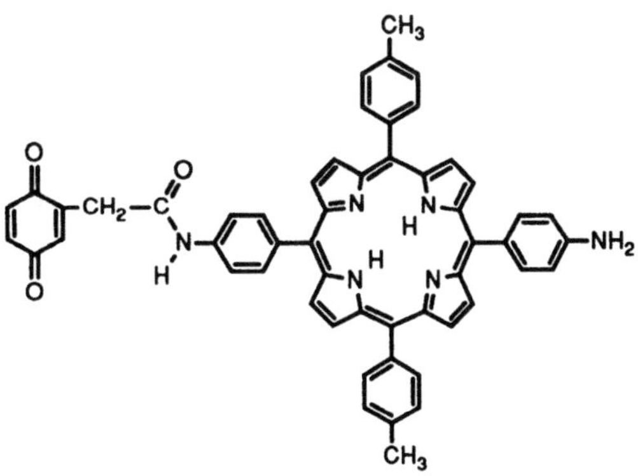

1

The study of compounds such as **1** has provided much information about the design of photochemical electron transfer systems based on biological molecules. Specifically, the roles of the linkage and the thermodynamics in making electron transfer competitive with fluorescence, intersystem crossing, and internal conversion are now qualitatively understood. However, a dyad system cannot have a branch point at an intermediate state in the electron transfer path and therefore cannot be used to study the competition between forward and recombination electron transfer. Also, the dyad system fails to mimic the long lifetimes of hundreds of milliseconds characteristic of

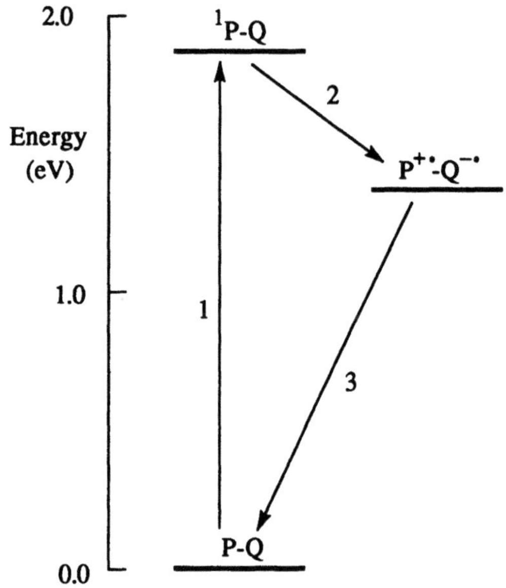

Figure 3. Transient states of dyad **1** and electron transfer pathways.

the charge separated states involving $B_2^{+\bullet}$ and $Q^{-\bullet}$ in reaction centers. Indeed, the charge separated state in compounds similar to **1** lives at most a few hundred ps.

b) .TRIADS

In order to explore the multistep strategy and to determine whether such a strategy could give rise to a long lived charge separated state in a synthetic system, we designed the first of a series of supramolecular compounds christened triads. In 1983 (Gust et al. (1983), Moore et al. (1984)), we reported the synthesis and photochemical studies of triad **2** which has nearly the same porphyrin quinone structure as **1**, but now the porphyrin bears a carotenoid polyene in a 5, 15 relationship to the quinone. The absorption spectrum of **2** is essentially the superposition of the quinone, porphyrin, and carotenoid components, indicating that the chromophores retain their identities.

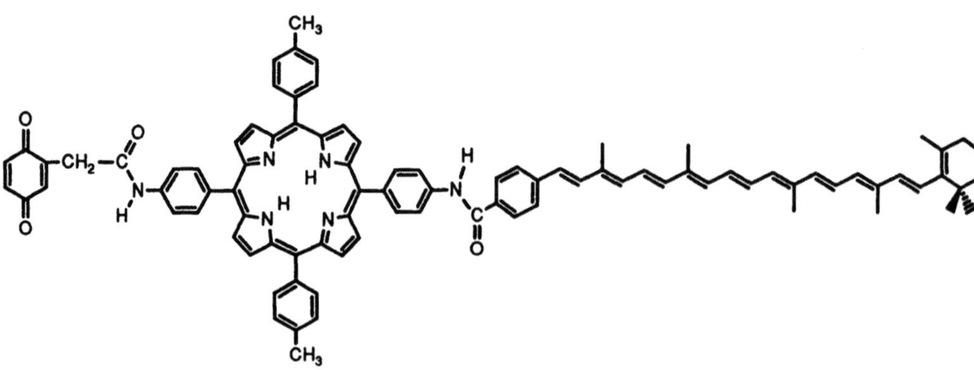

2

Excitation of **2** in solution populates the lowest excited singlet state of the porphyrin which then relaxes in much the same way as in **1** to give the species C-$P^{+\cdot}$-$Q^{-\cdot}$ in high yield. However, with **2** the course of events from this point on is fundamentally different. Transient absorption spectroscopy at 960 nm signaled the presence of a species that lived for nearly 300 ns. This signal was not observed in **1** or in a version of **2** in which the quinone was present as the hydroquinone form. By comparing the spectrum of this long lived transient with that of carotenoid radical cations generated by pulse radiolysis, the transient was assigned to the carotenoid radical cation species. These results can be interpreted as illustrated in Fig. 4. In competition with step 3, which is recombination from the intermediate state, is step 4, in which the carotenoid species donates an electron to the porphyrin radical cation. This is a forward electron transfer process in the sense that the charges are further separated and stabilized.

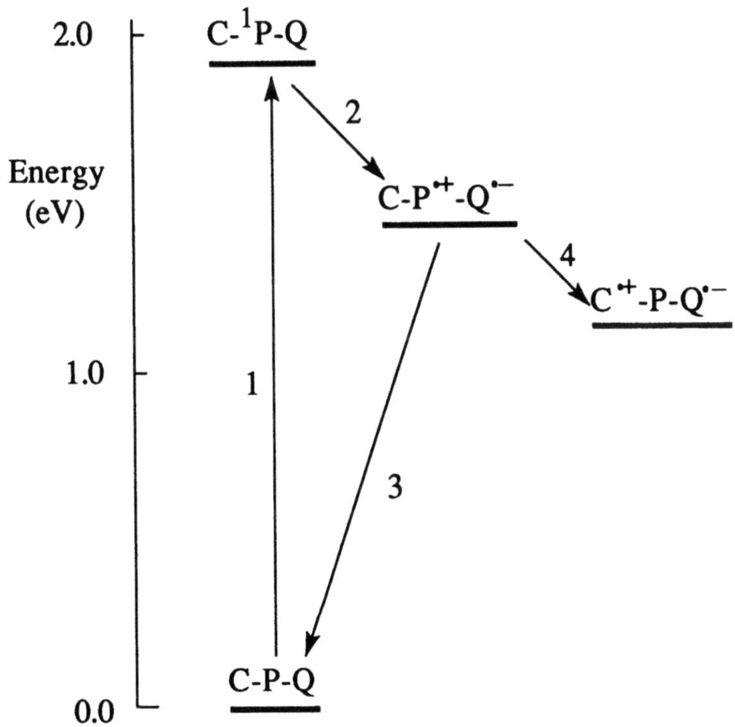

Figure 4. Transient states of C-P-Q triads such as **2** and related electron transfer pathways.

The overall yield of the final charge separated species in **2** is 4%, and the lifetime is ca. 300 ns in methylene chloride and a few μs in acetonitrile. The compound is stable for thousands of turnovers, even in the presence of O_2. Charge recombination from the final state involves an activated step which is the reverse of step 4 to repopulate the intermediate state which then undergoes recombination *via* step 3. We assigned the porphyrin singlet state as the precursor to the final charge separated state since the yield of triplet porphyrin in the 100 ps singlet lifetime (step 2) is less than the 4% overall yield. More recently, we have observed the final charge separated species in a related tetrad system by time resolved epr techniques and have confirmed this assignment (Hasharoni et al. (1990)). From the standpoint of energetics, triad **2** preserves ca. 1.1 eV of the 1.9 eV porphyrin excited singlet energy as intramolecular redox potential in the final charge separated state. Triad **2** demonstrates that it is possible to design a system in which the multistep electron transfer strategy seen in photosynthesis can be modeled. It has served as a starting point for investigating the factors controlling the forward and recombination processes at the branch point.

Turning to the Marcus theory of electron transfer (Marcus and Sutin (1985)) for guidance in efforts to improve the 4% overall yield of charge separated species, we note that in other studies of dyad systems in solvents such as methylene chloride the total reorganization energy, λ, is around 1 eV ((Wasielewski et al. (1985), Joran et al. (1987), Harrison et al. (1987)). This means that the rate of electron transfer reactions having $-\Delta G^o < 1$ eV will be increased by an increase in driving force, while those having $-\Delta G^o > 1$ eV will be slowed by such an increase. We can use this information in the following way. Examination of Fig. 4 reveals that the branch point from the intermediate charge separated species, C-$P^{+\cdot}$-$Q^{-\cdot}$, involves the recombination reaction having $-\Delta G^o > 1$ eV in competition with the forward electron transfer having $-\Delta G^o < 1$ eV. Therefore, increasing the energy of the intermediate species would be expected to increase the rate of the forward reaction and decrease the rate of the recombination reaction. This would certainly favor the yield of the final charge separated species. On the other hand, an increase in the energy of the intermediate would likewise slow step 2 which would mitigate against a high overall yield. Since in **2** the rate of step 2 is sufficiently fast to ensure a yield of ca. 97%, our strategy was to sacrifice some of the rate of step 2 by increasing the energy of the intermediate and thereby to improve the ratio of step 4 to step 3.

In order for the above argument to be valid, the changes to the structure to increase the energy of C-$P^{+\cdot}$-$Q^{-\cdot}$ must not appreciably modify the electronic coupling between the donor and acceptor pairs. With this in mind, we synthesized triad **3** in which the two *meso* p-tolyl groups on the porphyrin are replaced by pentafluorophenyl groups (Moore et al. (1988)). Comparison of the 1H-NMR chemical shifts of the carotenoid and quinone protons of **2** and **3** indicated very little change, showing that the conformations of **2** and **3** are similar. Both have the carotenoid and quinone moieties extended out away from the porphyrin. The absorption spectrum of **3** indicates that there are no significant perturbations to the porphyrin excited singlet state. Most importantly, electrochemical experiments showed that the radical cation of the porphyrin moiety in **3** is ca. 0.2 eV above that of **2**.

Excitation of **2** and **3** by laser flash at 590 nm under the same conditions resulted in similar transient spectra in the near ir which signaled the production of $C^{+\cdot}$-P-$Q^{-\cdot}$. However, quantitative

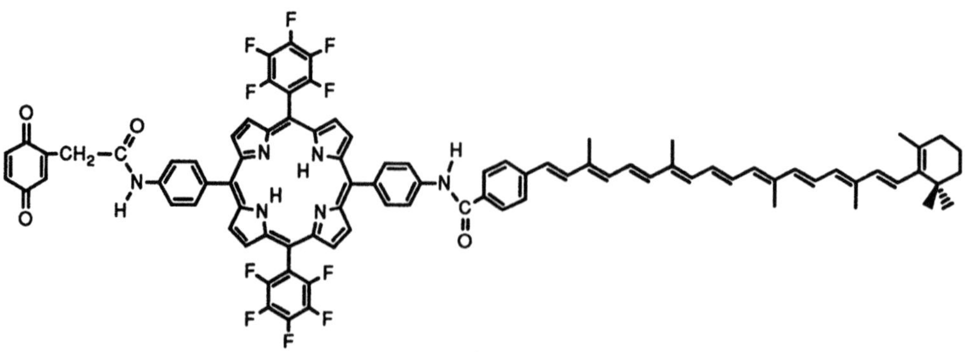

3

comparisons between **2** and **3** revealed that the quantum yield in **3** was 30%, which is more than a seven-fold increase over that in **2**. Furthermore, from fluorescence lifetime studies the yield of step 2 was found to be reduced from 97% in **2** to 41% in **3**. Therefore, the increase in overall yield was in spite of a decrease in the yield of the first step. Quantitatively, the yield of the forward step (4) at the branch point was improved from 4% in **2** to 73% in **3**, a factor of 18. Note that the energy of the final charge separated state is essentially the same in both compounds.

This result illustrates a strategy which seems to have been used by nature in the reaction center. Referring back to Fig. 4, imagine the intermediate level 0.2 eV above that shown and then refer to Fig. 2. Clearly, the first intermediate in Fig. 2 is likewise sufficiently energetic so that the recombination process would be well into the Marcus inverted region unless λ is very large. By using this thermodynamic factor to limit the rate of recombination, it is possible to have adequate electronic coupling between the donor and acceptor for a fast forward reaction without paying too high a price in terms of a fast recombination reaction.

c) TETRADS

In order to further explore the multistep strategy for achieving both long range electron transfer and a long lived charge transfer state, we have designed and synthesized tetrad **4** and triad **5** (Gust et al. (1988), Gust et al. (1989)). The plan here was to provide a second electron transfer step on the acceptor side of the porphyrin to compete with recombination from the first intermediate species.

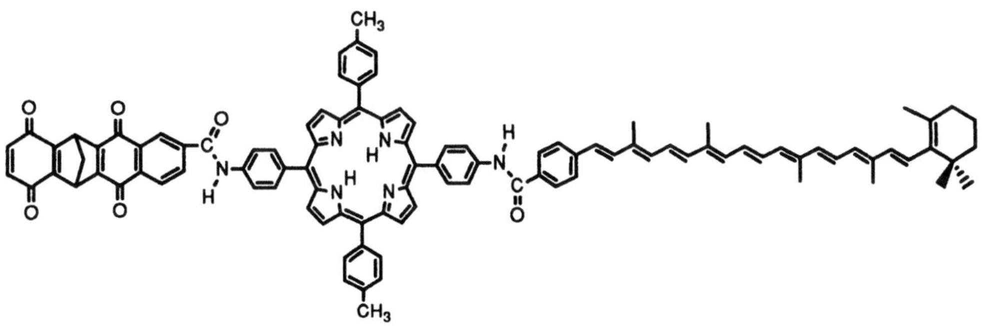

4

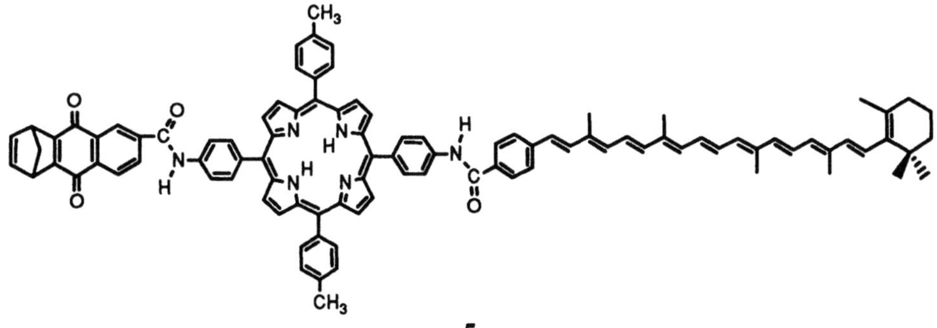

5

In **4** and **5** the carotenoid-porphyrin linkage is the same as in **2** and **3**, but the quinone has been changed. In **4** a diquinone species consisting of a naphthoquinone linked to a benzoquinone *via* a norbornadiene group was attached to the porphyrin. In order to compensate for the reduced driving force for electron transfer from the porphyrin excited singlet state to a naphthoquinone rather than to a benzoquinone, the methylene group was removed from the linkage between the porphyrin and the quinone. Electrochemical experiments with the diquinone species established that the two quinones act as independent electron acceptors; the first reduction occurred near the potential expected for this benzoquinone and the second reduction near the expected naphthoquinone potential. Compound **5** is a triad which serves as a reference for **4** in that the benzoquinone moiety has been removed.

Single photon timing techniques were used to measure the singlet lifetimes of **4** and **5**, and yielded 15 ps for **4** and slightly longer for **5**. This is faster than was found for **3** which means that the additional electronic coupling provided by removing the methylene group (and reversing the sense of the amide linkage (Connolly and Bolton (1988)) more than compensated for the decreased driving force for electron transfer to the naphthoquinone. In both cases the yield of step 2, the initial charge transfer process, is ca. 99%. Transient absorption studies to determine the lifetime and yield of the carotenoid radical cation species characteristic of the final charge transfer state produced very different results for **4** and **5**. In **4** the yield of the $C^{+\cdot}$-P-Q_A-$Q_B^{-\cdot}$ species is 23% and the lifetime is 460 ns, whereas in **5** the yield is only 4% and the lifetime ca. 70 ns.

These results can be interpreted from the electron transfer pathway scheme shown in Fig. 5. As before, the energies of these species have been determined from electrochemical measurements on model compounds and have not been corrected for coulombic attraction. As mentioned above, step

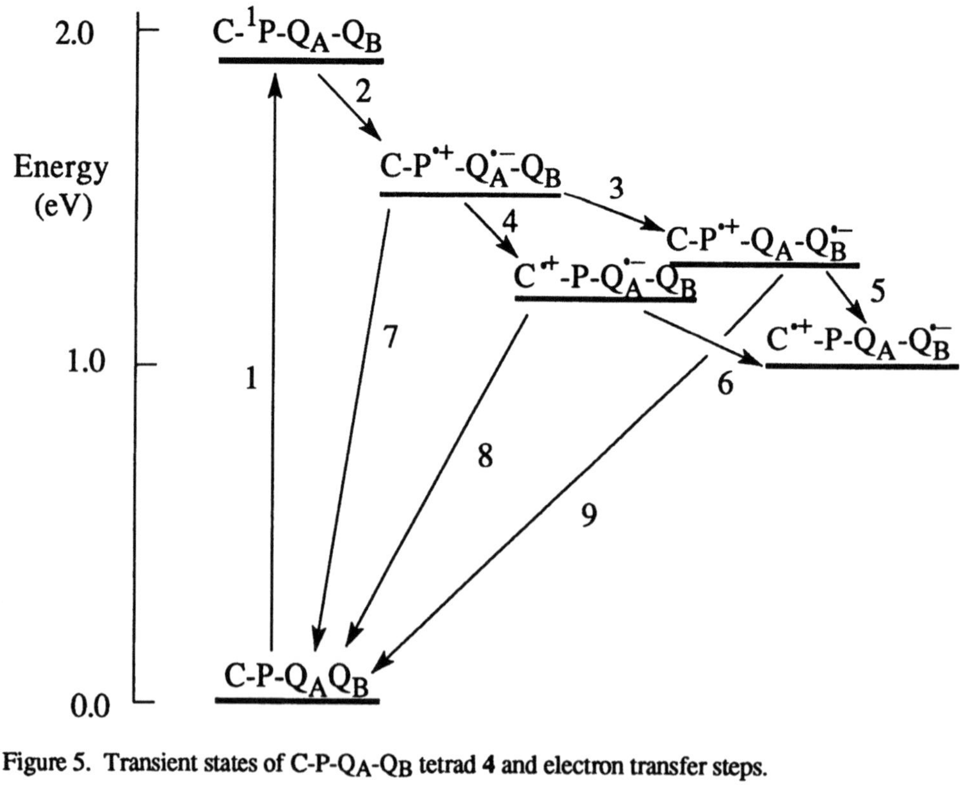

Figure 5. Transient states of C-P-Q_A-Q_B tetrad **4** and electron transfer steps.

2 is fast for both compounds and essentially 100% efficient. It is at the first intermediate species that the available pathways differ in a pivotal way. In **4**, there are two forward electron transfer pathways, steps 3 and 4, which compete with step 7, charge recombination. In **5**, since there is no benzoquinone moiety there is only step 4 to compete with recombination. Therefore, the increase in yield from 4% to 23%, almost a factor of six, can be attributed to step 3. This means that step 3 is five times faster than step 4 if the efficiency of the forward process is ca. 100% at the branch point between steps 5 and 9, and more than five times faster if step 9 is limiting the overall yield.

These compounds illustrate two points about multistep electron transfer. First, by designing the compound so that there is a fast second step in which the electron is transferred to a secondary acceptor, (step 3) in series with the primary charge separating step (step 2) it is possible to effectively compete against the primary recombination reaction (step 7). Second, it can be seen from Fig. 5 that steps 3 and 4 together with steps 5 and 6 represent parallel pathways that ultimately converge to the desired final charge separated species. Although the parallel pathway strategy does not seem to be consequential in photosynthesis today, the near two-fold symmetry axis of the reaction center suggests that it could have been at some point. In terms of designing improved supramolecular compounds, note that both the series and parallel strategies for abetting the forward electron transfer process can be applied together with the strategy of thermodynamic control (exemplified by **3**) to enhance the overall yield of charge separated species.

Thermodynamic control of the yield of the final charge separated species can be demonstrated with tetrad **4** by inserting a Zn^{+2} ion in the porphyrin. As shown by the scheme in Fig. 6, the metalated porphyrin has its excited singlet state blue shifted from that of the free base by ca. 0.2 eV. In addition, the radical cation of the metalated porphyrin is ca. 0.2 eV more stable than that of the free base. The combination of these two effects is that the driving force for primary charge separation in the metalated species is ca. 0.4 eV greater than in the free base analogue. As expected for a process in the normal Marcus region, the electron transfer is faster. On our single photon timing fluorescence spectrometer which has an instrument response function of 35 ps, the lifetime comes out to about 5 ps which is significantly faster than the 15 ps determined for the free base version. With respect to the effect of the added stability of the radical cation of the porphyrin, the argument has the opposite sense to the one used before for triad **3**. Steps 7 and 8 should be enhanced since they are in the inverted region while steps 4 and 5 should be slowed since they are in the normal region. The combination of increasing recombination rates while decreasing forward rates is expected to lower the yield of the final charge separated species, and that is what we observed. Quantitatively, the yield of long lived charge separated species in the metalated porphyrin tetrad is 70% that of the free base compound.

These results illustrate an apparent paradox in the design of supramolecular structures for multistep electron transfer in that a faster first step is not necessarily beneficial. It is also interesting to compare the energy level scheme in Fig. 6 with that in Fig. 2 for the reaction center. It can be seen that the driving force between the steps is distributed very differently in these two cases. It seems that, as mentioned in the discussion of triad **3**, one of the strategies for ensuring a high yield of multistep charge separation is to site recombination of the first intermediate well into the Marcus inverted region.

d) PENTADS

A common feature of the triads and tetrads mentioned above is that in each case at least part of the competition against charge recombination at the first intermediate species was provided by electron donation from the carotenoid moiety to the porphyrin radical cation (this is sometimes known as

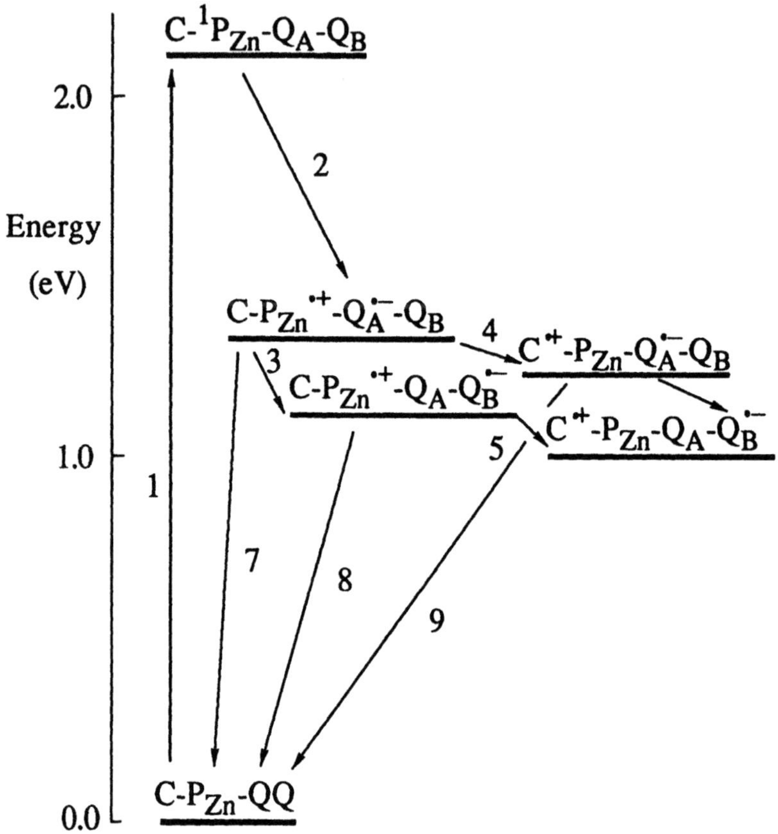

Figure 6. Transient states of a $C\text{-}P_{Zn}\text{-}Q_A\text{-}Q_B$ tetrad (the zinc analog of **4**) and electron transfer steps.

hole transfer since in the one electron approximation only the HOMO's of both species are involved). In order to explore the use of a tetrapyrrole species as the electron donor at this crucial step, and to include this tetrapyrrole in the supramolecular structure in such a way that the increased absorption cross section it provides would be available for charge separation, pentad **6** was synthesized (Gust et al. (1990)).

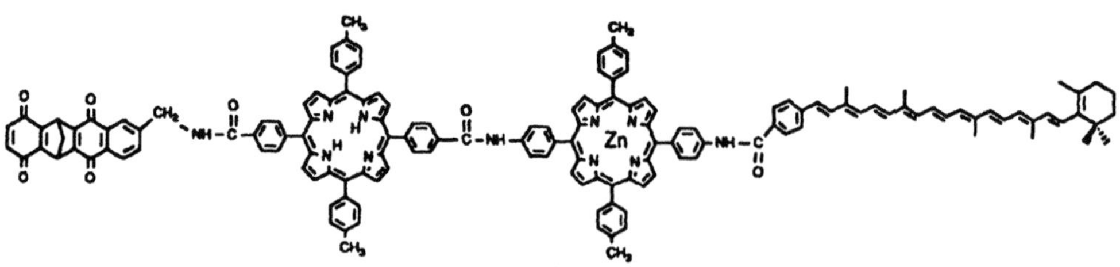

6

By metalating one of the porphyrins the degeneracy of excited singlet energy levels in a two porphyrin system is broken. This will insure that if the requirements for singlet energy transfer are satisfied, the singlet energy will localize on one porphyrin (the free base one). Furthermore, thinking in terms of providing a potential gradient for driving charge separation in the forward sense, metalating one of the porphyrins also lifts the near degeneracy of their radical cation states, thereby setting the stage for sequential hole transfer on the donor side of the system.

Pentad **6** retains the carotenoid as the ultimate electron donor and uses a diquinone species similar to that which provided the sequential electron transfer pathway on the acceptor side of tetrad **4**. Note that in **6** the linkage to the diquinone includes a methylene spacer and that the sense of the amide is reversed from that of **4**. These changes will reduce the rate of the initial charge separation step.

The experimental results for **6** were very encouraging. Excitation at 590 nm populates the two porphyrin excited singlet states, but steady state and time resolved fluorescence measurements revealed that singlet energy transfer from the P_{Zn} to the free base P occurs with a time constant of ca. 40 ps and an efficiency of 90% (step 1 in Fig. 7), resulting in the excitation being localized on P. This result confirmed the expectation that the added porphyrin would act as an antenna for the system. In a model dyad consisting of just the P_{Zn}-P species, the intrinsic lifetime of the free base

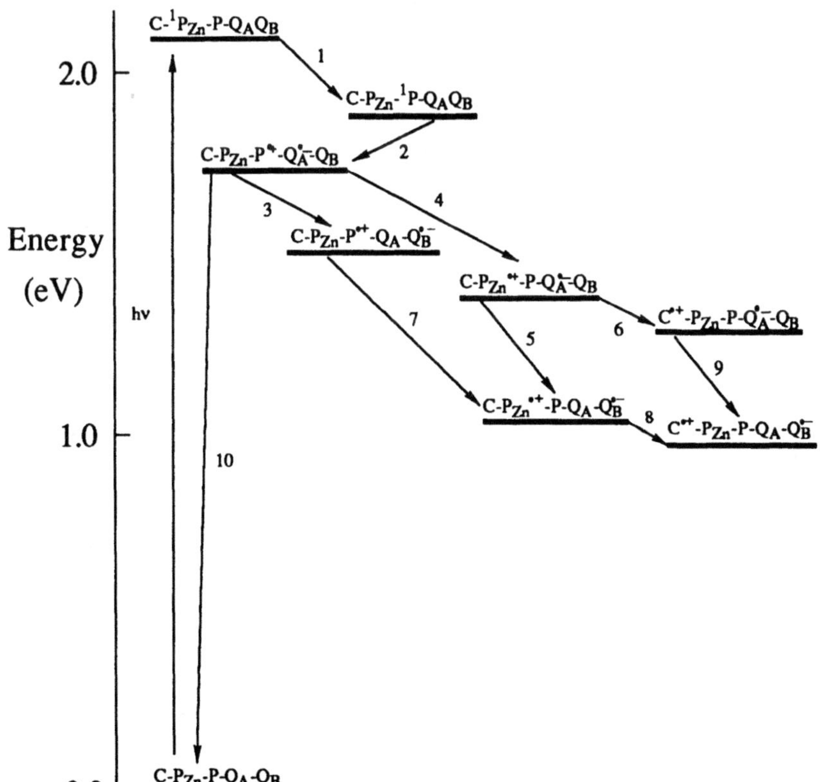

Figure 7. Transient states of C-P_{Zn}-P-Q_A-Q_B pentad **6** and singlet energy and electron transfer steps.

P was determined to be 7.8 ns, which is near that of P alone and indicates that in the absence of the diquinone species, P is not appreciably quenched. In **6**, P is quenched by electron transfer and decays in 1.2 ns to yield the primary charge separated species, $C\text{-}P_{Zn}\text{-}P^{+\cdot}\text{-}Q_A^{-\cdot}\text{-}Q_B$, as indicated by step 2. As mentioned above, the choice of linkage has reduced the rate of electron transfer from that observed for tetrad **4**. In fact, the rate calculated from equation 2 is $7.1 \times 10^8\ s^{-1}$, which results in a quantum yield of 0.85 for electron transfer in step 2.

Upon excitation of a chloroform solution of **6** at 650 nm in the laser flash spectrometer the carotenoid radical cation species was detected by its characteristic absorption spectrum with a maximum at 960 nm. This species is formed with a quantum yield of 83% and has a lifetime of 55 μs. The fact that the overall quantum yield is essentially the same as that of step 2 evinces that the combination of steps 3 and 4 have prevailed in the competition with recombination (step 10) for the decay of the primary charge separated intermediate, $C\text{-}P_{Zn}\text{-}P^{+\cdot}\text{-}Q_A^{-\cdot}Q_B$.

Pentad **6** introduces a supramolecular compound having sequential steps on both the acceptor and donor sides. As can be seen from Fig. 7, these sequential steps generate parallel pathways that converge on the final charge separated species. Although it is tempting to attribute the enhanced yield in **6** over **4** to the series of hole transfer steps on the donor side available in **6**, such an interpretation is precarious at this time since the acceptor side linkages are different and the ratios of the rates of step 4 in Fig. 7 to step 4 in Fig. 6 are not known. However, an indication of the importance of the donor side steps is given by the results for a derivative of **6** in which the two porphyrins are both present in the free base form. In this case, the rate of step 2 is unchanged but the quantum yield of step 2 (in dichloromethane) is 35% due to the equilibrium between the isoenergetic porphyrin singlet species. The overall yield is reduced to 15% which means that a significant driving force for moving the hole away from $Q_A^{-\cdot}$ and toward the carotenoid is crucial in the competition with recombination. This derivative demonstrates that for a high quantum yield of long lived charge separated species, it is imperative that the two porphyrins be spectroscopically and electrochemically distinct.

Conclusion

In conclusion, these results suggest that there are multiple factors to consider when choosing which combination of thermodynamic effects and electronic effects to use to control the rate of an electron transfer process. Several qualitative guidelines have emerged from this work. For example, increasing the rate by increasing the electronic coupling between the donor and the acceptor may also increase the rate of the recombination reaction, although not necessarily to the same extent. Because of this the rate of the first step should only be fast enough to compete efficiently with the other processes that deactivate the excited state. Tetrad **4** is an example where this step is probably too fast, resulting in a fast recombination step which limits the overall yield. In pentad **6** this step is too slow, resulting in the limiting yield of ca. 85% A reasonable plan to improve the situation for tetrad **4** would be to decrease the thermodynamic driving force for the first step by using the porphyrin from triad **3**. In this way the recombination step would be moved further into the inverted region and slowed. An alternative plan would be to decrease the electronic coupling between the porphyrin and the naphthoquinone, thus slowing the recombination step. Following the strategy for the reaction center illustrated in Fig. 2, the best choice would seem to be to use the thermodynamic control so that as much energy as possible would be stored at the first intermediate. The plans discussed above for improving the yield for **4** should apply to **6** as well.

Pentad **6** and its *bis*-free base relative illustrate that it is possible to design a supramolecular system that can photochemically move an electron over a distance of ca. 80 Å at a cost of slightly less than 1 eV of driving force which results in conserving just over 50% of the ^1S energy. With respect to the fraction of the ^1S energy that is conserved in the reaction center *vs.* that conserved in the pentad, the absolute drop in redox potential accompanying overall charge separation is similar in both systems; the higher fractional conservation in the synthetic system is simply a consequence of a higher ^1S level. At this point, further experiments on supramolecular structures in which the intricacies of electronic and thermodynamic control of electron transfer can be coupled with strategies of series and parallel steps will be necessary to further define the limits of energy conversion efficiency in these systems.

Acknowledgment

This research was supported by the National Science Foundation (CHE-8903216), and the U.S. Department of Energy (DE-FG02-87ER13791). This is publication 96 from the Arizona State University Center for the Study of Early Events in Photosynthesis. The Center is funded by the U.S. Department of Energy grant DE-FG02-88ER13969 as Part of the U.S. Department of Agriculture-Department of Energy-National Science Foundation Plant Science Center Program.

References

Antolovich, M., Keyte, P. J., Oliver, A. M., Paddon-Row, M. N., Kroon, J., Verhoeven, J. W., Jonker, S. A. and Warman, J. M. (1991) 'Modeling long-range photosynthetic electron transfer on rigidly bridged porphyrin-quinone systems', J. Phys. Chem. 95, 1933-1941.

Bixon, M., Jortner, J., Michel-Beyerle, M. E., and Ogrodnik, A. (1989) 'A superexchange mechanism for the primary charge separation in photosynthetic reaction centers', Biochim. Biophys. Acta 977, 273.

Connolly, J. S. and Bolton, J. R. (1988) 'Artificial photosynthesis', in M. A. Fox and M. Channon (eds.), Photoinduced Electron Transfer, Part A, Elsevier, Amsterdam, Ch. 6.2.

Feher, G., Allen, J. P., Okamura, M. Y., and Rees, D. C. (1989) 'Structure and function of bacterial photosynthetic reaction centres', Nature 339, 111-116.

Gust, D., and Moore, T. A. (1989) 'Mimicking photosynthesis', Science 244, 35-41.

Gust, D., and Moore, T. A. (1991a) 'Mimicking photosynthetic electron and energy transfer', Advances in Photochemistry 16, 1-65.

Gust, D. and Moore, T. A. (1991b) 'Photosynthetic model systems', Topics in Current Chemistry 159, 103-151.

Gust, D., Moore, T. A., Moore, A. L., Seely, G., Liddell, P., Barrett, D., Harding, L. O., Ma, X. C., Lee, S.-J., and Gao, F. (1989) 'A carotenoid-porphyrin-diquinone tetrad: Synthesis, electrochemistry and photoinitiated electron transfer', Tetrahedron 45, 4867-4891.

Gust, D., Moore, T. A., Moore, A. L., Lee, S.-J., Bittersmann, E., Luttrull, D. K., Rehms, A. A., DeGraziano, J. M., Ma, X. C., Gao, F., Belford, R. E., and Trier, T. T. (1990) 'Efficient multistep photoinduced electron transfer in a molecular pentad', Science 248, 199-201.

Gust, D., Moore, T. A., Liddell, P. A., Nemeth, G. A., Makings, L. R., Moore, A. L., Barrett, D., Pessiki, P. J., Bensasson, R. V., Rougée, M., Chachaty, C., De Schryver, F. C.,

Van der Auweraer, M., Holzwarth, A. R., and Connolly, J. S. (1987) 'Charge separation in carotenoporphyrin-quinone triads: Synthetic, conformational and fluorescence lifetime studies', J. Am. Chem. Soc. 109, 846-856.

Gust, D., Moore, T. A., Moore, A. L., Barrett, D., Harding, L. O., Makings, L. R., Liddell, P. A., De Schryver, F. C., Van der Auweraer, M., Bensasson, R. V., and Rougée, M. (1988) 'Photoinitiated charge separation in a carotenoid-porphyrin-diquinone tetrad: Enhanced quantum yields via multistep electron transfers', J. Am. Chem. Soc. 110, 321.

Gust, D., Mathis, P., Moore, A. L., Liddell, P. A., Nemeth, G. A., Lehman, W. R., Moore, T. A., Bensasson, R. V., Land, E. J., and Chachaty, C. (1983) 'Energy transfer and charge separation in carotenoporphyrins', Photochem. Photobiol. 37S, S46.

Harrison, R. J., Pearce, B., Beddard, G. S., Cowan, J. A., and Sanders, J. K. M. (1987) Chem. Phys. 116, 429.

Hasharoni, K., Levanon, H., Tang, J., Bowman, M. K., Norris, J. R., Gust, D., Moore, T. A., and Moore, A. L. (1990) 'Singlet photochemistry in model photosynthesis: Identification of charge separated intermediates by Fourier transform and CW-EPR spectroscopies', J. Am. Chem. Soc. 112, 6477-6481.

Holzapfel, W., Finkele, U., Kaiser, W., Oesterhelt, D., Scheer, H., Stilz, H. U., and Zinth, W. (1990) 'Initial electron-transfer in the reaction center from *Rhodobacter sphaeroides*', Proc. Natl. Acad. Sci. USA 87, 5168-5172.

Joran, A. D., Leland, B. A., Felker, P. M., Zewail, A. H., Hopfield, J. J., and Dervan, P. B. (1987) Nature (London) 327, 508.

Kirmaier, C. and Holten, D. (1987) 'Primary photochemistry of reaction centers from the photosynthetic purple bacteria', Photosyn. Res. 13, 225.

Kong, J., and Loach, P. A. (1978) 'Covalently-linked porphyrin quinone complexes as RC models', in P. C. Dutton, J. S. Leigh and H. Scarpa (eds.), Frontiers of Biological Energetics: From Electrons to Tissues, Vol. 1, Academic, NY, pp. 73-82.

Kugimiya, S., Lazrak, T., Blanchard-Desce, M., and Lehn, J.-M. (1991) 'Electron conduction across vesicular bilayer membranes induced by a caroviologen molecular curve', J. Chem. Soc., Chem. Commun. 1179.

Marcus, R. A. and Sutin, N. (1985) 'Electron transfers in chemistry and biology', Biochim. Biophys. Acta 811, 265 (and references therein).

Moore, T. A., Gust, D., Mathis, P., Mialocq, J.-C., Chachaty, C., Bensasson, R. V., Land, E. J., Doizi, D., Liddell, P. A., Lehman, W. R., Nemeth, G. A., and Moore, A. L. (1984) 'Photodriven charge separation in a carotenoporphyrin-quinone triad', Nature (London) 307, 630-633.

Moore, T. A., Gust, D., Moore, A. L., Seta, P., Bienvenue, E., and Bensasson, R. V. (1987) 'Transmembrane charge transfer in model systems for photosynthesis', in V. Balzani (ed.), Photoinduced Charge Separation and Energy Migration in Supramolecular Species, D. Reidel Press, Boston, pp. 283-297.

Moore, T. A., Gust, D., Hatlevig, S., Moore, A. L., Makings, L. R., Pessiki, P. J., De Schryver, F. C., Van der Auweraer, M., Lexa, D., Bensasson, R. V., and Rougée, M. (1988) 'Photoinitiated electron transfer in carotenoporphyrin-quinone triads: Enhanced quantum yields via control of reaction exergonicity', Israel J. Chem. 28, 87-95.

Moser, C. C., Keske, J. M., Warncke, K., and Dutton, P. L. (1991) 'The distance-rate relationship of intramolecular electron transfer', Biophys. J. 59, 521a.

Parson, W. W. (1991) 'Reaction centers', in H. Scheer (ed.), Chlorophylls, CRC Press, p. 1153.

Platt, J. R. (1959) 'Carotene-donor-acceptor complexes in photosynthesis', Science 129, 372.

Seta, P., Bienvenue, E., Moore, A. L., Mathis, P., Bensasson, R. V., Liddell, P., Pessiki, P. J., Joy, A., Moore, T. A., and Gust, D. (1985) 'Photodriven transmembrane charge separation and electron transfer by a carotenoporphyrin-quinone triad', Nature 316, 653-655.

Schenck, C. C., Parson, W. W., Holten, D., and Windsor, M. W. (1981) 'Transient states in reaction centers containing reduced bacteriopheophytin', Biochim. Biophys. Acta 635, 383.

Schenck, C. C., Blankenship, R. E., and Parson, W. W. (1982) 'Radical pair decay kinetics, triplet yields and delayed fluorescnece from bacterial reaction centers', Biochim. Biophys. Acta 680, 44.

Tabushi, I., Koga, N., and Yanagita, M. (1979) Tetrahedron Lett. 257.

Warman, J. M., Smit, K. J., Deltaas, M. P., Jonker, S. A., Paddon-Row, M. N., Oliver, A. M., Kroon, J., Oevering, H., and Verhoeven, J. W. (1991) 'Long-distance charge recombination within rigid molecular assemblies in nondipolar solvents', J. Phys. Chem. 95, 1979-1987, (and references therein).

Wasielewski, M. R., Niemczyk, M. P., Svec, W. A., and Pewitt, E. B. (1985) J. Am. Chem. Soc. 107, 1080.

Yeates, T. O., Komiya, H., Chirino, A., Rees, D. C., Allen, J. P., and Feher, G. (1988) 'Structure of the reaction center from *Rhodobacter sphaeroides* R-26 and 2.4.2: Protein-cofactor (bacteriochlorophyll, bacteriopheophytin, and carotenoid) interactions', Proc. Natl. Acad. Sci. USA 85, 7993-7997.

INTRAMOLECULAR ELECTRON TRANSFER IN DONOR ACCEPTOR SYSTEMS LINKED BY A FLEXIBLE CHAIN.

F.C. De Schryver*, M. Van der Auweraer, N. Boens, D. Declercq, N. Helsen, S. De Paemelaere, P. Van Haver, A. Onckelinckx

K.U.Leuven
Department of Chemistry
Laboratory for Molecular Dynamics and Spectroscopy
Celestijnenlaan 200F
B-3001 Heverlee, Belgium

and K. Iwai
Department of Chemistry Faculty of Science
Nara Women's University
Nara 630, Japan

Abstract. In this contribution some recent developments in the area of intramolecular excited state interactions will be discussed. In a first part the power of global compartmental analysis of fluorescence decays of intramolecular exciplex forming systems is exemplified. In a second part the use of photoacoustic spectroscopy in the study of intramolecular back electron transfer is illustrated and in a last part the alternative linking of a donor and an acceptor with a flexible silane chain will be compared with a similar donor and acceptor linked by an alkane chain.

1. Compartmental Analysis of Intramolecular Exciplex Formation .

The kinetics of excited-state reactions can be characterized from time-resolved measurements of the concomitant fluorescence emission. Current methodologies in time [1,2] and frequency domain [3-8] provide data with sufficient accuracy allowing an elaborate data analysis.
In many cases the time relaxation of the excited system can be described by a sum of exponentially decaying functions. The parameters of interest are not the relaxation times and

V. Balzani and L. De Cola (eds.), Supramolecular Chemistry, 313–323.

the corresponding preexponential factors (these may be called descriptive or empirical parameters) but the rate constants in the excited state and the species associated spectra. To determine these parameters of interest experiments have to be conducted under a variety of conditions so that a multidimensional fluorescence decay data surface is obtained. From the resulting set of relaxation times and preexponential terms the rate constants and the relative species associated spectra can be obtained. The species associated spectra can then be obtained from steady-state spectra. An accurate determination of the relaxation times and the pre-exponentials and a high model discrimination power is obtained by performing a simultaneous analysis of all related decay traces. [9-30]. In this so-called global analysis approach parameters can be linked over various decay curves. For example, the relaxation times can be linked over decay traces collected at various emission wavelengths. However, for decay curves collected at various concentrations of a reactant the relaxation times can generally not be linked. To benefit the most of the merits of a global analysis approach, one has to fit for the underlying parameters, i.e. the rate constants which can be linked indeed. This global analysis has the additional advantage that the parameters of interest are determined directly from the complete decay data surface in a single step [17-30]. The expressions for the relaxation times and the preexponential factors in terms of the rate constants and the species associated spectra for a causal, linear, time-invariant and concentration dependent fluorescent system consisting of n different excited-state species can be obtained in matrix notation, allowing a general computer program implementation [25-30]. In the first implementation of this direct approach the decay curves had to be properly normalized to obtain the species-associated spectra [25]. Recently, an extension of this method has been described in which the condition for normalization is no longer required [28]. The power of this approach has been exemplified in intermolecular excited state interactions in excimers [29] and in exciplexes [30].

This methodology has been extended to intramolecular excited state interactions [31-32]. In this contribution the fluorescence decay of 1-(4-N,N-dimethylaminophenyl)-3-(9-phenanthrylpropane) (9-Phen) measured using single photon timing is analysed at different wavelength of emission in the presence of a quencher. The mathematical analysis and simulation of such a system has shown that using this approach all rate constants of the kinetic scheme as well as the species associated emission spectrum can be derived and the presence of eventual ground state interactions can be unequivocally determined [32]. The emission spectrum of 9Phen (fig.1) in solvents of different polarity consists of an emission band of 9-ethylphenantrene, the locally excited state, in the range of 350 to 400 nm and a broad intense structureless band at longer wavelengths which is due to the emission of the intramolecular complex of 9Phen. The complex emission maximum is solvent dependent and shifts to longer wavelengths with increasing polarity of the solvent. In isooctane, toluene, diethylether and acetonitrile the emission maximum of the complex is situated respectively at 422 nm, 443 nm, 443 nm and 503 nm. The shift is an indication that the emission can be attributed to a state with charge transfer character.

The emission decays of 9Phen in isooctane at room temperature can be analyzed as two exponential with decay times of respectively 2.71 ns and 32.27 ns. Based on a identifiability study [32] a set of decay data of 9Phen was measured at three different emission wavelength and five different concentrations of the quencher 1,4-dibromobenzene at each wavelength. Two time increments were used for each set of quencher and emission wavelength. The fluorescence decay of 9-ethylphenantrene in absence and in presence of quencher was studied and the obtained values were used in the analysis . The set of data were analyzed according to the global compartmental scheme 1 and the results are reported in table 1.

Scheme 1

$$
\begin{array}{ccccc}
1^* + Q & \underset{k_{12}}{\overset{k_{21}}{\rightleftarrows}} & 2^* + Q \\
\downarrow k_{01} \quad \downarrow k_{q1} & & \downarrow k_{02} \quad \downarrow k_{q2} \\
1 \; + \; Q & & 2 \; + \; Q
\end{array}
$$

where compartment 1 represents the excited(*) and ground state when there is no interaction between linked donor and acceptor and compartment 2 the excited(*) and ground state of the complex , Q the quencher and k the respective rate constants.

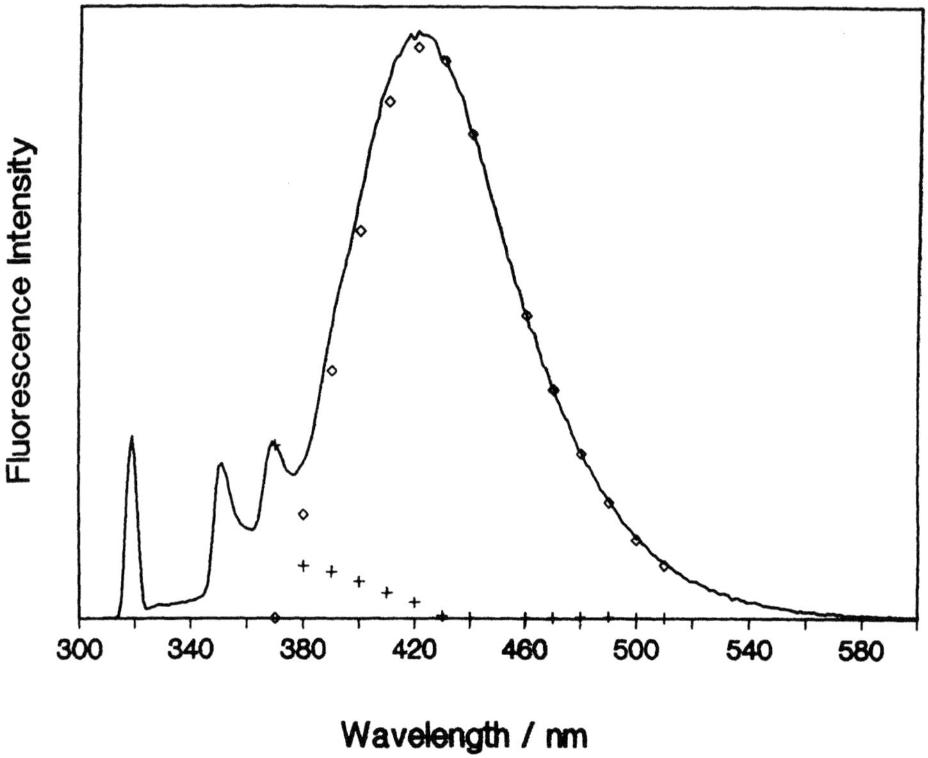

Figure 1 : Deconvoluted monomer (+) and exciplex (◊) fluorescence of 9Phen in isooctane (Excitation = 320nm)

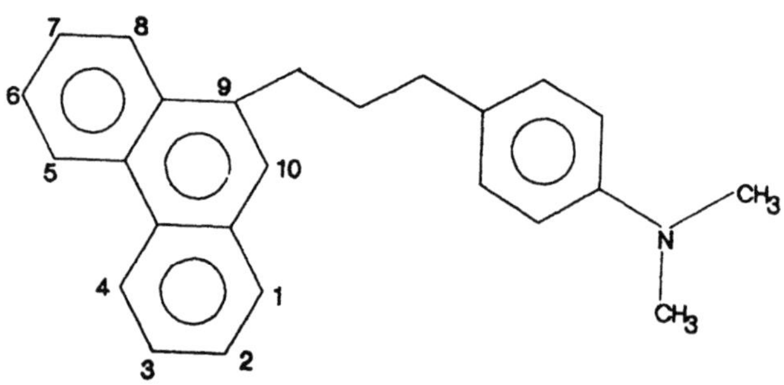

9Phen

Table I : Rate Constants and Relative Spectral Contributions of 9Phen in Isooctane at Room Temperature Estimated with Global Compartmental Analysis in Presence of the Quencher 1,4-dibromobenzene.

$k_{01} = 2.04 \pm 0.02\ 10^{7}\ s^{-1}$	$k_{21} = 3.32 \pm 0.67\ 10^{8}\ s^{-1}$
$k_{12} = 2.88 \pm 6.60\ 10^{7}\ s^{-1}$	$k_{02} = 3.13 \pm 0.22\ 10^{7}\ s^{-1}$
$\tilde{b}_1 = 0.99 \pm 0.01$	$\tilde{c}_1(370\ nm) = 1.00 \pm 0.18$
$\tilde{c}_1(400\ nm) = 0.35 \pm 0.02$	$\tilde{c}_1(460\ nm) = 0.02 \pm 0.01$

$\tilde{b}_1$ and $\tilde{c}_1$ are the normalised contribution of the absorption of the compartment 1 at 320 nm and of the normalized emission of compartment 1 at the respective emission wavelengths. The results indicate that in the excited state the equilibrium is shifted towards the complex and that in the ground state in this system no appreciable complex conformation is present. To determine in the species associated emission spectra, SAEMS, of the locally excited state and of the exciplex the fluorescence decay of the bichromophoric compound was measured every 10 nm throughout the emission band starting at 380 nm till 510 nm in absence of quencher. Based on the obtained $\tilde{c}_1$ and $\tilde{c}_2$ values the stationary emission spectrum can be deconvoluted in the SAEMS for the locally excited state and the complex as presented in fig. 1.

2. Influence of Polarity on the non radiative decay of the exciplex in a linked donor acceptor system.

It is interesting to obtain not only information on the rate of forward electron transfer in the exciplex formation but also on the rate of back electron transfer as occuring in the non radiative decay pathways of the exciplex.
Laser induced optoacoustic spectroscopy, LIOAS , is a very suitable method to study the radiationless decay processes of excited states. We have applied this method of LIOAS to access the total balance of energy in the intramolecular exciplex systems.[32]
The amplitude of the first optoacoustic signal H, the amount of prompt heat, is related to the energy of the incident laser pulse E_o by equation 1.

$$H = K\,\alpha\,E_o\,(1-10^{-A}) \quad (1) \qquad \text{eq 1}$$

where K is a constant containing the thermoelastic quantities of the medium and instrumental factors, A is the absorbance of the solution, and α is the fraction of the absorbed laser energy E_{abs} released as thermal energy E_{th}

within the response time of the detector, $\alpha = E_{th}/E_{abs}$. By plotting H as a function of E_0 for both the sample solution of bichromophoric compound and the 2-hydroxybenzophenone reference solution (α=1), α is calculated from the ratio of the two slopes directly., e.g. for 9Phen a value of $\alpha = 0.478$ in isooctane and of $\alpha = 0.898$ in acetonitrile is found.

The larger value of α in acetonitrile can be related to the smaller value of the quantum yield of fluorescence of locally excited state and of exciplex in a more polar solvent.

The amount of prompt heat for 9Phen can be calculated [33]. The ratio of the rate constant of internal conversion of the exciplex k_{ice} and the sum of the rate constants of all nonradiative pathways of the exciplex $k_{ice} + k_{isce}$ equal to ß, is calculated using the following equation.

$$\begin{aligned}\alpha \times A\ cm^{-1} &= B\ cm^{-1} + \Phi_{fm}\ C\ cm^{-1} \\ &+ \Phi_{iscm}\ E\ cm^{-1} \\ &+ \Phi_{fe}\ (-DH^{\circ} + E_{rep}) \\ &+ (\Phi_{ice} + \Phi_{isce})\ \beta\ F\ cm^{-1} \\ &+ (\Phi_{ice} + \Phi_{isce})\ (1 - \beta)\ E\ cm^{-1} \qquad \text{eq 2}\end{aligned}$$

The energy parameters in this equation are obtained in the following way:

A cm^{-1} = the excitation energy

B cm^{-1} = the energy difference between the Franck Condon excited state S_1^* and the relaxed locally excited state S_1 of the acceptor or donor.

C cm^{-1} = the energy difference between the Franck Condon ground state S_{0m}^* and the relaxed ground state S_0.

E cm^{-1} = the energy difference between S_1 and the locally excited triplet state T_1.

F cm^{-1} = the energy difference between S_1 and S_0.

$-DH^{\circ} + E_{rep}$ is the stabilization enthalpy of the exciplex and the sum of the ground state repulsion energy.

Φ_{fe} is the quantum yield of exciplex fluorescence, Φ_{fm} is the fluorescence quantum yield of the locally excited state, Φ_{iscm} is the quantum yield of intersystem crossing of the locally excited state, Φ_{icm} is the quantum yield of internal conversion of the locally excited state, Φ_{for} is the quantum yield of exciplex formation and $\Phi_{isce}+\Phi_{ice}$ is the sum of the quantum yields of intersystem crossing and internal conversion of the exciplex. Φ_{fm}° is the fluorescence quantum yield of a model compound. Φ_{icm} is supposed to be zero. Φ_{iscm}, Φ_{for}, and $\Phi_{isce} + \Phi_{ice}$ are calculated by the following equations.

$$\Phi_{iscm} = \Phi_{fm} (1 - \Phi_{fm}^{\circ}) \qquad \text{eq 3}$$
$$\Phi_{for} = 1 - \Phi_{fm}/\Phi_{fm}^{\circ} \qquad \text{eq 4}$$
$$\Phi_{isce} + \Phi_{ice} = (1 - \Phi_{fe}/\Phi_{for}) \qquad \text{eq 5}$$

From equation 2, ß values of 0.424 and 0.977 were obtained for 9Phen in isooctane and acetonitrile respectively. This can be compared with the values 0.21 and 1.00 obtained for 1-(1-pyrenyl)-3-(3-(N-methyl)indolylpropane and with the values 0.11 and 0.58 obtained for 1-(1-pyrenyl)-3-(2-(N-methyl)indolylpropane [33]. In all three examples does the branching ratio between internal conversion and intersystem crossing favours the former process upon increasing the solvent polarity. This can be explained within the framework of the current theory describing electron transfer [34] as the non radiative decay processes of the exciplex are actually charge recombination reactions.

3. Influence of the nature of the flexible linking chain on the intramolecular electron transfer process.

Where most of the flexible chains linking donor acceptor groups are based on carbon it is known that a silane chain has the capacity of extensive trough bond interaction [35]. To study the possibility of through bond interaction in such molecules 1-(4-NN-dimethylanilino)-3-(1-pyrenyl)hexamethyltrisilane PSi3D was synthesized [36] and its spectral properties compared with those of 1-(1-pyrenyl)heptamethyltrisilane PSi3 and of 1-(4-N,N-dimethylanilino)-3-(1-pyrenyl)propane PC3D [37]. Where the emission spectrum of PSi3 is independent of the solvent polarity excluding electron transfer from the silane chain to the excited pyrene does the emission spectrum as well of PSi3D (fig 2) and PC3D strongly depend on the polarity of the solvent. From this dependence using the Lippert Mataga equation the slope equals $2\mu^2/4\pi\epsilon_0 hc\rho^3$ where μ is the difference in dipole moment between ground and excited state and ρ is the radius of spherical cavity. A value of 7400 cm^{-1} and 12520 cm^{-1} is obtained for PC3D and PSi3D respectively. The absorption spectrum of Psi3D shows a slight shift of the ^1L_a transition and an broadening of the vibrational fine structure in comparison with PSi3. Furthermore the excitation spectrum of PSi3D measured in the locally excited state wavelength region and in the exciplex region differs suggesting an interaction between the donor and the acceptor. Additional evidence is found in the temperature behavior of donor acceptor linked by silane in solvents of medium polarity. On cooling the solution the emission of

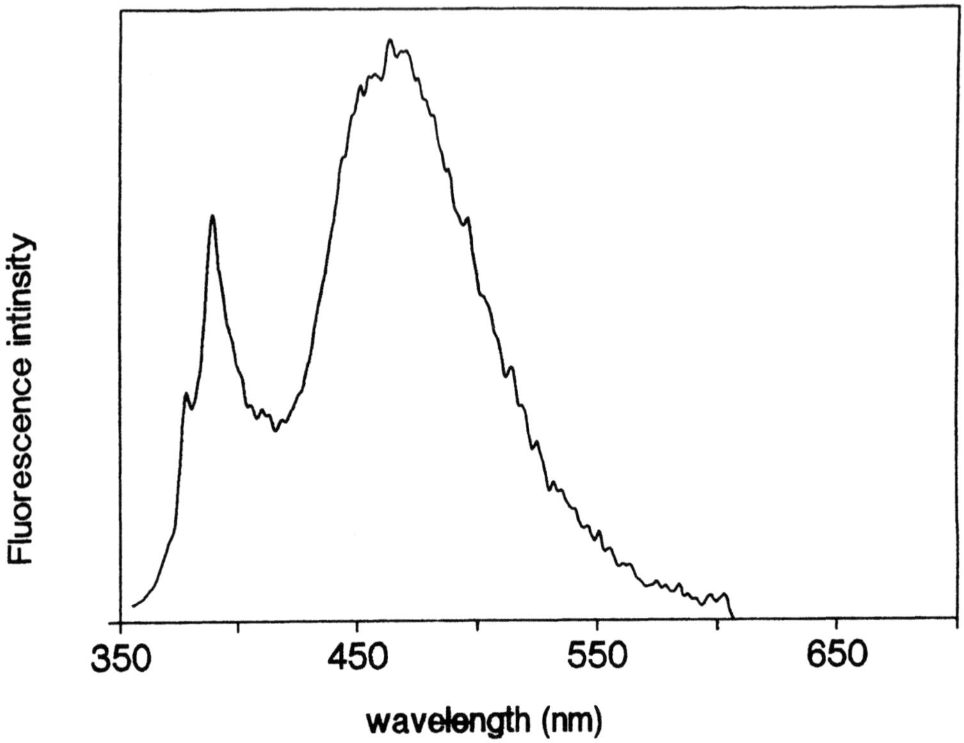

Figure 2 : Fluorescence spectrum of P3D in isooctane (Excitation = 350nm)

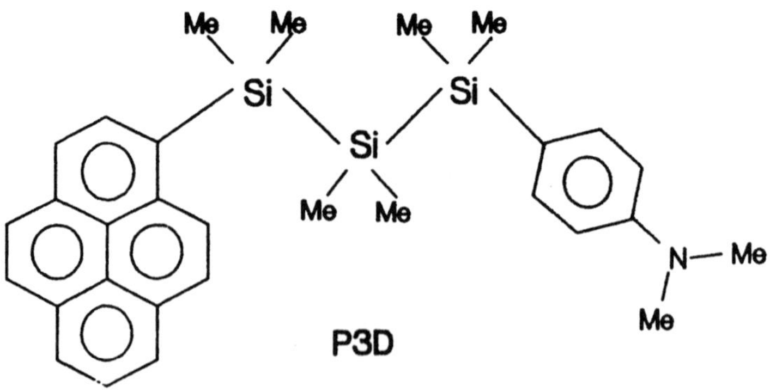

P3D

the locally excited state is reduced as is the total quantum yield of emission. This could indicate that on lowering the temperature the chain takes a conformation more favorable for through bond interaction. In the same temperature domain a donor acceptor system consisting of pyrene as acceptor but indole, a less good electron donor, shows opposite behavior and an increase of the emission intensity of the locally excited state and an increase in the total emission intensity. Further preliminary evidence for such an interaction is found in the fluorescence decay at longer wavelengths in the exciplex spectral region. The important deviation of the ratio of the preexponential terms of the decay from -1 in PSi3D not observed in PC3D could indicate interaction between the pyrenyl moiety and the anilogroup in the former compound and absence of this interaction in the latter one. As donor and acceptor are identical this could indicate that the interaction has to be mediated through the silane bonds. A further ongoing study of this and analogous systems will clarify this issue.

Acknowledgements. The authors are indebted to the Ministry of scientific Planning for financial support to the laboratory through UIAP-16. Dr. R. Miller is thanked for invaluable help in the synthesis of PSi3D. N.B. is Bevoegdverklaard Navorser of the Belgian Fonds voor geneeskundig Onderzoek, MVDA is Onderzoeksleider of The Belgian Fonds voor Kollektief Wetenschappelijk Onderzoek

References

(1) O'Connor, D.V.; Phillips, D. *Time-correlated single photon counting*; Academic Press: London, 1984.

(2) Boens, N. In *Luminescence Techniques in Chemical and Biochemical Analysis*; Baeyens, W.R.G.; De Keukeleire, D.; Korkidis, K. Eds.; Marcel Dekker: New York, 1991; pp 21-45.

(3) Lakowicz, J.R.; Laczko, G.; Gryczynski, I.; Szmacinski, H.; Wiczk, W. *J. Photochem. Photobiol., B: Biology* **1988**, *2*, 295-311.

(4) Laczko, G.; Gryczynski, I.; Gryczynski, Z.; Wiczk, W.; Malak, H.; Lakowicz, J.R. *Rev. Sci. Instrum.* **1990**, *61*, 2331-2337.

(5) Lakowicz, J.R.; Gryczynski, I.; Laczko, G.; Joshi, N.; Johnson, M.L. In *Luminescence Techniques in Chemical and Biochemical Analysis*; Baeyens, W.R.G.; De Keukeleire, D.; Korkidis, K. Eds.; Marcel Dekker: New York, 1991; pp 141-177.

(6) Gratton, E.; Limkeman, M. *Biophys.J.* **1983**, *44*, 315-324.

(7) Feddersen, B.A.; Piston, D.W.; Gratton E. *Rev. Sci. Instrum.* **1989**, *60*, 2929-2936.
(8) Gratton, E.; Alcala, J.R.; Barbieri, B. In *Luminescence Techniques in Chemical and Biochemical Analysis*; Baeyens, W.R.G.; De Keukeleire, D.; Korkidis, K. Eds.; Marcel Dekker: New York, 1991; pp 47-72.
(9) Eisenfeld, J.; Ford, C.C. *Biophys. J.* **1979**, *26*, 73-83.
(10) Knutson, J.R.; Beechem, J.M.; Brand, L. *Chem. Phys. Lett.* **1983**, *102*, 501-507.
(11) Löfroth, J.-E. *Eur. Biophys. J.* **1985**, *13*, 45-58.
(12) Beechem, J.M.; Brand, L. *Photochem. Photobiol.* **1986**, *44*, 323-329.
(13) Ameloot, M; Beechem, J.M.; Brand, L. *Biophys. Chem.* **1986**, *23*, 155-171.
(14) Weidner, R.; Georghiou, S. In *Time-Resolved Laser Spectroscopy in Biochemistry*; Lakowicz, J.R., Ed., Proc. SPIE 909, **1988**, 402-410.
(15) Han, M.K.; Walbridge, D.G.; Knutson, J.R.; Neyroz, P.; Brand, L. In *Fluorescent Biomolecules. Methodologies and Applications;* Jameson, D.M.; Reinhart, G.D. Eds.; Plenum Press: New York, 1989; pp 33-59.
(16) Janssens, L.D.; Boens, N.; Ameloot, M.; De Schryver, F.C. *J. Phys. Chem.* **1990**, *94*, 3564-3576.
(17) Löfroth, J.-E. *Anal. Instrum.* **1985**, *14*, 403-431.
(18) Beechem, J.M.; Ameloot, M.; Brand, L. *Anal. Instrum.* **1985**, 14, 379-402.
(19) Löfroth, J.-E. *J. Phys. Chem.* **1986**, *90*, 1160-1168.
(20) Beechem, J.M.; Gratton, E. In *Time-Resolved Laser Spectroscopy in Biochemistry*; Lakowicz, J.R., Ed., Proc. SPIE 909, **1988**, 70-81.
(21) Boens, N.; Malliaris, A.; Van der Auweraer, M.; Luo, H.; De Schryver, F.C. *Chem. Phys.* **1988**, *121*, 199-209.
(22) Boens, N.; Luo, H.; Van der Auweraer, M.; Reekmans, S.; De Schryver, F.C.; Malliaris, A. *Chem. Phys. Lett.* **1988**, *146*, 337-342.
(23) Boens, N.; Janssens, L.D.; De Schryver, F.C. *Biophys. Chem.* **1989**, *33*, 77-90.
(24) Weidner, R.; Georghiou, S. In *Time-Resolved Laser Spectroscopy in Biochemistry II*; Lakowicz, J.R., Ed.; Proc. SPIE 1304, **1990**, 717-726.
(25) Beechem, J.M.; Ameloot, M.; Brand, L. *Chem. Phys. Lett.* **1985**, *120*, 466-472.
(26) Ameloot, M.; Beechem, J.M.; Brand, L. *Chem. Phys. Lett.* **1986**, *129*, 211-219.
(27) Piston, D.W.; Bilash, T.; Gratton, E. *J. Phys. Chem.* **1989**, *93*, 3963-3967.
(28) Ameloot, M.; Boens, N.; Andriessen, R.; Van den Bergh, V.; De Schryver, F.C. *J. Phys. Chem.* **1991**, *95*, 2041-2047.

(29) Andriessen, R.; Boens, N.; Ameloot, M.; De Schryver, F.C. *J. Phys. Chem.* **1991**, *95*, 2047-2058.
(30) Khalil, M.M.H.; Boens, N.; Van der Auweraer, M.; Ameloot, M.; Andriessen, R.; Hofkens, J.; De Schryver, F.C. *J. Phys. Chem.*, **1991**, *95*, 9375.
(31) Andriessen, R.; Ameloot, M.; Boens, N.; De Schryver, F.C. *J. Phys. Chem.*, in press.
(32) Boens, N; Ameloot, M.; Andriessen, R.; De Schryver, F.C.; Hermans, B., in press
(33) Van Haver, Ph.; Helsen, N.; Depaemelaere, S.; Van der Auweraer, M.; De Schryver, F.C. *J. Am. Chem. Soc.* **1991**, *113*, 6849.
(34) Marcus, R.A. *J. Phys. Chem.* **1989**, 93, 3078
(35) Miller, R.D.; Michl, J. *J. Chem. Rev.*, **1989**, *89*, 1359.
(36) Declercq, D.; De Schryver ,F.C.;Miller, R.D. *Chem. Phys. Letters* **1991**, in press
(37) Masaki,S.; Okada,T.;Mataga,N.;Sakata, Y.; Misumi, S. *Bull. Chem. Soc. Jpn.*, **1976**, 49,1277

ATTEMPTS TO MIMIC BIOMEMBRANE PROCESSES: FUNCTION OF PHOSPHOLIPASE A_2 AT LIPID MONOLAYERS

A. Reichert, H. Ringsdorf, A. Wagenknecht
Institute of Organic Chemistry
Becher Weg 18-20
University Mainz
W-6500 Mainz, Germany

ABSTRACT. The interplay between molecular self-organization and molecular recognition leads to the construction of functional supramolecular systems in which order and mobility are combined and whose function is based on their organization. These fascinating phenomena, for which the living cell is a perfect example, can be understood only if a wide range of scientific disciplines work together. For this purpose, synthetic supramolecular systems can be used to simulate natural biomembrane processes. One example is the function of enzymes which is coupled to a recognition process. In the interaction of phospholipase A_2 with lecithin monolayers, specific recognition between enzyme and substrate is followed by an active phase of lipid cleavage and then by aggregation of the enzyme to give domains of regular morphology. Fluorescence microscopy can be used to follow this process directly. Spontaneous enzyme domain formation can be induced by using mixed monolayers of either DPPC/lysoPPC/palmitic acid or DPPC/palmitic acid alone. Binding studies with a cationic dye show that a locally high negative charge density is one of the prerequisites for enzyme binding. In addition however, small amounts of D- or L-α-DPPC headgroups within the monolayer seem to be necessary for the protein domain formation.

1. Functionalized Supramolecular Systems

One of the areas of organic chemistry now being intensively studied on an interdisciplinary basis in many laboratories, is that of supramolecular systems. Chemists are becoming more and more interested in investigating organic chemistry beyond the covalent bond. In nature, system functioning is achieved by interplay between different functional units. The smallest unit which demonstrates this perfectly is the cell [1]: it is the interplay between molecular self-organization and molecular recognition of its individual constituents (lipids, proteins, nucleic acids...) which leads to the construction of this natural supramolecular system. It combines order and mobility, and its functioning is based on its organization. This fascinating phenomenon can be understood only if a wide range of modern scientific disciplines work together. The special tasks of organic and macromolecular chemistry in this connection involve finding new syntheses and solving structural problems. One of the interesting questions which arises is whether it is possible to develop new materials which simulate natural supramolecular systems in that their functioning is determined by their organization. This will not be possible without going beyond the limits of classical thinking in

V. Balzani and L. De Cola (eds.), Supramolecular Chemistry, 325–341.

the individual disciplines. Where are examples of connecting links between apparently separated disciplines of science? Figure 1 describes common perspectives of life science and materials science relating to functional supramolecular systems.

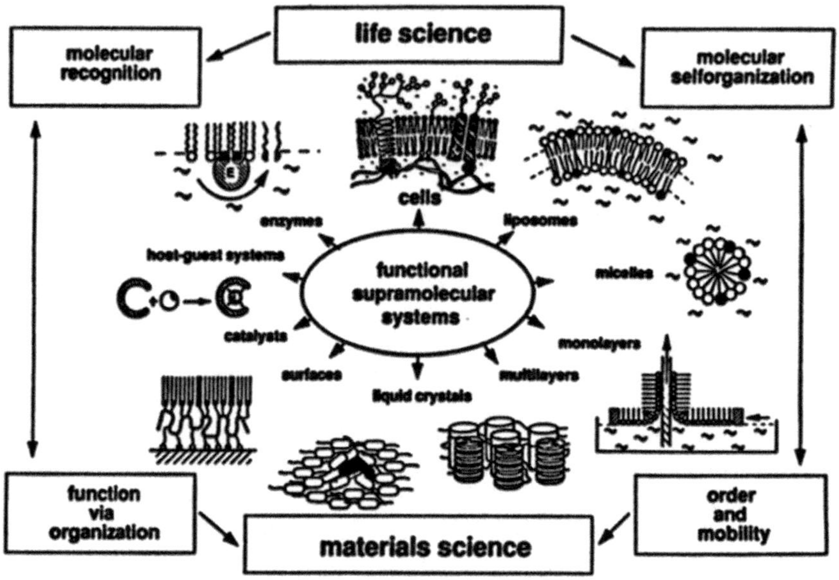

Figure 1. Functional supramolecular systems - connecting link between life science and materials science [2].

2. Attempts to Simulate Recognition Processes and Protein Functions on Model Membranes

The self-organization of lipids to form the plasma membrane [1] was a crucial step in the evolution of the earliest forms of life. Whereas the lipid bilayer determines the basic structure of biological membranes, most of their functions are provided by proteins embedded in the lipid bilayer. However, the structure of the cell membrane is so complex that it is diffcult to study specific individual processes in biological systems. This is why it is an advantage to choose model membranes for investigating biomembrane processes because they can be limited to a few variable factors. Figure 2 outlines various membrane processes which can be simulated with monolayers as model membrane systems.

The first important step in many surface reactions is the specific recognition taking place between a ligand on the membrane and a receptor protein (for example an antibody). An interaction of this type can be studied using fluoresceinlipids and the anti-fluorescein antibody [2]. As a consequence of specific recognition different processes can take place on the membrane. One example of this is the 2D crystallization of streptavidin after binding to biotin-lipid monolayers [2-4]. If the protein is an enzyme, the specific recognition can be followed by an enzymatic function being carried out. This will be illustrated in more detail by the example of phospholipid-monolayer hydrolysis by phospholipase A_2 [2,5,6].

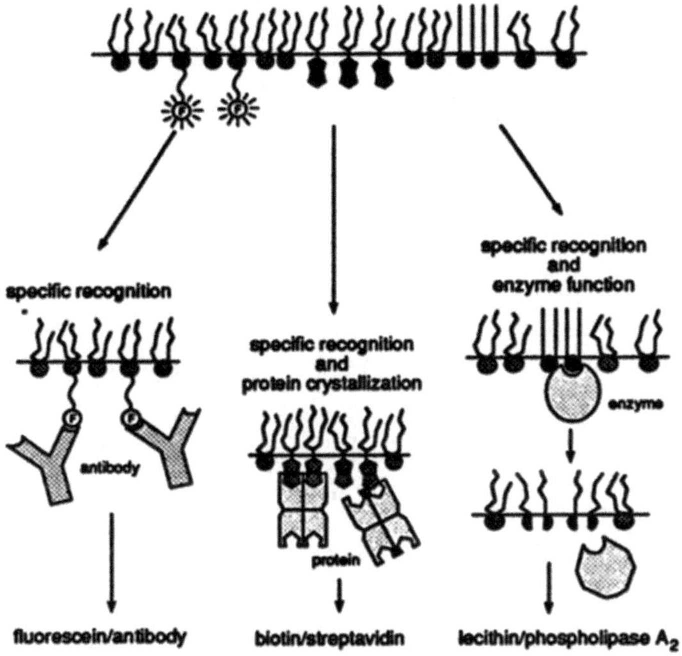

Figure 2. Examples of the simulation of specific recognition reactions at model membranes: recognition (fluorescein/anti-fluorescein antibody), recognition and protein crystallization (biotin/streptavidin) and enzyme function (lecithin, phospholipase A_2).

The technique used for the investigation of specific interaction of phospholipase A_2 with substrate monolayers is the fluorescence microscopy. This method is very well known for studying the formation of solid-analogous lipid domains during the main phase transition in lipid monolayers using a number of different fluorescence microscopy systems [7-9]. Recently it could be shown that fluorescence microcopy is also very suitable to study protein-lipid interactions at monolayers [2-6]. The fluorescence film balance used in all following experiments has been described elsewhere [7]. Materials and experimental procedure are briefly given in the text. A detailed description can be found in the cited papers [7,8].

3. Enzyme Substrate Interaction: Specific Recognition, Function and Domain Formation of Phospholipase A_2 at Lecithin Monolayers

3.1. PROPERTIES AND FUNCTION OF PHOSPHOLIPASE A_2 (PLA_2)

Phospholipases comprise an important class of enzymes [10]. They catalyze the hydrolysis of ester bonds in membrane-forming phospholipids. The enzyme from this class which has been investigated most thoroughly is phospholipase A_2 [11, 12]. One of its important functions is to rebuild and modify cell membranes. However, the detailed mechanism for regulating this enzyme is still unknown. Phospholipase A_2 also induces the release of arachidonic acid which leads to the biosynthesis of compounds which are involved in inflammation processes (prostaglandines, leukotrienes, thromboxanes) [13]. Thus, a way of regulating enzyme activity would provide a good method for controlling inflammatory processes.

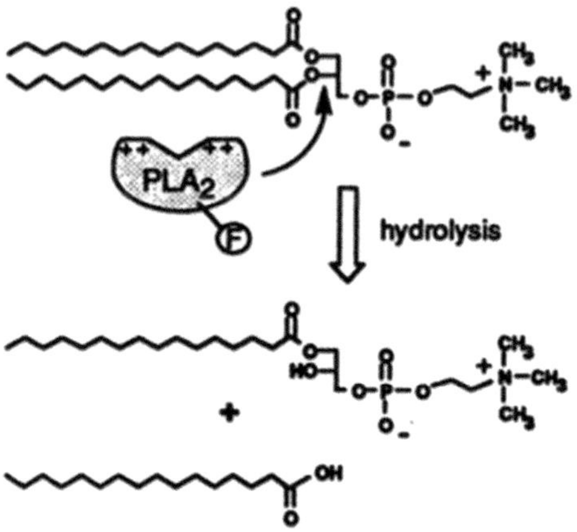

Figure 3. Cleavage reaction catalyzed by phospholipase A_2 (schematic): the enzyme hydrolyses the ester bond in C-2 position of a lecithin, resulting in the corresponding lysolecithin and the free fatty acid.

Phospholipase A_2 is a relatively small protein with a molecular weight of 12-14 kDa, it is soluble in water, and is very stable, having a large number of disulfide linkages. It catalyzes ester cleavage at the C-2 position in naturally occurring glycerophospholipids to give the corresponding lysolecithin and the free fatty acid (Fig. 3). The reaction takes place both regioselectively, that is, only the ester linkage in the C-2 position is cleaved, and stereospecifically, that is only the naturally occurring L-form of the lipid can be hydrolyzed, the D-form is inert [14]. Phospholipase A_2 activity depends very strongly on the physical state of the substrate [15]. It increases many times in going from a homogeneous solution to organized substrates such as micelles, liposomes or monolayers. The highest activity in liposomes is found when the lipid is in the phase transition region [16]. This property of phospholipase A_2 makes the enzyme especially interesting for investigations in monolayers because it is easy to bring about large changes in both the composition and the physical state of the membrane .

3.2. HYDROLYSIS OF SUBSTRATE LIPIDS OF PHOSPHOLIPASE A_2 IN MONOLAYERS

In order to visualize the hydrolytic action of phospholipase A_2 with the fluorescence microscope, two different fluorescence labels were needed. Thus the monolayer was doped with a sulforhodamine-lipid (SR-DPPE = N-(Texas Red sulfonyl)dipalmitoyl L-α-phosphatidyl-ethanolamine), and the phospholipase was labeled with fluorescein isothiocyanate (FITC). These two labels made it possible to observe processes in the lipid monolayer and the behavior of the enzyme separately. The experimental concept is outlined in Figure 4.

The lipid monolayer is compressed into the phase transition region. This leads to the formation of solid-analogous lipid domains in a liquid-analogous matrix (Fig. 4 A). Injection of the labeled phospholipase A_2 into the subphase (Fig. 4 B) is followed by specific recognition between the enzyme and its substrate. The attack by the protein takes place preferentially at the boundary between the solid-analogous and the liquid-analogous phase (Fig. 4 C). As hydrolysis progresses, the solid-analogous lipid domains are broken down and the hydrolysis products accumulate in the monolayer (Fig. 4 D). After a certain time, the phospholipase A_2 starts to aggregate to form domains of its own which have a specific shape (Fig. 4 E). This may indicate a change in the conformation of the protein due to interaction with the hydrolyzed monolayer. These processes can be observed directly under the fluorescence microscope [5,6], as will be illustrated by the example of L-α-dipalmitoylphosphatidylcholine (DPPC).

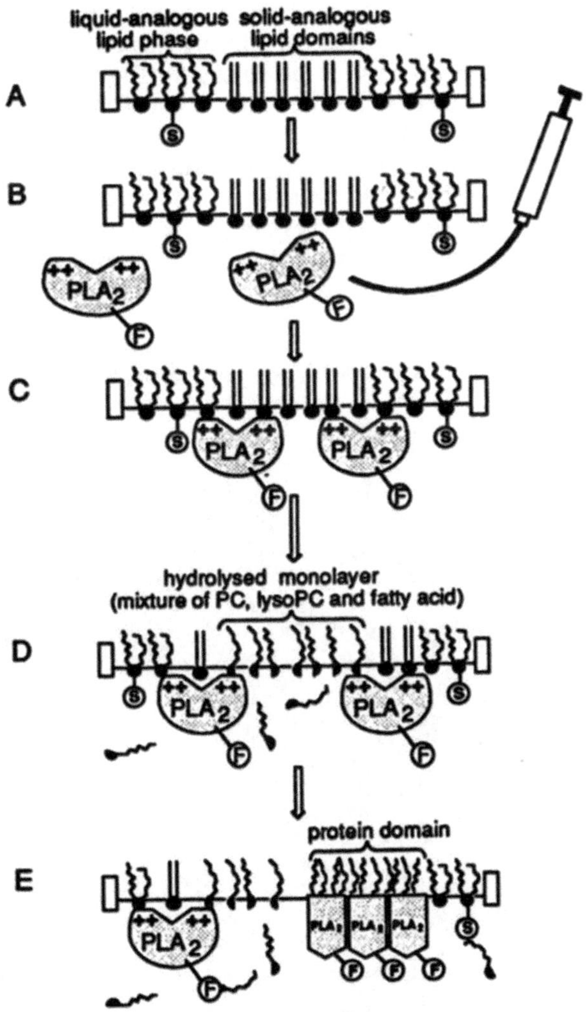

Figure 4. Hydrolysis of a lipid monolayer by phospholipase A_2 (schematic): (A) monolayer in the phase transition region with solid-analogous lipid domains in a liquid-analogous matrix mixed with SR-DPPE; (B) injection of FITC-labeled phospholipase A_2; (C) specific recognition of the lipids by the enzyme; (D) hydrolysis of the lipid domains and accumulation of the hydrolysis products in the monolayer; (E) aggregation of the enzyme. PC = phosphatidylcholine.

Hydrolysis of Solid-Analogous L-α-DPPC Domains [5,6]. As mentioned above, the enzyme is injected under a monolayer which is in the phase transition region. Viewed through the sulforhodamine filter, dark solid-analogous lipid domains which are typical of L-α-DPPC [17] are seen in the bright matrix of liquid-analogous lipid and sulforhodamine marker (see Fig. 5 A and scheme in Figs. 4 A, B). With the fluorescein filter, the homogeneous fluorescence of the protein is seen under the monolayer (see Fig. 5 B). The first signs that the enzyme has recognized its substrate in the monolayer and started to hydrolyze it appear after 10 minutes: firstly areas of erosion form specifically at the concavities previously present in the lipid domains (see Fig. 5 C and scheme in Fig. 4 C). The enzyme then starts to degrade the domains inwards from these points on (Fig. 5 C, E). Once the enzyme has bound to the phase boundary and started to hydrolyze the lipid, there is local accumulation of the acid and lysolecithin, which may lead to autocatalytic activation of the phospholipase A_2 at this point [18]. This would explain why further hydrolysis from an initial point of attack is preferred. After about 60 minutes, only slight remains of the lipid domains are left (Fig. 5 E, G, I, L).

Observation through the fluorescein filter reveals small bright dots after 30 minutes (Fig. 5 F): the protein is starting to aggregate. These bright dots increase in number and size (Fig. 5 F, H, K) until, at the end (after about 60 minutes), they have a typical kidney shape (see Fig. 5 L) . The monolayer now consists only of almost completely destroyed solid-analogous lipid domains and the protein domains in a mixed matrix of fluid lipid, lysolipid and fatty acid (see Fig. 5 L, M) .

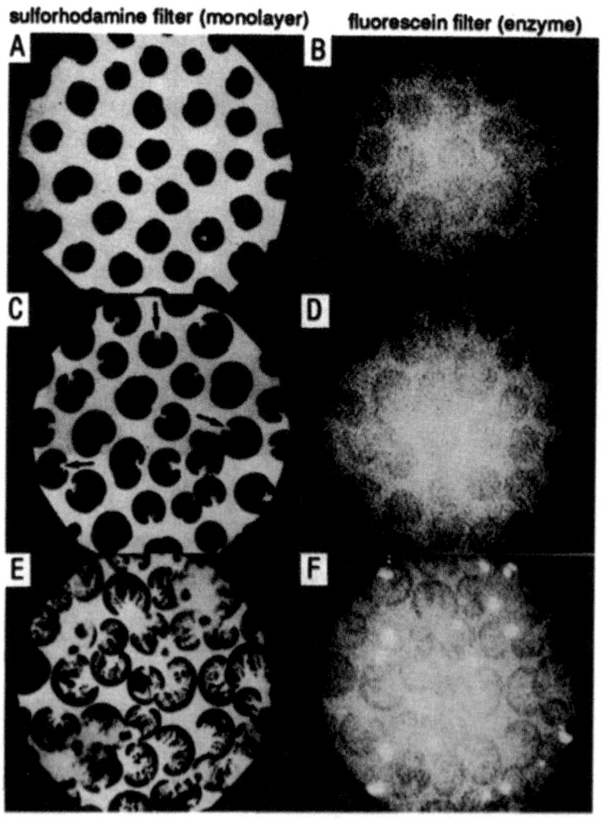

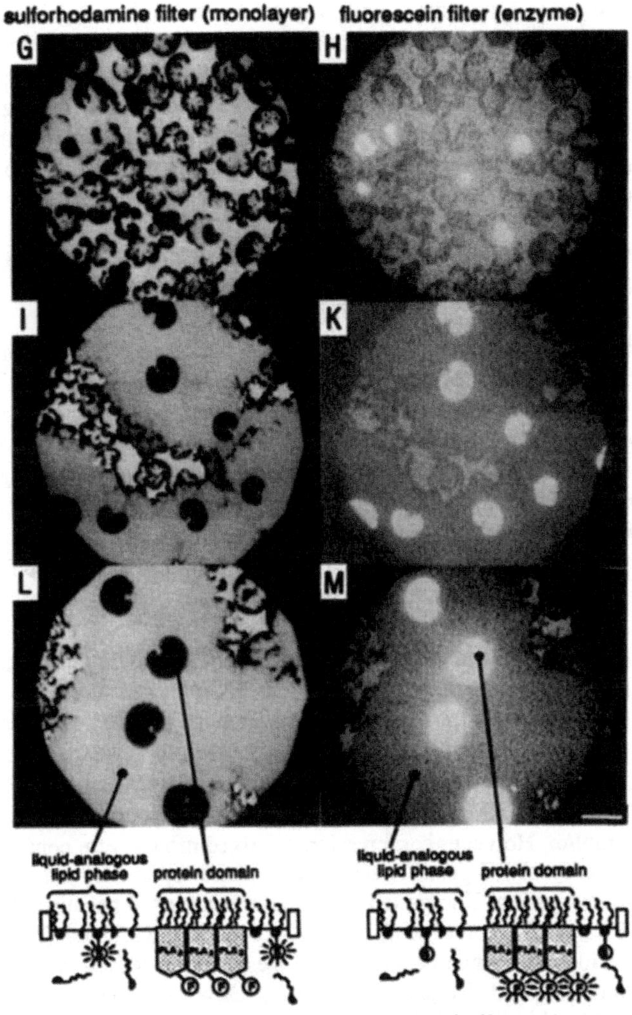

Figure 5. Hydrolysis of an L-α-DPPC monolayer by phospholipase A_2 observed by fluorescence microscopy; the monolayer is seen through the sulforhodamine filter (on the left) and the enzyme is seen through the fluorescein filter (on the right): (A) dark solid-analogous lipid domains in a bright liquid-analogous matrix; (B) homogeneous fluorescence of the enzyme in the subphase; (C) first signs of hydrolysis of the lipid domains (arrow); (D) homogeneous fluorescence of the enzyme in the subphase; (E), (F) progressive degradation of the lipid domains and first signs of the formation of enzyme aggregates below the monolayer (bright spots through the fluorescein filter (E), dark spots through the sulforhodamine filter (F)); (G)-(K) further hydrolysis of the lipid domains (G, I) with simultaneous growth of the protein domains (H, K); (L), (M) after almost complete degradation of the lipid domains there are now only protein domains in a fluid matrix composed of the hydrolysis products and DPPC; these enzyme domains have a specific shape and appear dark through the sulforhodamine filter (L) and bright through the fluorescein filter (M); (A), (B) t = 0 min (immediately after enzyme injection); (C), (D) t = 10 min; (E), (F) t = 30 min; (G), (H) t = 40 min; (I), (K) t = 50 min; (L), (M) t = 60 min; Π = 22 mN/m; T= 30°C; scale bar in (M) is 20 μm.

The first evidence that the lysolecithin and fatty acid produced during hydrolysis play an important part in the formation of the protein domains was gained from analogous experiments with D-α-DPPC: although the enzyme is able to recognize this lipid it cannot cleave it [14]. Unlike the hydrolysis of an L-α-DPPC monolayer, even after 2 hours there is neither degradation of the solid-analogous lipid domains nor formation of protein domains. Thus, because the existence of the hydrolysis products in the monolayer appears to be a necessary precondition for the formation of protein domains, investigations were carried out using mixed monolayers.

3.3. PHASE BEHAVIOR OF MIXED LECITHIN, LYSOLECIThIN AND PALMITIC ACID MONOLAYERS AND THEIR INTERACTION WITH PHOSPHOLIPASE A_2

In order to confirm that the existence of the hydrolysis products in the monolayer is crucial for the observed aggregation of phospholipase A_2 the effect of the hydrolysis products was investigated separatly from the hydrolysis process. This can be achieved by using mixed monolayers composed of lecithin, lysolecithin and fatty acid in various ratios. Such mixtures simulate the situation after different extents of monolayer hydrolysis. Assuming that only the cleveage products and not reactions during the hydrolysis (for example acylation [19]) induce domain formation of phospholipase A_2, there should be spontaneous development of protein domains when the composition of the mixed monolayers is suitable. In order to find the suitable composition, ternary as well as binary mixtures were investigated.

3.3.1. Investigations of Ternary Mixtures of DPPC, L-α-lysoPPC and Palmitic Acid [20]. In the mixtures of DPPC (L-form and D-form) with L-α-lysoPPC/palmitic acid, the amount of the hydrolysis products (1:1) was varied from 20% to 90% in steps of 10%. After spreading of the appropriate mixture doped with 0.5mol% DPPE-SR on buffer subphase the liquid-analogous monolayer appear homogeneously bright in the microscope (sulforhodamine filter, Fig. 6 A). Monolayer images of films with 20%, 30% and 40% of hydrolysis products do not change during compression up to 20 mN/m. However compression of mixed monolayers containing more than 40% of the 1/1 mixture of lysoPPC/palmitic acid lead to phase separation and formation of grey domains of unkown composition. The mixed domains appears grey because, unlike the black domains of pure DPPC, the fluorescence dye is not completely excluded (Fig. 6 B). Further compression leads to an increase in number and size of these regular shaped mixed domains.

After injection of fluorescein-labeled phospholipase A_2 below the domain containing monolayer (at 18mN/m), spontaneous binding of the enzyme onto these mixed grey domains can be visualized by switching the filter to view the fluorescein fluorescence. Domains which are grey in the sulforhodamine filter appear bright in the fluorescein filter already after 1-2 minutes (Fig. 6 C). This rapid binding of the enzyme to the monolayer is even more noteworthy when it is compared with the slow formation of protein domains during hydrolysis of L-α–DPPC monolayers [2,5,6] by phospholipase A_2, described in 3.2. The mixed domains therefore seem to represent a preferred environment for enzyme binding. Such a process is demonstrated in Figure 6 for a mixture of 10% L-α-DPPC with 90% lysoPPC/palmitic acid (1:1) simulating nearly complete hydrolysis of a lecithin monolayer. The shape of the formed mixed domains and therefore also of the protein domains is nearly the same as that observed for hydrolysis of L-α-DPPC after 60 minutes [6] (compare Fig. 5).

It is important to note that identical results (phase separation of mixed domains within the monolayer and fast binding of phospholipase A_2 to this domains) can be obtained with the non-hydrolyzable but recognizable D-α-DPPC instead of the L-α-DPPC in the mixtures [14]. Thus it appears that recognition of the headgroup of the lecithin, but not its cleavage, is necessary for rapid aggregation of the protein under the mixed monolayer. On the other hand the lecithin headgroup may not be needed at all to form the protein domains.

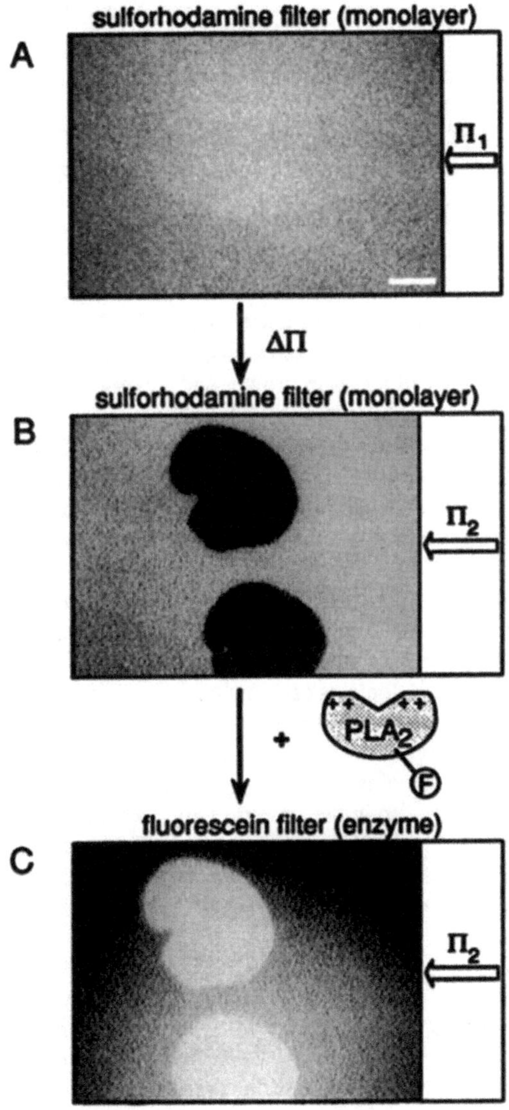

Figure 6. Interaction of fluorescein-labeled phospholipase A_2 with mixed monolayers composed of 90% lysoPPC/palmitic acid and 10% L-α-DPPC doped with 0.5mol% SR-DPPE: (A) homogeneous fluorescence of the liquid-analogous monolayer (Π_1 = 1 mN/m) ; (B) phase separated grey domains of unknown composition (Π_2 = 18mN/m); (C) fluorescein-labeled phospholipase A_2 binds very quickly to these mixed domains which thus appear bright in the fluorescein filter; Subphase: Tris buffer, pH 8.9; Temperature = 30°C; Scale bar in (A) is 20μm.

3.3.2. Investigations of Binary Mixtures [20]. In order to finf out which of the hydrolysis products in the ternary mixtures are necessary to induce phase separation followed by rapid enzyme aggregation at the monolayer, binary mixtures doped with 0.5mol% SR-DPPE were investigated.

In a mixed monolayer of lysoPPC and palmitic acid (1:1) without DPPC, neither phase separation within the monolayer nor protein domain formation after injection of the enzyme can be detected. Uncleaved lecithin thus seem to be necessary to induce domain formation protein aggregation.

All mixtures containing D-α-DPPC and palmitic acid (30%, 40% or 50%) show nearly the same behavior as the ternary ones. Directly after spreading on buffer subphase the monolayer looks homogeneously bright in the sulforhodamine filter. After compression small grey domains sometimes bright at the borders appear (Fig. 7 A). During further compression up to 18 mN/m these mixed domains grow in size and number (Fig. 7).

In contrast to ternary mixtures where only grey domains appear in the monolayer during compression up to 18 mN/m, use of the binary mixture of D-α-DPPC/fatty acid leads to an additional crystallization of DPPC (Fig. 7 B). Thus at 18 mN/m one finds a coexistence between lecithin domains (black in the sulforhodamine filter) and mixed domains (grey in the sulforhodamine filter). In the following experiments only the behavior of the mixed grey domains will be discussed. Size of these mixed domains depends strongly on the amount of fatty acid used in the mixture. In the case of 50% palmitic acid in the monolayer, the phase separated areas are of regular shape and clearly visible (diameter about 25 μm). Decrease of the amount of acid results in decreasing size of the domains and at a concentration of 30% palmitic acid no phase separation of mixed grey domains can be detected any more.

If fluorescein-labeled phospholipase A_2 is injected under a mixed monolayer of DPPC/palmitic acid which contains phase separated mixed domains (Fig. 7 B, C; arrow) as well as black DPPC domains the enzyme rapidly binds to the grey domains within 2 minutes as already shown for the ternary mixtures in Figure 6. These mixed domains can now be easily distinguished from the black lipid domains of pure lecithin by switching the filter to view the fluorescein fluorescence (Fig. 7 C). Only the mixed domains appear bright due to enzyme binding and no protein can be found below the black, probably pure DPPC domains. In addition, the fluorescein-labeled phospholipase A_2 allows detection of phase separation also in the mixture with 30% palmitic acid. The mixed domains are very small and therefore only visible in the fluorescein filter as small pinpoints of light after enzyme binding. The whole process of phase separation and enzyme interaction is shown in Figure 7 for two different monolayers which contain 50% palmitic acid (1) or 30% palmitic acid (2) respectively.

In mixed systems of lysoPPC/palmitic acid and DPPC/lysoPPC no phase separation of mixed domains can be observed. The mixed monolayers appear homogeneously bright in the sulforhodamine filter and nothing changes during compression up to 25 mN/m. For both mixtures, injection of fluorescein-labeled phospholipase A_2 at 18 mN/m results in homogeneous fluorescein fluorescence with no domain formation and thus no enzyme domain formation occuring even after a 60 minutes wait. This is a further hint that both palmitic acid as well as uncleaved lecithin, are necessary to induce enzyme aggregation.

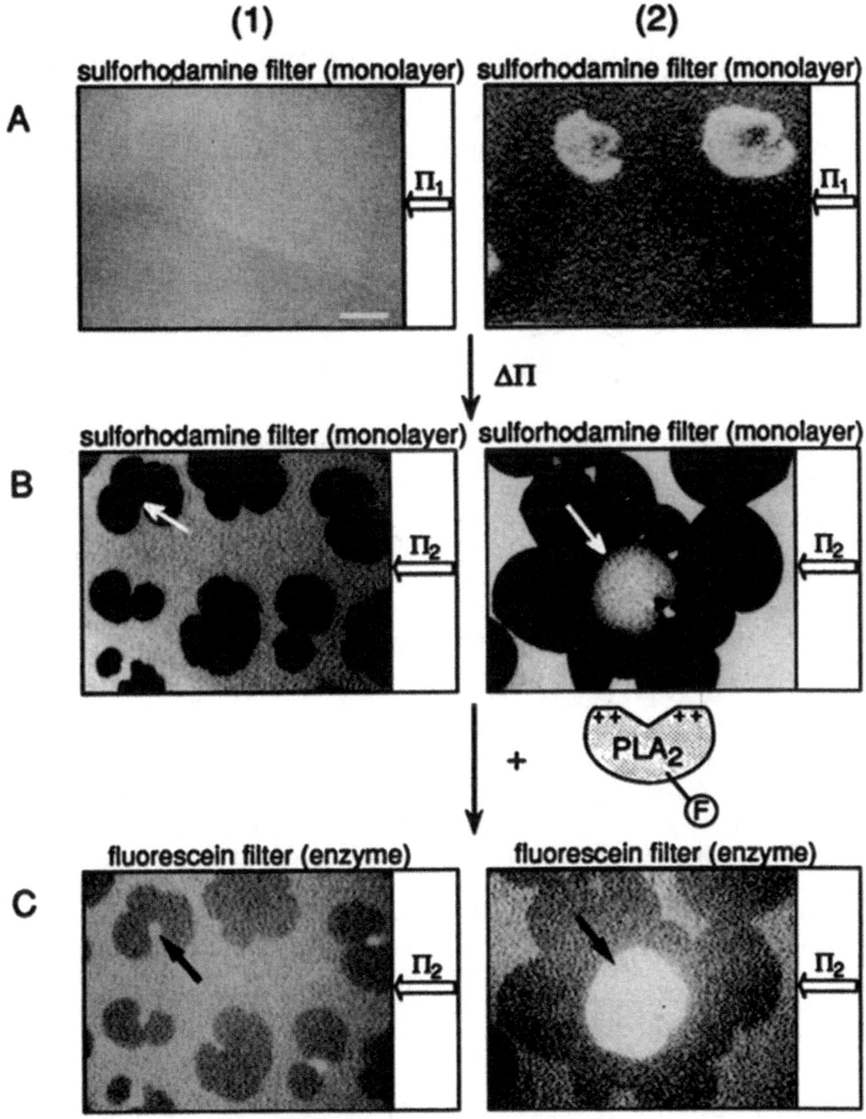

Figure 7. Interaction of fluorescein-labeled phospholipase A_2 with mixed monolayers consisting of (1): 50% DPPC/50% palmitic acid; (2): 70% DPPC/30% palmitic acid doped with 0.5 mol% SR-DPPE: (A) phase separated mixed domains of unknown composition (Π_1 = 12mN/m); (B) compression (Π_2 = 18mN/m) leads to an additional crystallization of solid analogous DPPC around the mixed grey domains (arrow); (C) fluorescein-labeled phospholipase A_2 injected under this monolayer binds very quickly (within 2 minutes) to these mixed grey domains which thus appear bright in the fluorescein filter (arrow); Subphase: Tris buffer, pH = 8.9; Temperature = 30°C, Scale bar in (A) is 20μm.

3.3.3. Interaction of Mixed Monolayers with a Cationic Dye. For further characterization of the mixed domains which induce preferential binding of phospholipase A_2, interaction of these mixed domains with 1, 1', 3, 3, 3', 3' -hexamethylindocarbocyanine-iodide **1** was examined [20]. This dye has a positive charge and binds very strongly to negatively charged interfaces [21]. Jain et al. have already used this dye in liposomal systems to study the influence of charges on phospholipase A_2 activity [22].

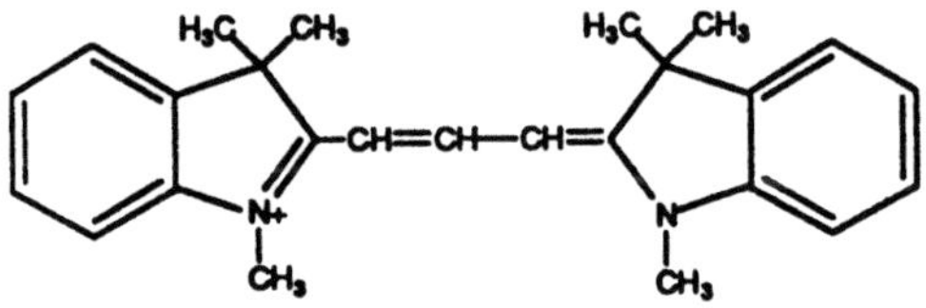

1

After spreading of the appropriate mixture (binary or ternary) on buffer subphase the monolayer looks homogeneously bright in the microscope (fluorescein filter). Compression again leads to phase separation of domains within the fluid matrix. Injection of 0.5 ml of the dye solution under the monolayer leads to direct binding of the dye to the phase separated areas indicating that negative charge (fatty acid) is concentrated in these areas.

Thus, the cationic dye **1** and phospholipase A_2 - as discussed in 3.3.1./3.3.2. - show rapid binding to the same phase separated areas suggesting so far that preferential binding of the enzyme is the result of charge-charge interactions. It is pertinent in this respect that the active site of phospholipase A_2 is surrounded by cationic amino acids [12] which are important for enzyme binding to interfaces [12, 23].

These findings are in agreement with results from Jain et al. showing that hydrolysis of DMPC vesicles with phospholipase A_2 leads to preferential binding of the cationic dye after a certain extent of hydrolysis [22], as well as to a higher amount of bound enzyme. All these results, explained by phase separation of acid are obtained by indirect evidences and domain formation of the enzyme is not mentioned. Fluorescence microscopy gives the first direct evidence of phase separation of negative charged domains. In addition one can directly visualize that these domains can specifically bind the cationic dye as well as the labeled phospholipase A_2. At this point of the discussion it is still unclear if the high negative charge density is the only reason for the fast docking of the enzyme to the phase separated mixed domains or if the lipid headgroup has an additional influence.

3.3.4. Hydrolysis of a Mixed Monolayer of L-α-DPPC/Pyrenedecanoic Acid. Visualization of the hydrolysis of a mixed film of L-α-DPPC with pyrenedecanoic acid (ratio 5:1) by phospholipase A_2 [20] offers the possibility to directly see what happens with the acid during hydrolysis reaction. It could be shown that hydrolysis of such a mixed monolayer in the phase transition region leads to accumulation of the acid above the enzyme domains. Whether the monolayer above the protein domains consists only of the acid, or of a mixture of acid and DPPC, can still not be distinguished with these measurements alone.

*3.3.5. Interaction of Phospholipase A_2 and the Cationic Dye **1** with Pure Carboxylic Acid Domains within a Liquid-Analogous D-α-DPPC Matrix.* To further study the influence of the lecithin headgroup on the rapid binding of phospholipase A_2 to the mixed grey domains, experiments were performed with a long chain carboxylic acid non-misscible with DPPC. For these investigations a diacetylene carboxylic acid **2** was used.

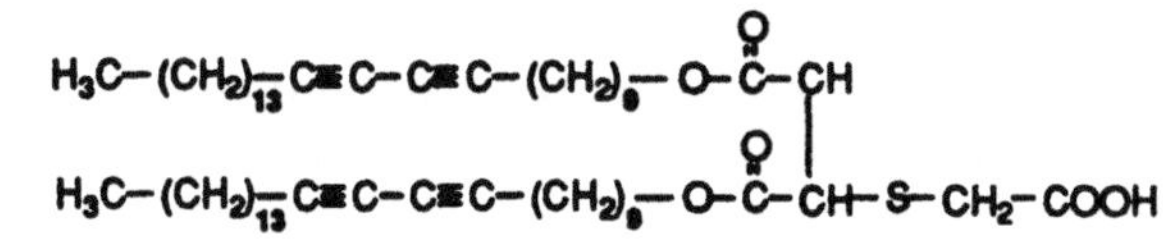

2

This acid has a high tendency to 2D-crystallize at the air/water interface: its pressure/area diagram shows only a solid-analogous phase [24].

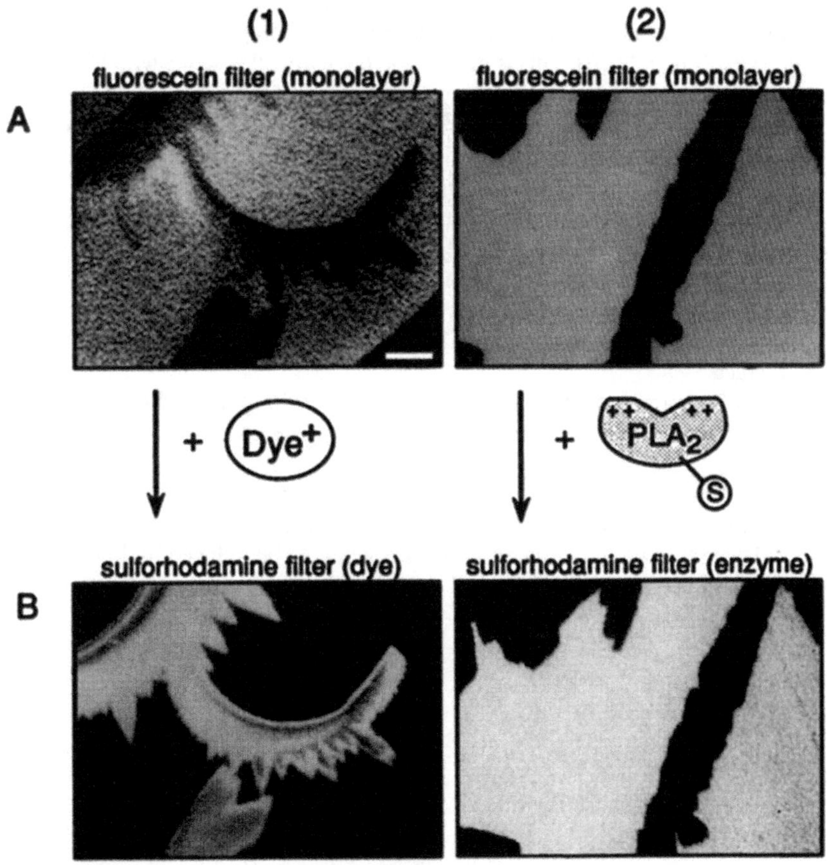

Figure 8. Interaction of the cationic dye (1) and sulforhodamine-labeled phospholipase A_2 (2) with a mixed monolayer of diacetylene carboxylic acid **2** and D-α-DPPC (1:1) doped with 0.5 mol% C6-NBD-PC: (A) black solid analogous acid domains in the liquid-analogous matrix of D-α-DPPC (Π = 18 mN/m), (B) interaction of the dye (1) / the enzyme (2) with the mixed monolayer: (1) after injection of the dye **1** the acid domains immediately appear bright in the sulforhodamine filter due to direct binding of the dye to the acid domains; (2) after injection of the enzyme no preferential binding of phospholipase A_2 to the acid domains can be detected; Subphase: Tris buffer, pH 8.9; Temperature = 30°C; Scale bar in (A) is 20μm

Directly after spreading of a mixture of D-α-DPPC/diacetylene carboxylic acid (1:1) doped with 0.5mol% C6-NBD-PC (3-palmitoyl-2-(6-(N-(7-nitro-bez-2-oxa-1,3-diazol-4-yl)amino)hexanoyl)-L-α-phosphatidylcholine), large, well defined characteristically shaped solid-analogous acid domains [25] are visible (Π = 1-2 mN/m). At this low pressure the DPPC is far away from any tendency to crystallize. Due to its immisibility with the diacetylene carboxylic acid one can assume that the black domains visible in the film consist of pure acid. Further compression leads to an increase in size of these domains and at a pressure of 18 mN/m the barrier is stopped to prevent co-crystallization of DPPC. Up to this pressure the monolayer thus consists only of solid-analogous acid domains in a liquid analogous matrix of DPPC mixed with C6-NBD-PC (Fig. 8 A). Injection of the cationic dye **1** leads to rapid binding (1min) of the dye to the acid domains showing the high negative charge density within the domains as expected for pure acid (Fig. 8 (1) B). After addition of sulforhodamine-labeled phospholipase A_2 no binding of the enzyme to the acid domains can be detected (Fig. 8 (2) B). This is in contrast to the mixed systems with DPPC/lysoPPC/palmitic acid and DPPC/palmitic acid, where dye and phospholipase A_2 show the same preferential and fast binding to the mixed domains. The different interaction of the cationic dye **1** and the sulforhodamine-labeled enzyme with the diacetylene carboxylic acid **2** domains within the fluid D-α-DPPC matrix are compared in Figure 8.

This direct comparison of dye and enzyme binding shows that not only the high negative charge density of the acid domains can be the reason for the rapid phospholipase A_2 domain formation. This points to the fact that besides ionic interactions the specific interaction of the lecithin headgroup with the enzyme seems to be necessary for the fast docking process at preformed acid and lecithin containing domains. This assumption is schematically shown in Figure 9, illustrating two facts:

(1) Interaction of labeled phospholipase A_2 with pure acid domains leads to an equilibrium between bound and free enzyme due to ionic interaction but no enzyme domain formation can be detected.

(2) Injection of the enzyme below monolayers containing acid domains with incorporated lecithin headgroups lead to fast binding of the protein to these domains..

Thus rapid protein domain formation seems to be induced by both, charge-charge interaction with the anionic monolayer as well as specific interaction with the headgroup.

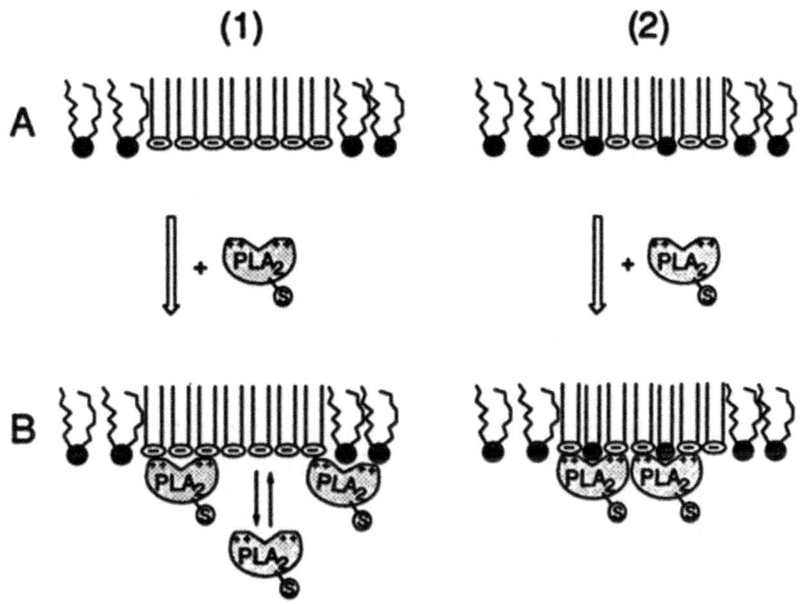

Figure 9. Schematic representation of the proposed influence of recognizable headgroups on the induction of phospholipase A_2 domain formation; (1) A: monolayer with pure acid domains; B: injection of labeled enzyme leads to an equilibrium between bound and free enzyme due to charge-charge interaction - no domain formation of the protein is detectable; (2) A: monolayer with acid domains which contain a small amount of recognizable headgroups; B: injection of labeled phospholipase A_2 leads to direct binding and thus domain formation of the enzyme due to specific recognition and charge-charge interaction.

4. Acknowledgements

The authors like to acknowledge the Deutsche Forschungsgemeinschaft and the New Energy and Industrial Technology Development Organization of Japan for research support. A. R. thanks the Fonds der Chemischen Industrie for support with a scholarship.

5. References

[1] Alberts, B., Bray, D., Lewis, J., Raff, M., Roberts, K. and Watsom, J. D. (1983) Molecular Biology of the Cell, Garland, New York.

[2] Ahlers, M., Müller, W., Reichert, A., Ringsdorf, H. and Venzmer, J. (1990) 'Specific interactions of proteins with functional monolayers - ways of simulating biomembrane processes', Angew. Chem. Int. Ed. Engl. 29, 1269-1285.

[3] Ahlers, M., Blankenburg, R., Grainger, D. W., Meller, P., Ringsdorf, H. and Salesse, C. (1989) 'Specific recognition and formation of two-dimensional streptavidin domains in monolayers: Applications to molecular devices', Thin Solid Films 180, 93-99.

[4] Darst, S. A., Ahlers, M., Meller, P. H., Kubalek, E. W., Blankenburg, R., Ribi, H. O., Ringsdorf, H. and Kornberg, R. D. (1991) 'Two-dimensional crystals of streptavidin on biotinylated lipid layers and their interactions with biotinylated macromolecules', Biophys. J. 59, 387-396.

[5] Grainger, D. W., Reichert, A., Ringsdorf, H. and Salesse, C. (1990) 'Hydrolytic action of phospholipase A_2 in monolayers in the phase transition region: direct observation of enzyme domain formation using fluorescence microscopy', Biochim. Biophys. Acta 1023, 365-379.

[6] Grainger, D. W., Reichert, A., Ringsdorf, H. and Salesse, C. (1989) 'An enzyme caught in action: direct imaging of hydrolytic function and domain formation of phospholipase A_2 in phosphatidylcholine monolayers', FEBS Lett. 252, 73-82.

[7] Meller, P. (1988) 'Computer-assisted video microscopy for the investigation of monolayers on liquid and solid substrates', Rev. Sci. Instrum. 59, 2225-2231.

[8] Knobler, C. M. (1990) 'Seeing phenomena in flatland: studies of monolayers by fluorescence microscopy', Science 249, 870-874.

[9] Brezesinsky, G., Dietrich, A. and Möhwald, H. (1991) 'Polymorphic domains in monolayers of isomeric triple-chain phospholipids', Macromol. Chem., Macromol. Symp. 46, 457-461.

[10] Waite, M. (1987) The Phospholipases, Plenum Press, New York; Dennis, E. A. (1991) Methods Enzymol. 197, Academic Press, New York.

[11] a) Berg, O. G., Yu, B., Rogers, J. and Jain, M. K. (1991) 'Interfacial catalysis by phospholipase A_2: determination of the interfacial kinetic rate constants', Biochemistry 30, 7283-7297; b) Gomashchi, F., O'Hara, T., Clary, D. and Gelb, M. H. (1991) 'Interfacial catalysis by phospholipase A_2: Evaluation of the interfacial rate constants by steady-state isotope effect studies', Biochemistry 30, 7298-7305; c) Jain, M. K., Yu, B., Ranadive, G. N. and Berg, O. G. (1991) 'Interfacial catalysis by phospholipase A_2: dissociation constants for calcium, substrate, products and competitive inhibitors', Biochemistry 30, 7307-7317; d) Gomashchi, F., Yu, B., Berg, O. G., Jain, M. K. and Jain, M. J. (1991) 'Interfacial catalysis by phospholipase A_2: substrate specificity in vesicles' Biochemisty 30, 7318-7329; e) Jain, M. K., Ranadive, G., Yu, B. and Verheij, H. M. (1991) 'Interfacial catalysis by phospholipase A_2: monomeric enzyme is fully catalytically active at the bilayer interface', Biochemistry 30, 7330-7340; f) Jain, M. K., Rogers, J. and Gelb, M. H. (1991) 'Interfacial catalysis by phospholipase A_2: activation by substrate replenishment', Biochemistry 30, 7340-7348.

[12] Ramirez, F. and Jain, M. K. (1991) 'Phospholipase A_2 at the bilayer interface', Proteins 9, 229-239.

[13] a) Chang, J., Musser, J. H. and McGregor, H. (1987) 'Phospholipase A_2: function and regulation', Biochem. Pharmacol. 36, 2429-2436; b) Scheuer, W. (1989) 'Phospholipase A_2: regulation and inhibition', Klin. Wochenschr. 67, 153-159; c) Nevalainen, T. J. (1989) 'The role of phospholipase A_2 in acute pancreatitis', Z. med. Lab. diagn. 30, 299-303.

[14] Bonsen, P.P., De Haas, G. H., Pieterson, W. A. and Van Deenen, L. L. M. (1972) 'Studies on phospholipase A and its zymogen from porcine pancreas: IV. The influence of the chemical modification of the lecithin structure on substrate properties', Biochim. Biophys. Acta 270, 364-382.

[15] a) Davies, D. E., Grainger, D. W., Lloyd, J. B., Reichert, A., Ringsdorf, H. and Salesse, C. (1990) 'Mixed monolayers of natural and polymeric phospholipids: structural characterization by physical and enzymatic methods', Biochim. Biophys. Acta 1022, 146-154; b) Romero, G., Thompson, K. and Biltonen, R. L. (1987) 'The activation of porcine pancreatic phospholipase A_2 by dipalmitoylphosphatidylcholine large unilamellar vesicles', J. Biol. Chem. 262, 13476-13482.

[16] Menashe, M., Romero, G., Biltonen, R. L and Lichtenberg, D. (1986) 'Hydrolysis of dipalmitoylphosphtidylcholine small unilamellar vesicles by porcine pancreatic phospholipase A_2', J. Biol. Chem. 261, 5328-5340.

[17] McConnell, L. K., Tamm, R. M. and Weis, R. M. (1984) 'Periodic structures in lipid monolayer phase transitions', Proc. Natl. Acad.Sci. USA 81, 3249-3253.

[18] a) Jain, M. K. and Jahagirdar, D. V. (1985) 'Action of phospholipase A_2 on bilayers. Effect of fatty acid and lysophospholipd additives on the kinetic parameters', Biochim. Biophys Acta 814, 313-318; b) Jain, M. K., Egmond, M. R., Verheij, H. M., Apitz-Castro, R., Dijkman, R. and De Haas, G. H. (1982) 'Interaction of phospholipase A_2 and phospholipid bilayers', Biochim. Biophys. Acta 688, 341-348; c) Plückthun, A. and Dennis, E. A. (1985) 'Activation, Aggregation, and product inhibition of cobra venom phospholipase A_2 and comparison with other phospholipases', J. Biol. Chem. 260, 11099-11109.

[19] Chow, W., Tomasselli, A. G., Heinrikson, R. L. and Kézdy, F. J. (1988) 'The chemical basis for interfacial activation of monomeric phospholipase A_2', J. Biol. Chem. 263, 11237-11241

[20] Reichert, A., Ringsdorf, H. and Wagenknecht, A. (1992) 'Spontaneous domain formation of phospholipase A_2 at interfaces: fluorescence microscopy of the interaction of phospholipase A_2 with mixed monolayers of lecithin, lysolecithin and fatty acid', Biochim. Biophys. Acta, submitted.

[21] Yu, B.-Z. and Jain, M. K. (1989) 'Segregation of anionic lipophiles in bilayers monitored by binding of cationic dye NK-529', Biochim. Biophys. Acta 980, 15-22.

[22] Jain, M. K., Yu, B.-Z. and Kozubek, A. (1989) 'Binding of phospholipase A_2 to zwitterionic bilayers is promoted by lateral segregation of anionic amphiphiles', Biochim. Biophys. Acta 980, 23-32.

[23] a) Volwerk, H. J., Jost, P. C., De Haas, G. H. and Griffith, O. H. (1986) 'Activation of porcine pancreatic phospholipase A_2 by the presence of negative charges at the lipid-water interface', Biochemistry 25, 1726-1733; b) Jain, M. K., Rogers, J. and De Haas, G. H. (1988) 'Kinetics of binding of phospholipase A_2 to lipid/water interfaces and its relationship to interfacial activation', Biochim. Biophys. Acta 940, 51-62.

[24] Bader, H. (1985) 'Functional amphiphiles in model membranes for surface recognition reactions and as membrane-forming compounds for polymerizable liposomes', thesis, Mainz.

[25] Meller, P., Peters, R. and Ringsdorf, H. (1989) 'Microstructure and lateral diffusion in monolayers of polymerizable amphiphiles', Colloid Polym. Sci. 267, 97-107.

PHOTOCHEMISTRY OF CROWN ETHER STYRYL DYES

M.V.ALFIMOV, S.P.GROMOV
Institute of Chemical Phisics Academy of Science
Novatorov st.7a 117421 Moscow USSR

ABSTRACT. The structure of crown ether styryl dyes representing a novel class of dyes is considered. Data on absorption and fluorescence spectra as well as photoisomerization and complexation of photochromic 15-crown-5-ethers are analyzed. The results are exhibited concerning dimerization of the complexes incorporating trans- and cis-isomers of the compounds obtained and stereospecificity of photoinduced cyclodimerization of the latter with formation of cyclobutane derivatives.

1. Introduction

Arylethylenes are a classical object of photochemistry. The products of their photolysis are: cis- and trans- isomers, condensed aromatic compounds and cyclobutane derivatives, each resulting in the course of photoisomerization, photocyclization and photocycloaddition, respectively [1-3].
One can imagine two basically different ways of the reaction control. The first one implies photochemical reactions in molecular organized microsystems, which influence the reaction course either incorporating one or many molecules into the matrix reaction cavity or building out of the reactants the following organized systems: glassy matrix. crystal, polylayers of LangmuirBlodgett, micelle, host-guest complex [4]. Influence of supramolecular organization of arylethylenes on their photochemical transformations in organized and disorganized systems was thoroughly studied by us [5]. The second way of arylethylenes phototransformation control can be based on the principles of molecular self-organization [6]. Spontaneous molecular self-organization occurs when the molecules capable of forming supramolecular structures wonted are obtained. The control over the process of self-organization on the molecular level, being of a considerable interest in terms of development and creation of the molecular system, becomes an object of scientific researches. As for the formation of supramolecular complexes in the course of photostructural and photoconformational changes in the recipient molecules, it is of special interest being based on the most promising method of molecular system control i.e. on light control.
Heretofore this approach has not found application in arylethylene photochemistry due to similarity of physical and chemical properties of trans- and cis- isomeric derivatives of ethylene studied [7] . A possibility that molecules

V. Balzani and L. De Cola (eds.), Supramolecular Chemistry, 343–358.

will be constructed whose photoisomers differ by their reactivity would mean new prospects for the chemical control. A considerable achievement in creating such systems would be a directed synthesis of arylethylenes where a change of molecule geometrical configuration under irradiation could lead to significant change of such parameters as, for instance, complexation constant. Such arylethylene derivatives can be of a considerable interest to supra-molecular photochemistry.

2. Presentation of the molecules.

At present there are intensive investigations on macroheterocyclic compounds which can serve as a basis for the construction of multifunctional compositions, possessing an ability to selectively complexate ions or neutral molecules according to host-guest type [8,9]. In this connection the photoactive compounds which can be constructed on the basis of styryl dyes containing macrohetero-cyclic fragment and possessing an ability to complexate metal cations are of interest.

We have supposed that bifunctional arylethylene derivatives which are crown ether styryl dyes comprising two structural fragments represented by dye and crown-ether residue, will serve as the efficient photochromic systems and photosensitive ionophores. On the other hand, ionophores containing a crown-ether cycle and anion group in a spatially accessible position are know to be more efficient receptors of cations of alkaline earth metals [10]. We could also expect that the synthesis of crown ether styryl dyes "covered" in cis-form by anion groups would allow us to obtain chromogenic crown ether capable of "taking off" and "putting on" its "anion cap" under photoirradiation, which in its turn provides an opportunity to achieve photocontrol of alkaline earth metal cations binding.

We have performed a synthesis of indolenine styryl dyes 1a,b [11,12] and styryl dyes of benzothiazole series 2a-d [13,14] . Compound 1b contains a fragment of crown ether, and perchlorate 2-[2- 3,4-dimethoxyphenyl)ethenyl]indolinium (1a) has in a benzene cycle two methoxygroups which reproduce in general terms electron-donating effect of a crown ether substituent on the dye conjugated double bonds. However, in contrast to crown ether styryl dye 1b, compound 1a should not be noticeably capable of complexation with ions of alkaline and alkaline earth metals.

Polyether macrocycle 15-crown-5 acting as pentadentate ligand at complexation with metal ions is built into compounds 2a-d. However, in contrast to cation dye 2a, compounds 2b - d are betaines containing sulfogroup which can form under certain conditions a bond with metal cation, thus supplementing its coordination sphere. Consideration of molecular models of the promising dyes from this series has shown that sulfopropyl and sulfobutyl groups are likely to be long enough as to form an additional coordination bond with a metal cation in cis-form of a dye.

According to the H-NMR data all styryl dyes obtained exist in a trans-form. This

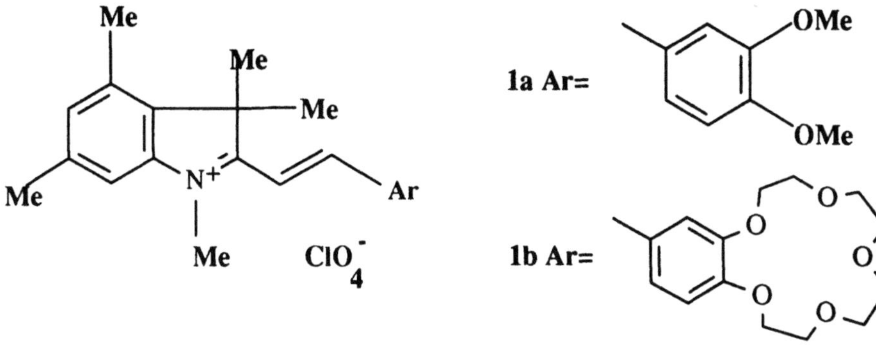

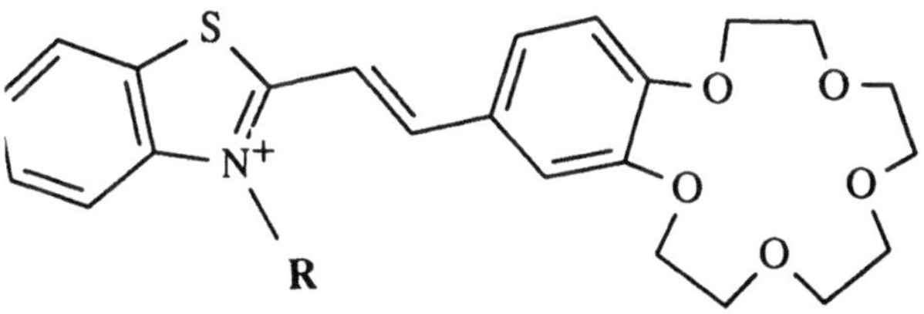

2a R = Et
2b R = $CH_2CH_2SO_3^-$
2c R = $CH_2CH_2CH_2SO_3^-$
2d R = $CH_2CH_2CH_2CH_2SO_3^-$

conclusion can be made on the basis of a high constant of spin-spin interaction $^3J_{trans}$= 15,6-16,7 Hz for olefinic protons of the dyes obtained. A performed X-ray analysis of monocrystals showed a molecule of crown ether styryl dye 2a being flat in a crystal lattice which is, evidently, attributed to a tight conjugation between orbitals of benzothiazole fragment, ethylene bond, carbon atoms of benzene cycle and some oxygen atoms of benzocrown ether [15].
The obtained data also allow the evaluation of conjugation chain length which spans a chain of bonds between nitrogen atom and oxygen atom, the latter being in para position to C=C bond. Since this oxygen atom participates in the electron pair conjugation, a considerable distortion of three-dimensional macrocycle structure occurs making the unpaired electrons of the rest oxygen atoms orientate to the exterior of the macrocycle cavity. All this indicates that before complexation crown ethers need preorganization by a series of conformational rearrangements.
The molecules of a crown ether styryl dye have been discovered to be arranged in monocrystal in pairs according to "head-to-tail" type so as to make ethylene bonds be situated one above the other. By analogy, the existence of dye molecules in dimeric form could be suggested for the solutions too. Later on we have discovered the susceptibility of crown ether styryl dyes to the formation of dimer in solutions [16].

3. Absorption and fluorescence spectra

Figure 1a and 2b show absorption and fluorescence spectra of the solution of trans 1b in water and the solution of trans-2a in acetonitrile, respectively. The absorption spectra of corresponding cis-isomers are represented in Figure 1a and 2b too. The quantum yields of fluorescence are exhibited in Table 1. A long-wave absorption band of trans-1a and trans-1b has a significant hypsochromic shift at an increase of solvent polarity. The band maximum of trans-1b is at λ=466 nm (ϵ=4.1 $\times 10^4$ l/mol×cm) in chlo- roform and at λ=432 nm (ϵ =3.9 $\times$ 10^4 l/mol×cm) in water.

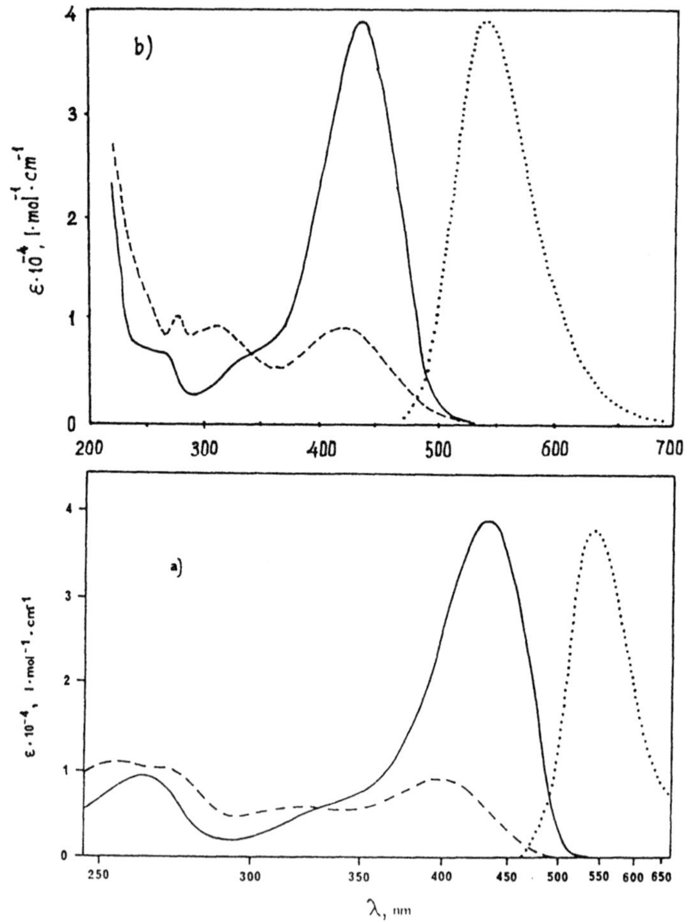

Figure 1. a) Absorption (—) and fluorescence (····) spectra of trans-1b and the calculated absorption spectrum of cis-1b (--) in water at T=295K. b) Absorption (—) and fluorescence (····) spectra of trans-2c and the calculated absorption spectrum of cis-2c (--) in acetonitrile at T=295K

In order to analyze the character of the absorption spectra we have simulated on a computer both the conformational changes and dye 2b electronic absorption spectrum shift. The calculation of conformation was performed by the MMX method using the program PCMODEL, and the calculations of the electronic absorption spectra were performed by CNDO/S method accounting for the conformational interaction (CI) [17].

According to our estimations trans-2b possesses flat configuration and exhibits only one long-wave electron transition which is in good agreement with the experimentally observed spectrum. It is mostly attributed about 90% to the configuration corresponding to electron transition from upper completed molecular orbit to lower uncompleted one. The second electron transition being close to the first one has relatively low intensity thus being seen as a shoulder against a long- wave absorption band.

Good agreement with the experiment is observed for both, energy and intensity of the electron transitions. Long-wave electron transition in cis- and trans isomers has the same nature and the character of their parameters change occurring at the disturbance of flat symmetry of trans- and cis-isomers are alike. At the same angle of inclination of the fragments the positions of long-wave absorption band is practically the same for both 2b isomers but their intensities differ.

4. Photoisomerization

Under the irradiation of trans-isomer solutions with light in the range of a long-wave absorption band the intensity of the latter decreases with a simultaneous rise of the absorption in the short-wave range which lasts till a photostationary state is achieved (Figure 1).

In all cases isobestic points are observed. These spectral changes are, apparently, explained by the reversible trans-cis-photoisomerization occurring in solution [12]. When the irradiation ceases in the moderate polarity solvents chloroform, acetone quick dark recovery of the spectra proceeds. It is, apparently, connected with the process of thermal cis-trans isomerization. In aqueous solutions the cis-1a,b are stable enough to allow the evaluation of the spectra of individual cis-isomers and their molar absorbance e by Fisher method [18]. Knowing e_c and e_t a standard method was used to evaluate the quantum yield of trans-cis (φ_{tc}) and cis-trans (φ_{ct}) photoisomerization. The obtained values of fluorescence and photoisomerization quantum yields are given in Table 1 (The error of φ_{tc} and φ_{ct} measurements is about 20% and that of v is about 40 %).

TABLE 1 Photochemical data for 1a,b and 2c at T=298K

Dye	φ_{tc}	φ_{ct}	φ_{fl}
11a (water)	0.47	0.51	0.0015
11b (water)	0.49	0.50	0.0020
12c (acetonitrile)	0.40	0.49	0.06

We studied quick stages of structural dynamics of crown ether styryl dyes by the method of picosecond fluorescence spectroscopy. It was found that the observed dependence of fluorescence lifetime on the solvent viscosity could be described by the barrier model of the trans-cis isomerization occurring in the existed singlet electron state [19].

5. Complexation of trans-2a-d with cations of alkaline and alkaline earth metals.

In the acetonitrile solution a compound 2c has in the absorption spectrum an intensive long-wave band with a maximum at λ=435 nm. On the addition of alkaline Na, K and alkaline-earth metals Mg, Ca, Ba into the solution of perchlorates a hypsochromic shift of the long-wave band is observed [13]. With the concentration of perchlorate $C_M \geq 10^{-3}$ M the absorption band maximum is at λ=392 nm ($\epsilon_{max}=3.69*10^4$ l/mol×cm), at λ=396 nm (ϵ_{max} = $3.62*10^4$ l/mol×cm) and at λ=407 nm ($\epsilon_{max}=3.66*10^4$ l/mol×cm) in case of Mg^{2+},Ca^{2+} and Ba^{2+}, respectively. Similar spectral changes were earlier observed when studying complexation of metal ions with the crown ethers not containing styryl chromogene [20]. Thus we attribute the observed spectral changes to the complexation.

From the absorption spectra of the solutions at the constant initial concentration of the ligand $C_0=2.05\times10^{-5}$ M and different concentration of the cation, a degree of the ligand transformation into the complex form with metal ions was determined by formula α= $(D_0-D)/(D_0-D)$, where D_0 is an absorption of the initial ligand solution at the fixed wavelength, D is an absorption of the solution upon complete complexation, D is an absorption of the solution at the different values of the cation concentration.

In the experiments on 2c complexation with Ba^{2+} ions the expenditure of the ligand at low degrees of transformation exceeded the concentration of the cation nearly two times. Thus, for the given compound the following scheme of the formation of the 2L:1M complex can be proposed:

$$L + M \underset{}{\overset{K_1}{\rightleftharpoons}} LM$$

$$LM + L \underset{}{\overset{K_2}{\rightleftharpoons}} L_2M$$

where M is a metal cation, L is a ligand molecule. The following equation corresponds to the scheme given above:

$$\frac{C_M}{\alpha} = \frac{1 + C_0K_1(1-\alpha) + C_0^2K_1K_2(1-\alpha)^2}{K_1(1-\alpha)(1 + 2C_0K_2(1-\alpha))} \qquad (1)$$

The experimental dependencies of the ligand concentration decrease $\alpha C_0 = C_0$-[L] on the total concentration of Mg^{2+}, Ca^{2+} and Ba^{2+} in the solutions are presented in Figure 2 by dots.

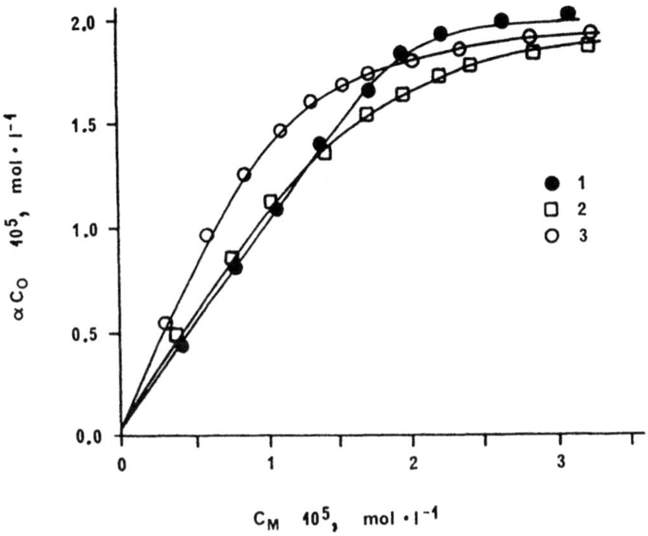

Figure 2. Experimental dependences of the necessary amount of ligand 2c for the total concentration of Mg^{2+}(1), Ca^{2+}(2) and Ba^{2+}(3).

The solid lines represent the theoretical dependencies corresponding to the equation (1) with constants $K_1 = 4\times10^6$ l/mol and $K_2=0$ for Mg^{2+}; 7×10^5 l/mol and 2×10^5 l/mol for Ca^{2+} and 6×10^5 l/mol and 1.5×10^5 l/mol for Ba^{2+}. A good agreement of the experimental data with the results of the calculation confirms the validity of the chosen scheme complexation. Besides, the decrease in value K_1, in the order of $Mg^{2+}>Ca^{2+}\cong Ba^{2+}$ correlates with the value of the long wave absorption band shift in the spectra of the complexes. The increase of value K_1 in the $Mg^{2+}<Ca^{2+}<Ba^{2+}$ order corresponds to the idea that the cations of large diameter tend to form complexes at the ratio of 2:1 with ethers which have small size cavities [21]. With a Mg^{2+} cation such a complex is apparently not formed.

We performed a comparative analysis of the complexation ability of 2b-d compounds differing solely by the length of sulfo- alkyl substituent. Spectral characteristics of both free trans- 2b-d and their complexes with Mg^{2+} differ insignificantly. In this case in order to reduce an error associated with the styryl dye sorb- tion occurring from a solution to the cell walls, the inner surface of the latter was modified with thrimethylchlorosilane. Except for this step, the method of estimation of the complexation constants was the same as the procedure mentioned above.

The constants of complexation of trans-2b-d with Mg^{2+} (K_1) obtained at $C_0=4.2*10^{-6}$ M have the following values:

compound	2b	2c	2d
K_1, l/mol	7×10^4	1×10^4	4×10^4

It is evident that an increase in number of methylene groups from 2 to 3 in sulfoalkyl N-substituent leads to a sharp increase of complexation constant. Further growth of their number up to 4 causes almost no changes [14].

6. Photoinduced formation of intramolecular coordination bond in 2b-d complexes with Mg^{2+}, Ca^{2+} and Ba^{2+} cations.

Compound 2c, like a crown-ether indolenine styryl dye 1b, can be subjected to a reversible photochemical reaction of trans-cis isomerization around the central double C=C bond. The absorption spectra of cis-2c in acetonitrile was calculated by means of Fisher method [18] using the spectra of trans-isomer and photostationary states obtained from the irradiation of a 2b solution with light at 365 nm and 436 nm.

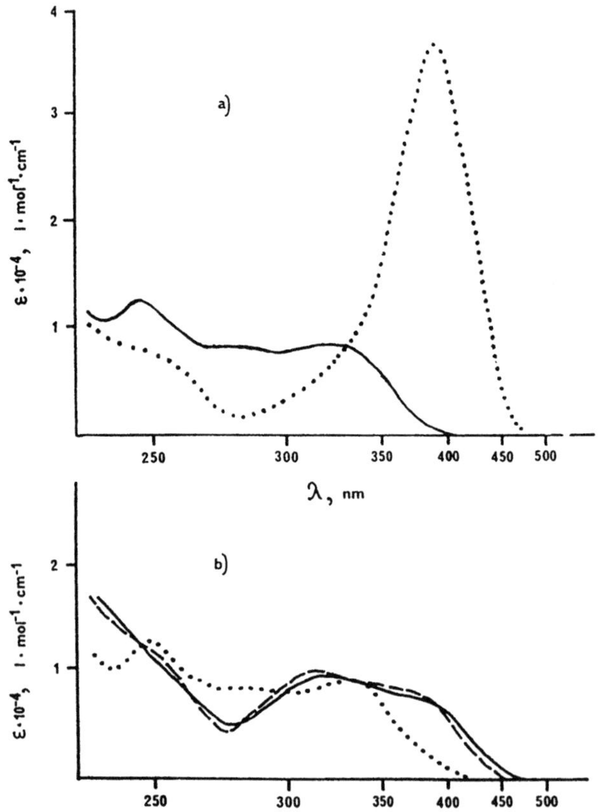

Figure 3. a) Absorption spectra of complexes of trans-2c (····) and cis-2c (—) with Mg^{2+} at $[Mg(ClO_4)_2]=1x10^{-3}M$. b) Absorption spectra of complexes of cis-2c with Mg^{2+} at $[Mg(ClO_4)_2]=3x10^{-3}M$ (····), 0.5M (—) and a cis-2a and Mg^{2+} complex at $[Mg(ClO_4)_2]=3x10^{-3}M$ (- -).

In Figure 3 trans- and cis-isomer absorption spectra of 2b are presented. The distance between the maxima of the long wave absorption bands of these isomers is only 15 nm which prevents complete transformation of trans-isomer into a cis-form through photoisomerization. Similar situation is characteristic for the majority of photochromic compounds [7]. Another situation is observed in the case of trans-2b complexation with cations of alkaline earth metals when the concentration of the latter is 10^{-3} M. The irradiation of the solutions with light at λ=436 nm leads to a photostationary state including more than 99% of cis-isomer. The absorption spectra of the cis-isomer complexes have a long- wave maximum at λ=321 nm ($\epsilon = 8.96\times10^3$ l/mol×cm), at λ=325 nm ($\epsilon = 9.51 \times 10^3$ l/mol×cm) and at λ=338 nm ($\epsilon = 8.05*10^3$ l/mol×cm) for Mg^{2+}, Ca^{2+} and Ba^{2+}, respectively. Figure 3a shows as an example the absorption spectra of cis- and trans-isomer of the complexes 2c with Mg^{2+}.
A strong hypsochromic shift by 70 nm of the cis-isomer complexes spectra in respect to the spectra of trans-isomers can be apparently explained by the formation in a cis-form of a coordination bond between a sulfo-group and a metal cation, located in a crown-cavity.

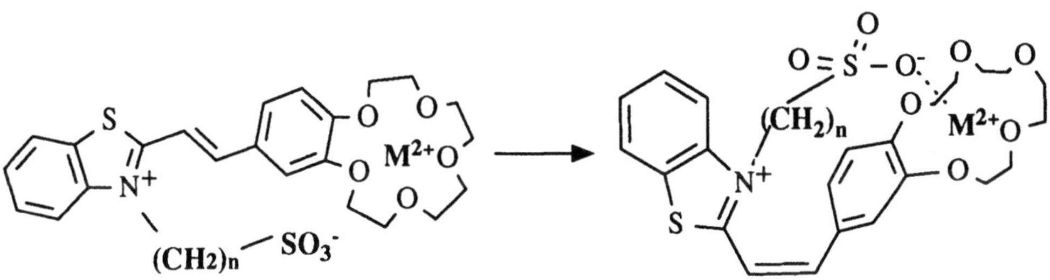

(Trans-2c,d)M^{2+} (Cis-2c,d)M^{2+}

In fact, according to our calculations by MMX method [17] while complexing with alkaline earth metal cation it is energetically advantageous for the cis-2c molecule to coil (calculated values for τ_1=90° and for τ_2=45°) and to form a coordination bond between a sulfo-group and a metal. Accordingly a conjugated chromophoric system undergoes full destruction and, as a result, long-wave absorption band not only shifts to a short-wave region but also absolutely loses its intensity. It occurs at any value of relative electric permittivity $\epsilon=\infty$ at the complete shielding of the cation electric field. Thus, in the experimentally observed spectrum it is the second absorption band of the molecule which corresponds to a long-wave absorption band. This circumstance accounts for the pronounced shift of a long-wave band in cis-2c electronic absorption spectrum observed at complexation. While comparing calculated values with various electric permittivity one can see the cation electric field influencing neither position nor intensity of the second absorption band of cis-2c. Thus, it can be concluded that the observed effect is due solely to the conformational change of cis-2c at complexation.
With an increase in the concentration of magnesium perchlorate in a cis-isomer solution of the 2c complex with Mg^{2+} (C_M > 0.01 M) the absorption in a

long-wave region of the spectrum gradually increases and at C_M > 0.2 M the spectrum takes an entirely different shape. The change of the spectrum shape is attributed probably to the destruction of intramolecular coordina- tion bond due to the association of SO_3^- group with an additional cation of Mg^{2+}.
Figure 3b shows the absorption spectra of a cis-2c complex with Mg^{2+} at concentrations of the $Mg(ClO_4)_2$ equal to 3×10^{-3} M and 0.5 M, as well as of a cis-2a complex with Mg^{2+} at concentration $[Mg(ClO_4)_2] = 3\times10^{-3}$ M. In a compound 2a an ethyl group instead of a sulfopropyl group is present, therefore the formation of an intramolecular coordination bond is impossible. The fact that the spectrum of the (cis-2a)Mg^{2+} at C_M=3x10^{-3} M and the spectrum of the (cis-2c)Mg^{2+} at C_M=0.5 M practically coincide, confirms the assumption that the destruction of the intramolecular bond in (cis-2c)Mg^{2+} at high concentrations of $Mg(ClO_4)_2$ occurs.
The presence of an additional coordination bond in cis-isomer complexes of bethaines of chromogenic crown ethers allows the suggestion of their higher stability in comparison with trans-isomer complexes. On the other hand, of interest is the modification of sulfoalkyl N-substituent of crown-ether bethaine styryl dyes by means of varying the number of methylene groups between a nitrogen atom of heteroaromatic residue and SO_3^- group. The stability of such chelate complexes should probably depend on the degree of steric tension in macrocycle resulting in cis-form due to the formation of the coordination bond and incorporating an ion of the metal. A decrease in the number of methylene groups will presumably lower the stability.
Constant of complexation of cis-2c with Mg^{2+} was estimated using the competitive reaction method [14]. The cis-isomer of the complex was formed by irradiating with light at λ=436 nm the solution of trans-2c and $Mg(ClO_4)_2$ in acetonitrile with concentration C =4.2×10^{-6} M and C =1.0×10^{-5} M, respectively. Then benzo-15-crown-5 (b15c5) was added to the solution and the degree of cis-2c complexation (α) was estimated by the absorption at λ=421 nm (the maximum of long-wave absorption band of cis-2c) [13]. The process of complexation in the presence of competitor can be represented as follows:

$$\text{cis-2c} + Mg^{2+} \underset{}{\overset{K_c}{\rightleftarrows}} \text{(cis-2c)}Mg^{2+}$$

$$\text{b15c5} + Mg^{2+} \underset{}{\overset{K_k}{\rightleftarrows}} \text{(b15c5)}Mg^{2+}$$

The scheme can be expressed by the following equation:

$$C_k = (1 + (K_c/K_k)(1-\alpha)/\alpha) \times (C_m - \alpha C_0 - \alpha/((1-\alpha)K_c)) \quad (2)$$

where C_k, C_M and C_0 is the total concentration of (b15c5), $Mg(ClO_4)_2$ and 2c respectively.
When $1/K_c << C_M = 1\times10^{-5}$ M the experimental dependence of the degree of cis-2 complexation on (b15c5) concentration

$C_k x10^3 M$	0	0.37	1.11	1.84
	1	0.72	0.51	0.39

corresponds to the equation (2) at K_c/K_k=140. Using the spectrophotometric titration we estimated K_k > 1×10^7 the small molar absorbance of (b15c5) prohibits the exact measurements , then K_c>1.4×10^9 l/mol. The fact that K_c considerably exceeds more than two orders the corresponding constant for trans-isomer see above confirms the afore made conclusion that an intramolecular coordination bond is formed in the complexes of cis-2c with cations of alkaline earth metals.

At photoisomerization of (trans-2d)Mg^{2+} at C_0= 10^{-6}- 10^{-4} M, as in the case with (trans-2c)Mg^{2+} an intramolecular cis-isomer complex with short-wave absorption spectrum is formed. The constant of complexation of cis-2d with Mg^{2+}, also estimated by the competitive reaction method, is eight times greater for cis-2c isomer. It is probably attributed to the fact that the larger the cycle the easier it satisfies the attempts of cis-2d to form an intramolecular coordination bond between SO_3^- group and the metal cation.

As for complexes of 2b with Mg^{2+} no experimental conditions provided the considerable hypsochromic shift in the absorption spectrum at trans-cis-photoisomerization. Moreover (cis-2b)Mg^{2+} spectrum calculated by Fisher method practically coincides with the corresponding absorption spectrum of 2a. The sulfoethyl chain is probably too short to form intramolecular complex.

Thus, the shorter is the sulfoalkyl N-substituent the weaker becomes the coordination bond in cis-isomer complexes and in the case of sulfoethyl derivative it is not formed at all.

7. Dimerization of trans- and cis-2c,d.

As we have shown above an intramolecular coordination bond in the complexes of cis-2c with alkaline-earth-metal cations breaks at a considerable excess of cations C_M≥0.1 M and a "closed" complex turns into an "open" form with longer-wave absorption spectrum. It turned out that the similar transition occurs when concentration of cis-2c complex increases under the lack of excessive metal. At C_0≥10^{-2} M cis-isomer complex is mostly in the "open" form. This effect is probably attributed to the dimerization of the complexes by "head-to-tail" type. intramolecular coordination bond breaks and corresponding intramolecular bonds form [16].

At the dilution of cis-isomer complex solution up to C_0<10^{-4} M the slow transition of an "open" form into a "closed" one occurs.

These process is easily observed by the absorption spectra Figure 4 . The kinetics of the transition which was followed by measuring the absorption of the solution at λ=393 nm are of the first order. The life time of an "open" form of (cis-2c)Mg^{2+} does not depend on the dilution degree of the solution within the concentration range from 10^{-5} to 10^{-4} M at the constant concentration of an excessive Mg^{2+} (1×10^{-4} M), and equals 550 sec at T=298 K.

The addition of $Mg(ClO_4)_2$ to the solution with low concentration of free cis-2c (C_0<10^{-4} M) causes almost immediate formation of a "closed" form of a

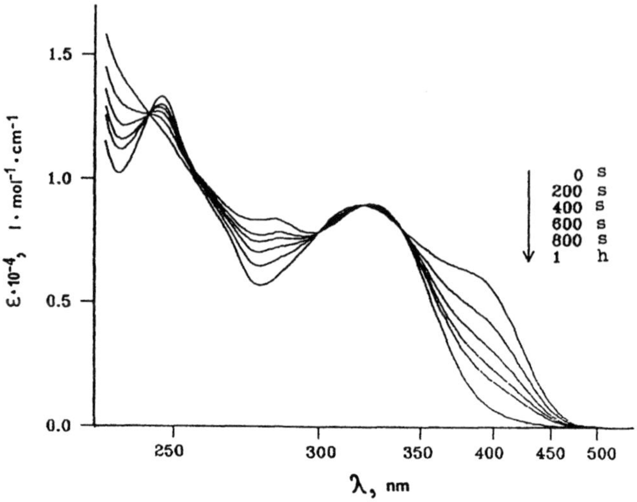

Figure 4. Absorption spectra of (cis-2c)Mg^{2+} measured at different moments after dilution (in 10^3 times) of initial solution ($2.2x10^{-2}$)

(cis-2c)Mg^{2+} complex. Consequently, according to our assumption the rate of the transition of a (cis-2c)Mg^{2+} "open" form to a "closed" one at the dilution of a solution is limited by the rate of the dimer dissociation, and the lifetime of an "open" form corresponds to the life time of dimers.

After a short-time irradiation of a diluted solution of (trans-2c)Mg^{2+} (5×10^{-6} M $\leq C_0 \leq 1x10^{-4}$ M, $C_M=1.2\times10^{-4}$ M) its absorption spectrum undergoes slow changes at dark. The character of these changes makes it reasonable to conclude that the transition of an "open" form of (cis-2c)Mg^{2+} into a "closed" form occurs, the duration of the transition being of the same order as the afore given lifetime of (cis-2c)Mg^{2+} dimers.

Basing on these data we assume that in the course of (trans-2c)Mg^{2+} photoisomerization in the diluted solution an "open" form of (cis-2c)Mg^{2+} is produced in the form of dimers. Such a situation is possible solely when the (trans-2c)Mg^{2+} complex at $C_0 > 5\times10^{-6}$ M is at least partially dimerized, and in the course of photoisomerization of dimers the intermolecular coordination bond is retained.

As the quantum yield of cis-trans-isomerization of 2cMg^{2+}, estimated at $C_0=1.6\times10^{-5}$ M and at $C_m=1.2\times10^{-4}$ M is high (φ_{tc} = 0.46) and the dissociation rate of 2cMg^{2+} dimers is low, the irradiation of the solution with high intensive light ($I_0 \cong 10^{16}$ $cm^{-2}sec^{-1}$) at λ=436 nm provokes the establishment of cis-trans-photostationary state, including (cis-2c)Mg^{2+} dimers which absorb light of the given wave length. The process of dimers dissociation with the formation of (cis-2c)Mg^{2+} "closed" form not absorbing light at λ=436 nm) gradually changes cis-isomer content in the mixture up to the full transition of a trans-isomer into a cis-form.

The processes occurring in the solution of 2cMg^{2+} at concentration < 10 M under the influence of light at =436 nm can be represented by the following scheme:

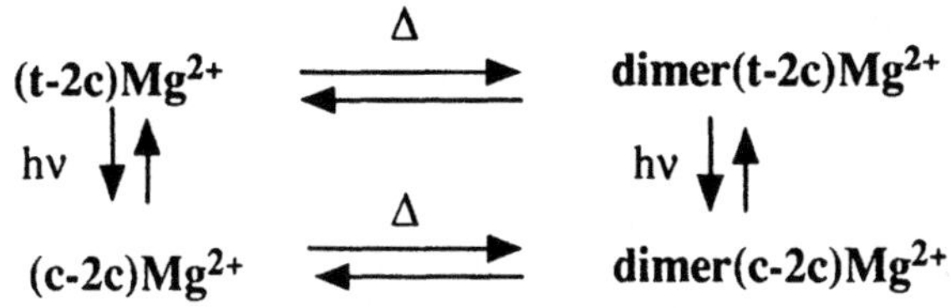

(3)

The picture changes when the solution of $2cMg^{2+}$ is irradiated with light at λ=365 nm which a "closed" form of (cis-2c)Mg^{2+} can absorb. Proceeding with a high quantum yield φ_{ct} =0.58 the reaction of cis-trans-isomerization of a (cis-2c)Mg^{2+} "closed" form leads to the closure of the cycle in this scheme.
A comparison of 2c,d complexes differing solely by the length of sulfoalkyl substituent reveals, as far as their ability to form dimers is concerned, that they are almost similar.
Thus, we can conclude that the formation of the complexes of 2c,d with Mg^{2+} cations in acetonitrile solution leads to the as- sociation of the 2c,d complexes in dimers.

8. Photochemical 1,2-cycloaddition of 2c,d.

It is known that the association of the molecules with double C=C bond in the solution can attribute to an increase in the quantum yield of photocycloaddition and influence its stereoselectivity. In our case it is reasonable to suggest that from (cis-2c)Mg^{2+} dimers the cyclodimers are photosynthesized. In fact in the solution of $2cMg^{2+}$ complex at concentration 1×10^{-2} M under the long-term influence of light at $\lambda\geq365$ nm a new photoproduct is formed. Its absorption spectrum is represented in Figure 5.

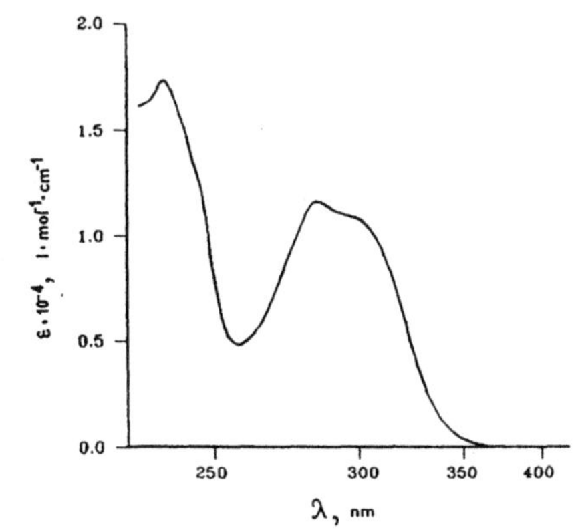

Figure 5. Absorption spectra of (cyclodimer-2c)Mg^{2+} solution ($6x10^{-3}$M) after photolysis (lR365 nm)

As in the solution irradiated with light both cis- and trans-isomers of a $2cMg^{2+}$ are present, the formation of various isomers of cyclobutane derivative is probable. The NMR ^{1}H and ^{13}C data, however, show undoubtedly that in the course of photolysis a single-type isomer is formed. It is obvious that it can be produced solely from two cis-2c molecules in the course of cycloadding by "head-to-tail" type. Thus, the existence of photocycloaddition and its high stereoselectivity give further evidence for the production of (cis-2c)Mg^{2+} dimers which form the intermolecular coordination bonds in acetonitrile solution under the high concentrations of dimers.

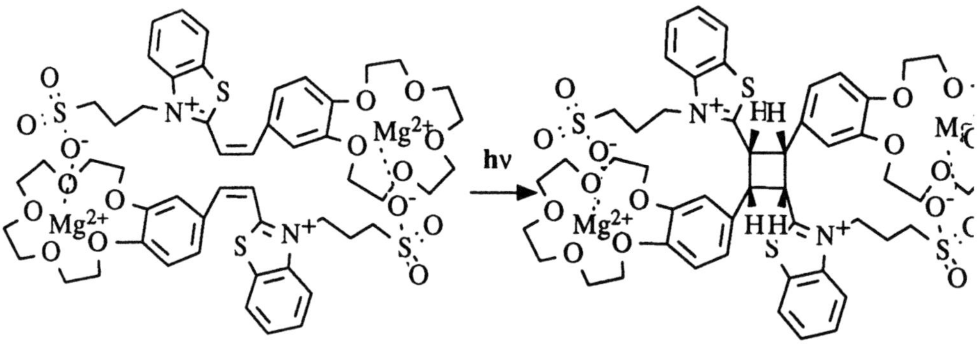

dimer (cis-2c)Mg^{2+} **cyclodimer $2cMg^{2+}$**

In the solution under the influence of constant irradiation the concentration of (cis-2c)Mg^{2+} dimers reaches some stationary level. Owing to this effect the reaction of photocycloaddition can achieve a grade of reactivity more than 90% even at concentrations of $2cMg^{2+}$ 10^{-5} M. The dependence of quantum yield of photo- cycloaddition φ_d on $2cMg^{2+}$ concentration estimated at constant intensity of light at λ=365 nm, $I_0=3.7\times10^{15}$ $cm^{-2}\times sec^{-1}$, looked as follows:

C_0, M	5×10^{-6}	2.4×10^{-5}	4.6×10^{-5}	2.1×10^{-4}	2.0×10^{-3}
φ_d	0.0022	0.0043	0.0052	0.0051	0.0055

A decrease in $2cMg^{2+}$ concentration from 2×10^{-3} M to 5×10^{-5} M pays practically no effect on the quantum yield of photocycloaddition. This fact as well as the structure of cyclodimers allow us to conclude that it is constructed solely from (cis-2c)Mg^{2+} dimers. A quantum yield within a following range of concentrations: 5×10^{-5} M -5×10^{-6} M becomes 2.5 times lower which can be explained by a decrease in the degree of (trans-2c)Mg^{2+} dimerization and accordingly by a decrease in the degree of (cis-2c)Mg^{2+} dimerization at photostationary state.

Using the data obtained we can represent the general scheme of the processes of 2c dimerization and 2c photocycloaddition proceeding in acetonitrile solution under the influence of light at λ= 365 nm as follows:

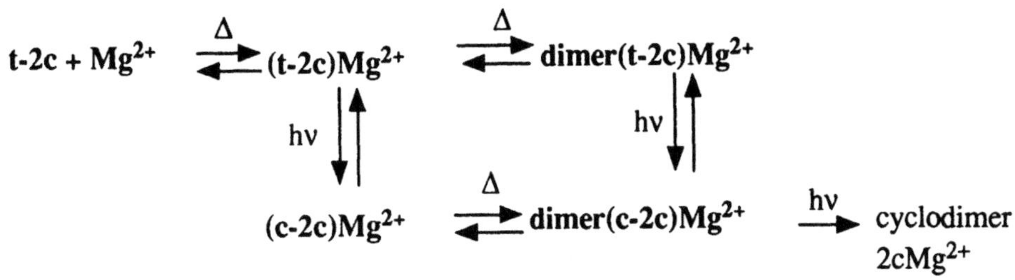

Thus the formation of dimers allows us to perform the reaction of photocycloaddition with high stereoselectivity and within a broad range of concentrations.

9. Conclusion

Spontaneous molecular self-organization is based on the molecules obtained capable of forming of themselves a desired supramolecular architecture. The studies on photochemical transformations of crown ether styryl dyes representing a novel class of bifunctional dyes and chosen as an example allow us to demonstrate for the first time two step molecular self-organization. The first step of the process includes self-organization of dye molecules into dimers in the course of complexation. The second step comprises photoinduced pre-organization of dye complexes (trans-cis-photoisomerization) serving as the basis for the formation of cis-form dimers, and followed by the process of photocycloaddition is completed with the formation of cyclobutane derivatives.

References

1. Brown, George H. (ed.) 1971 'Photochromism' 4, Wiley, New York.
2. Sheck, Yury B. and Alfimov, Michail V. 1978 'Photochemical reactions of rearrangement', Uspekhi nauchn. fotogr. 19, 216-229.
3. Sazhnikov, Vjacheslav A., Yakusheva, Olga B. and Alfimov, Michail V. 1978 "Photoaddition', Uspekhi nauchn. fotogr. 19, 207-215.
4. Ramamurthy, V. 1986 'Organic photochemistry in organized media', Tetrahedron 42, 5753-5839.
5. Alfimov, Michail V. 1990 'Photochemical reactions control principles in the molecular organized microsystems', Izv. Akad. Nauk SSSR, Ser. Khim. 10, 2197-2210.
6. Lindsey, Jonathan S. 1991 'Self-assembly in synthetic routes to molecular devices. Biological principles and chemical perspectives: a review', New J. Chem. 15, 153-180.

7. Eltsov, Andrei V. (ed.) 1982 'Organic photochromes' Khimiya, Leningrad.
8. Lehn, Jean-Marie 1988 'Supramolecular chemistry-scope and perspectives molecules, supermolecules, and molecular devices (Nobel lecture)', Angew. Chem. Intern. Ed. 27, 90-112.
9. Vogtle, F. and Weber, E. (eds.) 1985 'Host Guest complex chemistry macrocycles', Springer-Verlag, Berlin.
10. Wierenga, Wendell, Evans, Bruce R. and Woltersom, John A. 1979 'Synthesis of new, noncyclic ionophores exhibiting efficient Ca^{2+} transport' J. Am. Chem. Soc. 101, 1334-1336.
11. Gromov, Sergei P. and Bundel, Yury G. 1985 'New synthesis of indoles' Dokl. Akad. Nauk SSSR 281, 585-588.
12. Gromov, Sergei P., Fomina, Marina V., Ushakov, Evgeny N., Lednev, Igor K. and Alfimov, Michail V. 1990 'Synthesis, photoisomerization and complexation of crown ether dyes' Dokl. Akad. Nauk SSSR 314, 1135-1138.
13. Gromov, Sergei P., Fedorova, Olga A., Ushakov, Evgeny N., Stanislavsky, Oleg B., Lednev, Igor K. and Alfimov, Michail V. 1991 'Crown ether styryl dyes. 2. Synthesis, photoinduces and dark complexation of chromogenic 15-crown- 5-ether betaine', Dokl. Acad. Nauk SSSR 317, 1134-1139.
14. Gromov, Sergei P., Fedorova, Olga A., Ushakov, Evgeny N., Stanislavsky, Oleg B., and Alfimov, Michail V. 1991 'Crown ether styryl dyes. 4. Size of chelate macrocycle and complexation of cis-isomers of photochromic 15-crown-5-ether betaine', Dokl. Acad. Nauk SSSR 321, 104-107.
15. Gromov, Sergei P., Fedorova, Olga A., Alfimov, Michail V., Tkachev, Valery V. and Atovmyan, Lev O. 1991 'Crown ether styryl dyes. 3. Synthesis and X-ray crystal analysis of chromogenic 15-crown-5-ether', Dokl. Acad. Nauk SSSR 319, 1141-1144.
16. Ushakov, Evgeny N., Stanislavsky, Oleg B., Gromov, Sergei P., Fedorova, Olga A. and Alfimov, Michail V. 1992 'Crown ether styryl dyes. 5. Dimerization of chromogenic 15-crown-5-ethers', Dokl. Acad. Nauk SSSR, in press.
17. Baskin, Igor I., Burstein, Konstantin Y., Bagaturians, Alexander A., Gromov, Sergei P., and Alfimov, Michail V. 1992 'Computer simulation of the effect of complex formation on the conformation and electronic absorption of novel crown ether styryl dyes', Dokl. Acad. Nauk SSSR, in press.
18. Fischer, Ernst 1967 'The calculation of photo-stationary states in systems A--B when only A is known', J. Phys. Chem. 71, 3704-3706.
19. Alfimov, Michail V., Buyevich Oleg E., Gromov, Sergei P., Kamalov, Valei F., Lifanov, Alexander P. and Fedorova, Olga A. 1991 'Picosecond isomerization of crown-containing styryl dyes in excited electron state', Dokl. Akad. Nauk SSSR 319, 1149-1154.
20. Lohr, Haus-Gerd and Vögtle, Fritz 1985 'Chromo- and Fluoro- ionophores. A new class of dye reagents', Acc. Chem. Res. 18, 65-72.
21. Takaki, U. and Smid, J. 1974 "Complexes of crown ether and glymes with difluorenylbarium', J. Am. Chem. Soc. 96, 2588-2593.
22. Gromov, Sergei P., Fedorova, Olga A., Ushakov, Evgeny N., Stanislavsky, Oleg B., and Alfimov, Michail V. 1991 'Crown ether styryl dyes. Photocycloaddition of photochromic 15-crown-5-ether betaines', Dokl. Acad. Nauk SSSR, in press.

TOWARDS SYNTHETIC ENZYMES BASED ON PORPHYRINS AND STEROIDS

H. L. ANDERSON, R. P. BONAR-LAW, L. G. MACKAY,
S. NICHOLSON and J. K. M. SANDERS*
Cambridge Centre for Molecular Recognition
University Chemical Laboratory, Lensfield Road
Cambridge CB2 1EW, United Kingdom

ABSTRACT. Syntheses and binding properties of some supermacrocyclic hosts based on porphyrins and steroids are presented. It is shown that the ligand-binding properties of linear intermediates can be used to control the outcome of the synthesis; three distinct rôles for such ligand templates are identified and exploited in the synthesis of a linear porphyrin octamer. Some chemistry of cyclocholates and cholate-capped porphyrins is also described, with an emphasis on the control of ring size through steric effects.

1. Introduction

We want to construct model enzymes that are capable of binding, recognition and catalysis. The aim is *not* to mimic any particular natural enzyme. However, inspired by the example of nature, our long-term aim is to understand the principles of how to design and create homogeneous catalytic systems.

So far we have concentrated on devising efficient syntheses of new hosts based on porphyrins and steroids, and on trying to understand the binding properties of our systems. The justification for this approach, as opposed to using protein engineering or catalytic antibodies, is that the study of 'small', well-defined systems is likely to lead to the discovery of new principles of general utility; our first hosts, porphyrin dimers linked by flexible chains, unexpectedly led to a new understanding of π–π interactions,(1) while the second generation has illuminated and greatly expanded the rôle of templates in synthesis. The latter work forms the core of this present article.

Model enzymes should have the following features: rapid construction from large, rigid building blocks (the bond-by-bond approach of classical natural product chemistry will be too laborious); built-in spectroscopic probes for geometry, binding and catalysis; convergent binding sites for two or more substrates; and, eventually, chiral recognition properties. Diaryl porphyrins, **Zn1**, and cholic acid, **2**, each possess most of these features. The diaryl porphyrin unit can be made with a

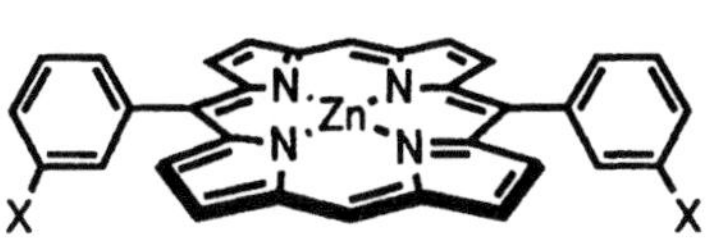

Zn1

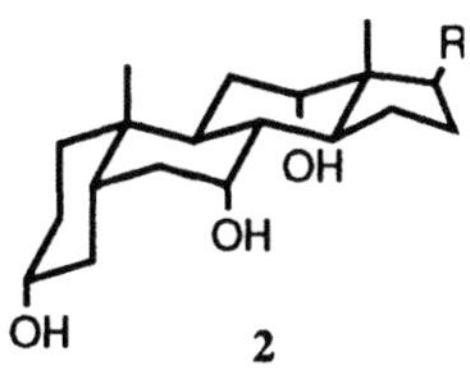

2

V. Balzani and L. De Cola (eds.), Supramolecular Chemistry, 359–374.

range of different side chains and linkers, has a π-system that tells us spectroscopically about coordination state (UV absorption) and geometry (NMR), and a central metal ion to act as binding site. We use zinc as the central metal ion because it is easy to insert and remove quantitatively, it is reliably four- or five-coordinate in porphyrins, and is spectroscopically convenient because it is diamagnetic. Cholate is cheap, occurs naturally as a single enantiomer, and is equipped with inward-facing hydroxyls that have the potential to act as binding or catalytic sites.

We have therefore sought the efficient syntheses of cyclic oligomers of **Zn1** and **2**, and of molecules combining these units. We have also been looking at their binding properties.

2. Synthesis of cyclic porphyrin oligomers

Our general synthetic strategy is outlined in Figure 1 below.(2–4) This synthesis converges rapidly, taking us from a single pyrrole ring to porphyrin trimer in three conceptual steps. We have used R = Et and CH_2CH_2COOMe; the latter can be saponified to give water-soluble hosts or transesterified to a range of sidechains with different solubility properties.

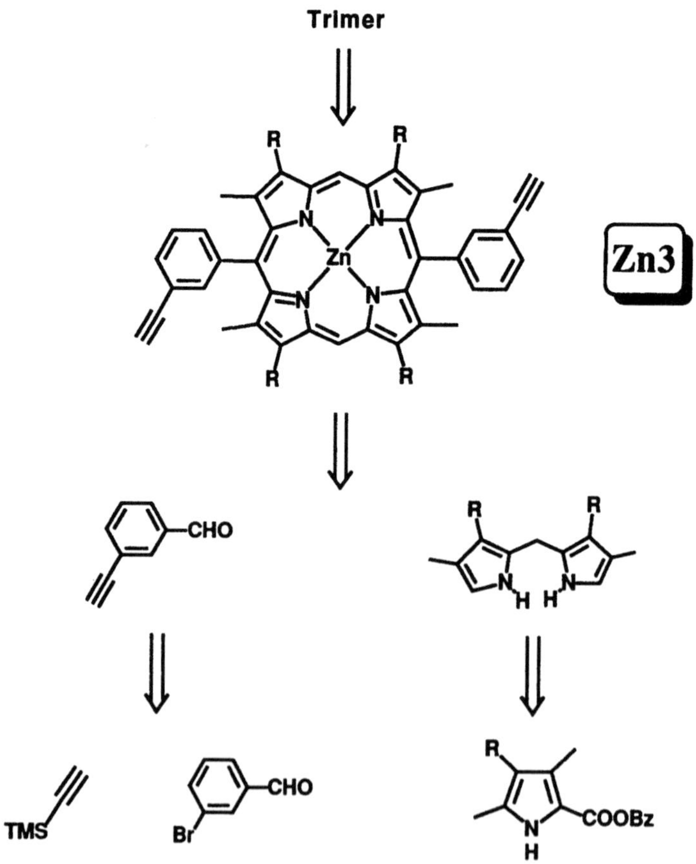

Figure 1. Convergent synthesis of porphyrin oligomers.

The key bis-acetylenic porphyrin monomer **Zn3** is available on a large scale using the route shown in Figure 1 and has been linked into two series of cyclic oligomers which have similar shapes but which enclose different sized cavities. The first, and most versatile, reaction is the Cu-catalysed Glaser–Hay coupling which yields an essentially quantitative mixture of cyclic butadiyne-linked oligoporphyrins as shown in Figure 2.(2, 3)

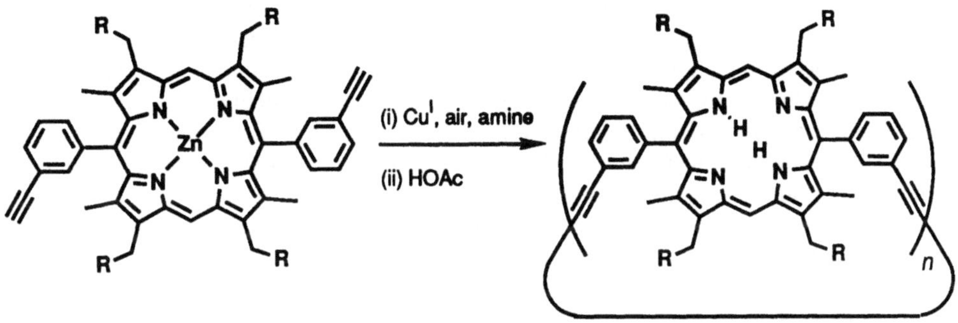

3

4 $n = 2$ (20–25%)
5 $n = 3$ (25–30%)
6 $n = 4$ (10–15%)

Figure 2. Glaser–Hay coupling of **3** to give cyclic porphyrin oligomers.

Dimer, $\mathbf{H_44}$, and trimer, $\mathbf{H_65}$, are readily isolated from this mixture. The tetramer, $\mathbf{H_86}$, is difficult to separate cleanly from the small quantities of higher oligomers formed in the reaction; a better synthesis is described in §4 below. These oligomers are well-behaved, surprisingly robust and more soluble that one might have expected.

Alternatively, monomeric **3** units can be linked through alkyne–Pt–alkyne links. The coupling can be carried out directly, as shown in Figure 3, or via the $SnMe_3$–alkyne.(4) In our experience these reactions are not as clean as the Glaser–Hay coupling, and the resulting oligomers are less stable, but it is possible to obtain 20% isolated yield of the cyclic trimer, **7**.

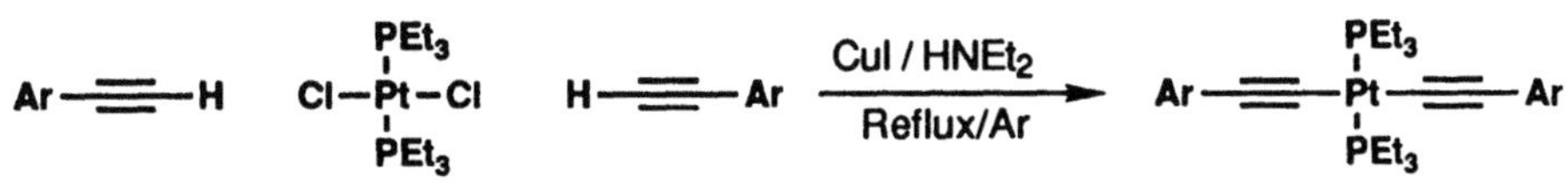

Figure 3. Alkyne coupling with a platinum linker.

UV and NMR spectroscopy of all the oligomers indicate that the component porphyrins are essentially isolated from each other electronically. They enclose open cavities that are accessible to solvent. The Zn–Zn distance in the butadiyne-linked trimer $\mathbf{Zn_35}$ is calculated to be around 15.8 Å, while in the Pt-linked trimer $\mathbf{Zn_37}$ it is around 18 Å.

3. Ligand binding

Diaryl zinc porphyrins bind pyridine ligands in chloroform and similar solvents with binding constants of around 3×10^3 M^{-1}. This binding is easily monitored as the change in zinc coordination state from four to five is accompanied by a shift in the intense porphyrin Soret absorption from 409 to 423 nm. The NMR resonances of bound ligands are shifted dramatically upfield by the porphyrin ring current, providing detailed information about adduct geometries.

Binding of simple pyridines to our cyclic oligomers occurs with the same microscopic binding constant as binding to monomer. So, each ligand binds independently: there is no barrier to binding several pyridines in close proximity within cavities. Our preliminary X-ray structure of the quinuclidine adduct of **$Zn_3$5** indeed shows three ligand molecules within each cavity.

Binding of a series of bipyridyls to dimer **$Zn_2$4** and trimer **$Zn_3$5** exhibits the pleasing selectivities summarised in Figure 4. The trimer binds a second molecule of the bipyridyls normally at its third site, confirming the independence of its component units.

The effective molarity (EM) for the second pyridine of bipyridyl itself (Bipy, $n = 1$) binding to dimer approaches 100 M, although modelling and NMR both suggest that the porphyrin units need to distort outwards to accommodate the ligand; the fact that Bipy binds as strongly as it does to the trimer implies that the host can distort significantly to assist ligand binding. Dipyridylpropane ($n = 3$) binding to trimer appears to be close to the optimum one could expect.

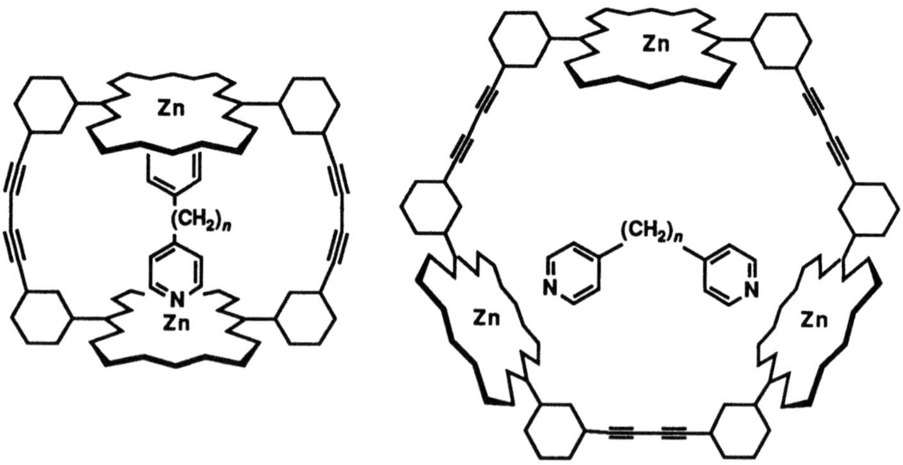

n	K	K
0	6×10^8	6×10^4
2	1×10^6	1×10^7
3	1×10^5	3×10^8

Figure 4. Binding constants for bipyridyl complexation with **$Zn_2$4** and **$Zn_3$5**.

The binding of tripyridyl guests can be even stronger. Tripyridyltriazine (TPT) is calculated to be a good fit, if slightly too small, for the butadiyne-linked trimer **Zn$_3$5**, while the Al(pyacac)$_3$ ligand was designed to fit the Pt-linked trimer, **Zn$_3$7**.(4) As Figure 5 shows, these expectations are correct. The second pyridine of TPT contributes a factor of around 10^5 to the binding constant, an EM of around 100, while the third pyridine contributes only 10^2 (EM ≈ 0.1) confirming that the ligand is slightly too small for a perfect fit.

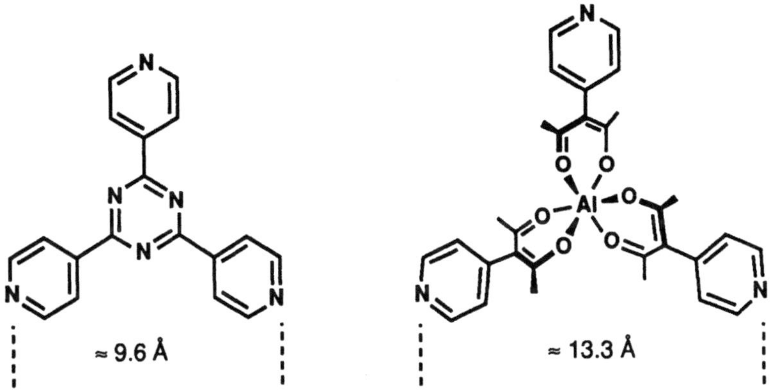

	TPT	***Al(pyacac)$_3$***
Trimer **Zn$_3$5**	Guest fits well $K \approx 10^{10}$ Slow exchange NMR	Guest too large $K \approx 10^7$ Broad NMR resonances
Trimer **Zn$_3$7**	Guest too small $K \approx 10^7$ Broad NMR resonances	Guest fits well $K \approx 10^{10}$ Slow exchange NMR

Figure 5. Binding of tripyridyl ligands to porphyrin trimers.

4. Templating: tetramer and octamer syntheses

The very different binding selectivities exhibited by dimer and trimer led us to ask whether it would be possible to control the outcome of the Glaser–Hay coupling by templating. Although the mechanistic details of the coupling itself are unclear, it seemed likely that the general course of the synthesis is as shown in Figure 6: once two monomers have coupled to form a linear dimer there is a competition between intramolecular cyclisation to give **4** and intermolecular coupling to give linear trimer. At this stage there is again competition between intramolecular cyclisation to **5** and further elongation by intermolecular coupling.

Would Bipy in the coupling mixture encourage dimer formation by binding linear dimer in the favourable conformation for intramolecular cyclisation? More interestingly, could TPT hold the linear dimer in an open conformation to inhibit cyclisation and so encourage intermolecular coupling? The answer is yes. After careful optimisation of conditions, the isolated yield of **4** is boosted to over 70% by Bipy and that of **5** to 50% by TPT.(3)

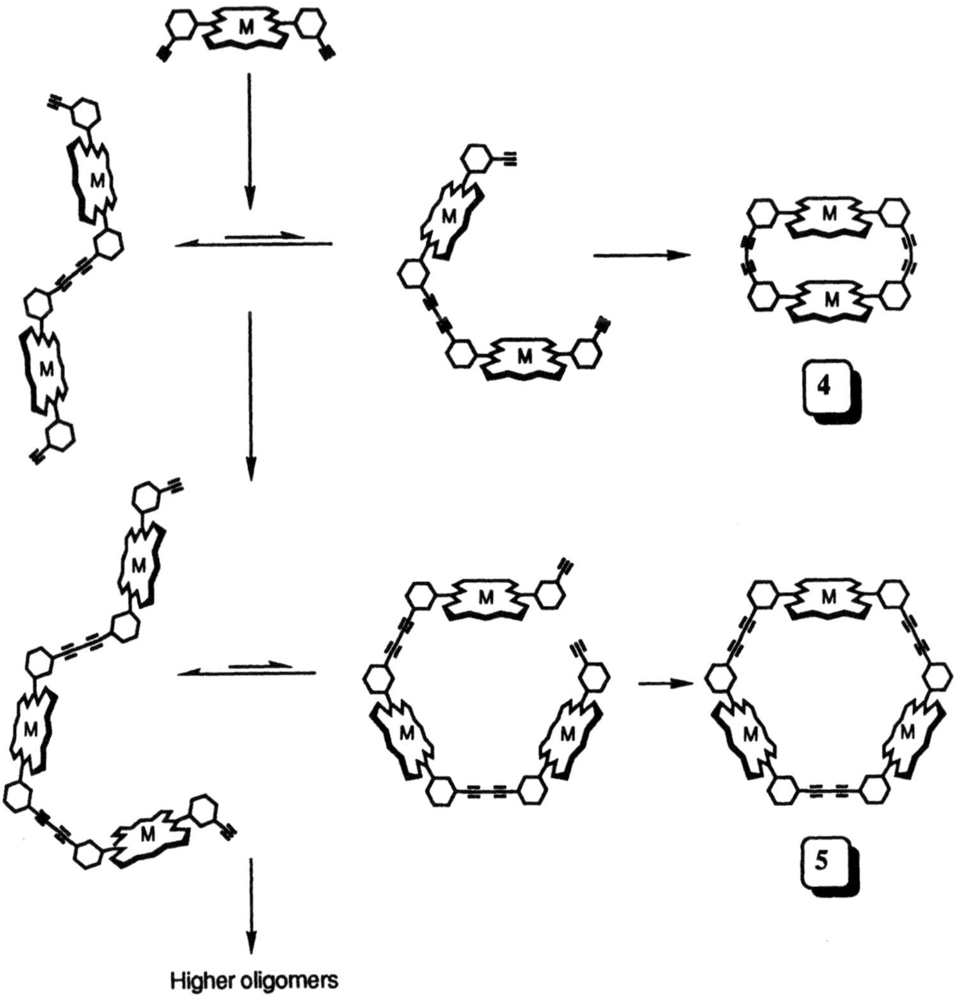

4

5

Figure 6. (Above) Reaction scheme for porphyrin oligomerisation

Figure 7. (Right) Positive templating brings reactive ends together; negative templating keeps reactive ends apart

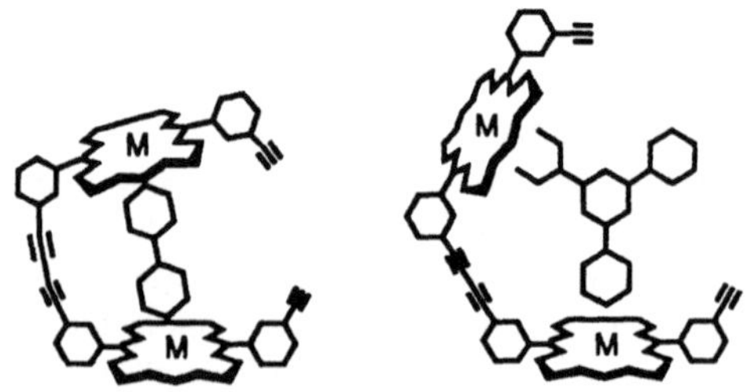

The main templating success of TPT resides in its ability to inhibit dimer formation as shown in Figure 7 — TPT is actually not very effective at catalysing trimer cyclisation. Dipyridyl-propane and other bifunctional templates of similar geometry are almost as good at templating trimer synthesis as TPT because they too can inhibit dimer cyclisation. We call this effect negative templating; the more conventional rôle of Bipy encouraging cyclisation of dimer we call positive templating.

The fact that these templating reactions work well does not prove that we understand how they they operate — it is not obvious, for example, that crucial intermediates will have the same ligand-binding properties as products. We therefore set out to synthesise some of the putative intermediate oligo-porphyrins and study their binding and cyclisation properties.(5)

The doubly-protected monomer **Si_2H_24** was prepared by trivial modification of the route in Figure 1, and partially deprotected with fluoride ion. Fully deprotected **H_24** was removed by chromatography and the resulting mixture of mono- (**SiH_24**) and di- (**Si_2H_24**) protected material was subjected to Glaser–Hay coupling. Unreacted **Si_2H_24** was readily separated chromatographically from the dimer **Si_2H_48** and recycled. Complete deprotection of **Si_2H_48** with fluoride yielded **H_48**, the presumed first intermediate in cyclo-oligomerisation.

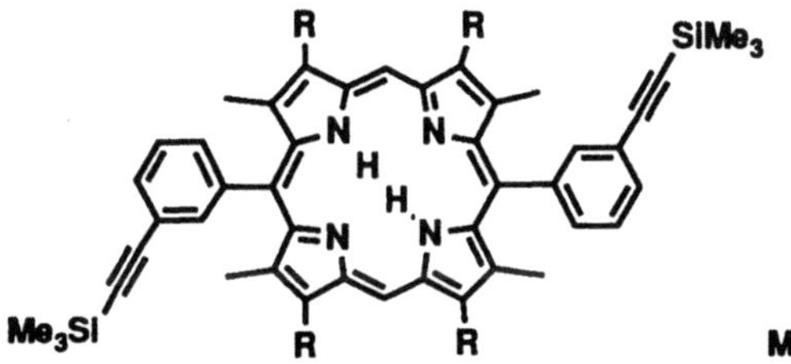

Si_2H_24

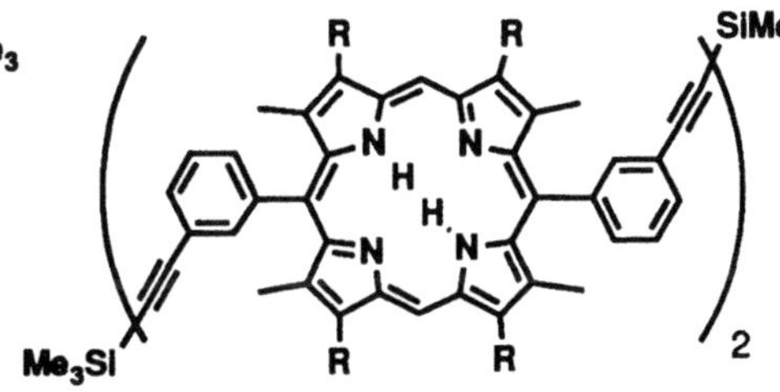

Si_2H_48

Glaser–Hay coupling of **Zn_28** in the absence of any template gave only 19% conversion to cyclic dimer **4**, together with many higher oligomers, while the Bipy-templated reaction gave 83% of **4**, providing support for the ideas in Figures 6 and 7. **Zn_28** does indeed bind strongly to Bipy in a conformation that facilitates intramolecular reaction.

The process can be extended in principle to linear and cyclic tetramer simply by repeating the sequence of partial deprotection, coupling, full deprotection and cyclisation steps. However, it is not practical to separate large amounts of the linear dimers **Si_2H_48**, **SiH_48** and **H_48**, or their zinc derivatives, and Glaser–Hay coupling of the mixture gives an intractable mixture of many different products because **Zn_28** mainly reacts intermolecularly.

Once again, templates come to our rescue, in yet another new rôle: as scavenger. Bipy effectively enforces the intramolecular cyclisation of **Zn_28** so this fully-deprotected molecule does not couple with monoprotected material in the reaction mixture. **Si_2Zn_28** will bind Bipy but not react, while **$SiZn_28$** will bind Bipy and then only be able to react with another similarly bound **$SiZn_28$** to give protected linear tetramer, **Si_2Zn_49**. As predicted, Glaser–Hay coupling of the partially deprotected dimer mixture in the presence of excess Bipy yields only three products: linear dimer, **Si_2Zn_28**; cyclic dimer, **Zn_24**; and linear tetramer, **Si_2Zn_49**. These are readily separated from each other after removal of the zinc.

The fully deprotected tetramer **Zn_49** reacts under untemplated Glaser–Hay conditions to give around 60% of the cyclic tetramer **Zn_410**, but it is very difficult to purify this from the mixture of other products. The NMR spectrum of **Zn_410** in the absence of any ligand is more complicated than one would expect, indicating the presence of more than one conformation in slow exchange: the large ring size allows the two aryl groups attached to one of the porphyrin units to adopt the *trans* arrangement as shown in Figure 8.

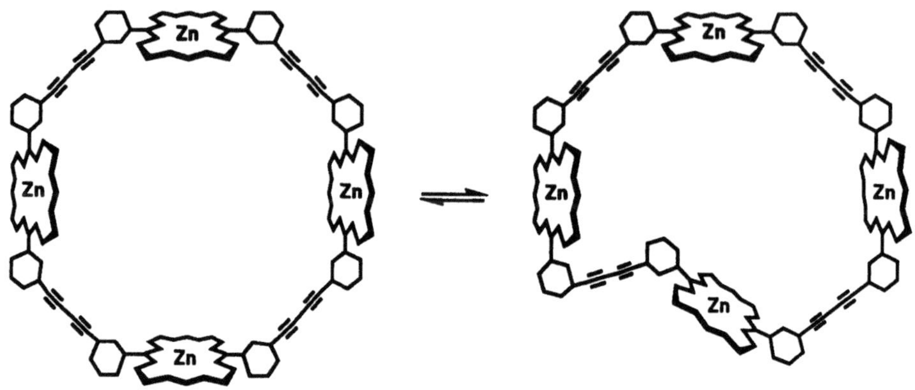

Figure 8. Conformational equilibrium of **$Zn_4$10**.

The cyclic tetramer binds very strongly to *meso*-tetrapyridylporphyrin, **11**, to yield a complex which is sufficiently stable to chromatograph as a single compound. Coupling of linear tetramer **$Zn_4$9** in the presence of **11** as template gave the **$Zn_4$10•11** complex as the only detectable product which is then readily isolated in a pure state.

On addition of **11**, the *meso* region of the **$Zn_4$10** spectrum simplifies in the way one would expect for complete conversion to the all-*cis* form required for binding. The complex is in slow exchange with either of the components at least up to 100° C, indicating its unusual strength. Molecular mechanics calulations and inspection of CPK models suggest that the complex takes up the tub-shaped geometry shown in Figure 9. The symmetry implied by this shape is supported by the NMR spectrum, which shows splitting of the signals Me_a and Me_b by 0.28 and 0.05 ppm respectively.

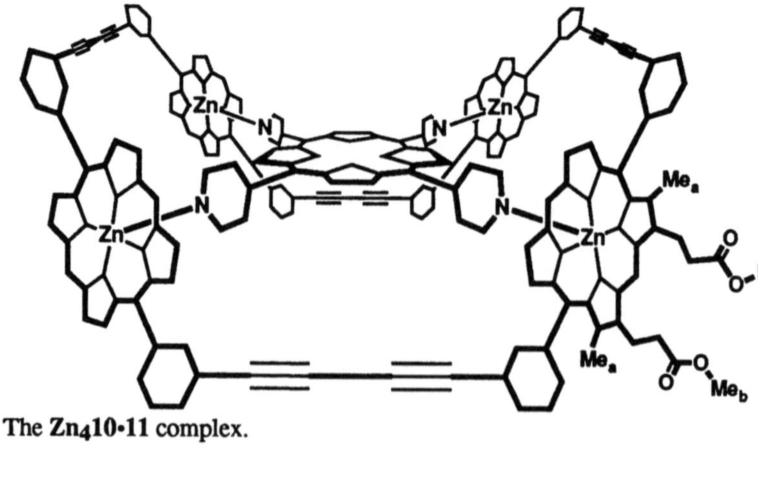

Figure 9. The **$Zn_4$10•11** complex.

The coupling–scavenging process can be taken one step further: **11** will clearly be an excellent scavenger for fully-deprotected linear tetramer **$Zn_4$9** in the presence of mono-protected **Si$Zn_4$9**. As expected, partial deprotection of **$Si_2Zn_4$9** followed by coupling in the presence of **11** led cleanly to the **$Zn_4$10•11** complex and linear octamer **$Si_2Zn_8$12** as sole products apart from recovered tetramer (Figure 10). Coupling in the absence of **11** gave an intractable mixture.

We have established three separate rôles for templates here: positive templates encourage intramolecular cyclisation; negative templates prevent intramolecular cyclisation and so encourage chain elongation; and scavenging templates allow efficient dimerisations in statistical mixtures by inhibiting intermolecular reactions of doubly-reactive monomers. The scavenging process is actually positive templating, but employed in rather a different way.

By this means we have developed syntheses of interesting oligomers that are practically inaccessible by conventional approaches. These principles should be applicable to other systems than porphyrins, although the templates and substrate molecules themselves may look very different.

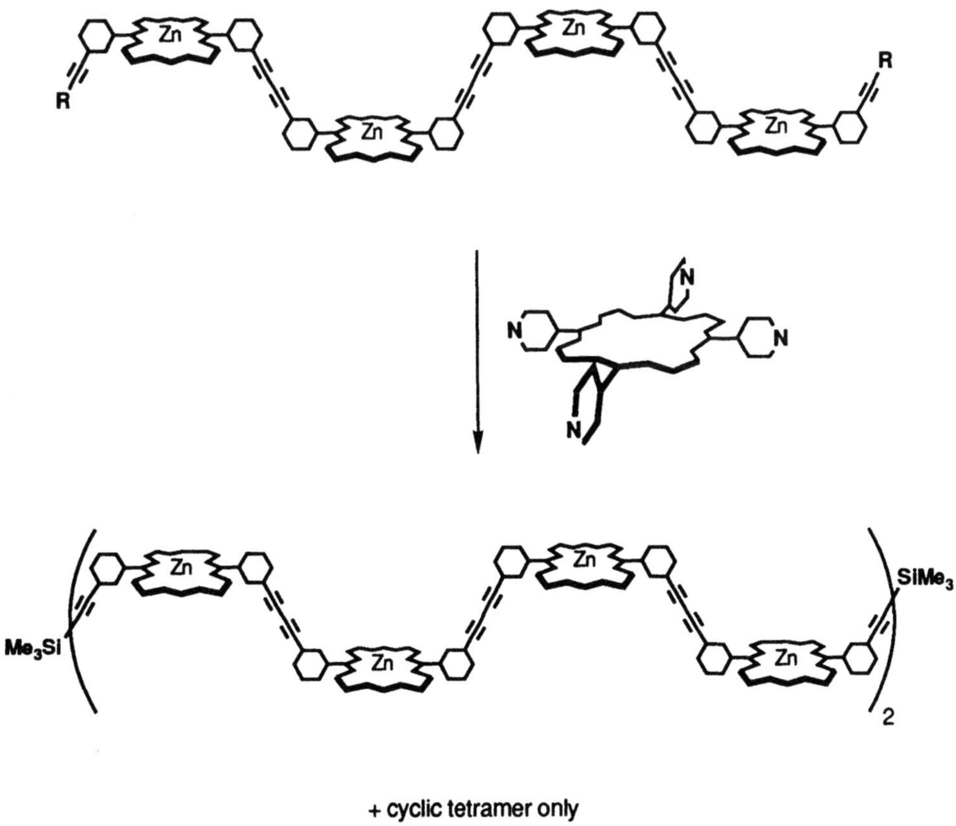

Figure 10. Synthesis of linear octamer **$Si_2Zn_8$12** by Glaser–Hay coupling of a mixture of linear tetramers (R = H, $SiMe_3$) using **11** as scavenging template.

5. Electrospray mass spectrometry

Fast atom bombardment mass spectrometry was an important tool for characterising our oligomers up to trimers, but molecular ions from tetramers are low intensity and we have seen no molecular ion at all for octamer. The electrospray technique became available at just the right time and promises to revolutionise the mass spectrometry of large molecules.(6) In essence, one protonates the molecule at several sites and injects the solution into the spectrometer; the conventional order of volatilisation followed by ionisation is reversed, so avoiding the problem of getting large molecules into the gas phase. Since each molecule carries many charges, its *m/z* is reduced, allowing the use of small magnets. Each molecular ion, MH_n^{n+}, is represented by several ions with different values of *n*, so it is a relatively simple piece of computing to find the true molecular weight.

We find that the sensitivity of the technique appears to be similar for different oligomers, and unambiguous molecular weights are readily obtained. Figure 11 shows a single example, the metal-free linear octamer, **Si_2H_{16}12**, which has a molecular weight of 7421.

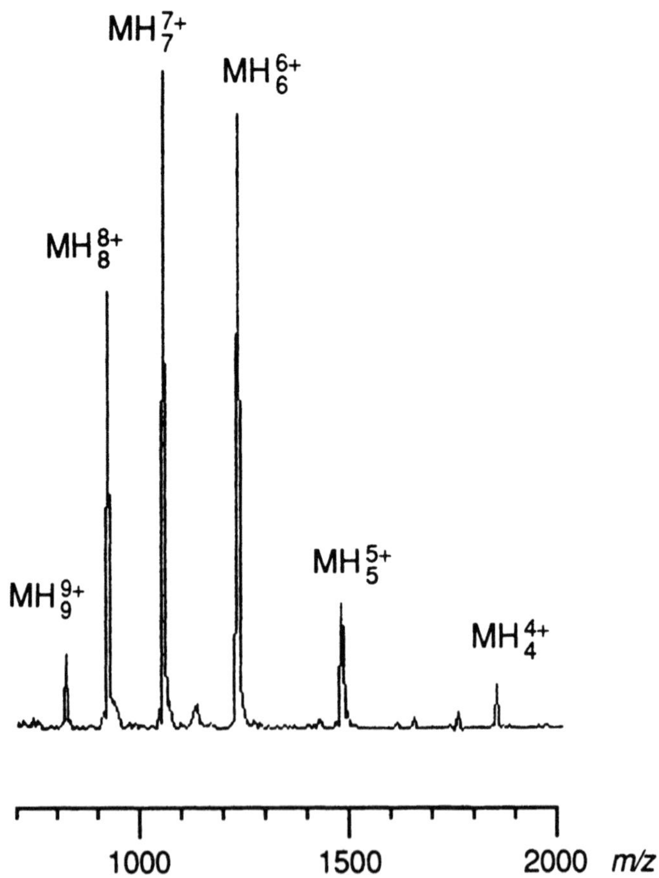

Figure 11. Electrospray mass spectrum of octamer **Si_2H_{16}12**; the sample was injected as a methanol solution containing 0.2%trifluoroacetic acid as proton source.

6. Cyclocholates

We turn now to a second building block, cholic acid, **2**. This can be incorporated into cyclic oligomers by formation of ester linkages between the C_3-hydroxyl of one unit and the carboxyl group of another. The C_{12}-hydroxyl group needs to be protected to prevent intramolecular lactonisation, while the C_7-hydroxyl can be also protected; versatile schemes are available for the protection and deprotection of each of the hydroxyl groups and the carboxylate.(7) Finally, the carboxyl-bearing sidechain can be shortened in four steps to give bisnorcholic acid, **13**; this has broadly the same chemistry but the shorter linkage between cholate units gives much greater rigidity to the whole assembly at the expense of a less reactive carboxyl resulting from steric congestion.

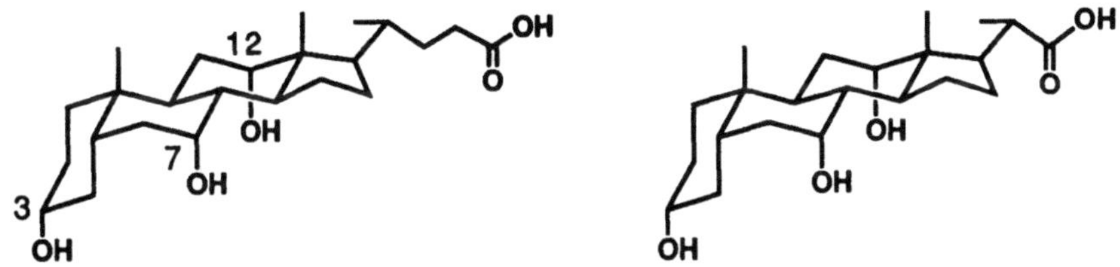

2 **13**

The shortest route to cyclocholates involves oligomerisation using mixed-anhydride chemistry, and leads to a range of ring sizes as shown in Figure 12.(8) Bulky substituents on the C_7-hydroxyl inhibit formation of 'small' rings: when R = H, some dimer is obtained; R = TFA (trifluoroacetate) gives trimer as the smallest oligomer, while bigger substituents increasingly favour formation of larger rings. Protecting groups such as TFA are readily removed with aqueous ammonia.

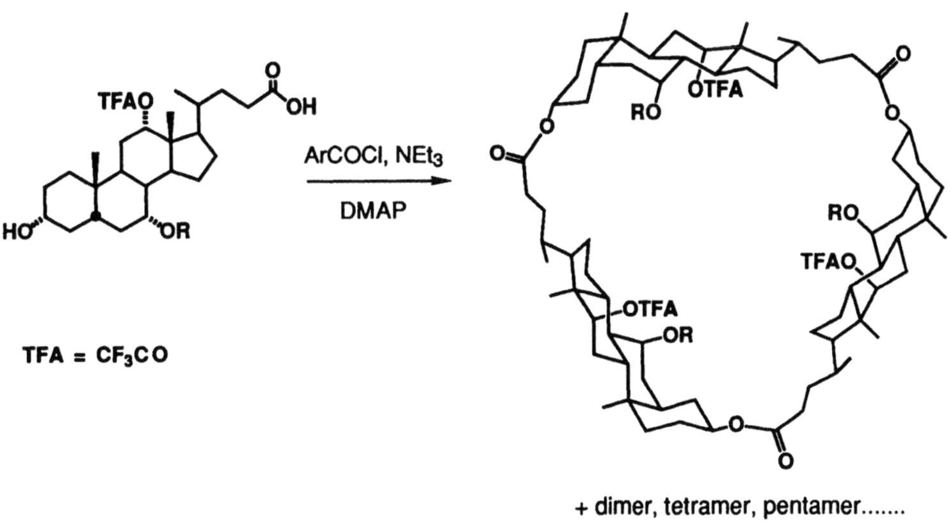

Figure 12. Synthesis of cyclocholates from monomer

A slightly longer but more controlled route is to synthesise linear oligomers by condensation of suitably protected monomers, and then cyclise these as shown in Figure 13.

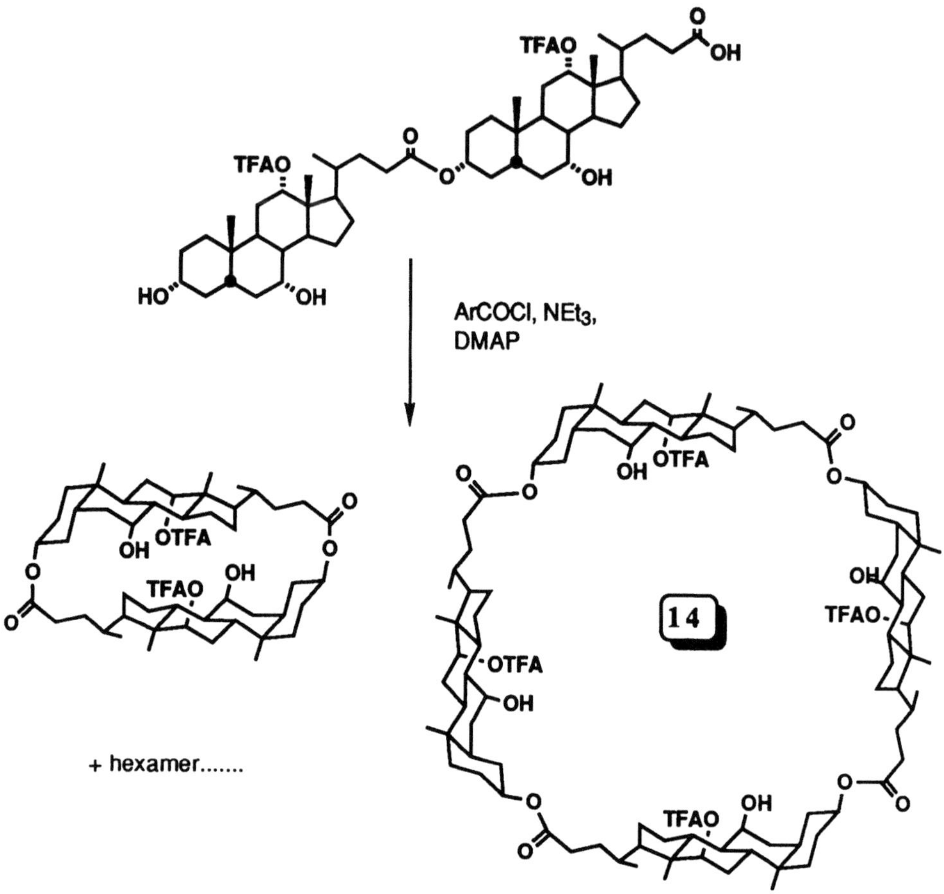

Figure 13.

The linear trimer derived from bisnorcholic acid, **13**, can be similarly synthesised and ring-closed to give cyclic trimer, **15**, or cyclic hexamer. Preliminary experiments suggest that steric effects from the C_7-substituent can again be used to direct the cyclisation either towards trimer (OH) or hexamer (OTFA) although the switch is not complete.

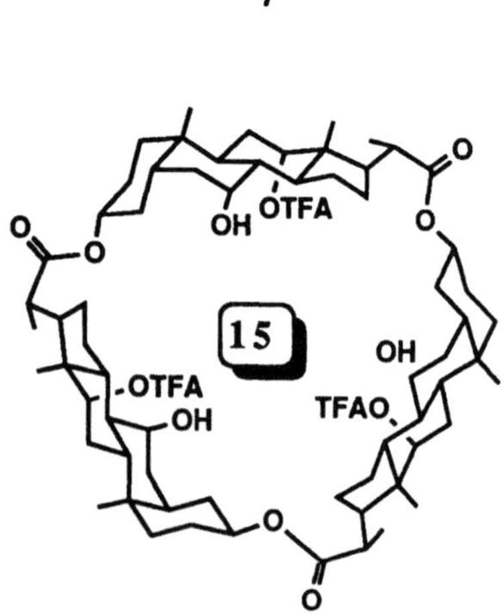

After deprotection, the bisnor cyclotrimer **15** weakly recognises several types of polar guest that are capable of donating and/or accepting hydrogen bonds in organic solution. For example, 1, 3, 5-trihydroxybenzene has a binding constant of around 100 M^{-1} in 5:1 $CDCl_3$–THF compared with only 3 M^{-1} for 1, 3-dihydroxybenzene in the same solvent mixture.

The cyclocholates are like 'inside-out' cyclodextrins, with hydrophobic exteriors and hydrophilic cores, and they may have interesting cross-membrane transport properties.

7. Porphyrin–Steroid combinations

The tetramer **14** can be converted in five steps to the bowl-shaped **Zn16** as shown in Figure 14.(8) This possesses a cavity lined with four hydroxyl groups and a floor containing the porphyrin metal ion as fifth binding site. As yet we know little of its binding properties.

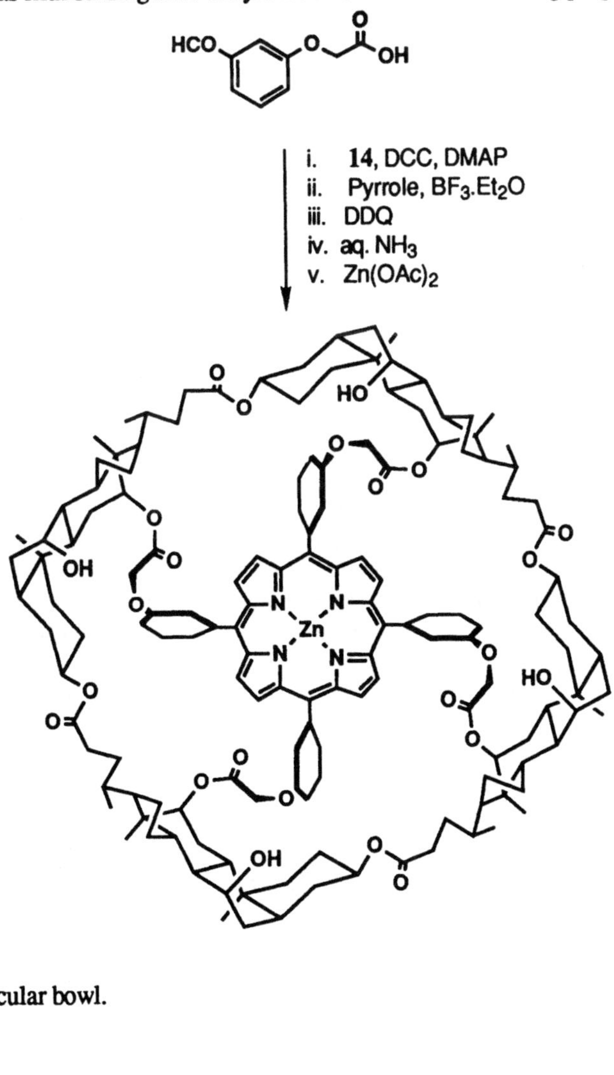

Figure 14. A molecular bowl.

A quite different geometry can be obtained by combining two cholates to give a bislactone as shown in Figure 15.(9) The steroid backbones form a rigid convex cap containing two inward-pointing hydroxyls.

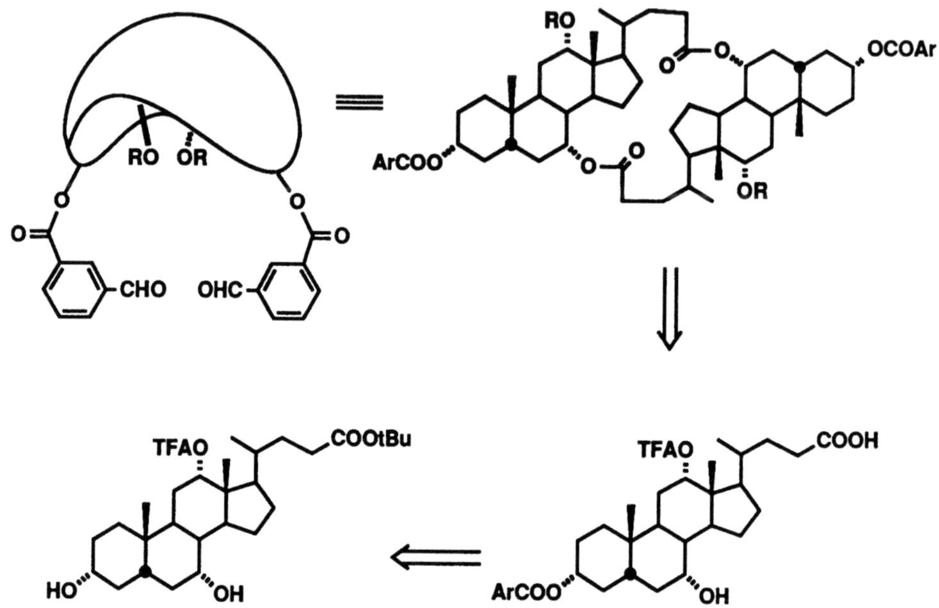

Figure 15. Cholate bis-lactone cap.

The bis-cholate lactone cap can be incorporated into a variety of porphyrins as shown in Figure 16.(9) Reaction with a bipyrromethane (as in Figure 1) gives the monomeric porphyrin **17** and dimer **18**. When the C_7-hydroxyls are free the main product is **17** but attachment of a bulky group such as TFA yields **18** as the dominant cyclic product; presumably this results from steric hindrance to intramolecular creation of the porphyrin. The asymmetry of the porphyrin environment in **Zn17** complicates its binding statistics so most of our binding studies have been carried out with the symmetric **Zn19** synthesised in 7% yield as shown in Figure 16 (R = H).

The three-dimensional cavity provided by the cholate cap has some interesting properties: relative to a standard porphyrin, binding of 4-ethyl pyridine is reduced by the expected steric hindrance but binding of 3, 5-dimethyl pyridine is enhanced by 3 kJ mol^{-1} because the methyl groups fit snugly into a cleft of the cap. Both 3- and 4-hydroxy-pyridine bind more strongly (by 6–9 kJ mol^{-1}) than to a standard porphyrin by virtue of an extra hydrogen bond to the C_7-hydroxyl(s) while purine binds 15 kJ mol^{-1} more strongly because it can make both donor and acceptor hydrogen bonds with the C_7-hydroxyl groups.(9)

Finally, 3-carboxypyridine selectively modifies **Zn19** under mixed anhydride esterification conditions as shown in Figure 17. The reaction is essentially quantititative, 100% selective for a single hydroxyl, and complete in 5 minutes at room temperature. The zinc-free porphyrin does not react at all under the same conditions, so binding in the appropriate geometry is an essential prerequisite for reaction.

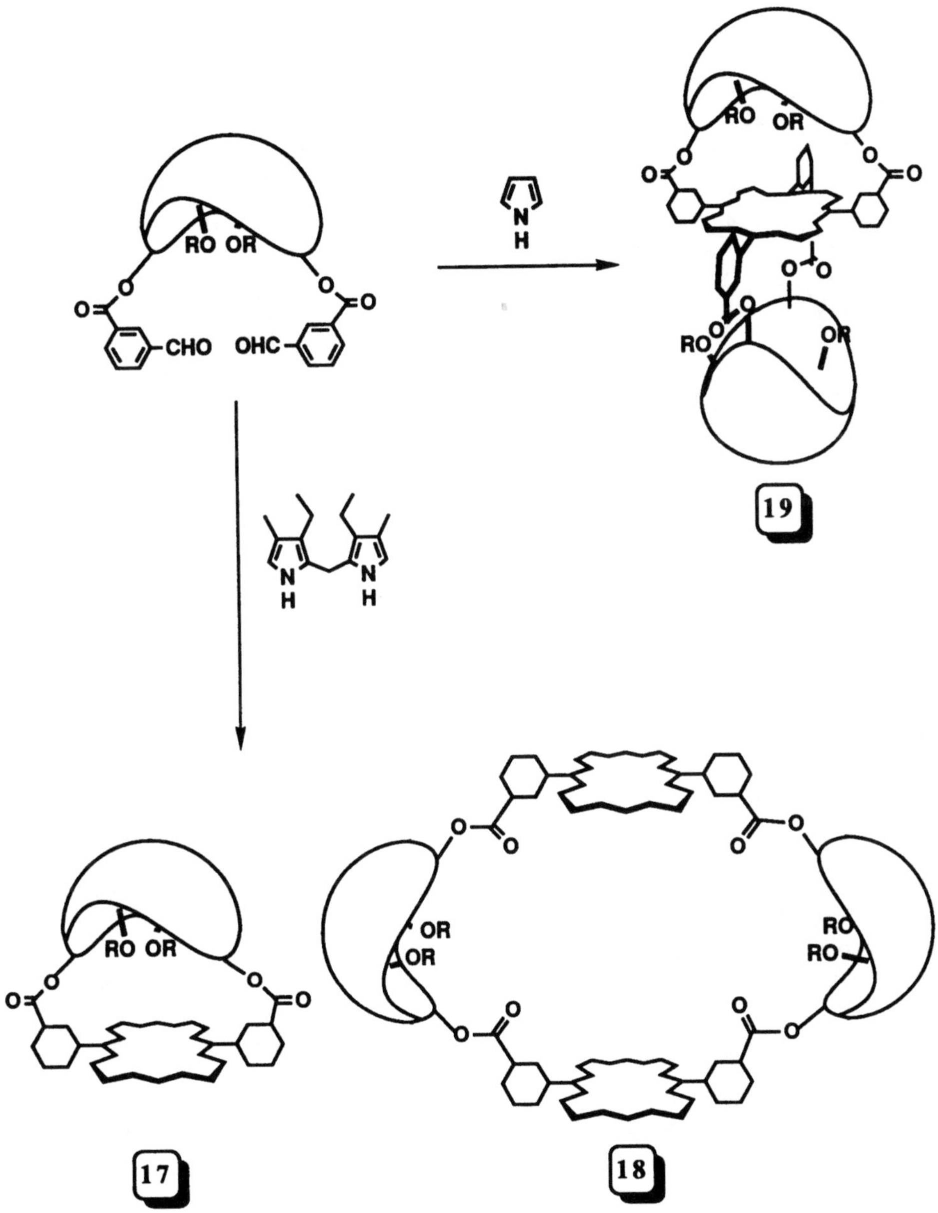

Figure 16. Synthesis of cholate-capped porphyrins

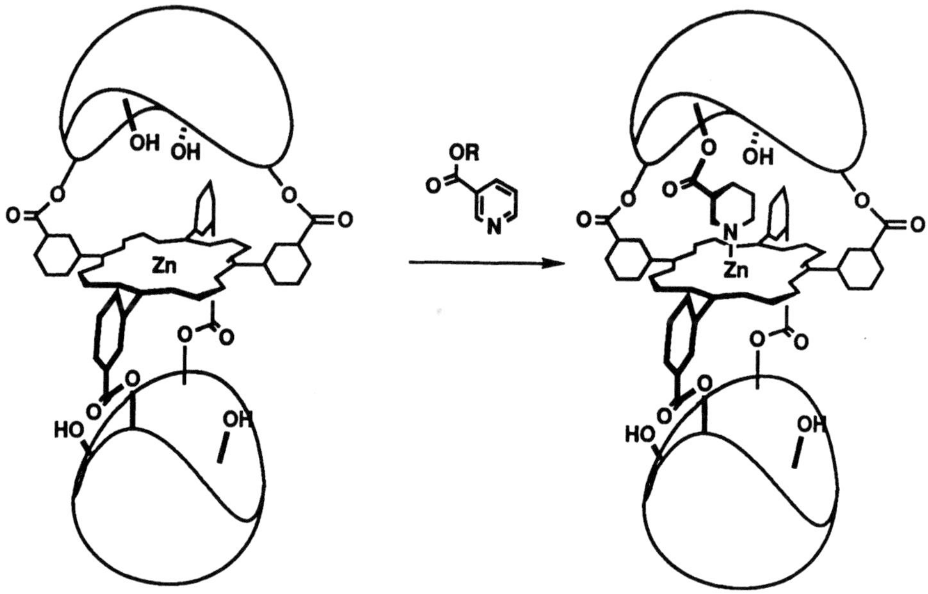

Figure 17. Binding-induced self-modification of a cholate-capped porphyrin.

8. References

1. C. A. Hunter and J. K. M. Sanders, *J. Am. Chem. Soc.*, 1990, **112**, 5525.
2. H. L. Anderson and J. K. M. Sanders, *J. Chem. Soc. Chem. Commun.*, 1989, 1714.
3. H. L. Anderson and J. K. M. Sanders, *Angew. Chem. Intl. Edn. Engl.*, 1990, **29**, 1400.
4. L. G. Mackay, H. L. Anderson and J. K. M. Sanders, *J. Chem. Soc. Chem. Commun.*, 1992, In press.
5. S. Nicholson, H. L. Anderson and J. K. M. Sanders, submitted.
6. J. B. Fenn, M. Mann, C. K. Meng, S. F. Wong, and C. M. Whitehouse, *Science*, 1989, **264**, 64.
7. R. P. Bonar-Law, A. P. Davis and J. K. M. Sanders, *J. Chem. Soc. Perkin Trans. I*, 1990, 2245.
8. R. P. Bonar-Law and J. K. M. Sanders, unpublished
9. R. P. Bonar-Law and J. K. M. Sanders, *J. Chem. Soc. Chem. Commun.*, 1991, 574.

9. Acknowledgements

We thank the SERC, Magdalene College, the Association of Commonwealth Universities, the New Zealand Vice-Chancellors' Committee and Rhône-Poulenc-Rorer for financial support, and the SERC Mass Spectrometry service for FAB mass spectra.

PHOTOPHYSICAL PROPERTIES OF THIONINE AND METHYLENE BLUE WHEN BOUND TO DNA, POLYNUCLEOTIDES OR NUCLEOTIDES

J.M. KELLY, E.M. TUITE, W.J.M. VAN DER PUTTEN
Chemistry Department
Trinity College
University of Dublin
Dublin 2
Ireland

G.S. BEDDARD, G.D. REID
Chemistry Department
University of Manchester
Manchester M13 9PL
UK

ABSTRACT. The photophysical properties of methylene blue and thionine in the presence of DNA, polynucleotides and mononucleotides have been studied by emission spectroscopy and transient absorption measurements. The singlet state fluorescence yield and lifetime for each of the dyes is reduced upon binding to guanine-containing polynucleotides or to deoxyguanosine monophosphate. The quenching, which is attributed to photo-induced electron transfer, is extremely rapid (4 ps and 0.8 ps for methylene blue and thionine respectively when bound to poly[d(G-C)].poly[d(G-C)]). Thionine, but not methylene blue, is strongly quenched by adenine derivatives, consistent with the higher oxidising potential of its singlet excited state. Transient absorption measurements show that even on a picosecond timescale yields of reduced dye and oxidised guanine formed are very low.

1. Introduction

Dye sensitised damage to DNA is of considerable interest both from a point of view of modifying the biological function of DNA and for developing procedures for site-specific reactions on DNA, which might be useful to molecular biologists. Thionine and its methylated analogues are of interest in this regard. They are known to cause photodynamic damage in bacteria, viruses, and some tumour tissues *in vitro* [*e.g.* 1-3] and photochemically-induced specific reactions with guanine have been suggested as methods for DNA-sequencing [4].

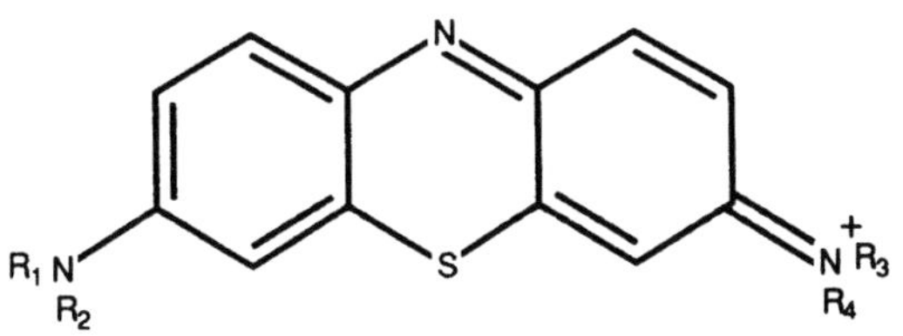

Thionine :	$R_1 = R_2 = R_3 = R_4 = H$	
Azure C :	$R_1 = CH_3$	$R_2 = R_3 = R_4 = H$
Azure A :	$R_1 = R_2 = CH_3$	$R_3 = R_4 = H$
Azure B :	$R_1 = R_2 = R_3 = CH_3$	$R_4 = H$
Methylene Blue :	$R_1 = R_2 = R_3 = R_4 = CH_3$	

Figure 1. Structural formulae of the phenothiazine dyes.

V. Balzani and L. De Cola (eds.), Supramolecular Chemistry, 375–381.

As it is expected that molecules of this family should bind to DNA in a well-defined manner, it is also of interest to consider the properties of the resultant supermolecule. In particular it is worthwhile to consider the interaction between the dye and the DNA bases. As the dyes' excited states are strongly oxidising, oxidation of guanine should be possible and this has been proposed as one mechanism for the damage to DNA photo-sensitised by these molecules. The thionine family of dyes are of interest as the redox and spectroscopic properties of the dyes vary systematically as thionine is methylated [*i.e.* from thionine, through azure C, azure A, azure B, to methylene blue] (Figure 1). In this paper we compare the properties of the triplet and singlet states of methylene blue and thionine when bound to DNA or synthetic double-stranded polynucleotides (poly[d(G-C)] or poly[d(A-T)]) or complexed to purine deoxynucleotides (dGMP or dAMP).

2. Binding of Dyes To DNA, Polynucleotides, and Mononucleotides and Study of Absorption and Fluorescence

We have studied the binding of the dyes to DNA by a number of techniques, especially absorption and fluorescence spectrometry and by dialysis. Results have been analysed using the Benesi-Hildebrand method for the mononucleotides and Scatchard-type methods for the polynucleotides. Both dyes bind strongly to dGMP and dAMP but no evidence could be found for binding to pyrimidine mononucleotides. A number of binding studies [5,6] have indicated that while the methylene blue intercalates into DNA at low ionic strength, at higher ionic strengths or in the presence of divalent ions other binding modes are operative. Under the latter conditions the dyes appear to bind more strongly to A-T rich sequences in DNA, whereas when intercalated the preference is for the G-C areas.

For thionine the absorption maximum at 599 nm shifts by 10-15 nm upon binding to the double-stranded polymers or in the presence of high concentrations of dAMP or dGMP (50 - 100 mM). Similar behaviour has been reported for methylene blue [7,8]. Marked red-shifts are also observed in the fluorescence spectra of thionine bound to poly[d(A-T)] or to dAMP at high concentrations. The shifts are diminished for the polynucleotide at higher ionic strengths. The fluorescence of methylene blue is quenched by DNA, poly[d(G-C)] or dGMP but not by poly[d(A-T)] or dAMP [7,8]. For thionine quenching occurs with each of the polymers and with both dAMP and dGMP, although the extent of quenching is not as great for the adenine-containing species as for those containing guanine moieties (Figure 2). The quenching of the fluorescence of dyes upon binding to DNA has has been attributed to photo-induced electron transfer [9] and the above behaviour can be rationalised by considering the redox potentials of the singlet excited states of the two dyes (see Table 1) and the expected higher oxidation potential for adenine compared to guanine. Oxidation potentials for adenine and guanine were determined polarographically at pH 6.5 by Kittler *et al.* [9] as 0.95 V and 0.85 V (*versus* 1 N sulphate electrode; 1.63 V and 1.53 V *versus* NHE). A more reliable value should be the one electron oxidation potential determined by pulse radiolysis [10,11]. The radical cations of adenine and guanine are both known to deprotonate at pH 7 (see Figure 3) and values of of >1.6 V and >1.35 V (*versus* NHE) respectively have recently been determined for the reduction potentials of these deprotonated forms [12].

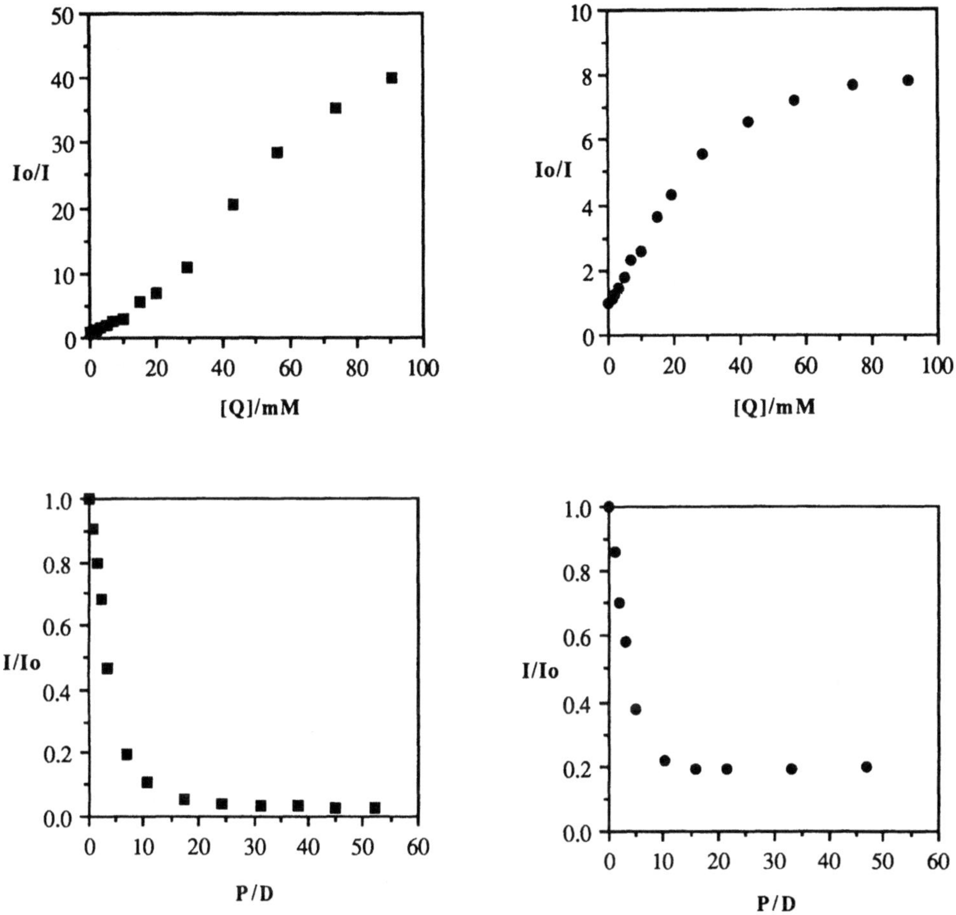

Figure 2. Quenching of thionine fluorescence by mononucleotides and polynucleotides. (A) dGMP, (B) dAMP, (C) poly[d(G-C)], (D) poly[d(A-T)}. $\lambda_f = 623$ nm. $[Th^+] = 2x10^{-6}M$.

TABLE 1. Estimated redox potentials for dye ground states and excited states at pH 7. (All values in volts *versus* NHE).

COMPOUND	E^0g.s.	E*s.s.	E^0s.s.	E*t.s.	E^0t.s.
Thionine	0.05 V	2.03 eV	2.08 V	1.69 eV	1.74 eV
Methylene Blue	0.04 V	1.83 eV	1.87 V	1.44 eV	1.48 eV

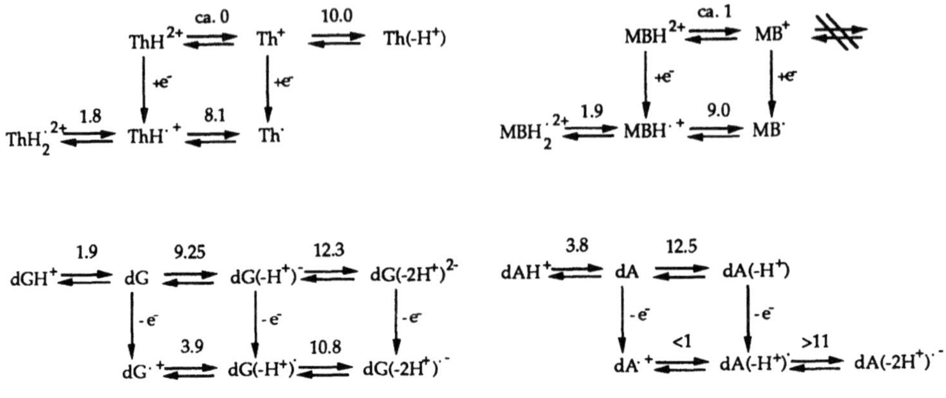

Figure 3. pH dependence of the dye and base species. At pH 7, phosphate units will exist as an equilibrium of mono- and di-anionic forms [11,16].

3. Transient Absorption Spectroscopy

3.1 NANOSECOND LASER FLASH PHOTOLYSIS

The triplet states of these dyes can be readily studied by transient absorption methods. Our previous studies indicated that the properties of the methylene blue excited state are modified on binding [7,13]. However unlike the singlet state, the triplet state lifetime is increased upon binding to DNA and polynucleotides. This can be taken to indicate that the electron transfer is disfavoured because of the lower redox potential of the triplet state. Enhancement of triplet state lifetimes has been observed in a number of other cases including acridine dyes [14]. The first order rate constants for decay of the triplet state for methylene blue in the presence of poly[d(G-C)] or poly[d(A-T)] in solutions of different ionic strengths, pH and aeration have been studied following excitation at 650 nm [13] (That first order kinetics are observed is probably indicative of exchange of the methylene blue triplet between different sites on the polynucleotide and possibly with molecules in solution within its lifetime.). It was noted that at low ionic strength the methylene blue triplet state bound to poly[d(G-C)] was less strongly quenched by oxygen than that bound to poly[d(A-T)] - as expected for an intercalated complex. However at higher ionic strength the greater protection is from the poly[d(A-T)], indicating that binding (presumably in the DNA grooves) to this polynucleotide is much stronger than to poly[d(G-C)] under these conditions. It was also found that binding to poly[d(A-T)] strongly retards the protonation of the triplet state, which has a pK_a of 7.1 in solution. These studies show that the triplet state is very sensitive to its environment and therefore has potential as a probe for DNA in chromatin and in other biologically relevant situations [15], although it is necessary to be cautious about the interpretation of data because of the sensitivity of the triplet state to several factors.

Flash photolysis studies can also be used to look for electron transfer products from either methylene blue or thionine with guanine-containing DNA. Our initial studies showed that the yield was small with either DNA or poly[d(G-C)] [7,13]. More recently Dunn *et al.* [8] have observed a low yield of the electron transfer products upon the flash photolysis of methylene blue in the presence of GMP. Analogous studies that we have carried out with thionine and dGMP at pH 7 reveal only very low concentrations of long-lived products - which we are unable to assign unambiguously to the electron transfer products. Pulse radiolysis experiments have shown that reduced thionine (Th$^{\cdot}$) protonates at neutral pH (pK_a = 8.1) [16]. The protonated species ($ThH^{\cdot +}$) has bands centred at 400 nm and 770 nm. Th$^{\cdot}$, which is expected to be formed initially on electron transfer but which will protonate at neutral pH, lacks the near infra-red band, making it more difficult to monitor its formation. It is interesting to note that at pH 7 guanine radical cation is known to deprotonate, so that electron transfer might be accompanied by a rapid proton transfer. However the deprotonated guanine radical cation absorbs only weakly in the visible and would be difficult to identify in the presence of thionine transient species. This lack of evidence for electron transfer may be contrasted to what is found with the ruthenium complex $Ru(TAP)_3^{2+}$ which is also quenched by guanine derivatives, such as dGMP. In this case clear evidence was found for both the reduced ruthenium complex and the deprotonated radical cation $G(-H^+)MP^{\cdot}$ [17]. Further with $Ru(TAP)_3^{2+}$ and poly[d(G-C)] or DNA the yield was significantly reduced but still measurable.

3.2 PICOSECOND LASER FLASH PHOTOLYSIS

In order to search for the electron transfer products and to determine more about the reactivity of the singlet state of the dyes we have carried out picosecond laser flash photolysis experiments. These used a colliding pulse mode-locked (CPM) R6G/DODCI dye laser, producing pulses of 60 fs (fwhm), centred at 625 nm. These were then amplified in four stages to 650 μJ at 15 Hz. Transient absorption decays were measured by a pump-probe method using a continuum generated from 50% of the amplified beam. Probing wavelengths were selected from the continuum by 10 nm bandwidth interference filters positioned in front of a pair of matched photodiodes.

The lifetimes of the singlet states of methylene blue and of thionine were found to be 380 ps and 360 ps respectively. Upon excitation with sub-picosecond pulses the transient bleaching and transient absorption signals can be observed (Figure 4 gives an example). As shown in Table 2 the singlet state lifetime is strongly reduced when MB^+ is bound to poly[d(G-C)] or in its complex with dGMP. Lifetimes of the short-lived species are identical at all wavelengths measured (*e.g.* 405, 436, 554, 590 nm). The slightly shorter lifetime with the poly[d(G-C)] may be a consequence of a better overlap between the dye and the guanine in the polynucleotide, where it is expected to be intercalated. For the thionine excited state the quenching with dGMP or poly[d(G-C)] is about 5 times faster than with methylene blue. This extremely rapid reaction can be attributed to the more favourable electron transfer energetics of the thionine compared to those for methylene blue.

TABLE 2. Lifetimes of short-lived species for thionine and methylene blue in the presence of polynucleotides and mononucleotides.

	QUENCHER			
COMPOUND	**poly[d(G-C)]**	**dGMP**	**poly[d(A-T)]**	**dAMP**
Thionine	0.8 ps	1 ps	100 ps	55 ps
Methylene Blue	4 ps	6 ps	290 ps	350 ps

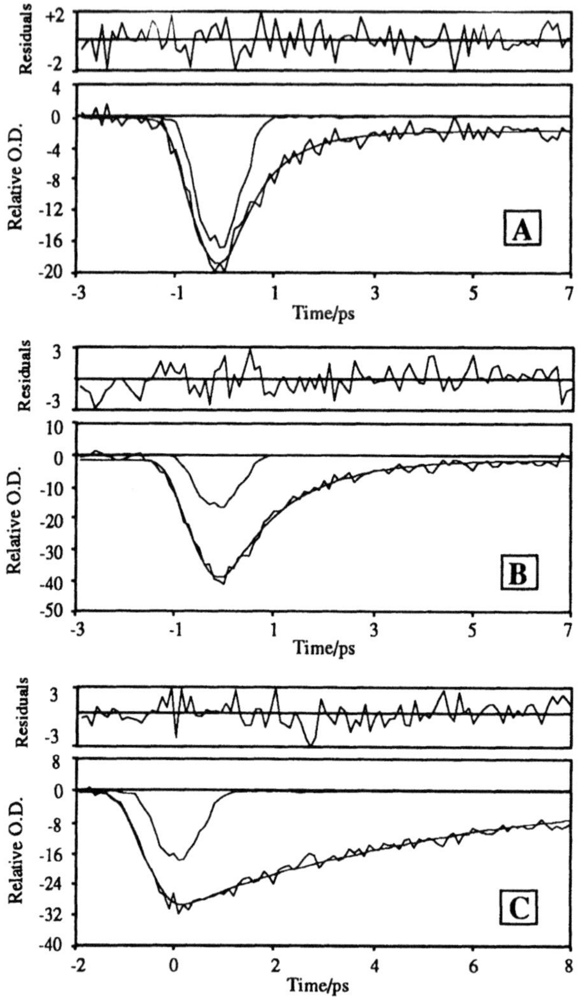

Figure 4. Picosecond transient absorption changes for thionine and methylene blue. (A) thionine & poly[d(G-C)]; P/D = 30, (B) thionine & dGMP (100 mM), (C) methylene blue & poly[d(G-C)]; P/D = 30. λ_{ex} = 620 nm; λ_{tr} = 605 nm. [Th^+] = $2x10^{-5}$M. 5mM phosphate buffer.

A striking observation from the thionine experiments is that the ground state bleaching is effectively reversible with guanine-containing compounds. Although in principle this might be explained by a fortuitous combination of extinction coefficients, it is more reasonable to assume almost complete recovery of the ground state. If this is the case then what we have so far assumed to involve a forward and backward electron transfer appears as a single step. This could mean that the initial electron transfer is essentially instantaneous and that what is being observed is the reverse process or, alternatively, that the reverse process is faster than the forward electron transfer.

When complexed to dAMP or bound to poly[d(A-T)] the lifetime of the MB^+ singlet state is only slightly reduced (350 ps and 290 ps respectively). However with thionine in the presence of adenine derivatives the lifetime of the singlet state is much shorter, consistent with electron transfer quenching due to the higher oxidising power of its excited state. It is interesting to observe that in this case the quenching with dAMP is faster than with poly[d(A-T)], possibly a further indication that under the conditions used the dye is not intercalated between the base pairs of the polynucleotide and therefore the electron transfer is less rapid than in the nucleotide complex. In contrast to what was observed with dGMP or poly[dG-dC] there is a substantial amount of longer-lived species formed in this case. This is most probably the triplet state.

4. Conclusion

Despite the lack of observation of electron transfer products following the quenching of thionine singlet excited state by guanine derivatives, the best explanation for the efficient deactivation of this species is that it participates in a very rapid electron transfer reaction. The contrast between its behaviour with adenine-containing compounds and that of the singlet excited state of methylene blue is consistent with its higher oxidation potential.

5. References

1. Martin, J.P. and Logsdon, N. (1987) J. Biol. Chem., 262, 7213-7219.
2. Swartz, M.R., Schnipper, L.E., Lewin, A.A. and Crumpacker, C.S. (1979) Proc. Soc. Exp. Biol. Med., 161, 204-209.
3. Gill, W.B., Huffman, J.L., Lyon, E.S., Bagley, D.H., Schoenberg, H.W., Straus, F.H. (1987) J. Urol., 138, 1318-1320.
4. Friedmann T. and Brown, D.M. (1978) Nucleic Acids Res., 5, 615-622.
5. Norden, B. and Tjerneld, T. (1982) Biopolymers, 21, 1713-1734.
6. OhUigin, C., McConnell, D.J., Kelly, J.M. and van der Putten, W.J.M. (1987) Nucleic Acids Res., 15, 7411-7427.
7. Kelly, J.M., van der Putten, W.J.M. and McConnell, D.J. (1987) Photochem. Photobiol., 45, 167-175.
8. Dunn, D.A., Lin, V.H. and Kochevar, I.E., (1991) Photochem. Photobiol., 53, 47-56.
9. Kittler, L., Lober,G., Gollmick, F.A. and Berg, H. (1980) J. Electroanal. Chem., 116, 503-511.
10. Jovanovic, S.V. and Simic, M.G. (1989) Biochim. Biophys. Acta, 1008, 39-44.
11. Steenken, S. (1991) Free Rad. Res. Commun., in press.
12. Steenken, S. personal communication.
13. Van der Putten, W.J.M and Kelly, J.M. (1989) Photochem. Photobiol., 49, 145-151.
14. Geacintov, N.E. and Brenner, H.C. (1989) Photochem. Photobiol., 50, 841-858.
15. Hogan, M.E., Rooney, T.F. and Austin, R.H. (1987) Nature, 328, 554-557.
16. Guha, S.N., Moorthy, P.N., Kishore, K., Naik, D.B. and Rao, K.N. (1987) Proc. Ind. Acad. Sci., Chem. Sci., 99, 261-271.
17. Kelly, J.M., Feeney, M.M., Tossi, A.B., Lecomte, J.-P. and Kirsch-de Mesmaeker, A. (1990) Anti-Cancer Drug Design, 5, 69-75.

6. Acknowledgements

We acknowledge support from the EC science program (SCI 009) and would like to thank Dr. Michael Waring for providing dialysis facilities in his laboratory.

STRUCTURE AND BINDING FOR CYCLOPHANE-ARENE COMPLEXES IN WATER FROM MONTE CARLO SIMULATIONS

WILLIAM L. JORGENSEN and TOAN B. NGUYEN
Department of Chemistry
Yale University
New Haven, Connecticut 06511-8118

ABSTRACT. Monte Carlo statistical mechanics simulations have been used to study the complexation of disubstituted benzenes by Diederich's octamethoxy tetraoxaparacyclophane host, <u>1</u>. The calculations were carried out in the NPT ensemble at 25 °C and 1 atm in the presence of 768 water molecules. Relative free energies of binding were obtained for p-xylene, benzene, p-cresol, and hydroquinone from statistical perturbation theory. The computed preference of 2.8 ± 0.3 kcal/mol in ΔG_b for binding p-xylene over hydroqinone compares well with the experimental result of 2.5 ± 0.3 kcal/mol. The computed results for benzene (2.0 ± 0.2 kcal/mol) and p-cresol (2.4 ± 0.2 kcal/mol) relative to p-xylene were predictions; however, the experimental data are now available and are lower by 1 - 2 kcal/mol. The computed structures for the complexes reveal interesting details. For example, hydroquinone protrudes somewhat from one side of the complex and participates in hydrogen bonds between one hydroxyl group and 1-2 water molecules and in an intracomplex hydrogen bond between the other hydroxyl group and ether oxygens. Benzene is predicted to be bound even more off-center, while p-cresol is centered and has an intracomplex hydrogen bond as in the case of hydroquinone.

1. INTRODUCTION

There is great activity in studying intermolecular interactions or molecular recognition in organic host/guest systems[1-4] and in the emerging area of supramolecular chemistry.[5] It is critical to have theoretical means to complement the experimental work by helping to clarify the strurtures and solvation of complexes, energetic contributions to binding, and necessary features for designing new systems with desired function. The realm is solution chemistry so Monte Carlo and molecular dynamics simulations with explicit treatment of the solvent are the current methods of choice. The efforts from other research groups have largely been focused on either enzyme-ligand pairs or complexes of alkali or halide ions with crown ethers, cryptands, and calixarenes.[6-13] Our own emphasis has been on computing structures and relative and absolute free energies of binding for complexes with neutral guests, as in our studies of amide and benzene dimerization,

V. Balzani and L. De Cola (eds.), Supramolecular Chemistry, 383–394.

Rebek's acridine diacid binding pyrazine and pyridine, Zimmerman's tweezer with adenine, imide and lactam association, and nucleotide base pairing.[14-16]

Water soluble hosts are of particular interest as models and mimics of biological systems and as important starting points for the design of artificial enzymes. Important contributions have been made in this area by many researchers, as reviewed by Diederich.[4] Two primary structural themes have been based on cyclodextrins and cyclophanes that bind aliphatic and aromatic guests largely through hydrophobic association. We wish to focus some theoretical studies on the cyclophanes in view of the potential structural diversity, the availability of extensive thermodynamic data on binding, the challenge of expanding our host/guest studies to aqueous media and to macrocyclic systems, and as a controlled route to better understanding the factors that contribute to binding in water, while avoiding the complexities of protein-ligand interactions.

The first effort has been aimed at Diederich's octamethoxy tetraoxaparacyclophane, **1**. The diphenylmethane spacers help enforce a slot-like cavity that is suitable for inclusion of aromatic guests. A crystal structure is available for the host along with much binding data

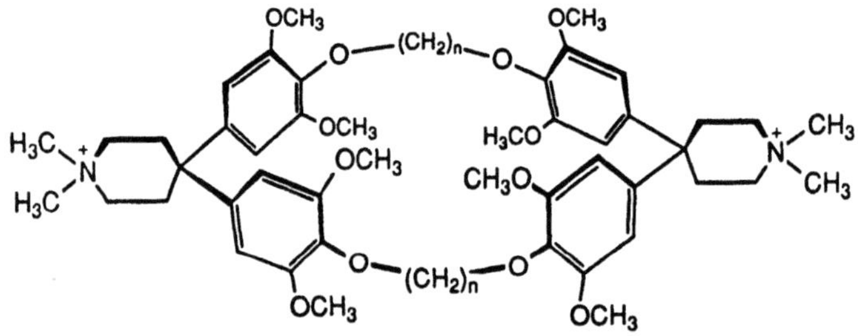

1 (n = 3) **2** (n = 4)

for para-disubstituted benzenes.[17-19] Naturally, low water solubility enhances binding for the guests, so ΔG_b for p-xylene is -4.2 kcal/mol, while it is only -1.7 kcal/mol for hydroquinone. The first challenge is to attempt to reproduce observed binding data and to characterize the structures of the complexes; the initial targets are the free energies of binding of hydroquinone, p-cresol, and benzene relative to p-xylene. The ΔG_b for p-cresol and benzene were unknown when the calculations were completed, but have recently been determined by Diederich and Sanford.[18b]

2. COMPUTATIONAL PROCEDURE

The calculations have been performed with version 3.1 of the BOSS program which permits sampling for any designated bonds, bond angles, and dihedral angles in the solutes.[20] The stretching and bending force constants have been adopted from the AMBER force field,[21] while the non-bonded interactions are described by the OPLS potential functions[22] and the torsional energy is given by a Fourier series for each dihedral angle with coefficients obtained by fitting experimental data or reliable gas-phase calculations. The solutes are represented in Z-matrix form and the sampling of internal coordinates involves making changes, Δr, $\Delta\theta$, and $\Delta\Phi$, to the designated elements in the Z-matrix. In the current simulations, bond lengths were not varied, though there is one ring-closure bond in the macrocycle that has a stretching term since it is varied implicitly. Internal coordinates within the benzene and piperidinium rings were not sampled; however, the remaining bond angles (11) and dihedral angles (16) in the macrocycle were varied along with the dihedral angles for the methoxy groups (8) and hydroxy groups of p-cresol and hydroquinone (1-2). All atoms are explicit except CH_2 and CH_3 groups, which were treated as united atoms.

Statistical perturbation theory was applied to compute the relative free energies of binding according to the thermodynamic cycle in Figure 1. The difference, $\Delta G_1 - \Delta G_2$, is obtained by computing $\Delta G_G - \Delta G_C$. This requires interconversions for the guests unbound and bound to the host in water. The perturbations are performed in small increments

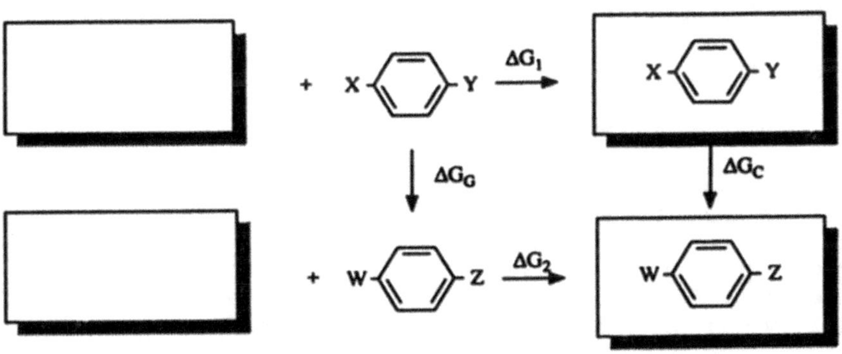

$$\Delta G_1 - \Delta G_2 = \Delta G_G - \Delta G_C$$

Figure 1. Thermodynamic cycle for computing the relative free energies of binding. The open box represents the host.

with the free energy change for the perturbation from increment i to j computed via eq 1 where the average corresponds to sampling configurations for the reference state i and E_i and E_j are the total

$$G_j - G_i = -k_BT \ln\langle\exp(-(E_j - E_i)/k_BT)\rangle_i \qquad (1)$$

energies.[23] The total energy consists of the solvent-solvent, host-solvent, guest-solvent, host-guest, bond stretching, angle bending, tosional and intramolecular non-bonded components. However, these cancel in the present case for $E_j - E_i$ except for the difference in host-guest, guest-solvent, and guest torsional energies. Independent translations and total rotations are attempted for all molecules during the simulations with the ranges for these motions and the variations in internal degrees of freedom adjusted to give 40-50% acceptance rates for new configurations. The Monte Carlo simulations were performed in the NPT ensemble at 25 °C and 1 atm using Metropolis and preferential sampling. Attempted moves for the host or guest were attempted every 20 configurations. Earlier papers can be consulted for additional details.[22]

The initial simulation was for the mutation of p-xylene to hydroquinone. The starting structure of the host was taken from the crystal structure,[17] standard geometries were used for the guests, and the p-xylene was placed in a low-energy configuration in the cavity. 768 TIP4P water molecules[24] were added in a rectangular, periodic cell providing a minimum of 8.5 Å of solvent between any solute atom and a face of the box. The mutations could be carried out with good precision in 6 simulations with double-wide sampling to yield 12 incremental free energy changes. Each simulation consisted of an equilibration phase for at least 10^6 configurations, followed by averaging over at least an additional 2×10^6 configurations. The water-water interactions were spherically truncated at an O-O separation of 8.5 Å, while solute-water interactions were included for water molecules within ca. 10.5 Å of any non-hydrogen atoms of the solutes. In both cases, the interactions were feathered quadratically to zero over the last 0.5 Å.

3. RESULTS AND DISCUSSION

3.1. Thermodynamic Results

It should be noted that the OPLS potential functions for substituted benzenes relevant to the present study were validated in an extensive series of interconversions for benzene, toluene, p-xylene, phenol, p-cresol, anisole, hydroquinone, and benzonitrile in the presence of 500 TIP4P water molecules.[25] These results combined with the previous calculation of the absolute free energy of hydration of our all-atom model of benzene yielded the absolute free energies of hydration for the substituted benzenes, as summarized in Table 1. The statistical uncertainties are $\pm 1\sigma$ and were obtained in the usual way by considering fluctuations in the averages over subsets of configurations.[22] The close accord with the experimental free energies of transfer[26] of the substituted benzenes from the gas phase to aqueous solution provides further support for the appropriateness of the OPLS potential functions for the present applications.

The key thermodynamic results from the simulations of the host/guest systems are given in Table 2. The calculations were well-behaved and yielded small satistical uncertainties (±0.1-0.2 kcal/mol) for both ΔG_G and ΔG_C. A second independent calculation for the bound p-xylene to hydroquinone mutation that used slightly different potentials for the COOH and COOC tosions in the phenolic and anisole-like units yielded a ΔG_C within 0.3 kcal/mol of the value in Table 2. The rejection rate for attempted moves of the host was high owing to the

Table 1. Free Energies of Hydration (kcal/mol) for Aromatic Molecules[a]

Solute	$-\Delta G_{hyd}$ calcd.	$-\Delta G_{hyd}$ exptl.	Solute	$-\Delta G_{hyd}$ calcd.	$-\Delta G_{hyd}$ exptl.
benzene	0.9 ± 0.4	0.9	benzonitrile	3.1 ± 0.5	3.5
toluene	1.0 ± 0.4	0.9	p-cresol	6.1 ± 0.4	6.1
p-xylene	0.9 ± 0.4	0.8	phenol	6.1 ± 0.4	6.6
anisole	1.3 ± 0.5	1.0, 2.5[b]	hydroquinone	12.3 ± 0.5	12.0

[a] For transfer of the solute from the gas phase to aqueous solution at 25 °C. Experimental data from ref. 26. [b] Ref. 26d.

Table 2. Calculated and Experimental Free Energy Changes (kcal/mol)

p-xylene to	ΔG_G	ΔG_C	$\Delta\Delta G_b$ calcd.	$\Delta\Delta G_b$ exptl.[18]
benzene	0.0	2.0	-2.0 ± 0.2	-0.6 ± 0.3
p-cresol	-5.2	-2.8	-2.4 ± 0.2	-0.4 ± 0.2
hydroquinone	-11.4	-8.6	-2.8 ± 0.3	-2.5 ± 0.3

numerous internal degrees of freedom, so the completeness of the sampling is unlikely in that case. Nevertheless, doubling the length of the simulations for some test increments had negligible effect on the computed free energy changes. Though the utilized method of estimating the statistical uncertainties is not conservative enough, it is unlikely that the intrinsic uncertainties in $\Delta\Delta G_b$ (= $\Delta G_1 - \Delta G_2$) are much greater than 0.5 kcal/mol. However, there may be more complex sources of error (vide infra).

The ΔG_G results for the conversion of p-xylene to benzene, p-cresol, and hydroquinone are also reflected in the absolute free energies of hydration in Table 1. Thus, the respective ΔG_G values in Table 2 of 0.0, -5.2, and -11.4 kcal/mol agree well with the experimental data, -0.1, -5.3, and -11.2 kcal/mol. Clearly, water solubility increases significantly from p-xylene and benzene to p-cresol to hydroquinone, so the binding to host 1 might be expected to decrease along the series. This is reflected in the computed $\Delta\Delta G_b$ results in Table 2, though the predicted free energies of binding for benzene, p-cresol, and hydroquinone are compressed into a 1 kcal/mol range. The experimental and computed $\Delta\Delta G_b$ for hydroquinone relative to p-xylene are in excellent agreement. However, the weaker binding for benzene and p-cresol relative to p-xylene is overestimated by 1.4 and 2.0 kcal/mol, respectively. These differences are subject to experimental and theoretical uncertainties of ca. 0.3 and 0.5 kcal/mol in $\Delta\Delta G_b$. Nevertheless, the computations do find the strongest and weakest binding for p-xylene and hydroquinone, and similar binding constants for benzene and p-cresol in accord with the experimental trends.

The conversion of p-xylene to benzene in the host is computed to be unfavorable by 2.0 kcal/mol, while the free energies of hydration are the same. As discussed below, the predicted positioning of benzene in the host's cavity is unusual and may contribute to the discrepancy with the experimental difference of 0.6 kcal/mol. For p-cresol and hydroquinone, the conversions from p-xylene are computed to be 2-3 kcal/mol more favorable in water than in the complexes which leads to the weaker binding. Hydrogen bonding of water to the phenolic hydroxyl groups in the solvent is presumably not fully compensated for in the complexes. It may also be noted that the average host-guest interaction energies (E_{XX}) in these systems are highly attractive. For the mutation of p-xylene to hydroquinone, E_{XX} declined from -14 to -19 kcal/mol; however, this is not enough to cancel the greater decline in the solute-water interaction energy for unbound p-xylene versus hydroquinone. For the benzene complex, E_{XX} (-12 kcal/mol) is 2 kcal/mol less attractive than for the p-xylene complex and appears to be the chief contributor to the $\Delta\Delta G_b$.

3.2 Structures of the Complexes

Figures 2-5 present structures of the complexes taken from the Monte Carlo simulations. The illustrated structures are the last ones from the different simulations. Though their choice is effectively random, they are good representations of typical structures sampled during the calculations. Relatively little motion was obtained for the host except for torsional changes for the eight methoxy groups. However, the guests rattled in the cavity with some twisting and translations of 2-3 Å.

As shown in Figure 2, p-xylene is well inserted in the cavity with two parallel and two perpendicular aryl-aryl interactions. The guest is not perfectly centered in the cavity which is reasonable since the face-to-face π-stacking interactions favor an offset geometry.[22c] There is no room for water in the cavity when the p-xylene is bound.

The structure of the hydroquinone complex in Figure 3 is particularly interesting. The host is too thick to allow both hydroxyl groups to be exposed to the solvent on opposite sides of the cavity. Consequently, the guest is shifted off-center so one hydroxyl group participates in 1-2 hydrogen bonds with water. There is always a hydrogen bond to the phenolic oxygen from a water molecule which has its other hydrogen typically hydrogen-bonded to an oxygen of a proximal methoxy group on the "frontside" of the host. The two water molecules that are hydrogen bonded to the hydroquinone in this configuration are shown in Figure 3, while the remaining 766 water molecules have been

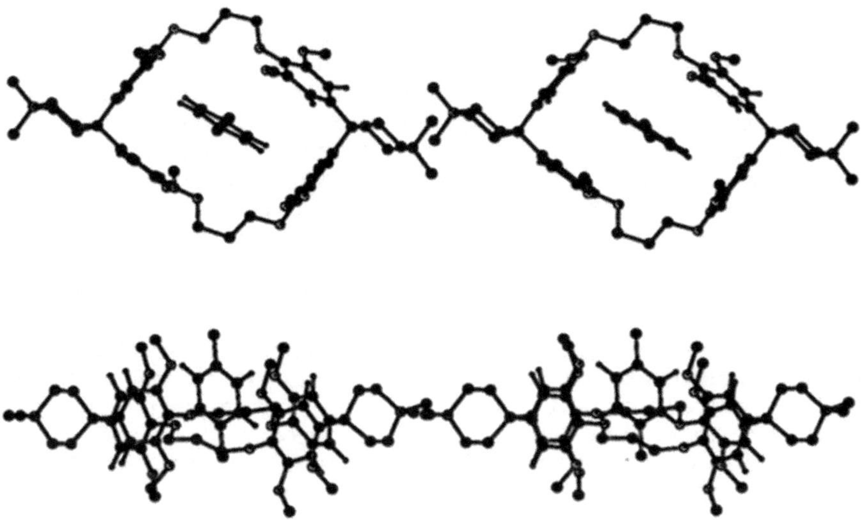

Figure 2. Two stereoviews of the complex of host 1 with p-xylene.

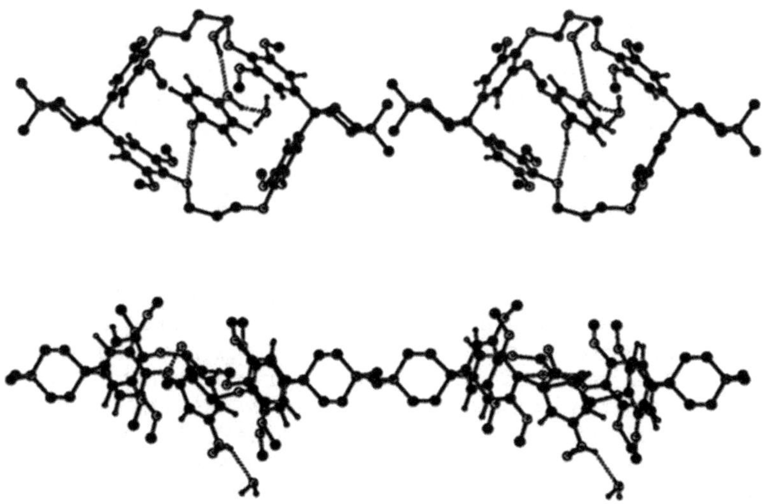

Figure 3. Two stereoviews of the complex of host 1 with hydroquinone.

deleted for clarity. The figure also illustrates that there is a hydrogen bond between the non-hydrated hydroxyl group of the guest and an ether oxygen of the macrocycle. This hydrogen bond persisted throughout the simulation. Though the distance to the macrocyclic ether oxygen is shorter, there is also frequently a longer hydrogen bond between the phenolic hydrogen and a nearby methoxy oxygen on the "backside" of the host. It is important to note that to achieve the intracomplex hydrogen bonding, the phenolic hydrogen must rotate 40-50° out of the plane of the aromatic ring at a cost of 1 kcal/mol in torsional energy. Remarkably, the intracomplex hydrogen bonding was presaged by Diederich and co-workers for complexes of 2 with p-cresol and hydroquinone.[18a] Their prediction followed from the observation that these complexes are particularly weakly bound, but show the most favorable enthalpies of binding (and unfavorable entropies of binding).[18a] Thus, the present results support their conclusion; the strong enthalpic attraction undoubtedly benefits from the intracomplex hydrogen bonding which is also reflected in the 5 kcal/mol lower E_{XX} for the hydroquinone than the p-xylene complex noted above.

Top views of the complexes for p-cresol and benzene are shown in Figure 4. The intracomplex hydrogen bond between the hydroxyl group of p-cresol and ether oxygens of the host has been retained rather than the hydrogen bonding with water. The perturbation proceeded from p-xylene to p-cresol and developed into this geometry. The computed structure is consistent with the thermodynamic results for the complex of 2 that led to the proposal of the intracomplex hydrogen bond.[18a] However, concern can be expressed that the alternative structure with the hydroxyl group hydrated has not been sampled, which might account for the 2 kcal/mol discrepancy between the calculated and experimental $\Delta\Delta G_b$. The simulation is now being repeated starting from the p-cresol end with the hydroxyl group hydrated in the complex. This issue was not problematic for the perturbation of p-xylene to hydroquinone owing to the symmetry of hydroquinone.

The computed structure for the benzene complex is striking since the guest is more off-center than in the other three cases. The off-center shift in Figure 4 is close to the maximal one sampled; extension of the simulation revealed no tendency for the benzene to leave the complex. The dissymetric placement of the guests in the host implies that a dynamic process likely exists in which the guest shuttles from one side of the cavity to the other. Such events did not occur during the present simulations; however, they would be interesting to investigate both theoretically and experimentally. Representations as in Figure 5 show that the central region of the benzene ring is being clasped by the four methoxy methyl groups on one side of the host. These interactions are electrostatically reasonable since the benzene carbons have partial negative charges, -0.115 e, and the methoxy methyl groups have charges of +0.25 e in the OPLS model. This type of methyl-aryl interaction has also been noted in other host/guest systems recently and is emerging as a common motif.[27] Further analysis of the NMR data for the complex should shed light on the validity of the

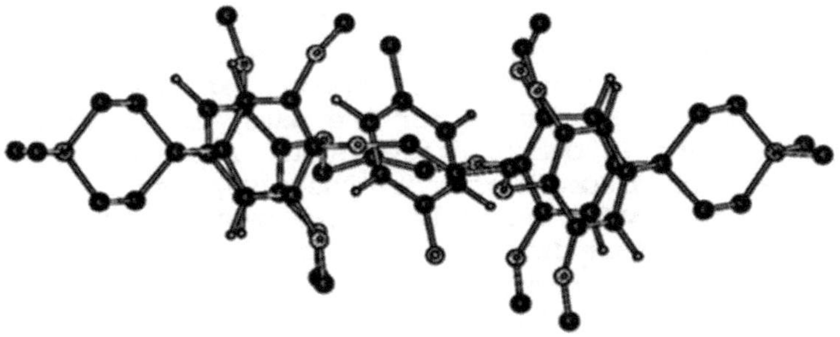

p-CRESOL

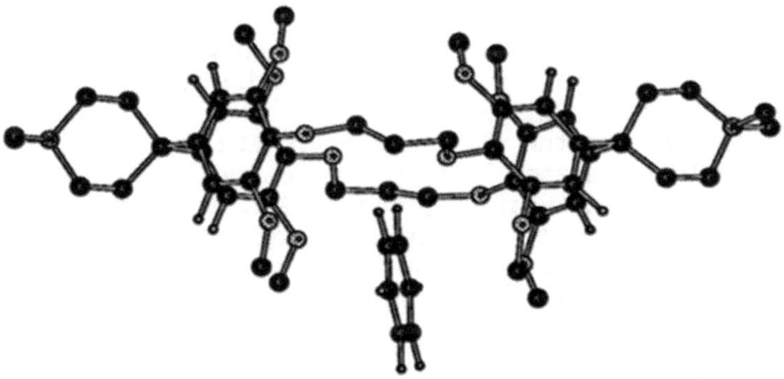

BENZENE

Figure 4. Calculated structures for the complexes of host 1 with p-cresol and benzene.

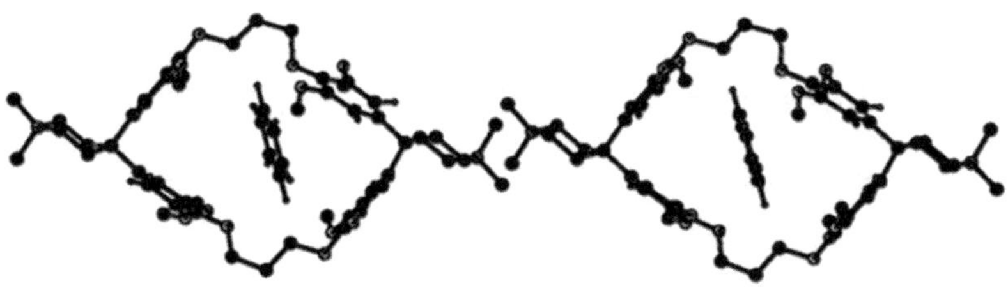

Figure 5. Stereoview for the complex of host 1 with benzene.

calculated structure. The calculated reduction in binding for benzene relative to p-xylene does overestimate the experimental result by 1.4 kcal/mol. Though this difference is approaching the range of the combined statistical uncertainties for the data, it is possible that the computed structure is overly dissymmetric. Another source of the too weak computed binding could arise if the dissymmetric structure is correct and there is a water molecule or two remaining in the cavity. In simulations of the bare host in water and in the X-ray struture, there are at least two water molecules in the cavity. In the mutation from p-xylene to benzene, no water slipped into the cavity, though there is room. If, in fact, there should be bound water, a higher net binding constant would be implied.

4. CONCLUSION

The present results illustrate the current capabilities of statistical mechanics simulations to examine the structure and binding in host/guest complexes. The average error in computed relative free energies of binding for the three cases is 1.2 kcal/mol. The larger errors for binding benzene and p-cresol may have geometricaaly interesting origins that require further study. The structural details that accompany the thermodynamic results are extensive, provocative, and complement the experimental studies. In the present case, the postulated intracomplex hydrogen bonding for hydroquinone and p-cresol with the cyclophane hosts received support. The assymmetric placement of the guests in the host's cavity also encourages further investigations of dynamic shuttling through the cavity.

ACKNOWLEDGEMENTS

Gratitude is expressed to the National Science Foundation for support of this work. Professor Francois Diederich, Ito Chao, and Elizabeth M. Sanford provided valuable comments and the unpublished binding data.

REFERENCES

(1) Lehn, J.-M. Angew. Chem., Int. Ed. Engl. **1988**, 27, 89.

(2) Cram, D. J. Angew. Chem., Int. Ed. Engl. **1988**, 27, 1009.

(3) Rebek, J., Jr. Acc. Chem. Res. **1990**, 23, 399.

(4) Diederich, F. Angew. Chem., Int. Ed. Engl. **1988**, 27, 362.

(5) Lehn, J.-M. Angew. Chem., Int. Ed. Engl. **1990**, 29, 1304.

(6) McCammon, J. A.; Harvey, S. C. "Dynamics of Proteins and Nucleic Acids"; Cambridge Univ. Press: Cambridge, 1987.

(7) van Gunsteren, W. F.; Berendsen, H. J. C. Angew Chem., Int. Ed.

Engl. 1990, 29, 992.

(8) Kollman, P. A.; Merz, K. M., Jr. Acc. Chem. Res. 1990, 23, 246.

(9) Brooks, C. L., III; Karplus, M.; Pettitt, B. M. Adv. Chem. Phys. 1988, 71, 1.

(10) Lybrand, T. P.; McCammon, J. A.; Wipff, G. Proc. Natl. Acad. Sci. USA 1986, 83, 833.

(11) Mazor, M. H.; McCammon, J. A.; Lybrand, T. P. J. Am. Chem. Soc. 1989, 111, 55.

(12) Dang, L. X.; Kollman, P. A. J. Am. Chem. Soc. 1990, 112, 5716.

(13) Grootenhuis, P. D. J.; Kollman, P. A.; Groenen, L. C.; Reinhoudt, D. N.; van Hummel, G. J.; Ugozzoli, F.; Andreetti, G. D. J. Am. Chem. Soc. 1990, 112, 4165.

(14) Jorgensen, W. L. Acc. Chem. Res. 1989, 22, 184.

(15) Jorgensen, W. L. Chemtracts - Organic Chemistry 1991, 4, 91.

(16) Jorgensen, W. L.; Boudon, S.; Nguyen, T. B. J. Am. Chem. Soc. 1989, 111, 755.

(17) Ferguson, S. B.; Seward, E. M.; Diederich, F.; Sanford, E. M.; Chou, A.; Inocencio-Szweda; Knobler, C. B. J. Org. Chem. 1988, 53, 5593.

(18) Ferguson, S. B.; Sanford, E. M.; Seward, E. M.; Diederich, F. J. Am. Chem. Soc. 1991, 113, 5410.

(19) Smithrud, D. B.; Wyman, T. B.; Diederich, F. J. Am. Chem. Soc. 1991, 113, 5420.

(20) Jorgensen, W. L. BOSS, Version 3.1 1991, Yale University, New Haven, CT.

(21) Weiner, S. J.; Kollman, P. A.; Case, D. A.; Singh, U. C.; Ghio, C.; Alagona, G.; Profeta, S., Jr.; Weiner, P. J. Am. Chem. Soc. 1984, 106, 765.

(22) (a) Jorgensen, W. L.; Tirado-Rives, J. J. Am. Chem. Soc. 1988, 110, 1657. (b) Jorgensen, W. L.; Briggs, J. M.; Contreras, M. L. J. Phys. Chem. 1990, 94, 1683. (c) Jorgensen, W. L.; Severance, D. L. J. Am. Chem. Soc. 1990, 112, 4768.

(23) Zwanzig, R. W. J. Chem. Phys. 1953

(24) Jorgensen, W. L.; Madura, J. D. Mol. Phys. 1985, 56, 1381.

(25) Nguyen, T. B.; Jorgensen, W. L., to be published.

(26) (a) Hine, J.; Mookerjee, P. K. J. Org. Chem. 1975, 40, 292. (b) Ben-Naim, A.; Marcus, Y. J. Chem. Phys. **1984**, 81, 2016. (c) Riddick, J. A.;Bunger, W. B.; Sakano, T. K. "Organic Solvents: Physical Properties and Methods of Purification, 4th Edition"; Wiley: New York, 1986. (d) Kuyper, L. F.; Hunter, R. N.; Ashton, D.; Merz, K. M., Jr.; Kollman, P. A. J. Phys. Chem. 1991, 95, 6661.

(27) (a) Andreetti, G. D.; Ori, O.; Ugozzoli, F.; Alfieri, C.; Pochini, A.; Ungaro, R. J. Inclusion Phenom. 1988, 6, 523. (b) Hamilton, A. D., et al. J. Chem. Soc. Chem. Commun., in press.

PHOTOINDUCED ELECTRON TRANSFER BETWEEN INTERCALATED MOLECULES: USING POLYNUCLEOTIDES TO SELF-ASSEMBLE DONOR-ACCEPTOR SYSTEMS

ANNE M. BRUN AND ANTHONY HARRIMAN
Center for Fast Kinetics Research
The University of Texas at Austin
Austin
Texas 78712
U.S.A.

ABSTRACT. Intercalation of ethidium bromide (EB^+) and N,N'-dimethyl-2,7-diazapyrenium dichloride (DAP^{2+}) into calf-thymus DNA is highly efficient and gives stable supramolecular systems. Selective excitation of intercalated EB^+ results in electron transfer to a nearby intercalated DAP^{2+} molecule. The transient products formed by this charge-shift reaction are readily observed by laser flash photolysis techniques. The rate of forward electron transfer decreases with increasing number of interspersed nucleic acid bases and, for the closest approach (≈10 Å), $k_f = (1.3 \pm 0.2) \times 10^9\ s^{-1}$ whereas the reverse electron transfer reaction occurs on a much slower timescale ($k_r = (5.0 \pm 1.1) \times 10^7\ s^{-1}$). Treating the kinetic data according to $k_f = A\exp(-\beta R)$, where R is the separation distance, gave an average β value of (0.88 ± 0.08) Å^{-1}.

1. Introduction

A major topic of interest in present day photochemistry concerns the construction of supramolecular systems capable of effecting photoinduced electron transfer over 10-20 Å. Many diverse approaches are being considered for mutual positioning of the donor and acceptor subunits, including the use of electrostatic and/or van der Waals interactions, coordination through transition metal chelates, hydrogen bonding, and covalent linkages. As an alternative to these avenues, we have considered using polynucleotides to self-assemble donor-acceptor systems by way of intercalation of planar polycyclic aromatic species between base-pairs. The resultant system should consist of a unidimensional ensemble in which donor and acceptor species are separated by a nonpolar, rigid medium of high density and, in such cases, electron tunneling must proceed through the interspersed nucleic acid base-pairs. An important parameter in understanding the mechanism of DNA-mediated electron transfer, therefore, concerns the distance dependence of electronic coupling within the duplex. As a first step toward this goal, we describe

V. Balzani and L. De Cola (eds.), Supramolecular Chemistry, 395–402.

the dynamics of photoinduced electron transfer between intercalated, but separate, donor and acceptor species.

Previous studies have shown the feasibility of electron transfer between intercalated reagents [1], between "partially" intercalated transition metal chelates [2], and between intercalated dyes and electron acceptors bound to the outer-surface of the DNA coil [3]. However, we are unaware of any prior studies in which direct spectroscopic evidence has been reported confirming photoinduced electron transfer between properly intercalated reagents. Furthermore, there is a complete scarcity of quantitative data relating to reorganization energies, electronic coupling or attenuation factors for electron transfer occurring within a polynucleotide matrix.

2. Experimental

Highly-polymerized deoxyribonucleic acid, (DNA), from calf-thymus, was purchased from Sigma Chem. Co. and used as received. Stock solutions of DNA were prepared by dissolution overnight in 5 mM phosphate pH 7 buffer and were stored at 4°C in the dark for short periods only. Concentrations of DNA per nucleotide phosphate were determined by absorption spectroscopy using a molar extinction coefficient of 6,600 M^{-1} cm^{-1} at 260 nm [4]. N,N'-Dimethyl-2,7-diazapyrenium dichloride (DAP^{2+}) was prepared and purified according to the method of Hünig et al. [5] whereas ethidium bromide (EB^+) was purchased from Sigma Chem. Co. and used as received.

Absorption spectra were recorded with a Hitachi U3210 spectrophotometer and fluorescence spectra were recorded with a Perkin-Elmer LS5 spectrofluorimeter. Excited singlet state lifetimes were determined by the time-correlated, single photon counting technique [6] using a mode-locked Nd-YAG synchronously-pumped, cavity-dumped Rhodamine 6G dye laser. The excitation wavelength was tuned to selectively excite ethidium bromide at 575 nm and fluorescence was separated from scattered laser light by glass narrow bandpass filters. Data analysis was made after deconvolution of the instrument response function (FWHM = 60 ps). Titrations were carried-out by adding successive amounts of aqueous solutions of DAP^{2+} to a freshly-prepared DNA stock solution already containing a fixed concentration of EB^+.

Laser flash photolysis studies were made with a frequency-doubled Nd-YAG laser. For ns-timescale studies, the laser was Q-switched and a pulsed Xe arc lamp was used as monitoring beam. The laser intensity was attenuated with metal screen filters and solutions were purged thoroughly with N_2 before and during the experiment. Improved time-resolution was obtained with a Nd-YAG mode-locked laser operated in the pump-probe technique. Approximately 300 individual shots were averaged at each time delay, baseline-corrected, and analyzed by computer iterations.

3. Results and Discussion

3.1. FORWARD ELECTRON TRANSFER

Intercalation of ethidium bromide (EB^+) into calf-thymus DNA (5 mM phosphate buffer, pH 7) occurs with a saturation of 0.4/base-pair and, at $[EB^+]$ = 67 μM and [DNA-bases] = 1.4 mM, more than 99% of EB^+ will be intercalated. Assuming a nonrandom distribution, the average distance between chromophores is 10 base-pairs (≈34 Å). Fluorescence from intercalated EB^+ decays with a lifetime (τ_0) of (21.1 ± 0.1) ns, compared to (1.6 ± 0.1) ns in water. Upon addition of N,N'-dimethyl-2,7-diazapyrenium dichloride (DAP^{2+}), which intercalates with high affinity, fluorescence from EB^+ was quenched and decay profiles became markedly non-exponential, even at very low concentrations of quencher.

In the presence of intercalated DAP^{2+}, the complex fluorescence decay profiles could be analyzed satisfactorily in terms of a 3-exponential fit.

$$I_f(t) = A_1\exp(-t/\tau_1) + A_2\exp(-t/\tau_2) + A_3\exp(-t/\tau_3) \quad (1)$$

The analysis involved deconvolution by iterative reconvolution, background subtraction, and optional correction for shift of the instrument response function. Statistical tests included chi-square, the Durbin-Watson test, the covariance matrix, a runs test, and the autocorrelation function [6]. An alternative form of data analysis involves distributions of lifetimes rather than a series of exponentials. Differentiation of systems obeying a decay law made up of three discrete components from systems where there exists a continuous distribution of lifetimes, or a distribution plus one or more discrete components, is a nontrivial analytical problem. Methods involving the minimization of the chi-square parameter are commonly used, but recently the maximum entropy method (MEM) has gained popularity [7]. Inherent in the MEM method is the theoretical lack of bias and the potential for recovering the coefficients of an exponential series with fixed lifetimes which are free of correlation effects and artificial oscillations. Recent work has compared the MEM with a new version of the exponential series method (ESM) which allows use of the same size probe function as the MEM and found that the two methods gave comparable results [8].

The precision of the data and the presence or absence of nonrandom noise play a critical role when one is fitting decay data to preconceived models. We have found that, at high levels of precision, the most satisfactory analytical fit to our experimental data involves a 3-exponential fit using ESM fitting routines [8]. Improved reliability was obtained by using a global analysis procedure covering ten emission wavelengths [9]. The derived lifetimes (τ) and fractional amplitudes (A) are collected in Table 1.

TABLE 1. Time-resolved fluorescence studies carried-out with ethidium bromide (70 μM) intercalated into DNA (1.4 mM bases) and various concentrations of DAP^{2+}. Monochromatic excitation was made in the range 575-590 nm, where only EB^+ absorbs, and fluorescence was detected over narrow wavelength segments within the range 630-660 nm.

$[DAP^{2+}]$ (μM)	τ_1 (ns)	A_1 (%)	τ_2 (ns)	A_2 (%)	τ_3 (ns)	A_3 (%)
0	21.1	100				
7	0.72	3	8.2	4	20.7	93
11	0.74	8	8.4	8	20.5	84
19	0.73	12	8.4	14	20.2	74
24	0.73	15	8.3	17	20.4	68
42	0.72	19	8.4	19	20.3	62
72	0.72	31	8.3	28	19.8	41
83	0.73	31	8.4	27	19.8	42
95	0.73	34	8.2	26	19.6	40
105	0.72	37	8.4	24	19.4	39
117	0.72	44	8.2	24	19.4	32
128	0.72	51	8.4	22	19.3	27

The decay profiles gave no indication of displaced EB^+, provided the concentration of DAP^{2+} was kept below ca. 120 μM, although excitation and detection wavelengths were optimized in favor of preferentially observing intercalated dye. Furthermore, the fluorescence spectral records showed that added DAP^{2+} was present exclusively in the intercalated form and that, under the experimental conditions, none was bound to the polynucleotide surface or free in solution. Gated fluorescence spectra, recorded at different times after the excitation pulse, were identical within experimental error and remained in excellent agreement with those recorded by steady-state methods.

Inspection of the parameters extracted from our analysis of the time-resolved fluorescence decay profiles (Table 1) shows that the entire titration data can be resolved into three discrete lifetimes. Thus, the lifetimes remain fixed, regardless of the amount of intercalated DAP^{2+}, as might be expected for a static system, rather than undergoing a progressive change, as might be indicative of a dynamic quenching process. This is further indication that only properly intercalated reagents are present. The two shorter lifetimes (τ_1 = (0.72 ± 0.03) ns; τ_2 = (8.4 ± 0.3) ns) increased in significance with increasing [DAP^{2+}] and contributed almost equally to the total initial fluorescence intensity except at higher DAP^{2+} loadings, where the shortest-lived component predominated (Table 1). The longest lifetime (τ_3) remained similar to that recorded in the absence of DAP^{2+}. This value decreased systematically, in both magnitude and significance, with increasing [DAP^{2+}] but appeared to reach a lower limit of (19.4 ± 0.5) ns at high loadings of DAP^{2+} (Table 1).

The time-resolved fluorescence behavior is explained as follows: Intercalation of DAP^{2+} can occur at different distances from the incumbent dye molecule, incremental according to the number of interspersed base-pairs, but it is quantized. The separation cannot be less than 3 base-pairs, because of the mutual exclusion volume of the intercalated molecules, but it can be 3, 4, 5, and so on. At low loadings, there will be a statistical distribution of separation distances, allowing for any prefered binding regions, whereas a high [DAP^{2+}] will raise the probability of effecting the minimum separation (i.e. 3 base-pairs).

As the concentration of intercalated DAP^{2+} approaches that of the dye, few excited states will escape quenching and, under such conditions, τ_3 can be assigned to dye molecules having a DAP^{2+} molecule separated by 5 (i.e. 17.0 Å) base-pairs. At lower loadings of DAP^{2+}, τ_3 will refer to a distribution of unquenched and quenched species and contains no useful information. Larger separations may abound but, in such cases, the "quenched" and "unquenched" fluorescence lifetimes will be too similar to be resolved.

Each intercalated DAP^{2+} molecule can quench EB^+ fluorescence via an electron transfer mechanism.

$$^*EB^+ + DAP^{2+} \rightarrow EB^{2+} + DAP^+ \quad (2)$$

The rate constant for the forward electron transfer reaction (k_f) is expected to decrease with increasing reactant separation distance (R)

$$k_f = A\exp[-\beta R] \quad (3)$$

where β quantifies the effectiveness of the intervening base-pairs in coupling donor and acceptor species. Assuming R is quantized, τ_1, τ_2, and τ_3 can be assigned, respectively, to dye molecules having a DAP^{2+} molecule separated by 3 (i.e. 10.2 Å), 4 (i.e. 13.6 Å), or 5 (i.e. 17.0 Å) base-pairs. In this manner, k_f ($k_f = [1/\tau - 1/\tau_0]$) can be calculated for separations of 10.2, 13.6, and 17.0 Å and the derived values are collected in Table 2.

TABLE 2. Derived rate constants for forward electron transfer processes observed with intercalated ethidium bromide (EB^+) or acridine orange (AO) at implied distances (R).

R (Å)	EB^+ k_f (s^{-1})	AO k_f (s^{-1})
10.2	1.3×10^9	4.3×10^9
13.6	7.2×10^7	2.3×10^8
17.0	2.5×10^6	1.3×10^7

Expressing the derived rate constants in terms of equation 3 gave a pre-exponential factor (A) of $(1.1 \pm 0.2) \times 10^{13}$ s^{-1} and a β value of (0.91 ± 0.04) $Å^{-1}$. Replacing EB^+ with acridine orange gave similar behavior with $\beta = (0.86 \pm 0.04)$ $Å^{-1}$. The average value of β [$\beta = (0.88 \pm 0.07)$ $Å^{-1}$] is comparable to values observed for modified proteins [10] and is suggestive of electron tunneling occurring through the intervening base-pairs rather than via the phosphate or ribose functions [11].

Laser flash photolysis studies provided firm evidence for the formation of charge separated products consistent with reaction 2. Thus, DNA (1.4 mM bases) was loaded with EB^+ (120 μM) and DAP^{2+} (170 μM). Immediately after excitation with a 30-ps laser pulse at 532 nm the excited singlet state of EB^+ was observed. This species decayed rapidly (τ_{decay} = 0.58 ns) and there was a concomitant growth (τ_{rise} = 0.64 ns) of a transient possessing the characteristic absorption spectral features of the charge separated products. Under these conditions, there is a high probability of finding EB^+/DAP^{2+} couples separated by only 3 base-pairs and the derived excited state lifetime is comparable to that observed in the time-resolved fluorescence studies (τ_1 = 0.72 ns) at much lower loadings.

3.2. REVERSE ELECTRON TRANSFER

Excitation of DNA (1.4 mM bases) loaded with EB^+ (67 μM) and DAP^{2+} (33 μM) with a 10-ns laser pulse at 532 nm gave rise to one-electron transfer products consistent with equation 2. These transient species decayed via first-order kinetics, presumably due to reverse electron transfer.

$$EB^{2+} + DAP^{+} \rightarrow EB^{+} + DAP^{2+} \tag{4}$$

Decay profiles were complex and the shortest-lived component, which dominated the decay records, corresponded to a lifetime of (20 ± 3) ns. This reaction is tentatively assigned to charge recombination across 3 base-pairs. The decay records at very long times may be contaminated with triplet state absorptions from unquenched EB^+ or displaced radical cations. Even allowing for such uncertainties, it is clear that charge recombination occurs slowly. The ps-laser flash photolysis studies showed no significant decay between 0-5 ns whereas ns-laser flash photolysis studies showed that the most rapid charge recombination process for the EB^{2+}/DAP^+ couple has a lifetime of 20 ns. The large amount of energy that has to be dissipated during charge recombination, 187 kJ mol^{-1}, could place the process within the Marcus "inverted" region [12] and, thereby, account for the slow rates. Alternately, charge separation and recombination could occur by different routes or involve different extents of electronic coupling between the reactants [10,11].

4. Conclusion

This work has demonstrated that long-distance (>10 Å) electron transfer processes are possible within a DNA matrix; this finding is not to be confused with earlier work concerned with electron transfer occurring along the outside of the helix [1-3]. Forward electron transfer appears to occur through the interspersed base-pairs and, for a photosensitizer possessing a long-lived excited state, reaction could occur over extremely long distances. This finding has important consequences for understanding the mechanism of DNA damage and repair processes. Studies employing covalently-linked bis-intercalators, comprising EB^+ and DAP^{2+} subunits, are in progress aimed at refining our understanding of these reactions.

We thank the National Science Foundation (CHE 9102657) for financial support. The Center for Fast Kinetics Research is supported jointly by the Biomedical Research Technology Program of the Division of Research Resources of the National

Institutes of Health (RR00886) and by The University of Texas at Austin.

5. References

1. (a) Baguley, B.C.; Le Bret, M. Biochem. **1984**, 23, 937. (b) Davis, L.M.; Harvey, J.D.; Baguley, B.C. Chem.-Biol. Interactions **1987**, 62, 45.
2. (a) Barton, J.K.; Kumar, C.V.; Turro, N.J. J. Am. Chem. Soc. **1986**, 108, 6391. (b) Purugganan, M.D.; Kumar, C.V.; Turro, N.J.; Barton, J.K. Science **1988**, 241, 1645.
3. (a) Fromherz, P.; Rieger, B. J. Am. Chem. Soc. **1986**, 108, 5361. (b) Atherton, S.J. Light in Biology and Medicine Douglas, R.H.; Moan, J.; Dal'Acqua F.: Eds. (Plenum, New York, 1988) Vol. 2, p. 77.
4. Reichmann, M.E.; Rice, S.A.; Thomas, C.A.; Doty, P. J. Am. Chem.Soc. **1954**, 76, 3047.
5. Hünig, S.; Gross, J.; Lier, E.F.; Quast, H. Liebigs Ann. Chem. **1973**, 339.
6. O'Connor, D.V.; Phillips, D. In Time Correlated Single Photon Counting Academic Press: London, 1984.
7. Smith, C.; Grady, W. Maximum Entropy and Bayesian Methods in Inverse Problems Reide: Boston, 1985.
8. Siemiarczuk, A.; Wagner, B.; Ware, W. J. Phys. Chem. **1990**, 94, 1661.
9. Knutson, J.; Beechem, J.; Brand, L. Chem. Phys. Lett. **1983**, 102, 501.
10. (a) Axup, A.W.; Albin, M.; Mayo, S.L.; Crutchley, R.J.; Gray, H.B. J. Am. Chem. Soc. **1988**, 110, 435. (b) Wallin, S.A.; Stemp, E.D.A.; Everest, A.M.; Nocek, J.M.; Netzel, T.L.; Hoffman, B.M. J. Am. Chem. Soc. **1991**, 113, 1842. (c) Isied, S.S.; Vassilian, A.; Wishart, J.W.; Creutz, C.; Schwarz, H.; Sutin, N. J. Am. Chem. Soc. **1988**, 110, 635.
11. (a) Beratan, D.N.; Betts, J.N.; Onuchic, J.N. Science **1991**, 252, 1285. (b) Jacobs, B.A.; Mauk, M.R.; Funk, W.D.; MacGillivray, R.T.A.; Mauk, A.G.; Gray, H.B. J. Am. Chem. Soc. **1991**, 113, 4390.
12. Marcus, R.A. J. Chem. Phys. **1956**, 24, 966.

SYNTHESIS AND PROPERTIES OF HETEROMETAL CUBANE-TYPE CLUSTERS IN FERREDOXINS

José J. G. Moura, Anjos L. Macedo, Cristina Moreno and Isabel Moura

Centro de Tecnologia Química e Biológica and Universidade Nova de Lisboa
Apartado 127, 2780 Oeiras, Portugal

ABSTRACT

Incorporation of different metals into the vacant site of a cuboidal [3Fe-4S] core in ferredoxins results in the formation of hetero metal clusters of the type [M,3Fe-4S]. These novel clusters show interesting magnetic properties, contain localized and delocalized valences and present a wide range of spin and oxidation states. These cores have been synthetized by chemical independent methods and natural occuring hetero metal cubane type clusters have been indicated to be present in proteins.

INTRODUCTION

Iron-sulfur proteins contain prosthetic groups composed of iron and sulfur atoms. They play a diversified functional role in biological systems being involved in electron transfer, in direct interaction with substrates and recently non-redox roles have been emphasized (1-5). Extensive spectroscopic investigations have been performed on a variety of iron-sulfur proteins and a wealth of information accumulated on their physical properties (1-10). Four basic structures have emerged: the FeS_4 center, the [2Fe-2S], the [3Fe-4S], and the [4Fe-4S] clusters (Figure 1). A common structural feature shared by these four centers is that each iron atom is tetrahedrally coordinated by four sulfur atoms. With the exception of the FeS_4 center found in rubredoxin, where the iron atom is coordinated by four cysteinyl sulfur atoms (11), all the other cores contain bridging inorganic sulfur atoms. The terminal ligands for the clusters are cysteinyl sulfur atoms (12-16). Possible distortion from the tetrahedral arrangement caused by structural constraints has been suggested for the FeS_4 center (17).

V. Balzani and L. De Cola (eds.), Supramolecular Chemistry, 403–410.

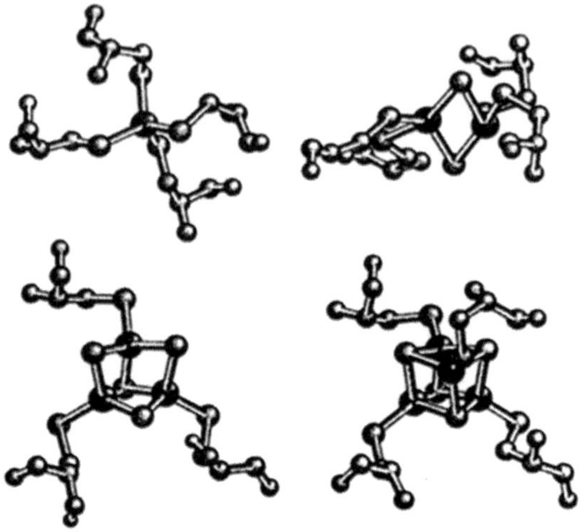

Figure 1
Simple iron-sulfur structures: FeS_4 , [2Fe-2S], [3Fe-4S] and [4Fe-4S] cores.

A close structural relation is apparent between the three and the four iron atom containing clusters and will be relevant for the following considerations.

The [3Fe-4S] cluster can be stabilized in two oxidation states, 1+ and 0. The $[3Fe\text{-}4S]^{1+}$ cluster contains three high-spin ferric ions spin-coupled to form a S=1/2 state (18-21) and the one electron reduced $[3Fe\text{-}4S]^0$ cluster has a S=2 (21, see text). A further two electron reduction was recently suggested to occur originating a all ferrous trinuclear core (22).

Three stable oxidation states, 3+, 2+ and 1+, have been observed for the [4Fe-4S] cluster. The 3+ and 2+ states are stable in high potential iron-sulfur proteins, while the 2+ and 1+ states are detected for bacterial ferredoxins and other 4Fe cluster-containing proteins (23). Both the 3+ and 1+ states are paramagnetic with S=1/2. The 2+ state is diamagnetic. In all three states, valence delocalized pairs are observed.

Theoretical investigations indicate that the concept of valence localization, as well as delocalization, is important for the understanding of the iron-sulfur core electronic states (24,25).

Facile interconversion between [4Fe-4S] cluster and [3Fe-4S] cluster has been reported (26-28) and shown to be physiologicaly significant (29,30).

More recent investigations on iron-sulfur proteins and model compounds (8), however, indicate that the above mentioned physical properties constitute only a limited understanding of iron-sulfur clusters, which in fact are susceptible to ligand and metal substitutions and are flexible in their electronic properties and structures. The presence of non cysteinyl ligands has been suggested (31,32). Different electronic ground states with higher spin manifold are observed in naturaly occuring systems (33-36) and observed after metal incorporation (37-39) and after selenium substitution (40,41). Unusual stoichiometries (six-iron containing cores) are being revealed (42,43).

1. Localized and delocalized valences in the [3Fe-4S] core

The percursor for the synthesis of hetero metal cuboidal type cores in proteins has been the [3Fe-4S] cluster. This structure has unusual electronic and magnetic properties which were explored in detailed in the cluster contained in ferredoxin II (FdII) isolated from *Desulfovibrio (D.) gigas* , a sulfate reducer. This protein is an electron transfer protein isolated as a tetramer of identical subunits of know amino acid sequence (6 kDa) and consists of 57 amino-acids, including six cysteinyl residues (44). Each subunit contains a trinuclear iron-sulfur cluster of the [3Fe-4S] type.

The X-ray structural analysis indicated that the cluster is bound to the polypeptide chain by three cysteinyl residues (Cys 8, Cys 14 and Cys 50) (45). Cys 11, usually identified as the forth ligand of the cubane, is not bound and was found to be tilted towards the solvent away from the cluster. Cys 18 and Cys 42 form a disulfide bridge. Only three cysteinyl residues are required to bind the trinuclear core to the polypeptide chain (Figure 2).

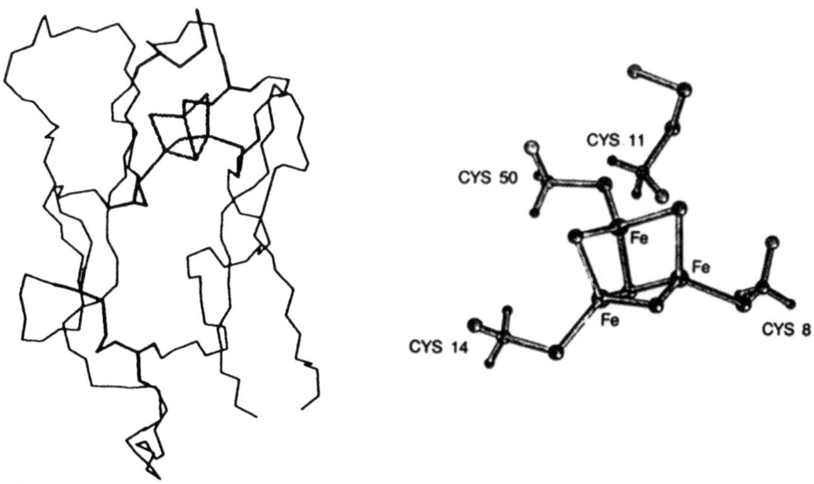

Figure 2

Folding of the polypeptide chain of D.gigas Fd II and localization of the [3Fe-4S] core and of the disulfide bridge (15). Details on the cysteinyl binding to the core.

In the oxidized form the cluster has an electronic spin S=1/2 and exhibits a correspondent fairly isotropic EPR signal centered around g=2.02 (19). Mössbauer spectra taken at T>40K (in the limit of fast electronic spin relaxation) exhibit one sharp quadrupole doublet with quadrupole splitting ΔE_Q=0.54 mm/s and isomer shift δ=0.27 mm/s. These parameters are typical of Fe^{3+} ions in a tetrahedral environment of thiolate ligands. The effective S=1/2 results from anti-ferromagnetic coupling of three high-spin ferric rubredoxin type (S_1=S_2=S_3=5/2) ions (21).

Reduction by one electron (E_m=-130 mV, vs NHE) yield a state with integer cluster spin Magnetic Circular Dichroism (MCD) studies showed that the reduced state has a spin S=2 (46) in agreement with the Mössbauer predictions and the detection of a Δm_s=4 EPR transitions, at around g =12 (21). At 4.2 K [zero-field] the Mössbauer spectrum of the reduced cluster exhibits two quadrupole doublets with intensity ratios 2:1. The more intense doublet, representing two iron sites (site I) has ΔE_Q (I)=1.47 mm/s and δ(I)=0.46 mm/s. The second doublet (site II) has ΔE_Q(II)=0.52 mm/s and δ(II)=0.32 mm/s. These values suggest that the iron atom associated with doublet (II) is high-spin ferric in caracter, similar to the iron sites in oxidized FdII. The parameters of doublet I indicate a formal oxidation state between +2 and +3. The two iron atoms are in the oxidation level $Fe^{+2.5}$. The reduced core is an example of a mixed valence compound with one localized site Fe^{+3} in site II) and a delocalized iron-pair. Spectroscopic evidence is consistent with the delocalized pair having a spin of 9/2 coupled to the high spin ferric ion forming a cluster spin of 2 (21).

This is the simplest iron cluster to exhibit this equal sharing of an electron by more than one iron site - a feature in commom with [4Fe-4S] cluster.

2. Hetero metal metal clusters-cubane structures with a [M,3Fe-4S] core, M= Co, Ni, Cd, Zn, Ga, Tl, V, W, Mo, Re

A number of proteins containing [3Fe-4S] centers bind certain metals in a reversible way to form cubane-like clusters of the type [M,3Fe-4S]. The [3Fe-4S] core present in *D.gigas* FdII was the first percursor used for the synthesis of hetero metal cores inside a protein matrix. Similar synthetic products have also been derived and extended with the ferredoxins isolated from *Pyrococcus (P.) furiosus* (39,47) and *D.africanus* (48).

The [3Fe-4S] core present in *D.gigas* Fd II can be interconverted into a cubane structure [4Fe-4S]. This facile conversion occur upon incubation of the protein with an excess of Fe^{+2} in the presence of a reducing agent (e.g. dithiothreitol). A summary of the interconversion pathways as well as the potentialities of the method for specific labelling an iron-sulfur core was previously discussed (26,49). Combination of ^{57}Fe isotopic enrichment and specific introduction of a forth iron atom into the [3Fe-4S] core produces different isotopic labelled clusters.

The facility of the [3Fe-4S] / [4Fe-4S] conversions has suggested that one may be able to incorporate other metals in the vacant site of [3Fe-4S] core and then generate a series of novel clusters containing hetero metals using this core as a synthetic precursor. According to the X-ray crystallographic analysis of oxidized *D.gigas* FdII, the candidate ligand to coordinate to the fourth site, forming a [M,3Fe-4S] complex, is Cys 11 (45).

The first derivative synthetized was the [Co,3Fe-4S] core (37). Cobalt (II) ion was introduced into a hetero metal cluster, assuming a paramagnetic configuration (S=3/2) at the forth site of a cubane structure. The newly formed cluster was studied in the oxidized and reduced states. In the oxidized state the cluster exhibits an S=1/2 EPR signal with g-values at g_x=1.82, g_y=1.92 and g_z=1.98 (g-values obtained after spectral simulation). The well resolved ^{59}Co hyperfine at the g_z

line (A_z~4.4 mT) is also broadened by ^{57}Fe isotopic substitution indicating that Fe and Co share a common unpaired electron. Upon one electron reduction (E_m= -220 mV,vs NHE) the cluster as an integer spin (S=1) as indicated by Mössbauer and MCD measurements . The Mössbauer spectrum of the oxidized cluster exhibits at 4.2K two distinct spectral subsites with an intensity ration 2:1, indicating that the three iron atoms reside in the same cluster. The isomer shifts, δ (one iron site) = 0.35 mm/s and δ (two iron site)=0.44 mm/s suggest the presence of a Fe^{3+} site and again a delocalized pair (37). Upon one electron reduction the third iron is formally Fe^{2+}. The [Co,3Fe-4S] cluster yields highly resolved spectroscopic data and its formation was a usefull probe for testing this proposed synthetic approach. However the paramagnetism of the cobalt site does not simplify the overall problem of the four site cluster magnetism.

The introduction of a metal with a full d-shell (e.g. Cu^+, Zn^{2+},$Cd^{2+}$and Ga^{3+}) would be more advantageous in this respect. The procedure of converting a [3Fe-4S] core into a [Zn,3Fe-4S] was described (38). The incubated product is EPR-active (spin system, S=5/2) with E/D~0.25 and D=-2.7 cm^{-1} and shows ^{57}Fe hyperfine broadening.

The Mössbauer studies performed at 4.2K spectrum (at this temperature only the lowest Kramer doublet is populated) consists of two spectral components with an intensity ratio of 2:1 (38). Doublet I (one iron) has ΔE_Q=2.7 mm/s and δ=0.62 mm/s, suggesting a FeS_4 site with slightly smaller d-electron density than a typical ferrous FeS_4 complex. This site appears to be a trapped valence Fe^{2+}. Doublet II (two iron sites) has ΔE_Q=1.6 mm/s and δ=0.52 mm/s. These parameters suggests again a delocalized Fe^{2+}/Fe^{3+} pair, as observed for the reduced [3Fe-4S] core, but with a slightly higher ferrous character as indicated by the value of δ. The formal oxidation states involved were assigned: two Fe^{2+}, one Fe^{3+}, and, being a Kramer system (S=5/2), the zinc site must be divalent. Incorporation of the zinc atom at the vacant site of the [3Fe-4S] core implies a previous reduction step of the cluster plus an extra electron for metal incorporation.

We have now preliminary evidence for the incorporation of Cd^{2+}, which yields a similar S=5/2 system under reducing conditions (see also (50)).

Recently, a similar $[Ni,3Fe\text{-}4S]^{1+}$ was produced in *P.furiosus* Fd(39) and *D.gigas* FdII (51). The dithionite reduced FdII was anaerobicaly incubated with excess nickel(II) in the presence of dithiothreitol. The EPR spectral features are dominated by the presence of a S=3/2 species and can be explained by the appropriated Hamiltonian. Peaks at 5.60 and 4.89 (broad derivative at 2.82) and very weak and broad signals around 1.8 could be interpreted in terms of a S=3/2 ground state with E/D=0.21 and negative D value.

The redox properties of a series of hetero metal clusters were accessed by electrochemical mewasurements and EPR. The derivatives formed in *D.gigas* FdII were measured by direct square wave voltammetry promoted by Mg(II) at a vitreous carbon electrode and the following redox transitions were detected: Cd(-495 mV), Fe(-420 mV), Ni(-360 mV) and Co(-245 mV) [vs NHE]. The values agree well with independent measurements ((51) and our unpublished results). Similar derivatives generated in *D.africanus* FdIII measured by cyclic voltammetry in film and bulk solution yield: Cd(-590 mV), Zn(-490 mV) and Fe(-400 mV) (48,52). A novel addition to these series was obtained by the introduction of a mono valent ion, Tl^{+} (53). The $[Tl,3Fe\text{-}4S]^{2+}$ core shows a redox transition at -80 mV.

Table I collects the variability of spin states found in the hetero metal cubane-type cores so far formed in simple iron-sulfur proteins, indicating the effect of different d-level occupancy of the forth site.

TABLE I

Hetero atom clusters in proteins

Cluster	Spin state
$[Co,3Fe\text{-}4S]^{+2}$	1/2
$[Co,3Fe\text{-}4S]^{+1}$	1
$[Zn,3Fe\text{-}4S]^{+2}$	5/2
$[Zn,3Fe\text{-}4S]^{+1}$	2
$[Ni,3Fe\text{-}4S]^{+2}$	3/2
$[Cd,3Fe\text{-}4S]^{+1}$	5/2
$[Tl,3Fe\text{-}4S]^{+2}$	1/2

Inorganic synthesis produce several hetero metal cluster of this type, complementing and extending the described work. A VFe_3 ($S=3/2$) core was synthetized and the X-ray structure determined (54). The complex had the peculiarity of depicting a V site with a high coordination number (six) and mimics well the chemical environment of the VFe cofactor of nitrogenase (55). Examples of cuboidal type cores with hetero metal have been described containing Mo (56), W (57), Re (58) and Co (59). $NiFe_3$ cores containing S or Se were also reported (60) and shown to be iso-electronic with the ones formed within the protein matrix.

Naturally occurring hetero metal clusters have also been detected in proteins, such as, the [Zn, Fe] cluster in the red kidney bean purple phosphatase (61) and the M center (a putative [M,6Fe] cluster, M=Mo or V) in the metal cofactor of nitrogenases (62). Also the nickel derivative is an interesting addition to the series of transition metal substituted cubanes and speculated as a model for nickel enzymes indicated to have a Ni and Fe containing center, such as CO dehydrogenase (63). Most of these natural occuring structures remain to be proven.

CONCLUSIONS

The described protein matrix-assisted inorganic synthesis provides an interesting pathway for the formation of novel hetero metal containing clusters. A particular interestet in this field arises from the possibility of forming new biocatalysts with unexpected properties. Also the introduction of a differentiated site into a cubane-type structure with intrinsic properties enables to explore magnetism in valence - delocalized /localized systems as well as the simplification of the spectral analysis. Synthesis of relevant model compounds *via* the formation of hetero metal cubane-type structures containing molybdenum, vanadium and nickel are potentially interesting for mimic and specific modification a [FeS] core in a complex enzyme.

ACKNOWLEDGMENTS

We thank the contributions of Drs. J. LeGall, E. Münck, L. Sieker, B.H. Huynh, K. Surerus, T. Kent and V. Papaefthymiou. This work was supported by grants (I.M. and J.J.G.M.) from I.N.I.C. and J.N.I.C.T..

REFERENCES

1) Lovenberg, W. (1973 - 77) in *Iron-sulfur Proteins*, Vol. I - III, Academic, New York
2) Spiro, T. G. (1982) in *Iron-sulfur Proteins*, Vol. IV, John Wiley & Sons, New York
3) Ohnishi, T. (1987) in *Current Topics in Bioenergetics* (Lee, C. P., ed) vol. 15, pp. 37-65, Academic, New York
4) Beinert, H. (1990) *FASEB Journal* **4**, 2483
5) Switzer, R.L. (1989) Biofactors 2, 77
6) Backes, G., Mino, Y., Loehr, T. M., Meyer, T. E., Cusanovich, M. A., Sweeney, W. V., Adman, E. T., and Sanders-Loehr, J. (1991) *J. Am. Chem. Soc.* **113**, 2055
7) Bertini, I., Briganti, F., Luchinat, C., Scozzafava, A., and Sola, M. (1991) *J. Am. Chem. Soc.* **113**, 1237
8) Holm, R. H., Ciurli, S., and Weigel, J. A. (1990) in *Progress in Inorganic Chemistry: Bioinorganic Chemistry* (Lippard, S. J., ed) Vol. 38, pp. 1-74, John Wiley & Sons, New York
9) Moura, I., Macedo, A., and Moura, J. J. G. (1989) in *Advanced EPR and Applications in Biology and Biochemistry* (Hoff, A. J., ed) pp. 813-838, Elsevier, Amsterdam
10) Beinert, H., and Kennedy, M. C. (1989) *Eur. J. Biochem.* **186**, 5
11) Frey, M., Sieker, L., Payan, F., Haser, R., Bruschi, M., Pepe, G., and LeGall, J. (1987) *J. Mol. Biol.* **197**, 525
12) Fukuyama, K., Matsubara, H., Tsukihara, T., and Katsube, Y. (1989) *J. Mol. Biol.* **210**, 383-398
13) Stout, C. D. (1989) *J. Mol. Biol.* **205**, 545
14) Stout, G. H., Turley, S., Sieker, L. C., and Jensen, L. H. (1988) *Proc. Natl. Acad. Sci., USA* **85**, 1020
15) Kissinger, C. R., Adman, E. T., Sieker, L. C., Jensen, L. H., and LeGall, J. (1989) *FEBS letters* **244**, 447
16) Fukuyama, K., Hase, T., Matsumoto, S., Tsukihara, T., Katsube, Y., Tanaka, N., Kakudo, M., Wada, K., and Matsubara, H. (1980) *Nature* **286**, 522
17) Moura, I., Huynh, B. H., Hausinger, R. P., LeGall, J., Xavier, A. V., and Münck, E. (1980) *J. Biol. Chem.* **255**, 2493
18) Emptage, M. H., Kent, T. A., Huynh, B. H., Rawlings, J., Orme-Johnson, W. H., and Münck, E. (1980) *J. Biol. Chem.* **255**, 1793
19) Huynh, B. H., Moura, J. J. G., Moura, I., Kent, T. A., LeGall, J., Xavier, A. V., and Münck, E. (1980) *J. Biol. Chem.* **255**, 3242
20) Kent, T. A., Huynh, B. H., and Münck, E. (1980) *Proc. Natl. Acad. Sci., USA* **77**, 6574-6576
21) Papaefthymiou, V., Girerd, J.-J., Moura, I., Moura, J. J. G., and Münck, E. (1987) *J. Am. Chem. Soc.* **109**, 4703
22) Armstrong, F. A. (1991) *Abstr. Fifth Int. Conf. Bioinorg. Chem., J.Inorg. Biochem.* **43**, 238
23) Carter, C. W., Jr. (1977) in *Iron-Sulfur Proteins* (Lovenberg, W., ed) Vol. III, pp. 157-204, Academic, New York
24) (a) Noodleman, L. (1991) *Inorg. Chem.* **30**, 246, and (b) 256
25) Jordanov, J., Roth, E. K. H., Fries, P. H., and Noodleman, L. (1990) *Inorg. Chem.* **29**, 4288
26) Moura, J. J. G., Moura, I., Kent, T. A., Lipscomb, J. D., Huynh, B. H., LeGall, J., Xavier, A. V., and Münck, E. (1982) *J. Biol. Chem.* **257**, 6259
27) Kent, T. A., Dreyer, J.-L., Kennedy, M. C., Huynh, B. H., Emptage, M. H., Beinert, H., and Münck, E. (1982) *Proc. Natl. Acad. Sci., USA* **79**, 1096
28) George, S. J., Armstrong, F. A., Hatchikian, E. C., and Thomson, A. J. (1989) *Biochem. J.* **264**, 275
29) Kent, T. A., Emptage, M. H., Merkle, H., Kennedy, M. C., Beinert, H., and Münck, E. (1985) *J. Biol. Chem.* **260**, 6871

30) Kennedy, M. C., Werst, M., Telser, J., Emptage, M. H., Beinert, H., and Hoffman, B. M. (1987) *Proc. Natl. Acad. Sci., USA* 84, 8854
31) Fee, J. A., Findling, K. L., Yoshida, T., Hille, R., Tarr, G. E., Hearshen, D. O., Dunham, W. R., Day, E. P., Kent, T. A., and Münck, E. (1984) *J. Biol. Chem.* 259, 124
32) Cline, J. F., Hoffman, B. M., Mims, W. B., Lahaie, E., Ballou, D. P., and Fee, J. A. (1985) *J. Biol. Chem.* 260, 3251
33) Lindahl, P. A., Gorelick, N. J., Münck, E., and Orme-Johnson, W. H. (1987) *J. Biol. Chem.* 262, 14945
34) Hagen, W. R., Wassink, H., Eady, R. R., Smith, B. E., and Haaker, H. (1987) *Eur. J. Biochem.* 169, 457
35) Robinson, A. E., Richards, A. J. M., Thomson, A. J., Hawkes, T. R., and Smith, B. E. (1984) *Biochem. J.* 219, 495
36) Smith, J. P., Emptage, M. H., and Orme-Johnson, W. H. (1982) *J. Biol. Chem.* 257, 2310
37) Moura, I., Moura, J. J. G., Münck, E., Papaefthymiou, V., and LeGall, J. (1986) *J.Am.Chem. Soc.* 108, 349
38) Surerus, K., Münck, E., Moura, I., Moura, J. J. G., and LeGall, J. (1987) *J.Am.Chem.Soc.* 109, 3805
39) Conover, R. C., Park, J.-B., Adams, M. W. W., and Johson, M. K. (1990) *J.Am.Chem.Soc.* 112, 4562 and Conover, R. C., Finnegan, M. G., Park, J.-B., Adams, M. W. W. amd Johnson, M. K. (1991) *Abstr. Fifth Int. Conf. Bioinorg. Chem., J.Inorg. Biochem.* 43, 245
40) Auric, P., Gaillard, J., Meyer, J., and Moulis, J.-M. (1987) *Biochem. J.* 242, 525
41) Gaillard, J., Moulis, J.-M., Auric, P., and Meyer, J. (1986) *Biochemistry* 25, 464
42) Hagen, W. R., Pierik, A. J., and Veeger, C. (1989) *J. Chem.Soc.Faraday Trans. 1* 85, 4083
43) Moura, I., Tavares, P., Moura, J. J. G., Ravi, N., Huynh, B.H. and LeGall, J. (1992) *J.Am.Chem.Soc.* in press
44) Bruschi, M., Hatchikian, E.C., LeGall, J., Moura, J. J. G., Xavier, A. V. (1976) *Biochim.Biophys.Acta* 449, 275
45) Kisinger, C. R., Adman, E. T., Sieker, L. C., Jensen, L. H. and LeGall, J. (1989) *FEBS Lett.* 244, 447
46) Thomson, A. J., Robinson, A. E., Johnson, M, K., Moura, J. J. G., Moura, I., Xavier, A. V. and LeGall, J. (1981) *Biochim. Biophys.Acta* 670, 93
47) Conover, R. C., Kowal, A. T., Fu, W., Park, J.-B., Aono, S., Adams, M. W. W. and Johnson, M. K. (1990) *J.Biol.Chem.* 265, 8563
48) George, S. J., Armstrong, F. A., Hatchikian, C. C., Thomson, A. J. (1989) *Biochem.J.* 264, 275
49) Kent ,T. A., Moura, I. , Moura, J.J.G., Lipscomb, J.D., Huynh, B.H., LeGall, J., Xavier, A. V., Münck, E. (1982) *FEBS Lett.* 138, 55
50) Münck, E., Papaefthymiou, V., Surerus, K. and Girerd, J.-J (1988) in *Metal Ions in Proteins*, Que, L. ed., ACS Symposium Series 372, Washington D.C. , Chapter 15
51) Moreno, C., Macedo, A. L., Surerus, K., Münck, E., LeGall, J. and Moura, J. J. G. (1991) in *Abstr. 3th International Conference on Molecular Biology of Hydrogenases*, Tróia, Portugal
52) Butt, J. N., Armstrong, F. A., Breton, J., George, S. J., Thomson, A. J. and Hatchikian, E. C. (1991) *J.Am.Chem.Soc.* 113, 6663
53) Butt, J. N., Sucheta, A., Armstrong, F. A., Breton, J., Thomson, A. J. and Hatchikian, E. C. (1991) *J.Am.Chem.Soc.* 113, 8948
54) Kovacs, J. A. and Holm, R.H. (1986) *J.Am.Chem.Soc.* 108, 340
55) George, G. N., Coyle, C. L., Hales, B. J. and Cramer, S. P. (1988) *J.Am.Chem.Soc.* 110, 4057
56) Ciurli, S. and Holm, R.H. (1989) *Inorg.Chem.* 28, 1685
57) Holm, R. H. and Simhon, E. D. (1985) in "*Molybdenum Enzymes*", Spiro, T. G. ed., Wiley-Interscience, New York, Chapter 1
58) Ciurli, S., Carney, M. J., Papaefthymiou, G., C. and Holm, R.H. (1989) *Inorg. Chem.* 28, 2696
59) Roth, E. K. H., Greneche, J. M. and Jordanov, J. (1991) *J.Chem.Soc., Chem.Commun.*, 105
60) Ciurli, Yu, S.-B., Holm, R.H., S., Srivastava, K. K. P. and Münck, E. (1990) *J.Am.Chem.Soc.* 112, 8169
61)Beck, J.L., de Jersey, J., Zerner, B., Hendrich, M.P., and Debrunner, P.G. (1988) *J.Am.Chem. Soc.* 110, 3317
62) Johnson, W.H. (1979) *Biochim..Biophys.Acta* 576, 192
63) Lindhal, P.A., Ragsdale, S.W., and Münck (1990) *J.Biol.Chem.* , 265, 3880

BINDING AND CATALYSIS IN WATER

RONALD BRESLOW
Department of Chemistry
Columbia University
New York, N.Y. 10027
U. S. A.

ABSTRACT. Recent work on hydrophobic binding and on catalysis by enzyme mimics is reported.

1. Introduction.

Although other forces can play an important role, binding in water often involves **hydrophobic** interactions. These take advantage of the fact that water is a solvent with low polarizability and high self-association energy, so non-polar organic molecules or molecular segments are not well soluble in water. Thus in water solution such non-polar segments tend to bind to each other. This is the type of binding that produces the molecular association in micelles and membranes, and it is also a major force in the binding of substrates to enzymes, or of many antigens to antibodies.

In the area of supramolecular chemistry of synthetic molecules, hydrophobic binding can be used to promote molecular recognition by binding, and reactions within catalyst-substrate complexes that mimic enzymes. As we have shown [1], hydrophobic binding of one substrate to another can also promote many simple organic reactions in water solution. In this paper we will review briefly the recent achievements in this area from our laboratory, and will point the way to the future of this exciting field.

2. Shape recognition and catalysis by dimeric hydrophobic binding molecules.

Hydrophobic binding of well-fitting hydrocarbon segments into α-, β-, or γ-cyclodextrin and its derivatives typically shows binding constants in water of the order of 10^4 M^{-1} or so. While this is enough to permit formation of molecular complexes for mimicking enzymes, for instance, it is too low to permit the use of the binding energy in other ways. For instance, many

V. Balzani and L. De Cola (eds.), Supramolecular Chemistry, 411–428.

enzymes bind a substrate so as to distort its geometry toward that of the transition state (i.e. they bind the geometry of the transition state better). Furthermore, catalytic antibodies may also use binding energy in this way.

To make this kind of thing possible with synthetic cavities, we have explored some of the chemistry of cyclodextrin dimers. As hoped, they can show much stronger binding than do the monomers, and in appropriate cases they show shape recognition of the type that will be useful in catalysis. Furthermore, catalytic groups placed in the center of such dimers are particularly well situated to attack coordinated substrates.

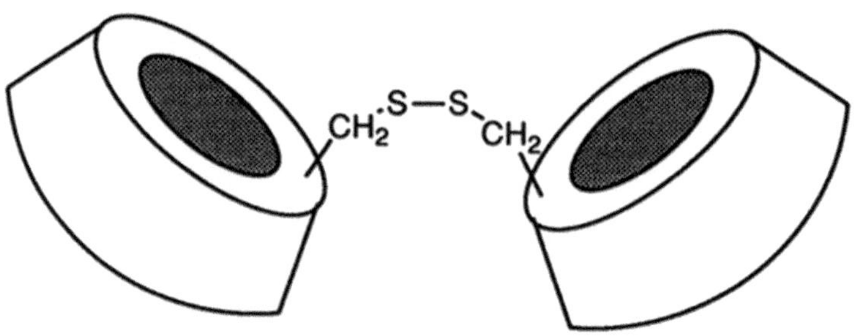

1

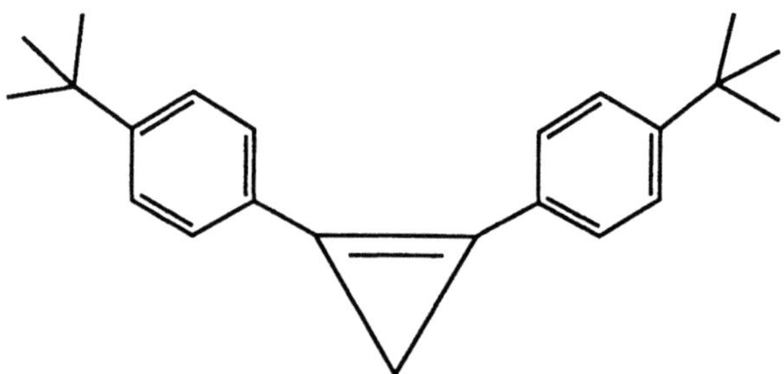

2

The first studies we did [2] focussed on compounds such as 1 with two β-cyclodextrin rings linked by a single chain, a type of system that we had initially prepared [3] many years earlier. We found that with appropriate rigid substrates such as 2, carrying two hydrophobic segments that could bind into both rings, binding constants as high as 10^8 were seen in water. However, with a single linking chain, attached at both ends to the primary C-6 carbon of the cyclodextrin, there was a lot of conformational flexibility in the host. Thus to improve the system, and add geometric specificity, we synthesized cyclodextrin dimers with two linking chains.

Two systems have been prepared so far. In one [4] we reacted the A,B isomer of cyclodextrin diiodide [5] with 6A,6B cyclodextrin bis-imidazole[5] (which is itself prepared from the same diiodide). The result was a mixture of two bis-imidazolium compounds, **3** which we call the "clam shell" type and **4**, the "love seat" type.

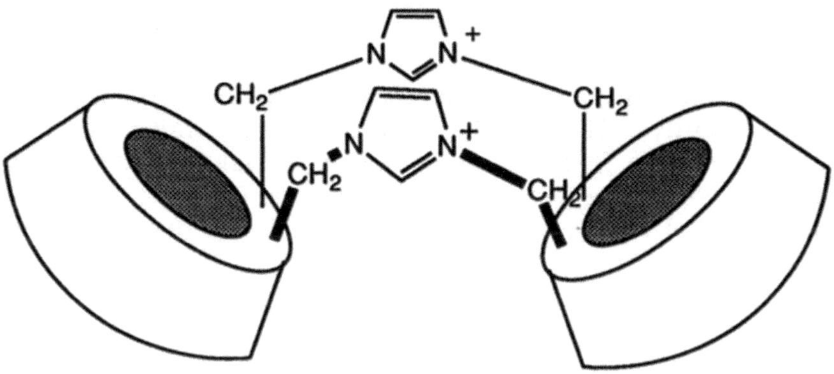

"Clamshell"

3

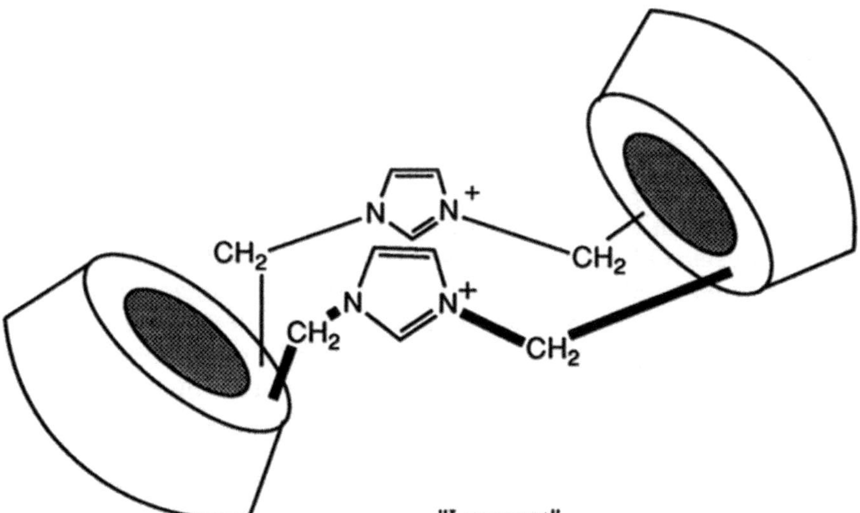

"Loveseat"

4

Labelling the cyclodextrin rings A to G clockwise, the clam shell has each imidazolium ring substituted by one 6A link of one cyclodextrin and one 6B link of the other. Thus the two imidazolium rings are equivalent, and show a single 2' hydrogen signal in the NMR. By contrast, the love seat arrangement has one imidazolium ring attached to two 6A sites, while the other is attached to two 6B sites. They are not equivalent in the NMR. We find that the clamshell arrangement is a strong binder of various substrates, while the love seat arrangement has only a normal 10^4 or so affinity for double-ended substrates since both ends cannot be bound with a love seat host.

These two isomers **3** and **4** can be separated by ordinary chromatography, but with difficulty. Thus we have prepared an affinity chromatography column, by attaching **5** to the leash of a commercial affinity gel material [4]. This easily separates the weakly binding love-seat isomer **4** from the strongly binding clamshell isomer **3**, and will probably be useful for other such separations as well.

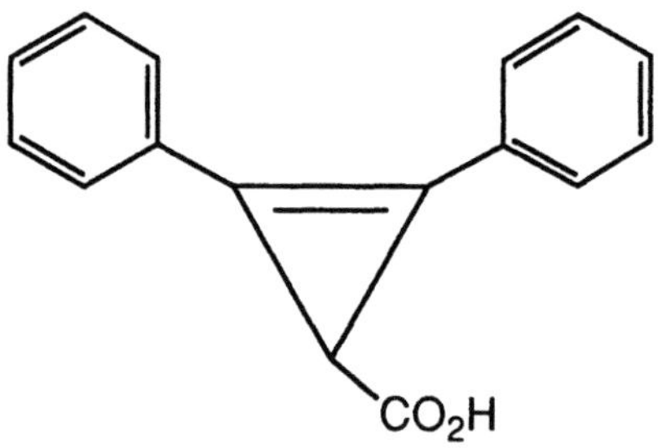

5

Reduction of the imidazolium rings in **3** with $NaBH_4$ affords N-methylethylenediamine units as linkers. These can bind metals, and are potential catalysts. Other such catalysts will be described below.

We were interested in constructing a clam shell dimer with two links of different lengths, so that the shell would not close all the way. Instead, with such uneven links we hoped for a host that would preferentially bind guests with angular geometry rather than linear geometry. Such a system might bind a linear amide system, for instance, but prefer to bind the tetrahedral intermediate formed during amide hydrolysis. Thus the difference in binding energy could be used to help drive the hydrolysis reaction. We were able to construct such a molecule, and it did show the desired geometric preference [6].

Reaction of the 6A,6B diiodide of β-cyclodextrin does not occur randomly at the two positions. Because the units of the cyclodextrin system are themselves chiral glucose units, the A and B carbons of a disubstituted cyclodextrin are diasterotopic, not enantiotopic. We find [6,7] that there is ca.

a 7:1 preference for reaction of a simple thiol nucleophile with the B position compared to the A position in the 6A, 6B diiodide, and this directly affects the selectivity for formation of the clamshell vs. love seat isomers. Thus reaction of the diiodide with one equivalent of naphthalene-2,6-dithiol occurs mostly with B-B linkage, with some A-B linkage and essentially no A-A linkage. Then reaction of the remaining iodides with sodium thioacetate, hydrolysis, and air oxidation affords two products. Compound 6 is the major love seat isomer, with the naphthalene bridging the B-B positions and the disulfide link at A-A; compound 7 is the clamshell, with a naphthalene and a disulfide each linking an A position to a B position.

NMR spectra confirm this assignment. The major isomer 6 indeed has a symmetric naphthalene ring, while the minor clamshell 7 has an unsymmetrical naphthalene. Again they were separated by ordinary chromatography, but affinity chromatography should now be easier. We found that the clamshell had high affinity for appropriate substrates. For example, with the linear disubstituted naphthalene substrate 8 the binding constant was of the order of 10^8, but as hoped it was even higher with the bent substrate 9 that can better fit the host. In this case the binding was so strong that we can so far only put a lower limit on it; the binding constant is at least 10^{11} M^{-1}. If this difference in binding energy can be turned into a driving force for bending a linear substrate toward a bent transition state structure, the rate contribution of this geometric preference should be at least 1000-fold.

We describe such compounds as having the clamshell geometry; in fact the resemblance is dynamic, not just static. In molecular models the hinge at the A and B positions permits the dimer to open and close like a clamshell, and it is likely that this is how substrates are bound. In models it is not possible to thread substrate 9 through the two cavities of a closed conformation of 7, since the bulky central naphthalene ring of the substrate doesn't make it through. Instead it seems possible to bind the substrate into the open conformation of the clamshell, and then close the shell up around the substrate to achieve maximum binding.

Of course we need some catalytic groups incorporated in such systems if they are to function as enzyme mimics. As mentioned above, we have prepared a clamshell dimer with two linking ethylenediamine groups, that can bind metal ions into the linker and has potential as a catalyst. Other metal-binding groups have also been put into the linker; the only one that we have started to examine is the bipyridyl linker in dimer 10. This indeed binds metal ions, and we find [8] that the resulting complex will hydrolyze bound esters reasonably well. The goal is of course to hydrolyze peptide links, for instance. In this regard, it is interesting that dimer 11 has a selective affinity for the binding of peptide 12, in which the two indole rings can bind in the cyclodextrin cavities. Thus we can hope for sequence selectivity, hydrolysis between two hydrophobic amino acid residues, in such catalysts.

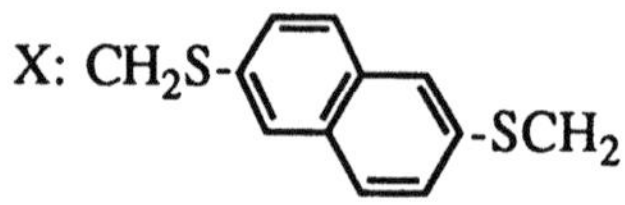

Y: $-CH_2-S-S-CH_2-$

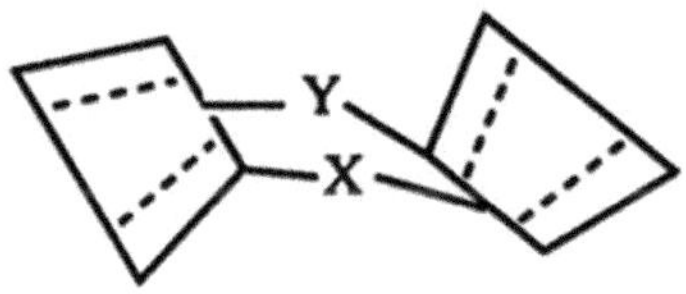

<u>6</u>

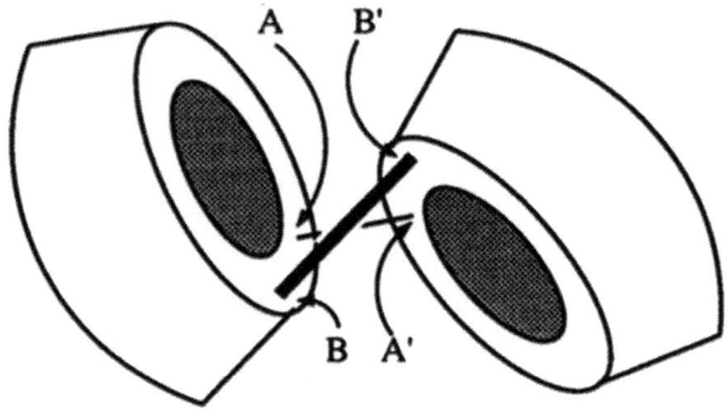

C_2 Aversive

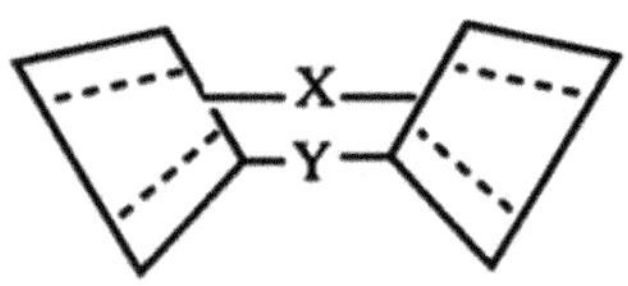

<u>7</u>

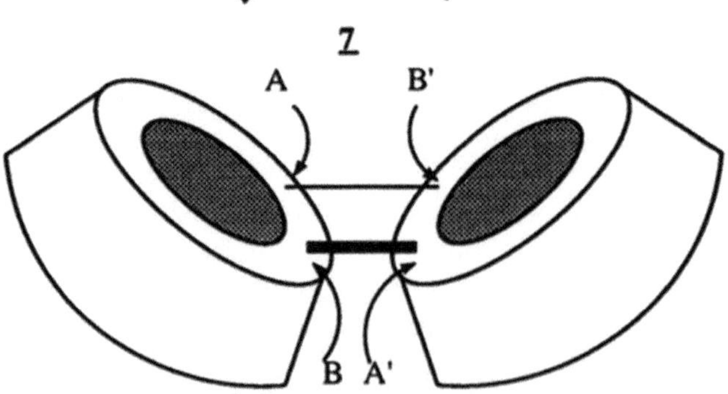

C_1 Occlusive

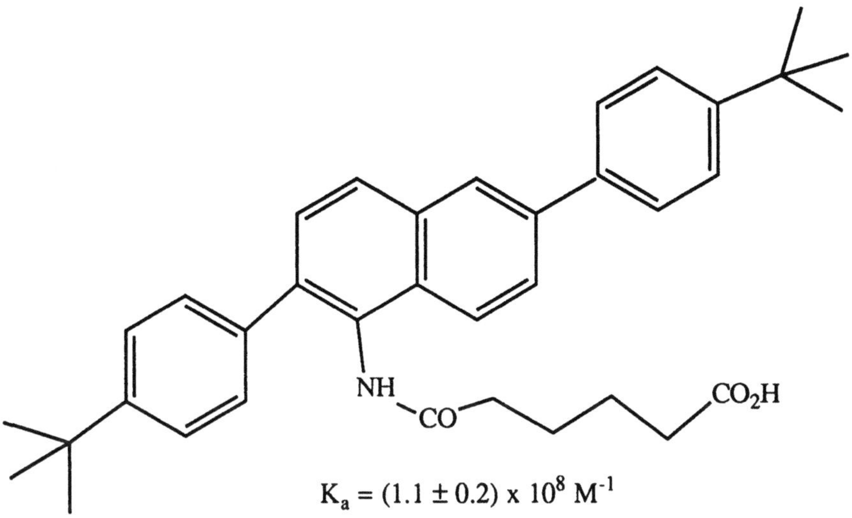

$K_a = (1.1 \pm 0.2) \times 10^8\ M^{-1}$

8

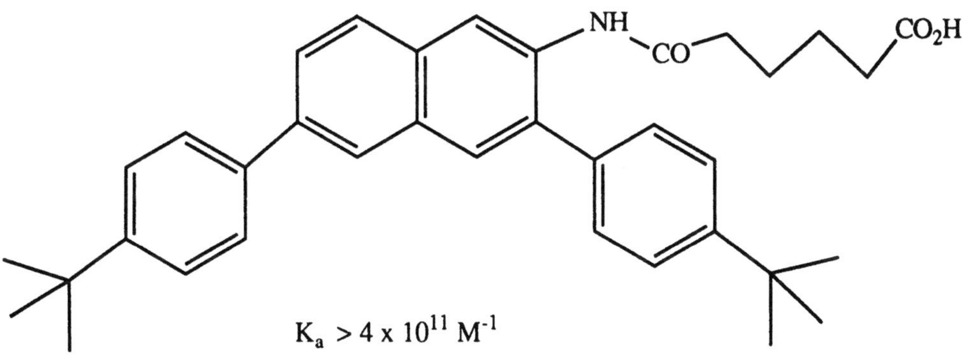

$K_a > 4 \times 10^{11}\ M^{-1}$

9

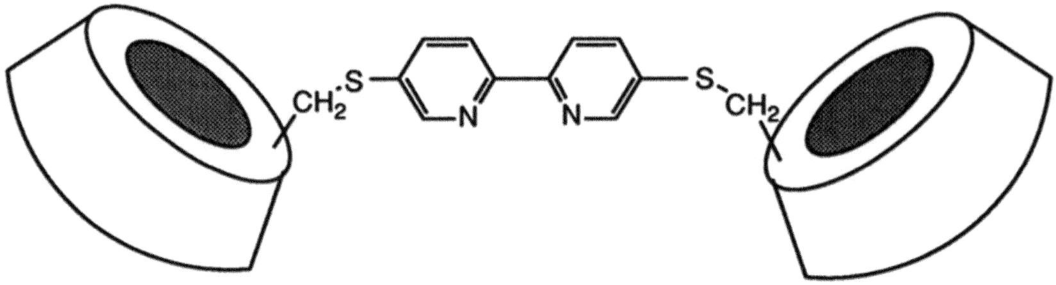

10

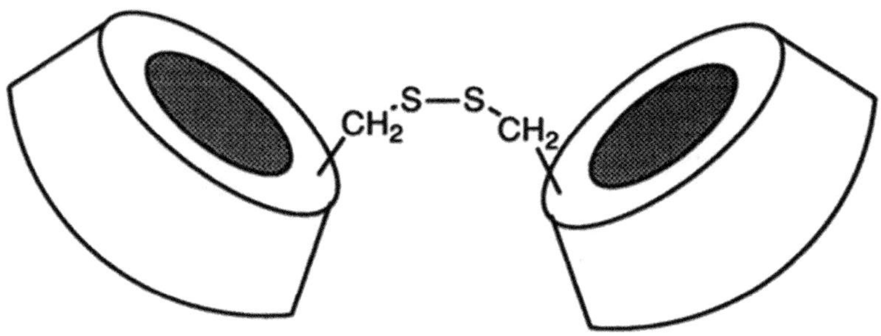

11

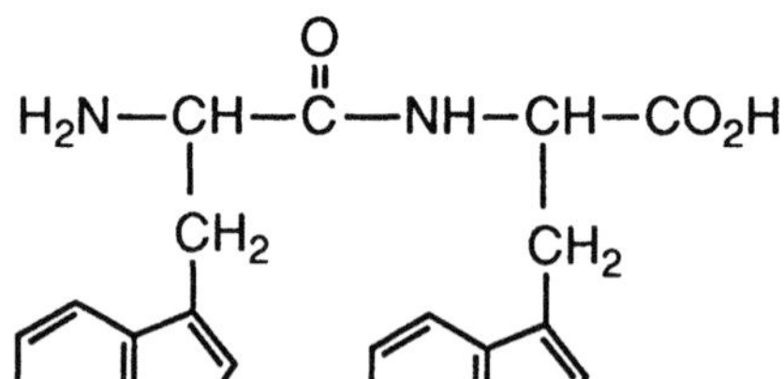

12

3. Salting in agents and the hydrophobic effect.

Protein chemists have long used certain substances as denaturants.

Compounds such as urea and guanidinium chloride (GnCl) cause unfolding of the proteins; they also cause unwinding of double helical DNA. In both cases the effect is caused by a decrease in the hydrophobic effect. This effect is a principal force that organizes protein structure, by packing the hydrocarbon sidechains of proteins into the interior, and it also contributes to the stability of DNA since the hydrophobic purines and pyrimidines can stack on each other away from water. The way in which such denaturants as urea act to overcome the hydrophobic effect has only recently been clarified, by studies we conducted aimed at distinguishing between the two likely mechanisms [8].

Additives such as GnCl decrease the hydrophobic effect, so they increase the solubility of hydrocarbons in water. They are commonly called "salting in" agents to indicate this. By contrast, ordinary salts such as NaCl decrease hydrocarbon solubility in water, and are "salting out" agents. We have studied both the reasons for this behavior and the general use of these effects to detect hydrophobic contributions to rates and equilibria. Recently we have initiated a program to put the effects on a more quantitative basis.

We reported some years ago [9] that certain Diels-Alder reactions are strongly accelerated in water, compared with other solvents. The contrast was so large that it seemed unlikely that we were seeing only the effect of increased polarity; instead, we invoked hydrophobic packing of the diene with the dienophile in water as part of the explanation. In support of this, we saw that the rate increased on addition of salting-out agents such as LiCl, which increase the hydrophobic effect, but decreased when GnCl was added. The contrast pointed to a hydrophobic contribution to the rates. Recently we saw [10] that other additives that have an even larger salting in effect slow the reaction even more; this finding confirmed that the additives were diminishing the hydrophobic effect, and ruled out other explanations.

As additional evidence for hydrophobic effects in the Diels-Alder reaction, there was a great increase in the preference for an endo geometry in the Diels-Alder reactions in water [11] , as expected if the transition state becomes more compact to minimize exposure of hydrocarbon segments to the water solvent. This endo preference was also increased to some extent when LiCl was added, and decreased when GnCl was the additive. Other evidence has also been cited for a hydrophobic effect on Diels-Alder reactions [12], and the water effect on geometric selectivity has been used in synthetic applications [13].

We also detected a hydrophobic effect in the benzoin condensation [14]. In this case, in contrast to the Diels-Alder reaction, it is not formally required that the two hydrophobic phenyl groups come together in the transition state, but our studies indicated that they do. Again there was a large increase in rate when water was the solvent, but in an ionic reaction of this sort such solvent effects could well be related only to the effect on the ions of the polar character of the medium. However, we saw that the reaction rate was increased with LiCl, but decreased when $LiClO_4$ was added. In this system $LiClO_4$ is a salting in agent, as shown by the fact that it increases the solubility of benzaldehyde in water, while LiCl decreases that solubility. A number of other salts were also examined, and there was a good correlation between their ability to modify the hydrophobic effect, as determined by changes in solubility of non-polar compounds, and their effect on the benzoin condensation.

The conclusion that the two phenyl groups pack together in the benzoin condensation transition state was confirmed by the finding that this reaction is accelerated by γ-cyclodextrin [14]. The cavity of this molecule is large enough to accomodate both rings, in contrast to that of β-cyclodextrin which with its smaller cavity, able to bind only one of the reactants, was an inhibitor of the reaction. As a further consequence of this picture, we attached a thiazolium salt to the γ-cyclodextrin and made an enzyme mimic that had both the hydrophobic cavity to organize the transition state and the thiazolium salt to furnish covalent catalysis for the reaction [15].

We were naturally curious as to the cause of the salting in and salting out effects. Salting in effects seem clear: NaCl or LiCl are strongly solvated by water, leading to electrostriction (volume decrease). This there is less room for the inclusion of a non-polar solute, and creating a cavity so as to make room would be energetically more expensive because of the electrostrictive effect.

There are two obvious possible explanations for the salting in effect of materials such as urea or GnCl. One possibility is that materials such as this are **water structure breakers**. If they set up alternative hydrogen bonding networks, they might make it easier for water to open up a cavity so as to accomodate a hydrocarbon solute. The other possibility is that such species as urea, Gn^+, or ClO_4^- bind to the hydrocarbon solute and help bridge its interaction with the water solvent. It might at first seem odd to invoke ions such as Gn^+ or ClO_4^- as solvaters of hydrocarbons preferred to solvation by neutral water, but the large ions are much more polarizable than is water and polarizability is important in solvating interactions. In fact, our evidence [16] points to this latter explanation.

If salting in materials make it easier to form a cavity in water, this should show up in a decreased surface tension when they are added. Instead, salts such as GnCl and $LiClO_4$ actually increase the surface tension of water, and so does urea; the effect is smaller than that of salting out materials such as LiCl, but it is in the same direction. As we have discussed in detail elsewhere [16], this really rules out the water structure breaking explanation of salting in, and leaves the solvation explanation.

It was not previously known whether salting in behavior was shown by materials such as GnCl in non-aqueous "water-like" solvents. We examined the situation in ethylene glycol and in formamide, and the effect was not seen [17]. In these media GnCl and $LiClO_4$ are now salting out additives. We believe that this simply means that in contrast to water such solvents are better hydrocarbon solvaters than are Gn^+ or ClO_4^-, so now the modest electrostrictive effect of these ions is seen, the same effect that makes them increase the surface tension of water. By contrast, organic salts such as $(Bu)_4^+$ Cl^- increase hydrocarbon solubility in glycol and formamide, probably by both a solvation and surface tension effect.

The important conclusion from this is that salting in materials are probably directly binding to the organic solutes. Thus we might see selective effects of different agents, depending on their ability to bind to a particular hydrophobic surface, while no such selective effects would be seen if the additives were simply affecting water structure. We have recently seen some phenomena that support this prediction.

A program is underway to put the salt effects on hydrophobic interactions on a more quantitative basis [4]. Instead of simply concluding that the two phenyl rings pack together during the benzoin condensation, for instance, we want to be able to say how much hydrophobic surface becomes covered. We think that we can use salting in agents to achieve such quantitative answers. For example, we find [4] that with a given concentration of GnCl (6 M) the solubility of p-methylbenzyl alcohol in water increases by 2.8 fold but that of p-t-butylbenzyl alcohol increases by 4-fold. Similarly, with 8 M urea the smaller methylbenzyl alcohol shows a 2.5-fold solubility increase, but the larger t-butyl derivative increases in solubility by 3.3-fold. Thus as expected, when the hydrophobic surface is larger the effect of salting in agents on solubility is also larger.

We have also compared this with the effect of such salting in agents on the binding of a hydrophobic molecule into cyclodextrin. In the molecule **13** that we call BANS, a t-butylphenyl group can bind into β-cyclodextrin, and this binding leads to an increase in the fluorescense intensity of the BANS. We find that binding of BANS into the cyclodextrin is decreased when 6M GnCl is added to the water, and by the same 4-fold by which the solubility of t-butylbenzyl alcohol was increased!

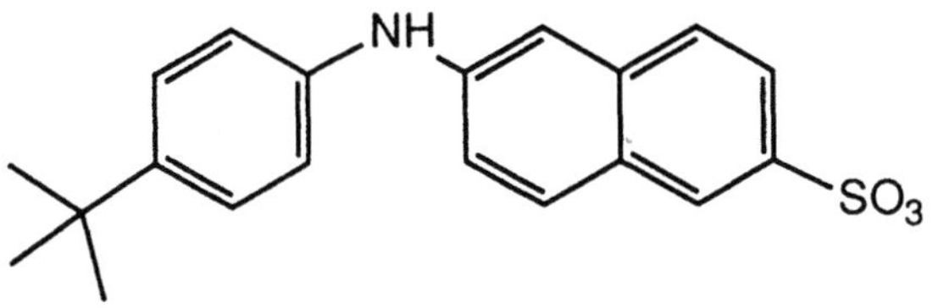

13

There are several possible interpretations of this coincidence. One attractive picture is that the entire effect on solubility is because of solvation of the t-butylphenyl group by Gn^+, that Gn^+ does not bind into cyclodextrin so it cannot bridge between water and the interior surface, and that solvation of the t-butylphenyl group by Gn^+ is completely blocked when the group binds into the cyclodextrin cavity. This is completely consistent with our previous picture that salting in compounds do not act by affecting water structure, but instead act by direct interaction with the hydrophobic surfaces.

If there were a generalized decrease in hydrophobicity because of a change in water structure, it would not be clear why this did not also affect the hydrophobicity of the cyclodextrin interior. If it had, the effect on binding should have been larger than the effect on solubility. In support of this argument, the situation is quite different when ethanol is added to the water solutions. With 10% ethanol the solubility of t-butylbenzyl alcohol in water is increased by only 1.3-fold, but the binding of BANS into β-cyclodextrin is decreased by 2.5-fold. The larger effect on binding must reflect the ability of small neutral ethanol to bind into the cyclodextrin cavity, in addition to solvating the substrate. As expected from this picture, it takes a 20% concentration of the less polar ethylene glycol to achieve the same 2.5-fold effect on binding.

This work is still in its infancy. However, the quantitative correlations have already given us additional support for our conclusions about the nature of the salting in effects, and promise to let us make very interesting conclusions about other hydrophobic phenomena.

4. Models for ribonuclease catalysis.

Since we have written a number of reviews of this work recently [18,19], it will be described only briefly here. In a series of studies, we determined the preferred mechanism for cleavage of RNA by imidazole buffer in aqueous solution, by the use of kinetics. We found that both the imidazole and the imidazolium ion played a catalytic role, but that they operated sequentially. That is, first one of them converted the RNA to an intermediate phosphorane monoanion, and then the other buffer component catalyzed the conversion of this intermediate to the cleavage product, the 2,3-cyclic phosphate ester.

A key part of this mechanistic study was the finding that with simple dinucleotides we could also see the isomerization of the 3,5' phosphate diester to the 2,5' ester, and vice versa. This isomerization was also catalyzed by a buffer component, but only the acidic ImH^+ component. A striking finding was that the rate of this isomerization was actually decreased as the basic buffer component Im was added (all data corrected for pH effects). We described this as negative catalysis, and pointed out how powerful this evidence was in pinning down the mechanism. Unfortunately, a distinguished journal permitted a fraudulent attack on this work, in which the previous findings and arguments were grossly misrepresented. This attack is an embarassment to science. It should be retracted, since it is without merit of any kind, so it will not be referenced here. It is more suitable to the Journal of Offensive Chicanery than to the journal which actually published it, after it had been rejected by JACS.

Of course, the finding that with imidazole buffer there is sequential bifunctional catalysis does not mean that the same sequence would be seen in the enzyme ribonuclease. In solution it is difficult to tie down two catalysts next to a substrate at once, but in an enzyme it is easy. Thus it seems likely that the enzyme uses a simultaneous version of our sequential mechanism. This has been discussed by us in detail [18,19], and the idea is supported by a lot of evidence on the enzyme itself.

With the evidence for our mechanism, it was of interest to construct a catalyst that could use it for cleavage of an RNA-like molecule. As we have described above, it is possible to attach imidazole rings to the primary carbons of cyclodextrins, and indeed we had prepared [20] two such catalysts that showed bifunctional catalysis of the hydrolytic cleavage of substrate **14**. However, we had designed our catalysts based on the previously accepted mechanism for RNA cleavage by ribonuclease. Since our mechanistic work revealed that a different mechanism was preferred, we redesigned the cyclodextrin bis-imidazole to place the imidazole rings in the correct position for the new mechanism. As hoped, this proved to be the best catalyst of all.

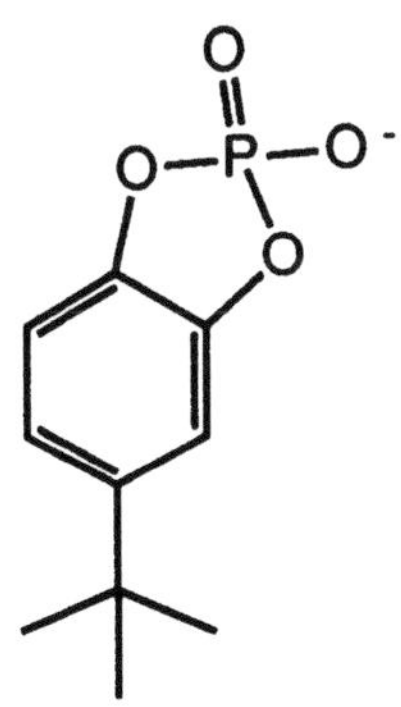

14

Work on the enzyme had included the study of its **proton inventory**, in which rates are determined in H_2O and D_2O mixtures. The results indicated that two protons move in the rate determining step, consistent with a simultaneous bifunctional mechanism. It was therefore gratifying that in our synthetic model enzyme the same technique also showed a two-proton simultaneous mechanism, as in the enzyme. The general lesson is that it is possible for simultaneous bifunctional catalysis to occur when the two catalytic groups are held in the same binding molecule, but unlikely when they are not. Bifunctional catalysts are not only better than monofunctional analogs, they also have mechanisms available to them that are distinct.

5. A model for enzymes that use coenzyme B-12.

We have made a number of enzyme mimics in which a coenzyme group is attached to a cyclodextrin. For instance, much work has gone into mimics for transaminase, with pyridoxal or pyridoxamine units attached to β-cyclodextrin [21]. In addition, as described above, a cyclodextrin with an attached thiazolium ring is a mimic of enzymes that use thiamine pyrophosphate as coenzyme. However, we thought that the attachment of a coenzyme B_{12} mimic to cyclodextrin would lead to a particularly interesting catalyst.

The biochemistry catalyzed by enzymes that use coenzyme B_{12} is particularly interesting. As an example, methylmalonyl CoA (**15**) can reversibly rearrange to succinyl CoA (**16**). Many other reactions are of this type, in which a hydrogen atom and another group change places along an ethano system. There is as yet no chemical process by which one can duplicate such reactions, regardless of rate. Thus in this case an enzyme mimic is needed not just to come up to the rate and selectivity of an enzyme, but indeed even to perform the chemical transformation itself.

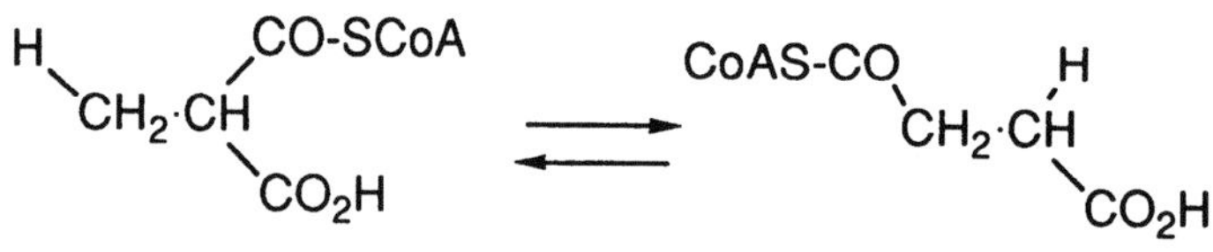

15 **16**

The mechanism used apparently involves [22] dissociation of the carbon-cobalt bond of coenzyme B_{12}, then hydrogen abstraction from the substrate by the resulting primary adenosyl radical, then rearrangement of the intermediate radical (in which the cobalt species may play a role [23]), and finally transfer of a hydrogen from deoxyadenosine to complete the substrate and regenerate the adenosyl radical, which is recaptured by the cobalt species. In models for some of these steps, we have studied transannular hydrogen transfers to radicals involving B_{12} species [24], and others have studied rearrangements of radicals that mimic some of the enzymatic steps [25]. However, there are yet no catalysts that perform the entire process, or indeed that perform an intracomplex hydrogen atom transfer related to the intramolecular reaction we studied.

It seemed to us that the previously unknown molecule cyclodextrin-B_{12} (17) would be particularly interesting. It resembles the coenzyme, with the cobalt attached to the primary carbon of a glucose instead of a ribose. At the same time, dissociation of the carbon-cobalt bond of 17 would generate a cyclodextrinyl radical 18 that might be able to abstract an atom from a bound substrate. If the resulting substrate radical would then rearrange while bound to the cavity, and retake the transferred atom from the cyclodextrin, there would be a complete model for the enzyme. Although there is still much to do, some of this has been successful. In addition, the properties of cyclodextrin-B_{12} were a pleasant surprise [26].

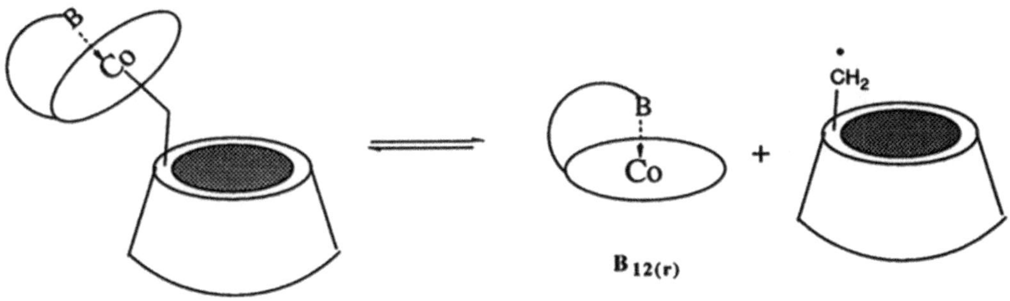

17 18

Synthesis of the compound [26] was straightforward, by reaction of cyclodextrin 6-bromide or iodide with the Co(I) B_{12} anion. The compound 17 could be handled at low pH, since protonation of the benzimidazole ligand prevents its coordination to the cobalt, a coordination that promotes carbon-cobalt bond fission. It had the expected optical spectrum, and the correct mass spectrum as well. However, on neutralization to pH 7.0 it underwent homolytic fission of the carbon-cobalt bond at 25 °C with a half-life of 150 minutes or so. Thus it readily generates the cyclodextrinyl radical under physiological conditions.

This was quite unexpected. Coenzyme B_{12}, with its bond to ribose, is completely stable under these conditions and dissociates readily only on heating or irradiation, or on binding to the enzyme. Thus cyclodextrin-B_{12} is a potential enzyme model in two respects: it combines the coenzyme with a mimic of the enzyme binding pocket, and it mimics the properties of the coenzyme bound to the enzyme in the sense that it now spontaneously dissociates under physiological conditions.

In the best mimic we would see transfer of a hydrogen atom from a bound substrate to the cyclodextrinyl radical, but we do not yet have such an example. However, halogen transfer to radicals is easier, so we examined the reaction of substrates such as t-butylbenzyl iodide **19** with the cyclodextrin-B_{12}. We found that indeed an iodine atom transfer occurs, to form iodocyclodextrin 20 and the benzyl radical. This radical is trapped to some extent by the cobalt of B_{12r}, the form of B_{12} with a Co(II) that dissociates from cyclodextrin B_{12}. With O_2 the product then forms t-butylbenzaldehyde, as is typical of benzyl-B_{12} compounds. Under the reaction conditions the t-butylbenzyl iodide should be bound into the cyclodextrin cavity, but we have not yet proven that reaction occurs only within such a complex.

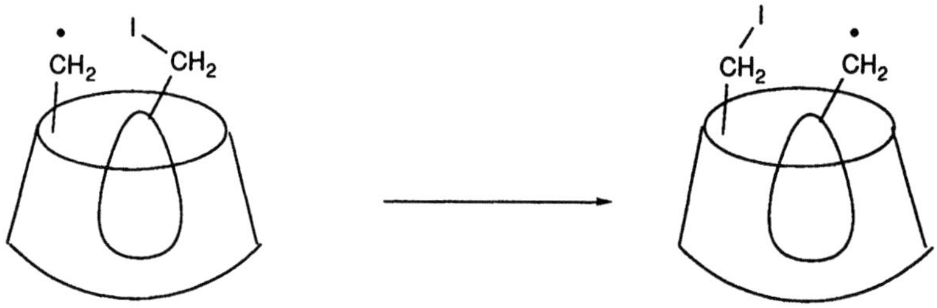

19 20

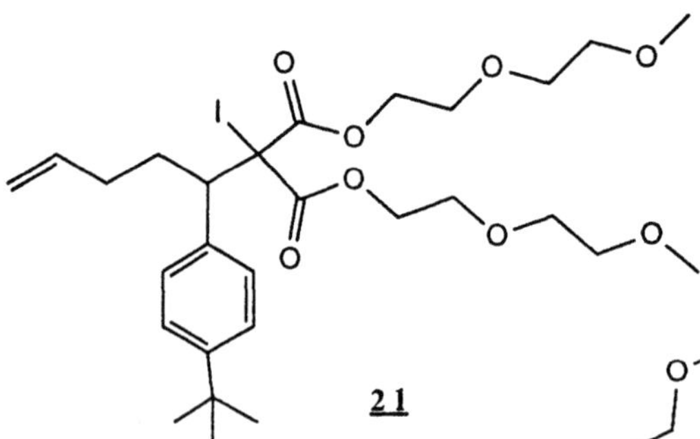

21

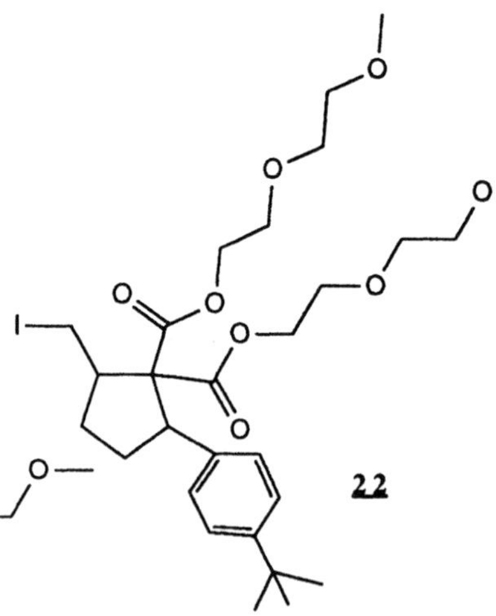

22

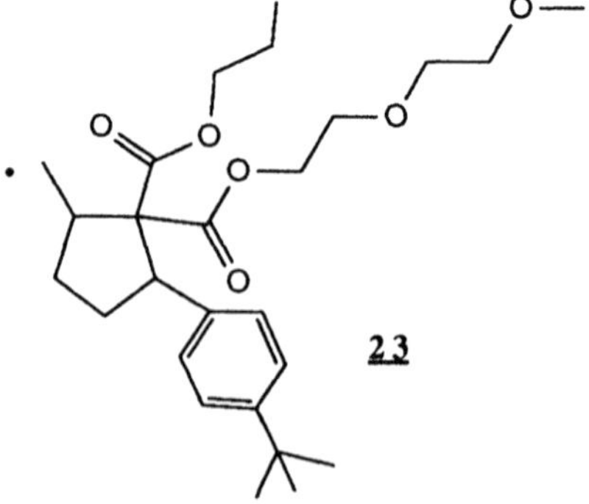

23

We have also looked at the ability of cyclodextrin-B_{12} to catalyze a free radical rearrangement. The reaction studied was the cyclization of the iodide **21** to the product **22**, which proceeds by iodine abstraction, cyclization of the resulting radical and then capture of the product radical [27]. We saw this process occurring, and again some of the product involved radical trapping by B_{12r}. However, there was also a lot of product in which X is iodine, and other studies made it clear that this occurs because of a free radical chain process that occurs outside the cavity, with product radical **23** abstracting the iodine from starting material **21** so as to continue the chain [27].

Even though some have argued [28] that there are enzymatic reactions in which coenzyme B_{12} acts as a simple free radical initiator for a chain process, this is not what we are hoping to mimic. Thus it is clear that some of the properties of cyclodextrin-B_{12} are quite promising, but that we must learn how to keep the radicals bound in the cavity so that external radical chain processes do not take over. There is still plenty to do, but the surprising properties of cyclodextrin-B_{12} furnish a very good start.

References

1. R. Breslow, "Hydrophobic Effects on Simple Organic Reactions in Water", Accts. Chem. Res. **24**, 159-164 (1991).
2. R. Breslow, N. Greenspoon, T. Guo and R. Zarzycki, "Very Strong Binding of Appropriate Substrates by Cyclodextrin Dimers" J. Am. Chem. Soc. **111**, 8296-8297 (1989).
3. Y. Chao, Ph.D. thesis, Columbia University (1972).
4. Halfon, Sherin, unpublished work.
5. E. Anslyn and R. Breslow, "Geometric Evidence on the Ribonuclease Model Mechanism", J. Am. Chem. Soc. **111**, 5972-5973 (1989).
6. R. Breslow and S. Chung, "Strong Binding of Ditopic Substrates by a Doubly Linked Occlusive C_1 "Clamshell" as Distinguished from an Aversive C_2 "Loveseat" Cyclodextrin Dimer", J. Am. Chem. Soc. **112**, 9659-9660 (1990).
7. Kumabe, Naofumi, Ph. D. thesis, Columbia University, 1990.
8. Zhang, Biliang, unpublished work.
9. D. Rideout, R. Breslow, "Hydrophobic Acceleration of Diels-Alder Reactions", J. Am. Chem. Soc., **102**, 7816 (1980).
10. R. Breslow and Carmelo J. Rizzo, "Chaotropic Salt Effects in a Hydrophobically Accelerated Diels-Alder Reaction", J. Am. Chem. Soc. **113**, 4340-4341 (1991).
11. R. Breslow, U. Maitra and D. Rideout, "Selective Diels-Alder Reactions in Aqueous Solutions and Suspensions", Tet. Letters, **24**, 1901 (1983).
12. H.-J. Schneider and N. K. Sangwan, J. Chem. Soc. Chem. Commun.

1787 (1986).
13. (6) P. Grieco, P. Garner, and Z. He, Tet. Letters **24**, 1897-1900 (1983), and later work.
14. E. T. Kool and R. Breslow, "Dichotomous Salt Effects in the Hydrophobic Acceleration of the Benzoin Condensation", J. Am. Chem. Soc., **110**, 1596 (1988).
15. R. Breslow and E. Kool, "A γ-Cyclodextrin Thiazolium Salt Holoenzyme Mimic for the Benzoin Condensation", Tetrahedron Lett., **29**, 1635 (1988).
16. R. Breslow and T. Guo, "Surface Tension Measurements Show that Chaotropic Salting-in Denaturants are not just Water-structure Breakers" Proc. Natl. Acad. Sci. USA **87**, 167-169 (1990).
17. R. Breslow and T. Guo, "Diels-Alder Reactions in Nonaqueous Polar Solvents. Kinetic Effects of Chaotropic and Antichaotropic Agents and of β-Cyclodextrin", J. Am. Chem. Soc., **110**, 5613 (1988).
18. R. Breslow, E. Anslyn, and D.-L. Huang, "Ribonuclease Mimics", Tetrahedron **47**, 2365-2376 (1991).
19. R. Breslow, Accts. Chem. Res. **24**, 317-324 (1991).
20. R. Breslow, J. Doherty, G. Guillot and C. Lipsey, "Beta-Cyclodextrinyl-bisimidazole, a Model for Ribonuclease", J. Am. Chem. Soc.,**100**, 3227-3229 (1978).
21. R. Breslow, J. W. Canary, M. Varney, S. T. Waddell, and D. Yang, "Artificial Transaminases Linking Pyridoxamine to Binding Cavities: Controlling the Geometry", J. Am. Chem. Soc. **112**, 5212-5219 (1990), and earlier references.
22. R. H. Abeles, Proc. Robert A. Welch Found. Conf. Chem. Res. **15**, 113 (1972) ; cf. also R. Breslow, *ibid.*, **15**, 119-120 (1972).
23. G. Choi, S.-C. Choi, B. Wilk and P. Dowd, Proc. Nat. Acad. Sci. USA **87**, 3174 (1990).
24. R. Breslow and P.L. Khanna, "An Intramolecular Model for the Enzymatic Insertion of Coenzyme B_{12} into Unactivated Carbon-Hydrogen Bonds", J. Am. Chem. Soc., **98**, 1297 (1976).
25. For a recent review see R. G. Finke in "*Molecular Mechanisms in Biological Processes*" C. Bleasdale and B. Golding, eds., Royal Society of Chemistry, Cambridge, England p. 344 (1990).
26. J. Light, P. Duggan and R. Breslow, submitted to J. Am. Chem. Soc.
27. Cf. D. P. Curran and C.-T. Chang, J. Org. Chem., **54**, 3140-3157 (1989).
28. J. Stubbe, Biochemistry **27**, 3893-3900 (1988).

FERROCENE AS THE CENTRAL UNIT IN NOVEL, REDOX-SENSITIVE LIGANDS, MONOMERS, AND RECEPTORS

GEORGE W. GOKEL,* JULIO C. MEDINA, CHENSHENG LI, TIMOTHY T. GOODNOW, MARIA T. ROJAS, JEANETTE C. HERNANDEZ-MEDINA, SERVANDO MUÑOZ, AKIO NAKANO,‡ JERRY L. ATWOOD,† AND ANGEL E. KAIFER,
Departments of Chemistry, University of Miami, Coral Gables, Florida 33124 and †University of Alabama, Tuscaloosa, AL 35487

ABSTRACT. The ferrocene system is remarkable for its structure, thermal stability, and chemical versatility. We describe in this article, diverse applications of ferrocene in the preparation of novel structures for use as cation binders, small molecule receptors, and in the formation of membranes.

1. Introduction

The remarkable "sandwich" structure known as ferrocene was discovered some four decades ago.[1] It has been studied extensively and it has been applied in a number of clever and remarkable ways.[2] Our own work over the last several years has involved the search for versatile molecules that might be used in cation binding,[3] in redox-switching,[4] and even in the development of molecules that could be used as receptors for other molecules.[5] In this brief article, we show several new applications of the ferrocene system that draw upon its unique structural and electrochemical properties.

2. Lariat Ethers

Although the lariat ethers[6] are not the focus of this article, it is informative to consider how their study led to the present developments. At the time the lariat ether program was initiated, both the crown ethers[7] and the cryptands[8] were well established cation binders. The naturally occurring mitochondrial transport agent valinomycin[9] was also known and it was apparent that a carrier required a combination of three-dimensionality and dynamics to be effective. The crown ethers show high cation binding rates but are generally two-dimensional. The cryptands are three-dimensional, but their cation binding rates are often slower than those observed for the crowns. Our solution to this difficulty was to develop crowns having one or more sidearms. The expectation was that these compounds would bind the cation primarily by the macroring but the sidearm would then add the third dimension of solvation. The latter step is essentially a conformational change and should therefore be rapid. The notion is shown schematically below.

The lariat ether notion has been confirmed in solution and by a variety of solid state structure studies.[10] One of us has recently summarized much of this effort.[11] For the present purpose, it is sufficient to consider the lariat ethers as compounds that have a crown ether-based binding site and a sidearm.

2.1 CLASSES OF LARIAT ETHERS

V. Balzani and L. De Cola (eds.), Supramolecular Chemistry, 429–442.

A Schematic of Lariat Ether Complexation

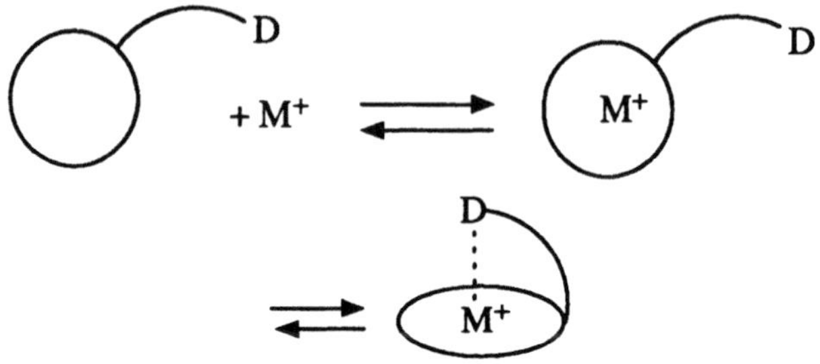

We have categorized the lariat ethers according to the means by which sidearms are attached (pivot point) and how many sidearms are present. The carbon-pivot compounds[12] are more chemically stable than the nitrogen-pivot compounds[13] but they are inherently less flexible. The ability of nitrogen to undergo facile inversion makes the nitrogen-pivot structures generally more adaptable to a cation than is the case with carbon-pivot derivatives. Examples of lariat ether compounds are shown below.

Examples of Lariat Ether Compounds

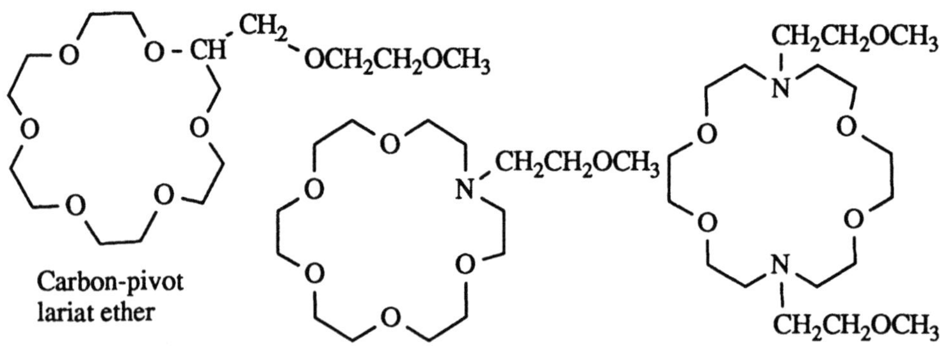

Carbon-pivot lariat ether

Nitrogen-pivot lariat ethers

As our understanding of lariat ether compounds increased, we generalized our working definition of lariat ethers from crowns having donor-group-containing sidearms to crowns having sidearms. In particular, we became interested to know if crown ethers having hydrophobic alkyl or steroidal sidearms would be amphiphilic. Examples of the compounds of interest to us are shown below as **1** and **2**.

We felt confident that the crown compounds would be amphiphilic. Indeed, a metal-cation--bound crown head group should certainly approximate a diffuse positive charge. At the time this work was undertaken, however, there was no evidence that either a crown or its complex could form such organized assemblies as vesicles. We were able to obtain direct evidence for the formation of vesicles from the steroidal lariat ether illustrated.[14] The essential characteristics of the vesicles produced is that they are unilamellar, their size is in the range 340-350 Å, and the volume entrapments were about 4%. Much of this work has been summarized in a recent review article.[15]

Amphiphilic Lariat Ethers

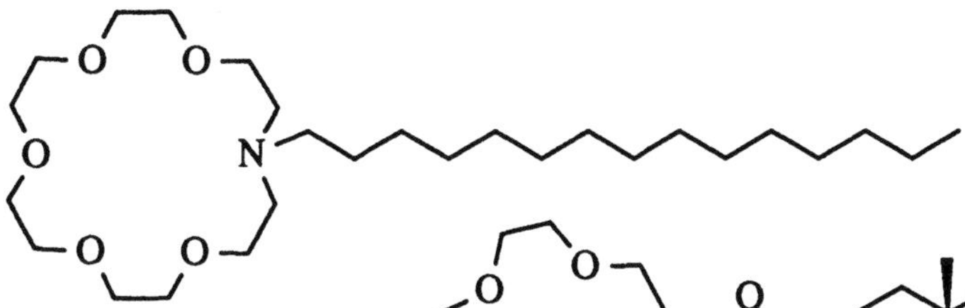

1

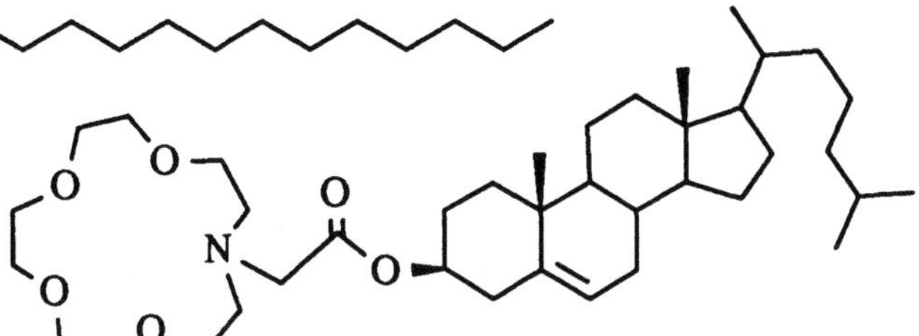

2

2.2 THE FORMATION OF MEMBRANES FROM LARIAT ETHERS

This success in the formation of membrane systems suggested to us that we might be able to form crown-ether-based bolamphiphiles inspired by the compounds isolated from archebacteria and their synthetic equivalents developed by Fuhrhop.[16] Indeed, vesicles were formed from aza-15-crown-5 and aza-18-crown-6 based dodecanes. A typical structure is shown below.

Preparation of Lariat Ether Bola-Amphiphiles

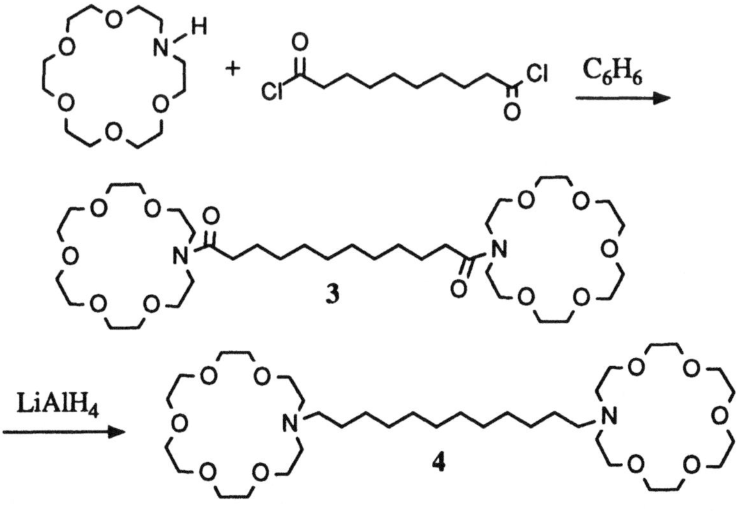

3

4

In all, we have prepared more than a dozen representatives of this class. The compound shown in the figure was studied in some detail. Aggregation studies relied upon transmission electron microscopy and static and dynamic light scattering. These techniques showed that the crown-bolyte surfactants in water formed spherical arrays. The assemblies appear from Debye light-scattering profiles, to be large vesicles (~2000Å in diameter) and probably multilamellar. These vesicles appear to be quite stable in acidic (pH~2) solution, a fact that suggests the formation of a hydrogen bonded network among the along both the endo- and exovesicular surfaces.[17]

3. The Concept of Redox Switching

3.1 THE PARADOX OF MEMBRANE TRANSPORT

A membrane is a physical barrier that separates two solutions that ordinarily have different character. Normally, these will be aqueous solutions separated by a hydrophobic lipid bilayer. In order for a cation to be transported across the hydrophobic barrier, it must be enveloped by a carrier that alters its character from polar and charge dense to non-polar and charge diffuse. In addition, the ideal carrier will completely surround the cation so that interactions between the cation and bilayer are minimized.

In addition to the structural requirement, there are dynamic requirements that must be considered. The binding of the cation on one side (source side) of the membrane should be rapid and the stability of the complex should be great. Inside the membrane, only the strength of the complex is important because decomplexation is unfavorable in the lipid environment and therefore not a significant factor. On the other (receiving) side of the membrane, complex stability should be weak and the decomplexation rate should be rapid so that cation release is facile. The kinetic and thermodynamic requirements for cation carriers on opposite sides of the membrane are thus incongruous.

If we consider that the stability constant for the complex is $K_S=k_{complex}/k_{decomplex}$, the requirement for a carrier is that the same molecule have a large and small decomplexation rate at the same time. Since this is impossible, compounds are chosen that achieve transport by a compromise of the complexation and decomplexation rate constants. One means of circumventing this problem is to use a ligand (binder) that can be made to form a more or less stable complex by a rapid chemical modification. This is usually referred to as "switching." Mechanisms to switch a ligand's binding ability based on ionization, thermal, photochemical, and redox properties have been described in the literature.

3.2 REDOX-SWITCHED CATION CARRIERS

There are several obvious mechanisms by which chemical switching can be achieved. These are (1) proton transfer (change in pH); (2) conformational or structural changes effected either by photochemical or thermal stimuli; (3) structural changes effected by oxidation-reduction chemistry; and (4) changes in the charge state of the donor array in a given ligand, effected by redox chemistry. Much of the recent work in this area has been reviewed by one of us.[18] It is the latter type of chemistry that has been of interest to us and the remainder of the present article is focused thereupon.

There are two basic cases that require consideration. In one, the ligand is switched from a neutral to a negative charge state. This transforms a weak, moderate, or strong cation binder into a strong, stronger, or very strong binder. At the source side of the membrane, cation binding is thus enhanced. On the opposite side of the membrane, oxidation converts the charged system to its previous, neutral state. The second case involves a binder that can accommodate a cation. Complexation strength is lowered by oxidation to convert the neutral or negative donor array to a neutral or positive system. The latter should be most effective since not only will binding be diminished, but charge repulsion will further affect the cation's departure from the ligand. The two possibilities are summarized schematically below.

Scheme: Redox-switching leading to cation binding enhancement

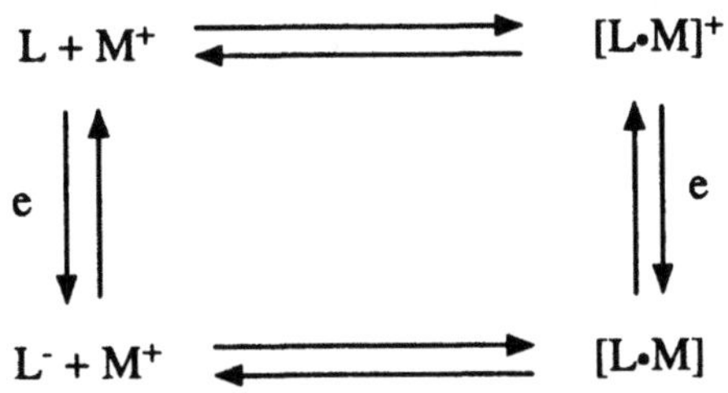

Scheme: Redox-switching leading to diminished cation binding

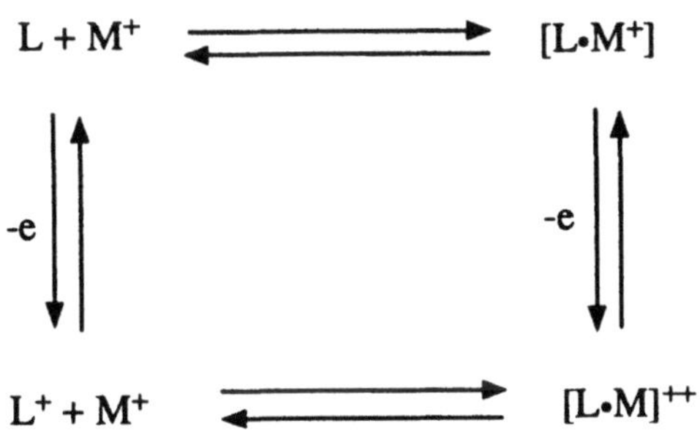

3.3 THE PROPERTIES OF FERROCENE

Ferrocene is also called "dicyclopentadienyl iron" which accurately describes its sandwich structure. What may not be apparent is that the molecule is quite nonpolar: it is soluble in hexane. The two cyclopentadienyl rings are formally aromatic, six-electron systems and as such are extremely electron rich. The ferrocene system can be functionalized by electrophilic aromatic substitution chemistry. The property of ferrocene most relevant to our efforts, however, is that an electron may be lost reversibly from iron [Fe(II) $\rightleftharpoons$ e + Fe(III)] so that the entire molecule becomes positively charged. The half wave potential for ferrocene is observed at about 400 mV (vs. calomel) in aqueous solution and the ferricinium species is stable in water for hours. This contrasts, for example with nitrobenzene or most quinones, which are generally less to much less stable than this upon reduction. These properties make ferrocene particularly attractive for applications in redox-altered chemistry.

3.4 FERROCENYL AMIDES AND CRYPTANDS

The first ferrocenyl cryptands were developed in England. Extensive studies from the groups of Hall, Beer, and others have produced various ferrocene-containing compounds that bind cations.[19] For the most part, these structures incorporate a macroring via one or more amide linkages. The amide functional group has both advantages and disadvantages. First, it is a relatively rigid, planar residue. Typical resonance energies for amides are 20 kcal per mole. Thus, a bis(amide) structure will have its conformations constrained by 40 kcal/mole or so. This may help hold a desirable conformation or it may hinder its formation. Second, the oxygen atoms of the amide residue, because of the same resonance that leads to planarity, are quite polar donor atoms. When present in a ligand, they may control cation binding selectivity. Third, and of critical importance for redox switching, is that an amide residue attached directly to the ferrocene will typically increase the half wave potential by more than 100 mV. This makes the ferricinium ion a better oxidant but a less stable chemical species.

The interesting amide structure, **9**, is the precursor to the compound that is discussed extensively in section 3.5. Note that it has two amide residues in addition to a diaza-18-crown-6 macrocyclic sub-structural unit. It appears that the amide residues dominate the chemistry of both ferrocenyl amides and ferrocenyl amides containing a crown ether subunit have. In the absence of the macrocycle, lithium is bound by the amide carbonyl groups. Remarkably, when the crown ether ring is present, lithium is no longer bound although other cations show interactions. The binding affinity generally increases in proportion to charge density. Thus Y^{3+} is bound more strongly than Ca^{2+} although both are similar in size. Our own data suggest that binding of Na^+, which is about the same size as the other two cations, is bound very weakly (log $K_S < 2$). Redox switching using these systems is possible, but the complexes become quite unstable when oxidized.[20]

Ferrocenyl Amides and Macrocycles

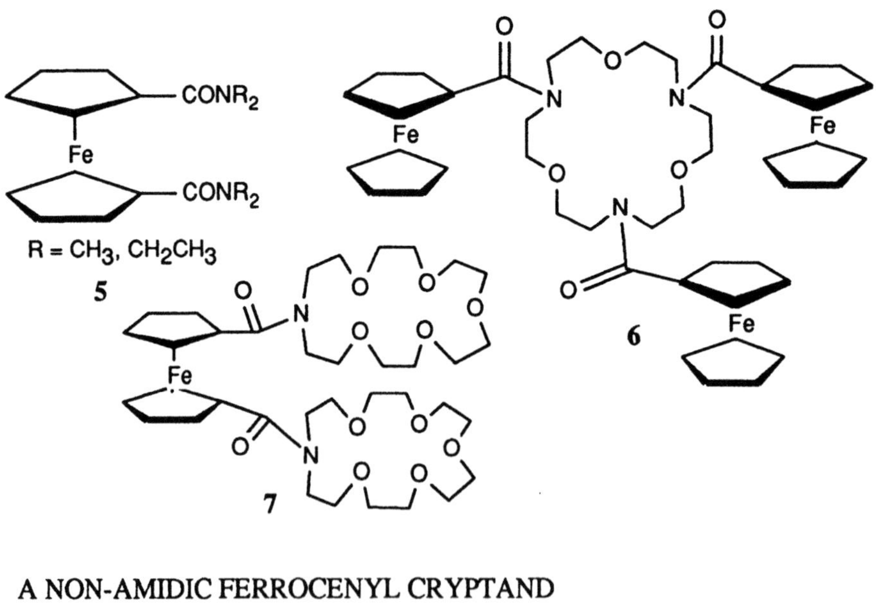

3.5 A NON-AMIDIC FERROCENYL CRYPTAND

Although some interesting properties are apparent in the amidic ferrocenyl cryptands, we were interested to see what chemistry could be developed for a system in which the donor group array was endocyclic. Thus, the much-studied cryptand **9** was reduced as described previously[21] and purified as shown in the scheme below. The product (**10**), as its hydrate, was obtained as a yellow-orange crystalline solid. The solid state structure was obtained and is shown below as well.[22] It is interesting that the water molecule bridges the external crown structure by forming two hydrogen bonds.

Synthesis of Ferrocenyl Cryptand 10

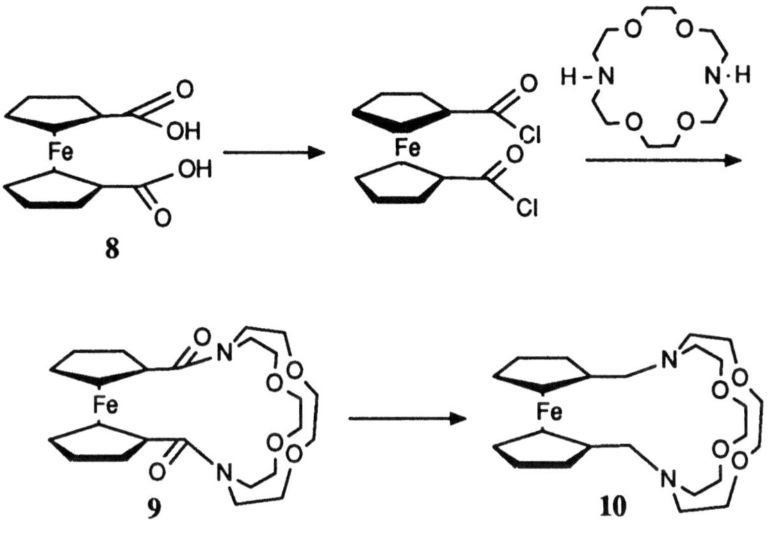

ORTEP Plot of Ferrocenyl Cryptand 10•H_2O

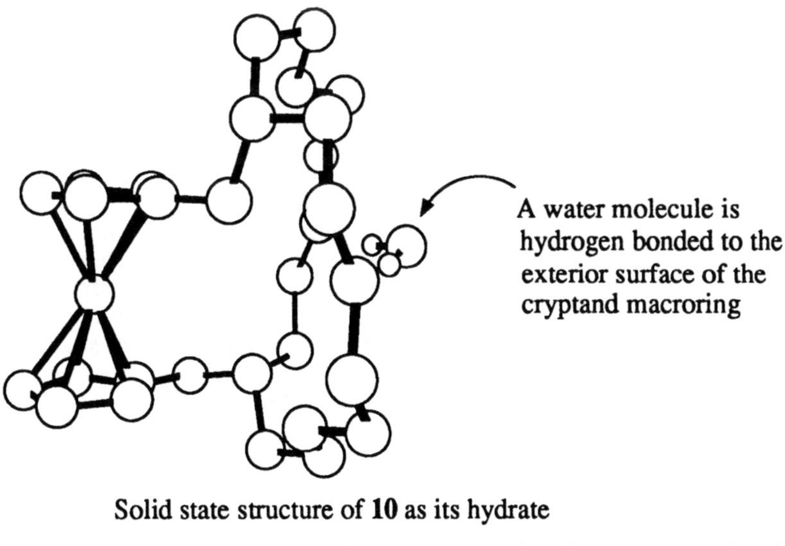

Solid state structure of **10** as its hydrate

3.5.1 Cation Binding. We have determined the cation binding constants for the ferrocenyl cryptand with a variety of cations in acetonitrile and methanol. The binding data are summarized in Table 1.

Table 1. Cation Binding by 10

Cation	*Solvent*	*Log K_S*
Li^+	acetonitrile	3.74
Na^+	acetonitrile	6.28
Na^+	methanol	3.72
K^+	acetonitrile	5.14
K^+	methanol	3.77
Ca^{2+}	acetonitrile	7.16
Ag^+	acetonitrile	8.35

Two trends are apparent. First, cation binding strength in acetonitrile is generally greater than in methanol. This is expected based upon the solvation enthalpies for Na^+ and K^+ in these solvents. Second, in acetonitrile, cation binding strength is in the order $Li^+ < K^+ < Na^+ < Ca^{2+} < Ag^+$. This is an unusual binding order. For example, in methanol, 18-crown-6 prefers K^+ over Na^+, Li^+ or Ca^{2+}. Likewise, [2.2.2]-cryptand (also in methanol) prefers K^+ over Na^+ or Ca^{2+} but Ag^+ is favored over the other three in this case. The trend is different in acetonitrile. In this solvent, [2.2.2] selects either Na^+ or K^+ (log K_S for either is about 10.6) over Ag^+ (log K_S ~9). The ferrocenyl cryptand selects Ag^+ over either alkali metal cation and the binding strength for Ag^+ by the ferrocenyl cryptand and [2.2.2] are not very much different.

3.5.2 Redox-switched Cation Binding. As noted above, the previously studied ferrocenyl amide compounds do not show evidence for redox switching in the presence of Na^+ or K^+. Cryptand **10**, on the other hand, shows evidence for such interactions and this is extensible to Ca^{2+} and even Ag^+ (see below). Redox switching ability was assessed by cyclic voltammetry in the presence of progressively increasing amounts of alkali metal cation salts. For example, cyclic voltammetry of **10** in acetonitrile containing 0.1 M Bu_4NPF_6, shows a reversible redox couple that may be ascribed to oxidation of ferrocene. In the presence of sub-stoichiometric amounts of Na^+ ion, a new redox couple (corresponding to the complex) appears at more positive potential. This redox couple shows currents that increase with increasing (0→1 equiv) cation concentration. After one equivalent of cation is present, only the new redox couple is in evidence. This means that one and only one equivalent of cation is bound by the ligand as expected for an inclusive cryptate. The fact that sub-stoichiometric quantities of cation afford two redox couples suggests that the equilibrium binding constant between the cation and the neutral ligand is large.

In order to confirm that the system was in dynamic equilibrium, 0.5 equivalent each of **10** and of the complex **10**•Na^+ was treated with 5 equivalents of [2.2.1]-cryptand. The original redox couple was restored immediately and no evidence of the second redox couple was any longer in evidence. This is because the electro-inactive cryptand completely sequesters the cation.

The goal of this effort is to prepare a ligand that binds a cation more strongly when neutral than when oxidized. The effectiveness of the ligand can be judged from the ratio of stability constants, K/K_+, which can be obtained from the half wave potential difference between the free and cation-complexed ligand. The K/K_+ values we obtained in acetonitrile solution are: Na^+, 1.5×10^3; K^+, 1.2×10^2; Ca^{2+}, 4.3×10^4; and Ag^+, 5.8×10^4. With the exception of the latter, these values can be understood in terms of cation charge density (see graph).

The silver cation does not fall on this line and binding is unusually strong between **10** and the cation. Some of the binding strength can be accounted for in terms of the

Relationship of Potential and Charge Density

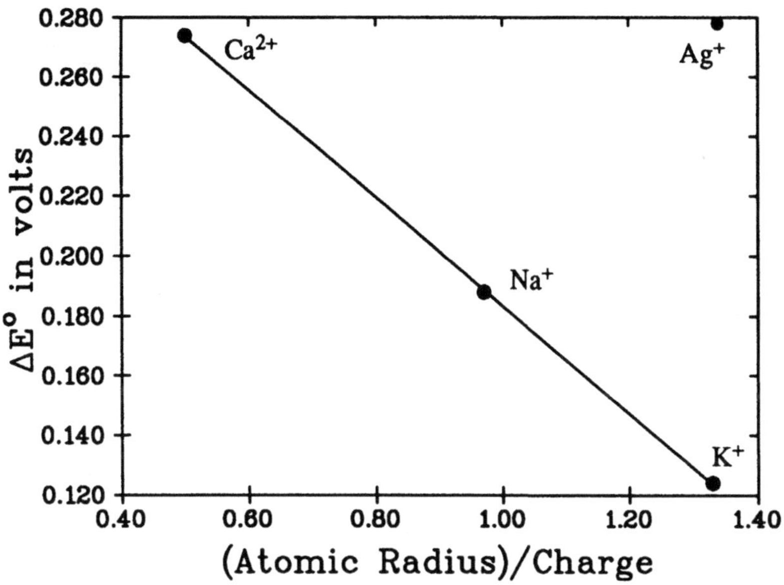

favorable N—Ag—N interaction. Nevertheless, the anomaly is large. We have accounted for this in terms of a direct Ag^{+}—Fe bond between the cation and the ferrocene ligand. We have documented the interaction by a combination of UV-visible and NMR spectral data. Although we were quite surprised to find this direct interaction, we have found a small number of literature reports[23] for other ferrocenyl systems and transition metal cations that exhibit this surprising, direct interaction. One of these precendents, a complex in which the ferrocene iron atom is in the palladium donor array, is illustrated in the figure below.

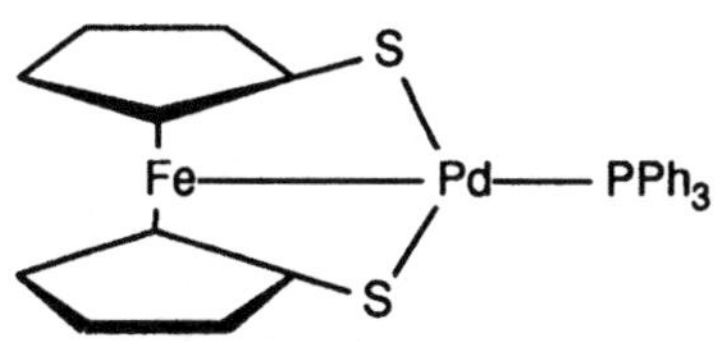

It should be noted that notwithstanding the precedents, ours is the first kinetically dynamic ligand system thus far reported that exhibits this type of direct, bonded interaction. The remarkable strength of this interaction, gave us hope that we might, with **10**, be able to observe redox switching behavior in aqueous solution. Recent cyclic voltammetric results confirm that, for the first time, redox switching can be observed in water.

4. Redox-Switched Membrane Formation

If cation complexing agents and carriers can be redox-switched, what about the formation of membranes themselves? An interesting candidate for this might again be ferrocene. As a relatively nonpolar material, a ferrocene substituted by a lipophilic sidechain should not be amphiphilic. On the other hand, if ferrocene is oxidized, the organometallic species would become a polar head group.

We thought to prepare several groups of structures. These would include polar alkyl sidechains and steroids. The attachment of these chains would also vary as shown in the figure below. The problem in this case is the same as that noted in section 3 above, namely that attachment of an amide carbonyl group directly to ferrocene, increases its redox potential and destabilizes the ferricinium cation.

Hydrophobic Ferrocene Derivatives

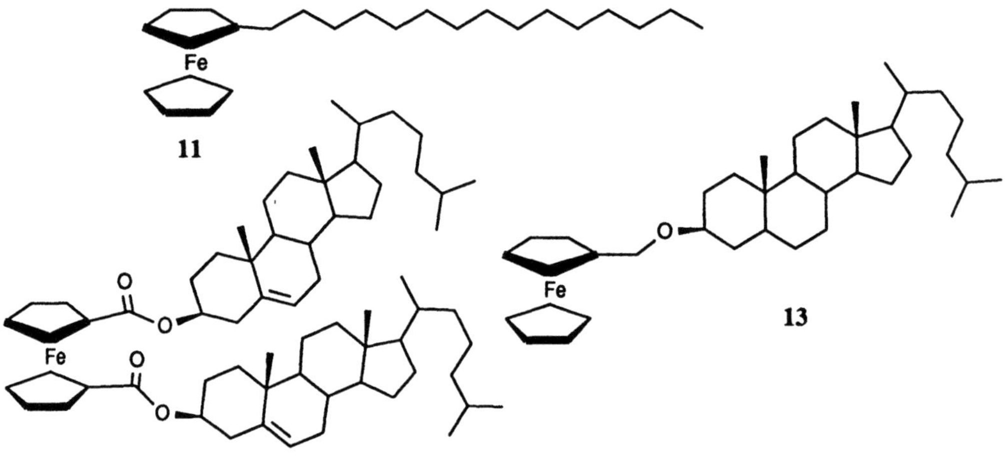

11 12 13

4.1 VESICLE FORMATION BY REDOX SWITCHING

A surprising observation is that certain alkyl ferrocenes are surface active, although we did not anticipate this.[24] None of our compounds formed assemblies that could be readily characterized in the neutral state and the properties of the non-oxidized systems were not investigated further. In any event, the mono-steroidal ferrocene shown above did form vesicles. In this particular study,[25] the half wave potentials for the ferrocenyl bis(amides) fell in the range 700-1000 mV. The mono-steroid differed from the other compounds in two important respects: it was the only monosubstituted system studied and it was the only compound lacking an amide carbonyl group bonded directly to the ferrocene structure. The fact that it formed vesicles may be due to its lower half-wave potential, a structural difference attributable to the ether, rather than amide, linkage, or the presence of one, rather than two, steroids. No doubt all of these variations contribute to the success of the system.

When the diamides were oxidized, each formed aggregates but the membranes generally lacked definition. In two cases, were able to obtain spherical particles that appeared to be vesicles. Light scattering indicated their sizes to be 670±190Å and 890±230Å. These values are reasonable but more work is necessary before these compounds can be considered to be characterized.

The steroidal ether was electrolyzed in a two-compartment cell using Pt working and Ag counterelectrodes. That the oxidized ferrocenylsteroid was a ferricinium cation was confirmed by UV-visible absorption (λ_{max} 628 nm). Controlled potential hydrolysis showed that approximately 80% conversion had taken place. Similar results were observed when either electrolysis of Ce^{+4} was used for oxidation, showing that the supporting electrolyte did not play a critical role. Moreover, after characterization, the vesicles were collapsed by addition of dithionite reducing agent. The monomer was recovered and characterized.

Electromicrographs of the monosteroidal vesicles showed them to be multilamellar and somewhat irregular in size. The membrane thickness appeared, again from EM, to be about 45Å. These systems await further characterization but represent the first known example of redox-switched formation of vesicles.[25]

5. Ferrocene in the Design of Small Molecule Receptors

The lovely and elegant design of molecular clefts reported by Rebek and coworkers[26] our evolving interest in the versatile ferrocene molecule led us to the design of small molecule receptor systems. The use of ferrocene was encouraged by two factors in particular. First, the inter-ring distance in ferrocene is 3.2-3.3Å. This is essentially an aromatic thickness. Thus, the top half of a ferrocene could be used as a mechanical link and any aromatic bound below would be in direct contact with the π-surface. Second, rotation of the cyclopentadiene rings about the iron atom requires, in ferrocene itself, no more than 5 kcal/mole. Thus, essentially free rotation in this system means that iron acts as an "atomic ball bearing." By using these notions, we schematically laid out the types of structures envisioned.

Schematic for Ferrocenyl Receptor Molecules

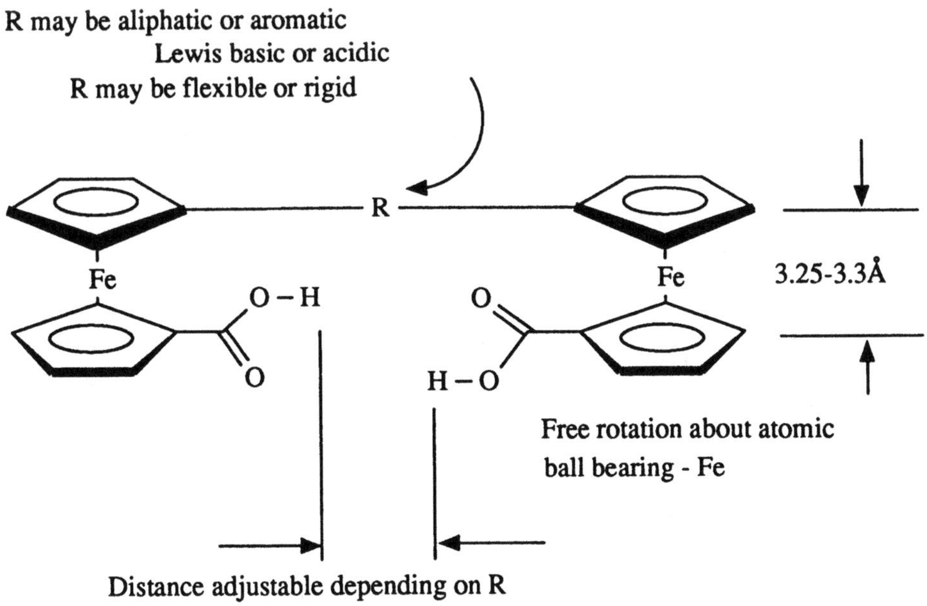

Several ferrocenyl receptors were brought to hand in this effort.[5,27] Two of them are shown in the figure below. Note that in each case, the two carboxyl groups on the bottom (1-position) rings may rotate away from each other to enlarge the molecular cavity. When rotation inward occurs, the two carboxyl groups are focussed towards each other at a distance fixed by the spacer used in the 1'-position. The two compounds shown were selected for specific reasons. The fluorenyl residue is capable of π-stacking but the bipiperidyl is not. We also obtained a solid state structure of the bipiperidyl compound that permitted us to confirm a number of our assumptions about these systems. Finally, the methoxyfluorenyl system is sufficiently de-symmetrized that all protons are discernible (at 400 MHz). This permits us to obtain binding constants in organic solvents appropriate for NMR. In the figure below, a general receptor structure is shown with the two R groups discussed. The ORTEP plot shown at the bottom of the figure is for the bipiperidyl structure (as its diethyl ester).

Receptors 14 and 15 and Solid State Structure of 15

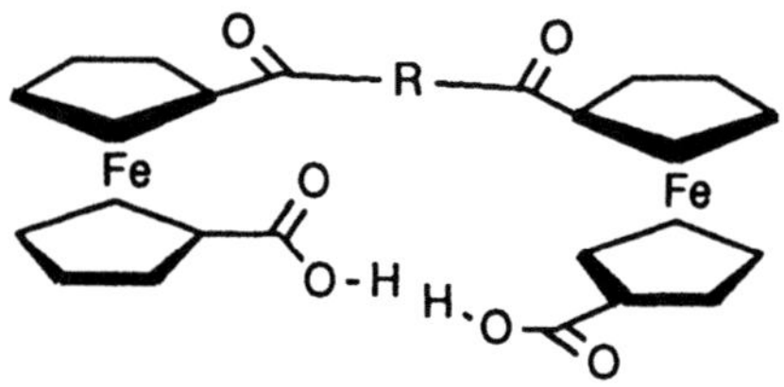

14, R = 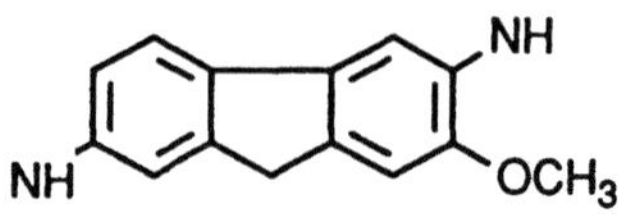

15, R = 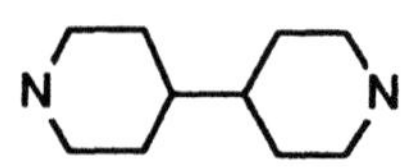

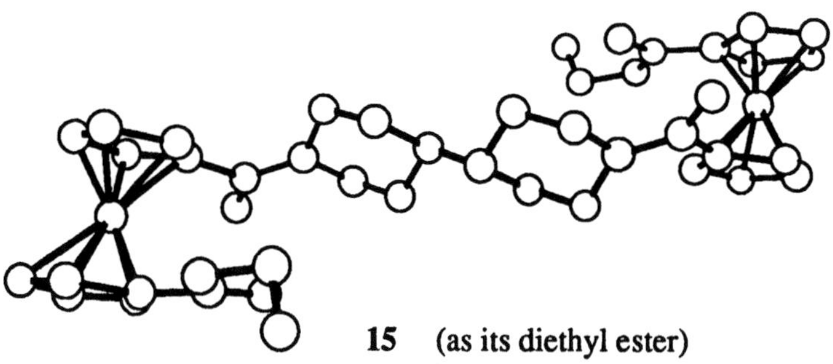

15 (as its diethyl ester)

The fluorenyl receptor showed good binding and selectivity. Thus, no significant binding was observed in THF solution between the fluorenyl receptor and pyridazine, 4,4'-bipyridyl, or 4-*t*-butylpyridine. In contrast, 4-aminopyridine and 4-aminoquinoline both bind strongly to the ferrocenyl receptor, the former having a binding constant of 3200±400.

6. Conclusion and Prospects

The ferrocene molecule is remarkable and versatile. We have demonstrated that it may be incorporated into redox-switchable cation binders, small molecule receptors, and we have demonstrated that the first example of redox-switched vesicles arises from an appropriately-substituted ferrocene. The prospect is that the rich redox chemistry of ferrocene will only augment its interesting structural qualities to afford a vast array of new compounds as it enters the second half of the century since its discovery.

Acknowledgment. We thank the NIH for a grant (GM-36262) that supported this work.

7. Notes and References

‡Present address: Toa University, Shimonoseki, Yamaguchi, Japan.

1. Kealy T. J.; Pauson, P. L. *Nature*, **1951**, *168*, 1039.
2. Rockett, B. W.; Marr, G. *J. Organomet. Chem.*, **1991**, *416*, 339.
3. Gokel, G.W.; Trafton, J.E.; in Inoue, Y.; Gokel, G.W. (Eds.), *Cation Binding by Macrocycles*, **1990**, Marcel Dekker, New York, p. 253.
4. Chen, Z.; Gokel, G.W.; Echegoyen, L.; *J. Org. Chem.* **1991**, *56*, 3369.
5. Gokel, G.W.; Medina, J.C.; Li, C., *Synlett*, **1991**, 677.
6. Gokel, G.W.; in Atwood, J.L.; Davies, E.; MacNicol, D.D. (Eds.), *Inclusion Phenomena, V. 4*, Oxford University Press, 1991, 287.
7. (a) Pedersen, C.J.; *J. Am. Chem. Soc.* **1967**, *89*, 2495. (b) Pedersen, C.J.; *J. Am. Chem. Soc.* **1967**, *89*, 7077. (c) Gokel, G.W.; *Crowns and Cryptands*, **1991**, Royal Society of Chemistry, London.
8. Lehn, J.-M.; *Angew. Chem. Int. Ed. Engl.* **1988**, *27*, 89.
9. (a) Brockmann, H.; Schmidt-Kastner, G.; *Chem. Ber.* **1955**, *88*, 57. (b) Grell, E.; Funck, T.; Eggers, F.; *Membranes*, **1975**, *5*, 1. (c) Pinkerton, M.; Steinrauf, L.K.; Dawkins, P.; *Biochem. Biophys. Res. Commun.* **1969**, *35*, 512.
10. Gandour, R.D.; Fronczek, F.R.; Gatto, V.J.; Minganti, C.; Schultz, R.A.; White, B.D.; Arnold, K.A.; Mazzocchi, D.; Miller, S.R.; Gokel, G.W.; *J. Am. Chem. Soc.* **1986**, *108*, 4078.
11. Gokel, G.W.; *Chem. Soc. Reviews*, **1992**, in press.
12. Dishong, D.M.; Diamond, C.J.; Cinoman, M.I.; Gokel, G.W.; *J. Am. Chem. Soc.* **1983**, *105*, 586.
13. Schultz, R.A.; White, B.D.; Dishong, D.M.; Arnold, K.A.; Gokel, G.W.; *J. Am. Chem. Soc.* **1985**, *107*, 6659.
14. (a) Echegoyen, L.E.; Hernandez, J.C.; Kaifer, A.; Gokel, G.W.; Echegoyen, L.; *J. Chem. Soc. Chem. Commun.* **1988**, 836. (b) Gokel, G.W.; *United States Patent* 4,783,528, November 8, 1988.
15. Gokel, G.W.; Echegoyen, L.; *Adv. Bio-organic Frontiers*, **1990**, *1*, 116.
16. a. Fuhrhop, J. H.; Mathieu, J. *J. Chem. Soc., Chem. Commun.*, **1983**, 144. b. Fuhrhop, J. H.; Liman, U. *J. Am. Chem. Soc.*, **1984**, *106*, 4643. c. Fuhrhop, J. H.; Liman, U.; David, H. H. *Angew. Chem. Int. Ed. Engl.*, **1985**, *24*, 339. d. Fuhrhop, J. H.; David, H. H.; Mathieu, J.; Liman, U.; Winter, H. J.; Boekema, E. *J. Am. Chem. Soc.*, **1986**, *108*, 1785. e. Fuhrhop, J. H.; Liman, U.; Koesling, B. *ibid*, **1988**, *110*, 6840.
17. Muñoz, S.; Mallen, J.V.; Gay, I.; Chen, Z.; Echegoyen, L.; Gokel, G.W.; Manuscript in preparation.
18. Kaifer, A.; Echegoyen, L.; in Inoue, Y.; Gokel, G.W. (Eds.), *Cation Binding by Macrocycles*, **1990**, Marcel Dekker, New York, p. 363.
19. Beer, P. D. *Chem. Soc. Rev.*, **1989**, *18*, 409.
20. a. Hall, C. D.; Sharpe, N. W.; Danks, I. P.; Sang, Y. P. *J. Chem. Soc., Chem. Commun.*, **1989**, 419. b. Beer, P. D.; Sikanyika, H.; Blackburn, C.; McAleer, J. F. *J. Organometal. Chem.*, **1988**, *350*, C15. c. Hall, C. D.; Sharpe, N. W. *Organometallics*, **1990**, *9*, 952. d. Hall, C. D.; Danks, I. P.; Lubienski, M. C.; Sharpe, N. W. *J. Organometal. Chem.*, **1990**, *384*, 139. e. Beer, P. D.; Keefe, A.

D.; Sikanyika, H.; Blackburn, C.; McAleer, J. F. *J. Chem. Soc. Dalton Trans.*, **1990**, 3289. f. Hall, C. D.; Danks, I. P.; Sharpe, N. W. *J. Organomet. Chem.*, **1990**, *390*, 227. g. Hall, C. D.; Sharpe, N. W. *ibid*, **1991**, *405*, 365. h. Hall, C. D.; Tucker, J. H. R.; Sharpe, N. W. *Organometallics*, **1991**, *10*, 1727.

21. Oepen, G.; Voegtle, F.; *Liebigs Ann. Chem.*, **1979**, 1094.
22. Medina, J.C.; Goodnow, T.T.; Bott, S.; Atwood, J.L.; Kaifer, A.E.; Gokel, G.W.; *J. Chem. Soc. Chem. Commun.* **1991**, 290.
23. a. Roberts, R. M. G.; Silver, J.; Morrison, I. E. G. *J. Organometal. Chem.*, **1981**, *209*. b. Sato, M.; Sekino, M.; Akabori, S. *ibid*, **1988**, *344*, C31. c. Watanabe, M.; Ichikawa, H.; Motoyama, I.; Sano, H. *Bull. Chem. Soc. Jpn.*, **1983**, *56*, 3291. d. Seyferd, D.; Hames, B. W.; Rucker, T. G.; Cowie, M.; Dickson, R. S. *Organometallics*, **1983**, *2*, 472. e. Akabori, S.; Kumagai, T.; Shirahige, T.; Sato, S.; Kawazoe, K.; Tamura, C.; Sato, M. *Organometallics*, **1987**, *6*, 2105.
24. (a) Fujihara, M.; Nishiyama, K.; Yamada, H.; *Thin Solid Films*, **1985**, *132*, 77. *Chem. Abstr.* **1986**, *105*, 106687f. (b) Nakahara, H.; Katoh, T.; Sato, M.; Fukuda, K.; *Thin Solid Films*, **1988**, *160*, 153.
25. Medina, J.C.; Gay, I.; Chen, Z.; Echegoyen, L.; Gokel, G.W.; *J. Am. Chem. Soc.* **1991**, *113*, 365.
26. (a) Rebek, J. Jr.; *Acct. Chem. Res.* **1984**, *17*, 258. (b) Rebek, J. Jr.; *Science* **1987**, *235*, 1478.
27. Medina, J.C.; Li, C.; Bott, S.G.; Atwood, J.L.; Gokel, G.W.; *J. Am. Chem. Soc.* **1991**, *113*, 366.

AUTHOR INDEX

SUBJECT INDEX

A bold-face type number corresponds to the first page of a chapter in which the subject is a major theme

GPSR Compliance

The European Union's (EU) General Product Safety Regulation (GPSR) is a set of rules that requires consumer products to be safe and our obligations to ensure this.

If you have any concerns about our products, you can contact us on ProductSafety@springernature.com

In case Publisher is established outside the EU, the EU authorized representative is:

Springer Nature Customer Service Center GmbH
Europaplatz 3
69115 Heidelberg, Germany

Batch number: 10607320

Printed by Printforce, the Netherlands